D0396774

Blind Speech Separation

Blind Speech Separation

edited by

Shoji Makino

NTT Communication Science Labs.
Japan

Te-Won Lee

UC San Diego, La Jolla
USA

and

Hiroshi Sawada

NTT Communication Science Labs.
Japan

A C.I.P. Catalogue record for this book is available from the Library of Congress.

ISBN 978-1-4020-6478-4 (HB)
ISBN 978-1-4020-6479-1 (e-book)

Published by Springer,
P.O. Box 17, 3300 AA Dordrecht, The Netherlands.

www.springer.com

Printed on acid-free paper

Preface

We are surrounded by sounds. Such a noisy environment makes it difficult to obtain desired speech and it is difficult to converse comfortably there. This makes it important to be able to separate and extract a target speech signal from noisy observations for both man–machine and human–human communication.

Blind source separation (BSS) is an approach for estimating source signals using only information about their mixtures observed in each input channel. The estimation is performed without possessing information on each source, such as its frequency characteristics and location, or on how the sources are mixed. The use of BSS in the development of comfortable acoustic communication channels between humans and machines is widely accepted.

Some books have been published on BSS, independent component analysis (ICA), and related subjects. There, ICA-based BSS has been well studied in the statistics and information theory fields, for applications to a variety of disciplines including wireless communication and biomedicine. However, as speech and audio signal mixtures in a real reverberant environment are generally convolutive mixtures, they involve a structurally much more challenging task than instantaneous mixtures, which are prevalent in many other applications.

The goal of this book is to provide a reference to the fascinating topic of BSS for convolved speech mixtures. The editors believe that this book is of particular value, as it comprises cutting edge reports by internationally recognized scientists on their findings and the state of the art. The individual chapters that we selected for this book were designed to be tutorial in nature with specific emphasis on an in-depth treatment of recent important results.

This book is organized into three sections that approximately follow the main areas of BSS.

Part 1 presents overdetermined or critically determined BSS where the number of sources is smaller than or equal to the number of microphones. Here, ICA is the main technology. ICA is a statistical method for extracting mutually independent sources from their mixtures. This approach utilizes spatial diversity to discriminate between desired and undesired components, i.e., it reduces the undesired components by forming a spatial null towards them. It is, in fact, a blind adaptive beamformer realized by unsupervised adaptive filtering.

Part 2 addresses underdetermined BSS, where there are fewer microphones than source signals. This is a challenging problem because of the inherently adverse conditions. ICA cannot be used for this case. Here, the sparseness of speech sources is very useful and time–frequency diversity plays a key role.

Part 3 presents monaural BSS where there is only one microphone. Even with a single microphone, we can separate a mixture by using the harmonicity and temporal structure of the sources. Here, we can build a probabilistic framework by assuming a source model, and separate a mixture by maximizing the a posteriori probability of the sources given the observations.

The authors and editors all hope that this book will serve as a guide for a large audience, inspire many readers, and be the source of new ideas to come. We hope this book will become a useful resource for readers ranging from students and practicing engineers to advanced researchers.

We would like to take this opportunity to express our deep gratitude to each and every one of our contributing authors for making this a unique work. Thanks to their cooperation, editing this book has turned out to be a very pleasant experience. Finally, we are very grateful to Dieter Merkle from Springer and his colleagues, for their encouragement and kind support.

Kyoto, Japan
San Diego, CA USA
July 2007

Shoji Makino
Te-Won Lee
Hiroshi Sawada

Contents

Part I. Multiple Microphone Blind Speech Separation with ICA

Part I

Multiple Microphone Blind Speech Separation with ICA

1 Convolutive Blind Source Separation for Audio Signals

Scott C. Douglas and Malay Gupta

Department of Electrical Engineering
Southern Methodist University
Dallas, Texas 75275 USA

Abstract. In this era of ever-improving communications technologies, we have become used to conversing with others across the globe. Invariably, a real-time telephone conversation begins with a microphone or other audio recording device. Noise in the environment can corrupt our speech signal as it is being recorded, making it harder to both use and understand further down the communications pathway. Other talkers in the environment add their own auditory interference to the conversation. Recent work in advanced signal processing has resulted in new and promising technologies for recovering speech signals that have been corrupted by speech-like and other types of interference. Termed *blind source separation* methods, or BSS methods for short, these techniques rely on the *diversity* provided by the collection of multichannel data by an array of distant microphones (sensors) in room environments. The practical goal of these methods is to produce a set of output signals which are much more intelligible and listenable than the mixture signals, without any prior information about the signals being separated, the room reverberation characteristics, or the room impulse response.

The main goal of this chapter is to describe several current approaches for the convolutive BSS task as they relate to speech processing applications. Novel methods and results that are described in this chapter include:

1. Block-based extensions of the natural gradient algorithm using robust gradient scaling.
2. Spatio-temporal extensions of the well-known FastICA algorithm to convolutive speech separation.
3. A performance comparison of several speech separation methods in real room environments and with different microphone array configurations.
4. Performance evaluations of various algorithms under the practical situation of an input signal power mismatch at the sensors.
5. A study of the importance of *a priori* room acoustics and direction-of-arrival information on algorithm convergence.

1.1 Historical Perspective

The BSS task has its roots in communication theory, in which man-made signals pass through a convolutive transmission medium that can alter the spectral and temporal characteristics of the signal to an extent such that

S. Makino et al. (eds.), Blind Speech Separation, 3–45.

the received signal shape is completely different from that of the originally-transmitted signal. It also has connections to human perception, in that human hearing system is able to focus attention on particular sound sources in the presence of significant noise and interference. It is not surprising that early contributions in the development of BSS techniques were made in two different areas of study: blind equalization/deconvolution in signal processing, and the analysis of neurological structures in the brain in medicine.

1.1.1 Blind Equalization and Deconvolution

Blind equalization and deconvolution originally arose in the context of communication systems involving symbols that are transmitted using modulated analog waveforms. Since the goal of communications is signal recovery, the channel effects must be removed from the observed signal, leading to the concept of channel identification to guide the design of a second system applied to the received signals that reverses the channel's effects. Training signals are often transmitted periodically to aid in this recovery, and these transmitted symbols lower the effective data rate of the communications link. In the *blind* counterpart to the above communications task, termed blind equalization or deconvolution, no training data is used, and the goal is to recover the original transmitted symbols based only on the received signal measurements.

Blind signal recovery is possible due to certain structural or statistical constraints that are often placed on the transmitted sources in communications tasks. This structural constraint can be on the amplitude of the signals being transmitted, as in the case of the constant modulus algorithm [2] and the Sato algorithm [1]. Alternatively, a constraint on the probability density function (p.d.f.) of the sources, such as their non-Gaussianity [3], can also be used, in which case the goal of signal recovery is to match the density of the equalized signal to that of the transmitted symbols. Further simplifications to this approach were suggested in [4], in which higher-order statistics are exploited for the equalization task. A similar method was originally proposed by Wiggins [5]. This work illustrates that non-Gaussian statistics have played an important role in many blind signal recovery algorithms.

In BSS, multiple unknown sources are transmitted through an unknown multiple-input, multiple-output channel. Recovery of the sources is hampered by an additional difficulty not present in the blind equalization or deconvolution task: the source distributions may differ from one another.

1.1.2 Neurological Structures and Information Representation

The study of brain function naturally leads one to consider the coding strategies employed by neurological structures such as the brains of higher animals in storing information [6]. Interestingly, concepts in information theory, concerned with study of randomness and structure in arbitrary data [7] became a key concept used in these studies, along with the specific concept of mutual information. These connections led Linsker [8, 9] to propose unsupervised information-theoretic learning rules to maximize the mutual information between the inputs and outputs of a nonlinear computational neural network. Previously, in the context of linear networks, it had already been shown that an application of a Hebbian-type learning rule extracts the principal component of stationary input data. Linsker observed that if linear units in the network were replaced by nonlinear units (neurons), similar learning rules extract independent components instead of principal components. This fact led several to postulate the information maximization (INFOMAX) principle [8, 10, 11] which states that, in a network of nonlinear computational units, maximum information flow can reduce the redundancy between the data at the output of the system. These studies were later linked to the term *independent component analysis*, which has as its goal the representation of a data set such that the components of the representation have a joint density that factors into products of marginal densities. Under linear mixing conditions and the constraint that the source signals are statistically-independent, ICA becomes a viable criterion by which to perform BSS, although technically these two fields of study are different, with different overall goals.

1.2 Mixture and Separation Models

The difficulty of the blind source separation task strongly depends on the way in which the signals are mixed within the physical environment. The simplest mixing scenario is termed *instantaneous mixing,* for which most early BSS algorithms were designed. Although useful for theoretical derivations, such algorithms have limited practical applicability in speech separation problems unless additional effort is spent on the system implementation. Real-world acoustical paths lead to *convolutive mixing* of the sources when measured at the acoustic sensors, and the degree of mixing is significant when the *reverberation time* (RT) of the room is large. Additional difficulties are encountered when the number of talkers is large; our experience indicates that separation of four or more talkers can be challenging in the presence of significant room reverberation.

1.2.1 Instantaneous Mixture and Separation Models

In instantaneous mixing, m unknown source signals $\{s_i(k)\},\ \ 1 \leq i \leq m$ are combined to yield the n measured sensor signals $\{x_j(k)\},\ \ 1 \leq j \leq n$ as

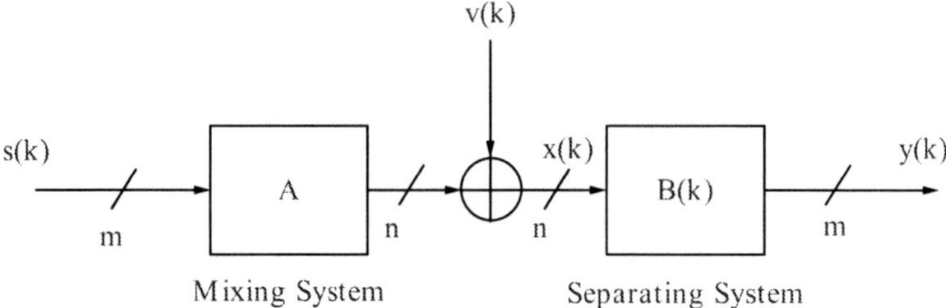

Fig. 1.1. Block diagram of the instantaneous BSS task.

$$x_j(k) = \sum_{i=1}^{m} a_{ji} s_i(k) + v_j(k) \tag{1.1}$$

where $\{a_{ji}\}$ are the coefficients of the linear time-invariant mixing system represented by the $(n \times m)$ matrix $\mathbf{A}$ and $v_j(k)$ is additive noise at the jth sensor.

The goal of BSS for instantaneous mixtures is to adjust the coefficients of an $m \times n$ separation or demixing matrix $\mathbf{B}$ such that

$$y_i(k) = \sum_{j=1}^{n} b_{ij}(k) x_j(k) \tag{1.2}$$

contains an estimate of a single original source $s_{i'}(k)$ for some index assignment $i \Rightarrow i'$ without repetition. The block diagram of this task is shown in Fig. 1.1.

While many algorithms have been developed for the instantaneous mixing case, practical algorithms for speech separation must take the convolutive mixing of the acoustic paths into account, a fact that has driven most recent research work in the BSS area.

1.2.2 Convolutive Mixture and Separation Model

Convolutive mixing arises in acoustic scenarios due to time delays resulting from sound propagation over space and the *multipath* generated by reflections of sound off of different objects, particularly in rooms and other enclosed settings. As a result, the m sources are mixed by a time-dispersive multichannel system as

$$x_j(k) = \sum_{l=-\infty}^{\infty} \sum_{i=1}^{m} a_{jil} s_i(k-l) + v_j(k) \tag{1.3}$$

where $\{x_j(k)\}$, $1 \leq j \leq n$ are the n sensor signals and $\{a_{jil}\}$ are the coefficients of the discrete-time linear time-invariant mixing system $\{\mathbf{A}_l\}_{l=-\infty}^{\infty}$, where each matrix $\mathbf{A}_l$ is of dimension $(n \times m)$. Here, we have

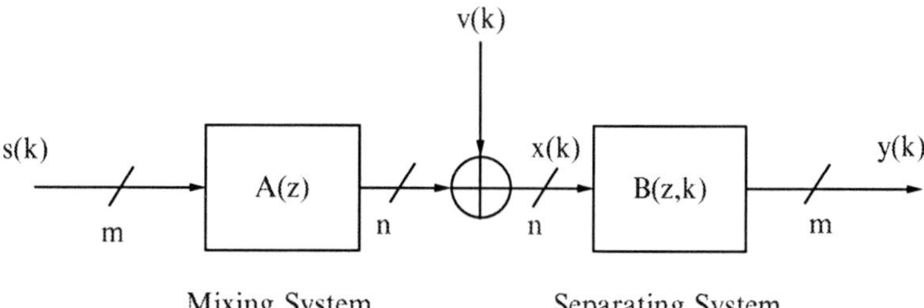

Fig. 1.2. Block diagram of the convolutive BSS task. $\mathbf{A}(z) = \sum_{l=-\infty}^{\infty} \mathbf{A}_l z^{-l}$ and $\mathbf{B}(z,k) = \sum_{l=-\infty}^{\infty} \mathbf{B}_l(k) z^{-l}$ denote the z transform of the sequence of the systems $\{\mathbf{A}_l\}$ and $\{\mathbf{B}_l(k)\}$, respectively.

assumed bandlimited signals such that sampled signals can take the place of continuous-times ones. The block diagram for the convolutive BSS in the discrete-time or z-domain is shown in Fig. 1.2.

Reverberant environments, in which sounds undergo hundreds or thousands of reflections as they propagate within the acoustical space, introduce time dependencies in the signal mixtures beyond those already present in the original source signals. For this reason, designers of convolutive BSS algorithms must consider both (i) spatial unmixing and (ii) the temporal changes introduced by the mixing system. Ideally, the temporal structure of the original sources would be preserved in the outputs of the separation system, although this feature requires some knowledge of the temporal structures of the source signals. Convolutive BSS algorithms must exploit both spatial and temporal signal characteristics to function properly, which is why they are sometimes referred to as *spatio-temporal* BSS algorithms.

Without any additional constraints, a convolutive BSS system processes the signal mixtures $x_j(k)$ such that

$$y_i(k) = \sum_{l=-\infty}^{\infty} \sum_{j=1}^{n} b_{ijl}(k) x_j(k-l) \tag{1.4}$$

contains a filtered estimate of one of the sources $s_{i'}(k)$ for some index assignment $i \Rightarrow i'$ without repetition.

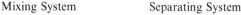

The relation in (1.4) is in the form of a multichannel convolution, where the coefficients $\{b_{ijl}\}(k)$ represent the impulse response of the multichannel separation system filter. Using well-known Fourier transform properties, it is possible to express the above demixing system in the frequency-domain, where the Fourier coefficients of the multichannel separation system filter are the unknown parameters. This parametrization of the separation system is discussed in more detail in a later section, where issues regarding the dimensionality of the model are dealt with explicitly.

The number of unknown sources m plays an important role in BSS algorithms in that, under reasonable constraints on the mixing system, the separation problem remains linear if the number of mixture signals n is greater than or equal to m ($n \geq m$). Of course, many practical situations involve signal mixtures where $n < m$, and thus the separation problem cannot be solved by linear filtering alone. Dealing with these scenarios, through preprocessing of the measurements, postprocessing of the separated signal outputs, or changes to the separation algorithm, is an important open problem. For the ease of our discussions in this chapter, we will assume $m = n$ in all subsequent discussions. We will also assume that the noise signals $\{v_j(k)\}$ are negligible as compared to the interference due to multiple mixed sources.

1.3 Ambiguities of BSS

Ideally, BSS algorithms assume as little knowledge as possible about the mixing system or the sources being mixed. This lack of knowledge leads to several ambiguities regarding the possible solutions provided by a BSS algorithm. To understand these ambiguities, consider the signal model in (1.1) in which we denote the true values of the mixing matrix $\mathbf{A}$ and source signal vector $\mathbf{s}(k) = [s_1(k) \ \cdots \ s_m(k)]^T$ as $\mathbf{A}_o$ and $\mathbf{s}_o(k)$, respectively. Clearly, the same signal model $\mathbf{x}(k) = [x_1(k) \ \cdots \ x_m(k)]^T$ is generated using the mixing matrix $\mathbf{A}_o\mathbf{M}$ and source signal vector $\mathbf{M}^{-1}\mathbf{s}_o(k)$ through

$$\mathbf{x}(k) = \mathbf{A}_0\mathbf{s}_0(k) = \mathbf{A}_0\mathbf{M}[\mathbf{A}^{-1}\mathbf{s}_0(k)] \tag{1.5}$$

for any nonsingular matrix $\mathbf{M}$. Thus, it is the structure of $\mathbf{M}$ which determines the feasibility constraints of BSS, as specified by the waveform preserving relations [12]. In this case, the ambiguities within the matrix $\mathbf{M}$ must be of the following form for separation to be possible:

$$\mathbf{M} = \mathbf{D}\mathbf{P} \tag{1.6}$$

where $\mathbf{D}$ is a nonsingular diagonal matrix and $\mathbf{P}$ is a permutation matrix. This form of $\mathbf{M}$ translates into scaling and permutation ambiguities that are potentially present in all solutions to the BSS task. As most of the information is carried by the shape of the waveform [12], these amplitude and permutation ambiguities do not pose a significant problem in practice.

In speech separation tasks, the overall goal is typically to improve speech intelligibility by removing interchannel interference. Permutation and scaling ambiguities do not pose a problem in this regard. Due to the convolutive nature of the mixing system and the lack of temporal knowledge of the sources, however, additional *phase* ambiguities are also present in the separated system outputs. These phase ambiguities can lead to undesirable situations, such as output signals that are separated from one another but with increased amounts of signal reverberation. Typically, phase ambiguities are kept in check through careful overall system design, in which the temporal extent of the separation system is constrained to be long enough to allow separation without additional degrees of freedom that allow arbitrary phase relationships to be imposed.

1.4 Criteria for BSS in Speech Separation

It is fortunate that speech signals are feature-rich and possess certain characteristics that enable BSS systems to be developed. Despite the wideband nature of the voiced speech spectrum, it is well-known that bandlimiting speech to frequencies between 300 Hz and 3500 Hz does not significantly harm its intelligibility [13]. The spectrum of unvoiced speech is not inherently bandlimited but instead tends to fall off rapidly at the upper and lower frequency edges of human hearing.

Speech is inherently a nonstationary signal, and amplitude modulations are largely responsible for this characteristic. Additional properties of speech signals that are relevant to speech separation include the following:

1. Speech signals originating from different talkers at different spatial locations in an acoustic environment can be considered to be statistically-independent.
2. Each speech signal typically has a unique temporal structure over short time frames (less than 1 second).
3. Speech signals are quasi-stationary for small time durations ($\approx$10 ms) but nonstationary over longer periods.

Theoretically, all of the above properties can simultaneously be exploited by a separation system [14], although it is possible to design systems that use only one of these features to achieve adequate separation.

Statistical independence of speech signals

BSS algorithms can be designed to exploit the statistical independence of different talkers in an acoustic environment. These algorithms attempt to make the output signals as statistically-independent as possible. The algorithms in this class are motivated from information-theoretic concepts and generally employ higher- or lower-order (*i.e.* non-second-order) signal statistics. Because of the spatio-temporal nature of the mixing process, the interchannel phases of the separate signals are important features, and the use of statistical independence alone as a criterion limits these methods to scenarios in which almost all of the source signals are non-Gaussian-distributed. It is known that the statistics of jointly-Gaussian random processes can be completely specified by their first- and second-order statistics (*i.e.* mean and covariance properties); hence, higher- and lower-order statistical features do not carry any additional information about Gaussian signals. Fortunately, speech is a non-Gaussian signal, in which its amplitude statistics have an exponential tail similar to that of a Laplacian distribution [15].

Temporal structure of speech signals

Speech can also be separated using second-order statistics (SOS) alone if the source signals have unique temporal structures with distinct autocorrelation functions. In such cases, statistical independence of the individual talker's signals need not be assumed, and the non-Gaussian nature of the speech signals are not important when these statistics are used. Algorithms that employ this criterion impose spatial decorrelation between the different outputs of the separation system through simultaneous diagonalization of output correlation matrices over multiple time-lags.

Nonstationarity of speech signals

Speech is also a highly nonstationary signal due to the amplitude modulations inherent in the voiced portions of speech and to the intermingling of voiced and unvoiced speech patterns in most dialects. Because the nonstationary characteristics of individual talkers is not likely to be similar, the second-order statistical properties of the individual speech signals can be expected to have different autocorrelation properties over short time periods, and thus second-order statistics can be used to perform separation In this general class of methods, the measured mixtures is divided into blocks corresponding to different amplitude modulations of the underlying source signals. A multichannel correlation matrix is then computed for each block, which is different from block to block due to the nonstationarity of the underlying sources. Source separation is then achieved by simultaneous diagonalization of these correlation matrices.

1.5 Time-Domain Convolutive BSS Using Spatial Independence

BSS algorithms exploiting source independence are widely-studied in the scientific literature under an instantaneous mixing model, leading to solutions for independent component analysis (ICA). It should be recognized that ICA and BSS are different goals. ICA is concerned about linear signal decomposition onto independent statistical bases, whereas BSS is concerned with the separation of signals that have been linearly mixed. Convolutive BSS directly addresses a particular structured linear mixing model in the latter task. Approaches for BSS that exploit the statistical independence of the sources can be classified into one of two types [16]:

1. Density-matching-based approaches
2. Contrast-function-based approaches

1.5.1 Density Matching BSS Algorithms Using Natural Gradient Adaptation

The concept of input–output density matching for inverting unknown single-input, single-output linear systems driven by independent, identically-distributed (i.i.d.) sources is a well-studied problem in the communications field [3]. Convolutive BSS techniques, however, must deal with multiple-input multiple-output (MIMO) systems and sources that do not have the same statistical distributions. Even so, the ideas presented in [3] are quite valuable in that they can be extended to MIMO systems when certain input signal conditions are satisfied. The strongest of these conditions is the requirement that all but one of the sources be non-Gaussian distributed in their amplitude statistics.

For instantaneous mixtures, density-matching methods can also derived using the so-called information maximization approach to ICA (*e.g.* see [11]). Under linear mixing conditions, both criteria assume that the joint probability density function (p.d.f.'s) of the sources can be expressed as

$$p(\mathrm{s}) = \prod_{k=1}^{m} p_{\mathbf{s}}(s_k) \tag{1.7}$$

where $p(\mathbf{s})$ is the joint p.d.f. of the source vector $\mathbf{s}(k) = [s_1(k)s_2(k)\cdots s_m(k)]^T$ and $p_s(s_i)$ for $i \in \{1, \ldots, m\}$ are the marginal p.d.f.'s of the sources.

The goal of density-matching BSS methods is to linearly transform the observed data in $\mathbf{x}(k) = [x_1(k) \ \cdots \ x_m(k)]^T$ under the mixing model in (1.1) using a system matrix $\mathbf{B}(k)$ with entries $\{b_{ij}(k)\}$ according to (1.2) such that the joint p.d.f. of the output vector sequence $\mathbf{y}(k) = \mathbf{B}(k)\mathbf{x}(k)$, denoted by $p_{\mathbf{y}}(\mathbf{y})$, closely matches a model density $\hat{p}_{\mathbf{y}}(\mathbf{y})$.

Calculating the difference between $\hat{p}_{\mathbf{y}}(\mathbf{y})$ and $p_{\mathbf{y}}(\mathbf{y})$ requires a criterion that is well-motivated by the underlying problem structure. A particularly-useful criterion common to all information-theoretic approaches [17] is the *Kullback–Leibler* (KL)-divergence measure $D(\cdot||\cdot)$ given by

$$D\left(p_{\mathbf{y}}||\hat{p}_{\mathbf{y}}\right) = \int p_{\mathbf{y}}(\mathbf{y}) \log\left(\frac{p_{\mathbf{y}}(\mathbf{y})}{\hat{p}_{\mathbf{y}}(\mathbf{y})}\right) d\mathbf{y} \tag{1.8}$$

where the shorthand $d\mathbf{y} = dy_1 \cdots dy_m$ in the m-dimensional integral has been used. Note that the KL-divergence is not a true *distance* measure as it is asymmetric with respect to $p_{\mathbf{y}}(\mathbf{y})$ and $\hat{p}_{\mathbf{y}}(\mathbf{y})$. Different choices of $\hat{p}_{\mathbf{y}}(\mathbf{y})$ yield different solutions to the BSS task. For example, if

$$\hat{p}_{\mathbf{y}}(\mathbf{y}) = C \prod_{i=1}^{m} p_s(y_i) \tag{1.9}$$

is chosen, where C is a normalization constant, a *maximum-likelihood* (ML) estimate of the demixing matrix $\mathbf{B}(k)$ is obtained under the linear mixing model in (1.1). Setting

$$\hat{p}_{\mathbf{y}}(\mathbf{y}) = C \prod_{i=1}^{m} \hat{p}_{y_i}(y_i) \tag{1.10}$$

where $\hat{p}_{y_i}(y_i)$ are the marginal p.d.f.'s of each output for the current value of $\mathbf{B}(k)$, yields the minimum mutual information (MI) [18] based approach to BSS. ML and MI-based approaches represent two extremes to this task, as the former approach has perfect knowledge of the inputs signal statistics (most restrictive) whereas the latter approach has no knowledge of the source statistics except their assumed independence (least restrictive). In practice, one typically trades off accuracy and computational complexity in a pragmatic choice by selecting marginal p.d.f.'s $\hat{p}_{y_i}(y_i)$ that are "close to" those of the underlying sources. How close is "close enough" can be answered formally in terms of local stability conditions [19] that depend both on the source distributions and the chosen p.d.f. models.

Convolutive mixing and demixing is also a linear process, and hence we can extend the density-matching criterion in (1.8) to extend to the convolutive BSS task. When $m = n$, the output-system-dependent portion of the KL-divergence cost function is given by

$$\hat{\mathcal{J}}(\mathbf{B}(z,k)) = -\sum_{i=1}^{m} \log p_s(y_i(k)) - \frac{1}{2\pi j}\oint \log|\det \mathbf{B}(z,k)|z^{-1}dz \qquad (1.11)$$

where $\mathbf{B}(z,k) = \sum_p \mathbf{B}_p(k)z^{-p}$ is the z-transform of the multichannel separation system's impulse response and $\det \mathbf{B}(z,k)$ denotes the determinant of $\mathbf{B}(z,k)$ [22]. The first term in (1.11) is affinely-related to the mutual information of the output vector sequence $\{\mathbf{y}(k)\}$ when $p_s(y)$ accurately models the source p.d.f.'s [20]. The second term on the R.H.S. of (1.11) is due to the input–output transformation of the separation system and prevents the trivial solution $\mathbf{B}(z,k) = \mathbf{0}$ for the minimization of (1.11).

Given the cost function in (1.11), we desire simple procedures to minimize it over all possible multichannel filters $\mathbf{B}(z,k)$. Stochastic gradient procedures are often useful for linear estimation tasks due to their simplicity and have proven to be highly popular when the criterion being minimized has a single minimum and nearly-isotropic gradient magnitudes around this minimum. Unfortunately, the cost function in (1.11) has neither of these properties. Moreover, the derivative of the constraint term in (1.11) with respect to each $\mathbf{B}_p(k)$ of the separation system is difficult to implement in practical situations. Fortunately, an alternative descent procedure, termed the *natural gradient* by Amari [21] and the *relative gradient* by Cardoso [18], is useful for this optimization task. Natural gradient adaptation alters the standard gradient direction according to the local Riemannian structure of the parameter space and provides isotropic convergence about any local minima of the cost function in this situation [21]. The complete natural gradient algorithm for minimizing (1.11) is given by [22]

$$\mathbf{y}(k) = \sum_{l=-\infty}^{\infty} \mathbf{B}_l(k)\mathbf{x}(k-l) \qquad (1.12)$$

$$\mathbf{u}_l(k) = \sum_{q=-\infty}^{\infty} \mathbf{B}_q^T(k)\mathbf{y}(k-l+q) \tag{1.13}$$

$$\mathbf{F}_l(k) = \mathbf{f}(\mathbf{y}(k))\mathbf{u}_l^T(k) \tag{1.14}$$

$$\mathbf{B}_l(k+1) = (1+\mu(k))\mathbf{B}_l(k) - \mu(k)\mathbf{F}_l(k) \tag{1.15}$$

where $\mathbf{f}(\mathbf{y}) = [f(y_1)\cdots f(y_m)]^T$ in (1.14) is a vector of source nonlinearities

$$f(y) = -\partial \log p_s(y_i)/\partial y_i \tag{1.16}$$

and $\mu(k)$ is a small positive step size constant that controls the convergence speed of the update procedure. The nonlinear function $\mathbf{f}(\mathbf{y})$ is applied element wise on the vector $\mathbf{y}$, which for simplicity is constructed from m identical scalar nonlinear functions $f(y)$. Typical choices for $f(y)$ are

$$f(y) = \operatorname{sgn}(y) = \frac{y}{|y|} \tag{1.17}$$

$$f(y) = \tanh(\alpha_1 y) \tag{1.18}$$

with $2 \leq \alpha_1 \leq 10$.

The algorithm in (1.12)–(1.15) remains impractical because it uses a doubly-infinite noncausal multichannel filter which cannot be easily implemented. A delayed and causal finite-impulse-response (FIR) approximate implementation was first proposed in [22]. Because this approximate version does not achieve spatial and temporal independence of the output signals $\{y_i(k)\}$, however, proper signal truncation and windowing become critical for the nonlinear correlations in the update terms to provide good gradient estimates. As a result, the algorithm in [22] generally requires long FIR filter lengths to work properly, and the increased number of parameters in the system hampers convergence. Recently, a new implementation of this approach that uses causal FIR approximations to the gradient calculations has been proposed [23]. The coefficient updates for this method are given by

$$\underline{\mathbf{B}}(k) = [\mathbf{B}_0(k) \;\; \mathbf{B}_1(k) \;\; \cdots \;\; \mathbf{B}_L(k)] \tag{1.19}$$

$$\underline{\mathbf{x}}(k) = [\mathbf{x}^T(k) \;\; \mathbf{x}^T(k-1) \;\; \cdots \;\; \mathbf{x}^T(k-L)]^T \tag{1.20}$$

$$\underline{\mathbf{z}}(k) = \underline{\mathbf{R}}(k)\underline{\mathbf{x}}(k) \tag{1.21}$$

$$\mathbf{y}(k) = \underline{\mathbf{B}}(k)\underline{\mathbf{x}}(k) \tag{1.22}$$

$$\underline{\mathbf{B}}(k+1) = (1+\mu(k))\underline{\mathbf{B}}(k) - \mu(k)\mathbf{f}(\mathbf{y}(k))\underline{\mathbf{z}}^T(k) \tag{1.23}$$

where

$$\underline{\mathbf{R}}(k) = \begin{bmatrix} \mathbf{R}_0(k) & \mathbf{R}_1(k) & \cdots & \mathbf{R}_L(k) \\ \mathbf{R}_{-1}(k) & \mathbf{R}_0(k) & \cdots & \mathbf{R}_{L-1}(k) \\ \vdots & & & \vdots \\ \mathbf{R}_{-L}(k) & \mathbf{R}_{-L+1}(k) & \cdots & \mathbf{R}_0(k) \end{bmatrix} \tag{1.24}$$

$$\mathbf{R}_l(k) = \begin{cases} \sum_{p=0}^{L-l} \mathbf{B}_p^T(k)\mathbf{B}_{p+l}(k) & \text{if } 0 \leq l \leq L \\ \mathbf{R}_{-l}^T(k) & \text{if } -L \leq l \leq -1 \end{cases} \tag{1.25}$$

In the above algorithm, the matrix $\underline{\mathbf{R}}(k)$ typically changes slowly over time, and thus it only needs to be computed periodically during adaptation. The calculation of $\underline{\mathbf{z}}(k)$ in (1.21) involves $m^2(L+1)^2$ multiply/adds, making it the most-complex data-dependent calculation in the update. An approximate recursive update for $\underline{\mathbf{z}}(k)$ was proposed in [23] as

$$\underline{\hat{\mathbf{t}}}(k) = [\hat{\mathbf{t}}_0^T(k) \;\; \hat{\mathbf{t}}_1^T(k) \;\; \cdots \;\; \hat{\mathbf{t}}_{2L-1}^T(k) \;\; \hat{\mathbf{t}}_{2L}^T(k)]^T \tag{1.26}$$

in which the last $m(L+1)$ entries of $\underline{\hat{\mathbf{z}}}(k)$ are used in place of $\underline{\mathbf{z}}(k)$. This recursive update is

$$\underline{\hat{\mathbf{z}}}(k) = \hat{\mathbf{t}}_{L+p}(k) \tag{1.27}$$

$$\hat{\mathbf{t}}_p(k) = \begin{cases} \mathbf{R}_L^T(k)\mathbf{x}(k) & \text{if } p = 0 \\ \hat{\mathbf{t}}_{p-1}(k-1) + \mathbf{R}_{L-p}^T(k)\mathbf{x}(k) & \text{if } 1 \leq p \leq L \\ \hat{\mathbf{t}}_{p-1}(k-1) + \mathbf{R}_{p-L}^T(k)\mathbf{x}(k) & \\ \quad - \mathbf{R}_{2L+1-p}^T(k-L-1)\mathbf{x}(k-L-1) & \text{if } L+1 \leq p \leq 2L \end{cases} \tag{1.28}$$

We have successfully used this approximate version of the causal FIR natural gradient algorithm in speech separation experiments involving rooms of various reverberation times from 150 ms to 450 ms for two- and three-source mixtures; see the simulations section for representative results. The main issue in using the algorithm is the choice of step size $\mu(k)$ due to the highly-nonlinear nature of the coefficient updates.

Block-based scaled natural gradient algorithm. In all natural gradient BSS procedures, convergence is governed both by the step size sequence $\mu(k)$ and the initial coefficient values of the separation system. It is often challenging to choose these quantities to obtain fast convergence without explosive divergence. Moreover, in the instantaneous mixing case, our experience with natural gradient BSS methods indicates that for larger-order mixtures (*e.g.* $m \geq 10$), it is nearly impossible to choose $\mathbf{B}(0)$ to obtain convergence for *any* constant step size μ. Although some work has been done in time-varying step size sequences for this algorithm class [24], initial convergence is still slow, and $\mu(k) \leq \mu_{max}$ must be maintained for some unknown μ_{max}. In a dynamic environment, the algorithm in (1.23) may not be robust to the positions of talkers to microphones, amplitude variations in the source data, and variations in the reverberation characteristics of the room.

We have recently proposed a novel, simple block-based modification to (1.23) that alleviates the aforementioned difficulties [25]. According to this modification, a scaling constraint is used to maintain a constant (natural) gradient magnitude for the algorithm. This is possible since the separation performance does not depend explicitly on the scaling of the separation system coefficients. The modification adds little computational complexity and achieves significant practical advantages. In particular, the algorithm exhibits fast convergence and excellent performance for a fixed step size μ, independent of the magnitudes of the measured signals and the separation system coefficients. When tested in various challenging situations, the procedure never diverges and typically converges in only 40 to 150 iterations under mixing conditions in which the original algorithm does not converge reliably for any fixed step size.

In the instantaneous mixing case, this algorithm attempts to solve the following optimization problem for the combined system via natural gradient descent:

$$\text{minimize} \quad -\log|\det(\mathbf{B}(k))| - \sum_{l=N(k-1)+1}^{Nk} \sum_{i=1}^{m} \log p_s(y_{ik}(l)) \tag{1.29}$$

$$\text{such that} \quad \frac{1}{mN} \sum_{l=N(k-1)+1}^{Nk} \sum_{i=1}^{m} \sum_{j=1}^{m} |f(y_{i(k+1)l}) y_{j(k+1)l}| = 1 \tag{1.30}$$

where $p_s(y)$ is a chosen source p.d.f. model. Equation (1.30) is a scalar L_1-norm constraint on the *a posteriori* gradient data matrix that is satisfied at algorithm convergence for separated sources. A scalar constraint is chosen to stabilize the algorithm as it does not alter the goal of making the output signals statistically-independent and is expected to have a minimal effect on the algorithm's local convergence behavior.

The adaptation procedure used to achieve (1.29)–(1.30) jointly adjusts $\mathbf{B}(k)$ in a natural gradient procedure to try to minimize (1.29) while imposing a multiplicative scaling of all of the elements of the updated matrix to satisfy

(1.30) exactly. The complete algorithm in the instantaneous mixing case is given by

$$\overline{\mathbf{B}}(k) = c(k)\mathbf{B}(k) \tag{1.31}$$

$$\overline{\mathbf{y}}_k(l) = \overline{\mathbf{B}}(k)\mathbf{x}_k(l) \tag{1.32}$$

$$\overline{\mathbf{G}}(k) = \frac{1}{N}\sum_{l=N(k-1)+1}^{Nk} \mathbf{f}(\overline{\mathbf{y}}_k(l))\overline{\mathbf{y}}_k^T(l) \tag{1.33}$$

$$\mathbf{B}(k+1) = (1+\mu)\overline{\mathbf{B}}(k) - \mu\overline{\mathbf{G}}(k)\overline{\mathbf{B}}(k) \tag{1.34}$$

where $c(k)$ is a scaling factor sequence. In order to impose (1.30), we compute $c(k)$ as

$$c(k) = \frac{1}{h\left(d(k)\right)} \tag{1.35}$$

$$d(k) = \frac{1}{m}\sum_{i=1}^{m}\sum_{j=1}^{m} |g_{ij}(k)| \tag{1.36}$$

where $h(d)$ is the inverse function of the magnitude of $yf(y)$; *i.e.* $h(yf(y)) = |y|$. To further simplify the algorithm we can absorb the scaling factor $c(k)$ into the update for $\mathbf{B}(k)$, such that little computational complexity is added. The simplified algorithm then becomes

$$\mathbf{y}_k(l) = \mathbf{B}(k)\mathbf{x}(l) \tag{1.37}$$

$$\mathbf{B}(k+1) = (1+\mu)c(k)\mathbf{B}(k) - \mu\frac{c(k)}{d(k)}\mathbf{G}(k)\mathbf{B}(k) \tag{1.38}$$

As for choices of algorithm nonlinearity $f(y)$, typical choices and their associated $h(d)$ functions are

$$f(y) = |y|^2 y \Rightarrow h(d) = d^{\frac{1}{4}} = \sqrt{\sqrt{d}} \tag{1.39}$$

$$f(y) = \frac{y}{|y|} \Rightarrow h(d) = d \tag{1.40}$$

$$f(y) = \tanh(\alpha|y|)\frac{y}{|y|} \Rightarrow h(d) = d \tag{1.41}$$

where $\alpha > 1$. The last choice uses an approximation for $h(d)$, as the inverse of $|y|\tanh(\alpha|y|)$ approaches $|y|$ as α gets large.

We have extended this method to the time-domain algorithm for multichannel blind deconvolution and convolutive BSS in [23]. The complete algorithm in this case is

$$\mathbf{y}_k(l) = \sum_{p=0}^{L} \mathbf{B}_p(k)\mathbf{x}(l-p) \tag{1.42}$$

$$\mathbf{F}_p(k) = \frac{1}{N} \sum_{l=N(k-1)+1}^{Nk} \mathbf{f}(\mathbf{y}_k(l))\mathbf{x}^T(l-p), 0 \leq p \leq L \tag{1.43}$$

$$\mathbf{G}_q(k) = \sum_{p=0}^{L} \mathbf{F}_{q-p}(k)\mathbf{B}_{L-p}^T(k), 0 \leq q \leq 2L \tag{1.44}$$

$$\mathbf{H}_p(k) = \sum_{q=0}^{L} \mathbf{G}_{L+p-q}(k)\mathbf{B}_q(k), 0 \leq p \leq L \tag{1.45}$$

$$d(k) = \frac{1}{m} \sum_{i=1}^{m} \sum_{j=1}^{m} \sum_{q=0}^{2L} |g_{ijq}(k)| \tag{1.46}$$

$$\mathbf{B}_p(k+1) = (1+\mu)c(k)\mathbf{B}_p(k) - \mu \frac{c(k)}{d(k)} \mathbf{H}_p(k) \tag{1.47}$$

In the above relations, both $\mathbf{B}_p(k)$ and $\mathbf{F}_p(k)$ are assumed to be zero outside the range $p \in [0, L]$.

Simulations of the scaled natural gradient algorithms above in instantaneous and convolutive BSS tasks indicate that the algorithms achieve fast convergence for a range of step sizes that are independent of the initial scaling of either the separation system or the input signal mixtures so long as the amount of data samples N in the gradient calculations are large enough. Data reuse is recommended, such that $\mathbf{x}(k+iN) = \mathbf{x}(k)$ for $i \in \{1, 2, \cdots\}$, and the entire data set $\mathbf{x}(k)$, $k \in \{1, N\}$ is used at each algorithm iteration. Typically, step sizes in the range $0.1 \leq \mu \leq 0.5$ have been chosen and appear to work well.

1.5.2 Contrast-Function-based Convolutive BSS

In some applications, extracting all of the sources present in the observed mixtures may not be the ultimate goal. Simultaneous extraction of all of the sources present in the signal mixtures is best performed using BSS algorithms that work in a *symmetric mode*, in which the coefficient solutions used to generate all of the system outputs are solved for jointly without regard for signal ordering. Systems that extract sources one by one, using the previous $(i-1)$ system outputs or coefficient solutions to obtain the ith estimated source, work in *deflation mode*. By definition, the density matching algorithms discussed in the previous section operate in a symmetric mode. In this section, we consider contrast-based algorithms that have the flexibility of operating in either symmetric or deflation modes. A *contrast* is a continuous-valued nonquadratic function depending only on a single extracted output signal whose maxima under suitable source or coefficient constraints correspond to separated sources. Contrast functions are used in a constrained maximization procedure over the demixing system coefficients to extract independent components from their observed mixtures. These algorithms rely on both the statistical independence and the non-Gaussianity of the source signals.

We first consider the instantaneous mixing case for notational simplicity and later develop appropriate algorithmic extensions to convolutive BSS. For the ith output signal computed as in (1.2), we define the contrast function $\boldsymbol{\Phi}[y_i(k)]$ as an operator on the sequence $y_i(k)$ that indicates a separating solution at its maximum under some suitable constraint on the coefficients $\{b_{ij}(k)\}$, $1 \leq j \leq m$, or the output signal $y_i(k)$. An example of such a contrast is the *squared kurtosis* given by

$$\boldsymbol{\Phi}_\kappa[y_i(k)] = |\kappa[y_i(k)]|^2 \tag{1.48}$$

$$\kappa[y_i(k)] = E\{|y_i(k)|^4\} - 3E^2\{|y_i(k)|^2\} \tag{1.49}$$

where $\kappa[y_i(k)]$ is the kurtosis of the output signal $y_i(k)$. In this case, the constraint $E\{|y_i(k)|^2\} = 1$ is often used. It can be shown that in the noise-free case, maximizing $\boldsymbol{\Phi}_\kappa[y_i(k)]$ across the ith output system coefficients under the constraint $E\{|y_i(k)|^2\} = 1$ results in separated sources in the instantaneous mixing case [26].

Because contrast-based approaches involve the optimization of separate cost functions for each extracted source, one must design appropriate *joint constraints* on the output signals to guarantee uniqueness of the extracted components, as extraction of the same source multiple times within the procedure is computationally wasteful. These joint constraints are usually chosen to be *whitening constraints*, in which

$$E\{y_i(k)y_j(k)\} = \delta_{ij} \tag{1.50}$$

where δ_{ij} is the Kronecker impulse function. Whitening constraints lead to a two-stage data processing strategy consisting of

(i) a prewhitening stage, in which observations are linearly transformed by an $m \times n$ prewhitening matrix $\mathbf{H}$ such that the output sequence $\mathbf{v}(k) = \mathbf{H}\mathbf{x}(k)$ satisfies $E\{\mathbf{v}(k)\mathbf{v}^T(k)\} = \mathbf{I}$; and

(ii) a separation stage, in which a separation matrix $\mathbf{W}(k) = [\mathbf{w}_1(k)\ \mathbf{w}_2(k)\ \cdots\ \mathbf{w}_m(k)]^T$ is computed such that each of the output signals in

$$\mathbf{y}(k) = \mathbf{W}(k)\mathbf{v}(k) \tag{1.51}$$

maximize the chosen contrast $\boldsymbol{\Phi}[y_i(k)]$ under appropriate signal magnitude constraints. The combined separation system is given by $\mathbf{B}(k) = \mathbf{W}(k)\mathbf{H}$.

Prewhitening the data allows us to impose *orthogonality constraints* on the separation system to guarantee output signal uniqueness, a fact we can easily show in the instantaneous case. Denoting the prewhitened mixing system by $\Gamma = \mathbf{HA}$, it is straightforward to show that $E\{\mathbf{s}(k)\mathbf{s}^T(k)\} = E\{\mathbf{v}(k)\mathbf{v}^T(k)\} = \mathbf{I}$ implies that Γ is orthonormal; *i.e.* $\Gamma\Gamma^T = \Gamma^T\Gamma = \mathbf{I}$. Hence, the separation matrix $\mathbf{W}(k)$ can also be constrained to be orthonormal. The orthogonality of $\mathbf{W}(k)$ ensures that once a source has been extracted by the ith row of $\mathbf{W}(k)$, the solutions to other rows of $\mathbf{W}(k)$ can be made different from that of the ith row by imposing

$$\mathbf{w}_i^T(k)\mathbf{w}_j(k) = 0, \quad j \neq i \tag{1.52}$$

The unit-power constraint $E\{|y_i(k)|^2\} = 1$ results in $\mathbf{w}_i^T(k)\mathbf{w}_i(k) = 1$. In the instantaneous mixing case, these coefficient orthogonality constraints can be imposed by various numerical procedures such as Gram–Schmidt orthogonalization of the rows of $\mathbf{W}(k)$ or a singular-value decomposition of $\mathbf{W}(k)$ [27]. The optimization procedure used to calculate any one row of $\mathbf{W}(k)$ in this formulation of the BSS task becomes

$$\text{maximize} \quad \mathcal{J}_C(\mathbf{w}_i(k) = \mathbf{\Phi}[y_i(k)]) \tag{1.53}$$

$$\text{such that} \quad \mathbf{w}_i^T(k)\mathbf{w}_j(k) = \delta_{ij}, \quad 1 \leq j \leq m \tag{1.54}$$

One of the most popular procedures for contrast based ICA and instantaneous BSS is the FastICA algorithm of Hyvärinen and Oja [28, 29]. For an m-element linear non-Gaussian signal mixture, the block-based procedure maximizes the kurtosis-based contrast of (1.48). The updates for an N-sample block of data are

$$\tilde{\mathbf{w}}_i(k) = \left[\sum_{l=0}^{N-1} y_i^3(k+l)\mathbf{v}(k+l)\right] - 3\mathbf{w}_i(k) \tag{1.55}$$

$$\mathbf{w}_i(k+N) = \frac{\tilde{\mathbf{w}}_i(k)}{||\tilde{\mathbf{w}}_i(k)||} \tag{1.56}$$

Equation (1.55) can be shown to be an approximate Newton-based procedure; in which the second step maintains the unit-length constraint on $\mathbf{w}_i(k)$ imposed by (1.54). Coefficient vector orthogonality is used to guarantee uniqueness of the extracted components. The algorithm enjoys a number of useful properties, including fast convergence, guaranteed global convergence for certain mixing conditions and contrasts, and robust behavior when noise is present. More forms of contrast function can be obtained by different choices of $\mathbf{\Phi}[y_i]$.

Spatio-temporal extensions. Spatio-temporal extensions of contrast-based methods involve additional constraints beyond spatial whitening and coefficient orthogonality. In particular,

- Signal prewhitening is spatio-temporal, such that

$$E\{v_i(k)v_j(l)\} = \delta_{ij}\delta_{kl} \tag{1.57}$$

- Under the above signal prewhitening constraints, the separation system must be constrained to be a *paraunitary filter bank* [30] to ensure component uniqueness.

Assuming that the source signals are spatially- and temporally-decorrelated, the combined system represented by $\{b_{ijl}\}$ satisfies the following paraunitary filter constraints:

$$\sum_{p=1}^{m}\sum_{q=-\infty}^{\infty} b_{ipq}b_{jp(l+q)} = \begin{cases} 1 & \text{if } i=j \text{ and } l=0 \\ 0 & \text{otherwise} \end{cases} \tag{1.58}$$

Paraunitary filters are the multichannel extension of allpass filters and have numerous applications in coding and system modeling [30]. The constraints in (1.58) are the spatio-temporal extension of orthonormality constraints on the rows (or columns) of a square matrix used by the FastICA procedure for instantaneous BSS.

To make paraunitary constraints useful, we need to develop ways to impose them on a multichannel filter. Given such methods, the extension of FastICA to convolutive BSS is straightforward. Suppose $\{h_{ijp}\}$ describes a multichannel prewhitening filter in the principal signal subspace, such that the m signals

$$v_i(k) = \sum_{j=1}^{n}\sum_{p=-\infty}^{\infty} h_{ijp}x_j(k-p) \tag{1.59}$$

are uncorrelated in space and in time with unit variance. Define

$$\mathbf{v}(k) = [\mathbf{v}_1^T(k)\ \mathbf{v}_2^T(k)\ \cdots\ \mathbf{v}_m^T(k)]^T \tag{1.60}$$

$$\mathbf{v}_j(k) = [v_j(k)\ v_j(k-1)\ \cdots\ v_j(k-L+1)]^T \tag{1.61}$$

Furthermore, define

$$\mathbf{w}_i = [\mathbf{w}_{i1}^T\ \mathbf{w}_{i2}^T\ \cdots\ \mathbf{w}_{im}^T]^T \tag{1.62}$$

$$\mathbf{w}_{ij} = [w_{ij0}\ w_{ij1}\ \cdots\ w_{ij(L-1)}]^T \tag{1.63}$$

as the separation system coefficient vector for the ith system output. Then, we compute the ith separated signal sequence as

$$y_i(k) = \sum_{j=1}^{m}\sum_{p=0}^{L-1} w_{ijp}v_j(k-p) \ = \ \mathbf{w}_i^T\mathbf{v}(k) \tag{1.64}$$

for $1 \leq k \leq N$, assuming a data record length of N samples, where $\{w_{ijp}\}$ are the adjustable system parameters. Due to the statistical orthogonality of the prewhitened signals $\{v_j(k)\}$ in space and in time, we can constrain $\{w_{ijp}\}$ to be jointly paraunitary, *i.e.*

$$\sum_{p=1}^{m}\sum_{q=-L}^{L} w_{ipq}w_{jp(l+q)} = \begin{cases} 1 & \text{if } i=j \text{ and } l=0 \\ 0 & \text{for } 0<|l|\leq Q \end{cases} \tag{1.65}$$

where Q is an integer less than L due to the finite length of the separation system and w_{ipq} is assumed to be zero outside of the range $0 \leq q \leq L-1$. Thus, our single-unit FastICA procedure for convolutive BSS has the following steps:

- *Step 1:* Compute $y_i(k)$ in (1.64) for $1 \leq k \leq N$.
- *Step 2:* Update the coefficient vector as

$$\mathbf{w}_i \longleftarrow \frac{1}{N}\sum_{k=1}^{N} f(y_i(k))\mathbf{v}(k) - f'(y_i(k))\mathbf{w}_i \tag{1.66}$$

  where $f(y)$ is the FastICA algorithm nonlinearity and $f'(y)$ is its derivative [28].
- *Step 3:* Impose all or a subset of the paraunitary constraints defined by (1.65) on the $\{w_{ijl}\}$ coefficients within $\mathbf{w}_i$ depending on the type of extraction method (*e.g.* symmetric mode or deflation mode).

It turns out that the choice of FastICA algorithm nonlinearity $f(y)$ in (1.66) is governed by the same rules and considerations as in the instantaneous BSS case. As has been observed, many choices of nonlinearity are possible, with the most popular choices being $f(y) = |y|^{q-1}y$ for $q \geq 1$ and $f(y) = \tanh(\alpha y)$, $\alpha > 1$.

Adaptive paraunitary constraints

Each update of the spatio-temporal FastICA procedure outlined in the previous section requires paraunitary filter constraints on a multichannel linear filter. Due to the FastICA coefficient updates, we are motivated to develop procedures that impose these constraints on multichannel FIR filters. While a direct projection method could be developed, such a technique would likely involve a significant computational overhead if conventional methods (*e.g.* Gram–Schmidt) were employed. We describe two simple adaptive procedures for imposing paraunitary filter constraints on the separation system coefficients $\{w_{ijl}\}$. Since these procedures are adaptive, several iterations of each procedure are needed after each coefficient update in (1.66) to maintain system paraunitariness.

Sequential orthogonalization

Our first procedure assumes a standard sequential implementation of the FastICA procedure. The procedure assumes that $\mathbf{w}_j$ for $1 \leq j \leq i-1$ have converged to separating solutions. The procedure is given as follows:

1. Normalize the length of the coefficient vector as

$$\mathbf{w}_i \longleftarrow \frac{\mathbf{w}_i}{\sqrt{\mathbf{w}_i^T\mathbf{w}_i}} \tag{1.67}$$

2. While $\mathbf{w}_i$ is not paraunitary with $\mathbf{w}_1, \mathbf{w}_2, \ldots, \mathbf{w}_{i-1}$,

$$\mathbf{w}_i \longleftarrow \frac{3}{2}\mathbf{w}_i - \frac{1}{2}\mathbf{g}(\mathbf{w}_i, \mathbf{w}_i) - \sum_{j=1}^{i-1}\mathbf{g}(\mathbf{w}_i, \mathbf{w}_j) \tag{1.68}$$

where

$$\mathbf{g}(\mathbf{w}_i, \mathbf{w}_j) = [\mathbf{g}_{ij1}^T \; \mathbf{g}_{ij2}^T \; \cdots \; \mathbf{g}_{ijm}^T]^T \tag{1.69}$$

$$\mathbf{g}_{ijk} = \mathbf{C}_{ij}\mathbf{w}_{jk} \tag{1.70}$$

and the (p,q)th element of $\mathbf{C}_{ij}$ is given by

$$[\mathbf{C}_{ij}]_{pq} = \begin{cases} \sum_{k=1}^{m}\sum_{l=0}^{L-1} w_{jkl}w_{ik(l+p-q)} & \text{if } |p-q| < \frac{L-1}{2} \\ 0 & \text{otherwise} \end{cases} \tag{1.71}$$

To understand the above method, define the $(n \times 1)$-dimensional ith system vector polynomial as

$$\underline{\mathcal{W}}_i(z) = \sum_{l=0}^{L-1} [w_{i1l} \; w_{i2l} \; \cdots \; w_{iml}]^T z^{-l} \tag{1.72}$$

Then, (1.68)–(1.71) can be expressed using polynomials as

$$\underline{\mathcal{W}}_i(z) \longleftarrow \frac{3}{2}\underline{\mathcal{W}}_i(z) - \Bigg[\frac{1}{2}\left[\underline{\mathcal{W}}_i^T(z^{-1})\underline{\mathcal{W}}_i(z)\right]_{-(L-1)/2}^{(L-1)/2}\underline{\mathcal{W}}_i(z) + \sum_{j=1}^{i-1}\left[\underline{\mathcal{W}}_j^T(z^{-1})\underline{\mathcal{W}}_i(z)\right]_{-(L-1)/2}^{(L-1)/2}\underline{\mathcal{W}}_j(z)\Bigg]_0^{L-1} \tag{1.73}$$

where $[\cdot]_J^K$ denotes truncating the polynomials of its argument to order $-J$ through $-K$. Extensive simulations of this iterative subprocedure indicate that (1.73) causes

$$\left[\underline{\mathcal{W}}_i^T(z^{-1})\underline{\mathcal{W}}_i(z)\right]_{-(L-1)/2}^{(L-1)/2} \longrightarrow 1 \tag{1.74}$$

$$\left[\underline{\mathcal{W}}_j^T(z^{-1})\underline{\mathcal{W}}_i(z)\right]_{-(L-1)/2}^{(L-1)/2} \longrightarrow 0 \quad \text{for } 1 \le j < i \tag{1.75}$$

The above constraints are a spatio-temporal extension of the orthonormality constraints imposed on $\mathbf{w}_i$ in the original FastICA procedure and imply that the separation system is paraunitary.

As further justification of the iterative procedure for enforcing paraunitary constraints, let $L \to \infty$, and define

$$\underline{\mathbf{w}}_i = \underline{\mathcal{W}}_i(z)\big|_{z=e^{j\omega}} \tag{1.76}$$

$$\underline{\mathbf{W}}_i = \left[\underline{\mathcal{W}}_1(z) \; \underline{\mathcal{W}}_2(z) \; \cdots \; \underline{\mathcal{W}}_{i-1}(z)\right]\big|_{z=e^{j\omega}} \tag{1.77}$$

Then, (1.73) can be rewritten as

$$\underline{\mathbf{w}}_{i,new} = \tfrac{3}{2}\underline{\mathbf{w}}_i - \tfrac{1}{2}||\underline{\mathbf{w}}_i||^2\underline{\mathbf{w}}_i - \underline{\mathbf{W}}_i\underline{\mathbf{W}}_i^H\underline{\mathbf{w}}_i \tag{1.78}$$

where $\cdot^H$ denotes Hermitian transpose. Define the variables

$$a_{0i} = ||\underline{\mathbf{w}}_i||^2 - 1 \tag{1.79}$$
$$a_{1i} = ||\underline{\mathbf{W}}_i^H \underline{\mathbf{w}}_i||^2 \tag{1.80}$$

where $||\underline{\mathbf{w}}_i||^2 = \underline{\mathbf{w}}_i^H \underline{\mathbf{w}}_i$. The condition $|a_{0i}| = a_{1i} = 0$ implies that $[\mathcal{W}_1(z) \cdots \mathcal{W}_i(z)]$ form an m-dimensional paraunitary sequence if $[\mathcal{W}_1(z) \cdots \mathcal{W}_{i-1}(z)]$ is already paraunitary. Then, (1.78) implies that these state variables evolve as

$$a_{0i,new} = \frac{1}{4} a_{0i}^2 (a_{0i} - 3) + (a_{0i} - 1) a_{1i} \tag{1.81}$$
$$a_{1i,new} = \frac{1}{4} a_{0i}^2 a_{1i} \tag{1.82}$$

This pair of nonlinear coupled scalar equations can be easily simulated for different initial conditions, and such simulation studies show that a_{0i} and a_{1i} converge to zero for a wide range of initial value pairs. Empirically, we have observed convergence of this system if

$$\begin{array}{c} -1 < a_{0i} < 2 \\ a_{1i} < a_{0i} + 1 \end{array} \quad \text{or} \quad \begin{array}{c} 0 < ||\underline{\mathbf{w}}_i||^2 < 3 \\ ||\underline{\mathbf{W}}_i^H \underline{\mathbf{w}}_i||^2 < ||\underline{\mathbf{w}}_i||^2 \end{array} \tag{1.83}$$

which are typically satisfied in practice.

Tables 1.1 and 1.2 provide MATLAB code for implementing the above STFICA1 algorithm with sequential orthogonalization, where `v` is the $m \times N$ prewhitened signal matrix, `L` is the separating system filter length, and **`numiter`** is the maximum number of iterations allowed for each FastICA procedure.

Symmetric orthogonalization

The above adaptive sub-procedure imposes paraunitariness in a sequential fashion to the system vectors $\{\mathbf{w}_i\}$. In other words, the value of $\mathbf{w}_1$ affects the solutions of $\mathbf{w}_i$, $2 \le i \le m$, the value of $\mathbf{w}_2$ affects the solutions of $\mathbf{w}_i$, $3 \le i \le m$, and so on. This effect is similar to that observed in the original FastICA algorithm applied in a sequential fashion with Gram–Schmidt coefficient orthogonalization. While this type of constraint is robust in terms of source acquisition, it causes error accumulation in the separated system outputs, such that sources extracted "later" in the separation process have worse signal-to-interference ratios.

We now describe a symmetric method for imposing adaptive paraunitary constraints jointly on all $\mathbf{w}_i$, $1 \le i \le m$. Complete details regarding this procedure, including a proof of convergence and an analysis of its convergence speed, can be found in [31]. The procedure is given as follows:

1. Normalize all vectors $\{\mathbf{w}_i\}$, $1 \le i \le m$, as

$$\mathbf{w}_i \longleftarrow \frac{\mathbf{w}_i}{\sqrt{\mathbf{w}_i^T \mathbf{w}_i}} \tag{1.84}$$

Table 1.1: MATLAB implementation of the STFICA1 algorithm update.

```
function [y,W] = stfica1(v,L,numiter);

L = L + rem((L+1),2);
[m,N] = size(v);
W = kron(eye(n),[zeros((L-1)/2,1);1;zeros((L-1)/2,1)]);
V = zeros(m*L,N);
for i=1:m
  V((i-1)*L+1:i*L,:) = toeplitz([v(i,1);zeros(L-1,1)],v(i,:));
end
y = zeros(m,N);
for i=1:m
  Wold = zeros(m*L,1);
  k = 0;
  y(i,:) = W(:,i)'*V;
  crit = 1;
  while (crit*(k<numiter))
    Wold = W(:,i);
    k = k+1;
    f = y(i,:)'.^3;             % f = tanh(20*y(i,:)');
    fp = 3*sum(y(i,:).^2);  % fp = 20*sum(sech(20*y(i,:)).^2);
    W(:,i) = V*f - fp*W(:,i);
    W(:,i) = orthW(W(:,1:i),m,L,10);
    y(i,:) = W(:,i)'*V;
    crit = (abs(abs(W(:,i)'*Wold)-1)>0.0001);
  end
end
```

Table 1.2: MATLAB implementation of functions called by STFICA1 algorithm update.

```
function [Wi] = orthW(W,m,L,numorth);

i = size(W,2);
Wi = W(:,i)/norm(W(:,i));
for k=1:numorth
  Wt = zeros(m*L,1);
  for j=1:i-1
    Wt = Wt + gfun(Wi,W(:,j),m,L);
  end
  Wi = 3/2*Wi - 1/2*gfun(Wi,Wi,m,L) - Wt;
end
```

```
function [G,C] = gfun(U,V,m,L);

Wi = zeros(L,m);     Wi(:) = U;
Wj = zeros(L,m);     Wj(:) = V;
Ct = zeros((3*L-1)/2,1);
Z = zeros((L-1)/2,1);
ll = (L+1)/2:(3*L-1)/2;     llr = L:-1:1;
for i=1:m
  Ct = Ct + filter(Wi(llr,i),1,[Wj(:,i);Z]);
end
C = Ct(ll);
Gt = filter(C(llr),1,[Wj;zeros((L-1)/2,m)]);
Gt = Gt(ll,:);
G = Gt(:);
```

2. While the $\{\mathbf{w}_i\}$ are not jointly paraunitary, do for all $1 \leq i \leq m$:

$$\mathbf{w}_i \longleftarrow \frac{3}{2}\mathbf{w}_i - \frac{1}{2}\sum_{j=1}^{m} \mathbf{g}(\mathbf{w}_i, \mathbf{w}_j) \tag{1.85}$$

where $g(\mathbf{w}_i, \mathbf{w}_j)$ is as defined in (1.69).

In order to better see the structure of this algorithm, define the z-transform matrix

$$\underline{\mathbf{W}}(z) = \sum_{l=0}^{L-1} \begin{bmatrix} w_{11l} & \cdots & w_{1ml} \\ \vdots & & \vdots \\ w_{m1l} & \cdots & w_{mml} \end{bmatrix} z^{-l} \tag{1.86}$$

Then, this algorithm can be written as

$$\underline{\mathbf{W}}(z) \longleftarrow \frac{3}{2}\underline{\mathbf{W}}(z) - \frac{1}{2}\left[\left[\underline{\mathbf{W}}(z)\underline{\mathbf{W}}^T(z^{-1})\right]_{\frac{-(L-1)}{2}}^{\frac{(L-1)}{2}} \underline{\mathbf{W}}(z)\right]_0^{L-1} \tag{1.87}$$

For $L = 1$, this procedure is a Newton-based adaptive orthogonalization scheme; see [32] for more details as well as a convergence analysis of the procedure. Typically, between 10 and 20 iterations of this procedure are needed at each FastICA update to obtain a system that is sufficiently close to paraunitariness for speech separation applications.

The method we have described, is particularly useful for speech separation tasks, and has a number of advantages over competing approaches:

1. No step size needs to be selected.
2. Knowledge of the source distributions is not needed, so long as their statistics imply a nonzero contrast value.
3. For the STFICA1 algorithm, the number of non-Gaussian sources within the mixture need not be known *a priori*.
4. Convergence of these procedures appears to be as fast as its spatial-only counterparts. The single-unit STFICA1 procedure usually requires fewer than 20 iterations per unit when the sources are spatially- and temporally-independent, whereas more iterations are needed for temporally-correlated sources such as speech signals.

1.6 Frequency-Domain Convolutive BSS

Convolutive BSS as applied to speech signal mixtures involves relatively-long multichannel FIR filters to achieve separation with even moderate amounts of room reverberation. While time-domain algorithms can be developed to perform this task, they can be difficult to code primarily due to the multichannel convolution operations involved. One way to simplify the conceptualization of convolutive BSS algorithms is to transform the task into the frequency-domain, as convolution in time becomes multiplication in frequency. Ideally, each frequency component of the mixture signal contains an instantaneous mixture of the corresponding frequency components of the underlying source signals.

Transformation of time-domain signals into the frequency-domain is usually done via the sliding window discrete Fourier transform (DFT) or the short-time Fourier transform (STFT). The window length is chosen such that the data is quasi-stationary within the window. Let us denote the time-domain data over T sample points as the matrix $\mathbf{X}(k)$, where

$$\mathbf{X}(k) = [\mathbf{x}(k), \mathbf{x}(k+1), \cdots, \mathbf{x}(k+T-1)]$$

The DFT of this frame of data is given as

$$\mathbf{x}(\omega, k) = \sum_{\tau=0}^{T-1} \mathbf{x}(k+\tau)\mathrm{e}^{-j2\pi\omega\tau/T} \tag{1.88}$$

If we assume that the mixing system impulse response is of length P samples, then so long as $(P << T)$, we have

$$\mathbf{x}(\omega, k) \approx \mathbf{A}(\omega)\mathbf{s}(\omega, k) \tag{1.89}$$

where $\mathbf{x}(\omega, k)$ denotes the multichannel DFT of the current mixed vector frame at time instant k and frequency ω and $\mathbf{s}(\omega, k)$ denotes the corresponding DFT of the source vector frame. The matrix $\mathbf{A}(\omega)$ is a $(n \times m)$ time-invariant mixing matrix at frequency ω. For a particular frequency bin ω, (1.89) represents an instantaneous mixing system.

A simplified block diagram of a generic frequency-domain convolutive BSS procedure is shown in Fig. 1.3. The separation system at frequency ω and

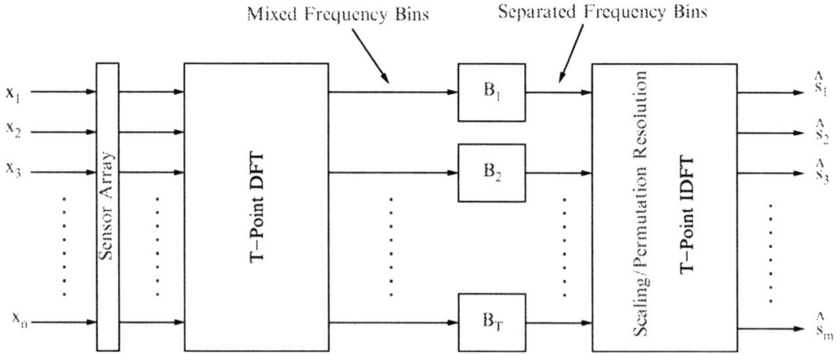

Fig. 1.3. Block diagram of frequency-domain blind source separation.

time k is represented as $\mathbf{B}(\omega, k)$ such that

$$\mathbf{y}(\omega, k) = \mathbf{B}(\omega, k)\mathbf{A}(\omega)\mathbf{s}(\omega, k) \tag{1.90}$$

Ideally, the separation system is adapted such that $\mathbf{B}(\omega, k)\mathbf{A}(\omega) = \mathbf{I}$ after a sufficient number of data frames have been obtained. Due to permutation and scaling ambiguities, however, it is the case that $\mathbf{B}(\omega, k)\mathbf{A}(\omega) = \mathbf{P}(\omega)\mathbf{D}(\omega)$, where $\mathbf{P}(\omega)$ and $\mathbf{D}(\omega)$ are frequency-dependent permutation and scaling matrices, respectively. Because of these ambiguities, the signals that are reconstructed via inverse DFT after the separation step may have spectral components corresponding to multiple sources, and they may suffer from distortions due to spectral errors.

1.6.1 Criteria for Frequency–Domain Convolutive BSS

As frequency-domain BSS methods assume instantaneous mixing conditions for each frequency bin, the algorithms employed for separation are closely-related to instantaneous BSS methods. One chief difference is the complex-valued nature of the mixing process, as DFT processing yields complex-valued data. Complex-valued BSS and ICA methods are active research subjects, and some details of these algorithms are still being developed. Note that such algorithms *generally cannot* be correctly developed by replacing symmetric and orthogonal matrices by Hermite-symmetric and unitary matrices, respectively.

BSS algorithms using temporal structure. One class of criteria for which frequency-domain techniques have been widely studied are those based on second-order statistics, yielding frequency-domain extensions of instantaneous BSS methods first proposed in [34] and later refined by the authors of [35]. Since speech is both correlated and nonstationary, it has the appropriate statistical structure for this formulation. It is assumed that

$$E\{s_i(\omega, k)s_j^*(\omega, k+l)\} = 0 \quad \text{for} \quad i \neq j \quad \text{and all} \quad \omega \text{ and } l \tag{1.91}$$

Moreover, it is assumed that the normalized correlation coefficients

$$\rho_i = \frac{E\{s_i(\omega, k)s_i^*(\omega, k+l)\}}{\sqrt{E\{s_i(\omega, k)^2\}E\{s_i(\omega, k+l)^2\}}} \tag{1.92}$$

are distinct for all ω and $1 \leq i \leq m$ and at least one value of $l \neq 0$. These constraints on the correlation statistics of the sources are due to certain identifiability conditions that must be satisfied for this formulation to work properly.

Consider the normalized cross-correlation matrix of the input signals at frequency ω and lag l, given by

$$\begin{aligned}\mathbf{R}_x(\omega,l) &= \left[E\{\mathbf{x}(\omega,k)\mathbf{x}^H(\omega,k)\}\right]^{-1}E\{\mathbf{x}(\omega,k)\mathbf{x}^H(\omega,k+l)\}\\ &= \left[\mathbf{A}(\omega)E\{\mathbf{s}(\omega,k)\mathbf{s}^H(\omega,k)\}\mathbf{A}^H(\omega)\right]^{-1}\\ &\quad\times\mathbf{A}(\omega)E\{\mathbf{s}(\omega,k)\mathbf{s}^H(\omega,k+l)\}\mathbf{A}^H(\omega)\end{aligned}\tag{1.93}$$

Under the assumption of unit variance sources such that $E\{\mathbf{s}(\omega,k)\mathbf{s}^H(\omega,k)\}=\mathbf{I}$, we have

$$\begin{aligned}\mathbf{R}_x(\omega,l) &= \mathbf{A}^{-H}(\omega)E\{\mathbf{s}(\omega,k)\mathbf{s}^H(\omega,k+l)\}\mathbf{A}^H(\omega)\\ &= \mathbf{A}^{-H}(\omega)\mathbf{R}_s(\omega,l)\mathbf{A}^H(\omega)\end{aligned}\tag{1.94}$$

Defining the eigenvalue decomposition of $\mathbf{R}_x(\omega,l)$ as

$$\mathbf{R}_x(\omega,l) = \mathbf{Q}(\omega)\mathbf{\Lambda}(\omega,l)\mathbf{Q}^{-1}(\omega)\tag{1.95}$$

we have that

$$\mathbf{\Lambda}(\omega,l) = \mathbf{P}(\omega)\mathbf{R}_s(\omega,l)\mathbf{P}^H(\omega)\tag{1.96}$$

with $\mathbf{Q}(\omega)=\mathbf{A}^{-H}(\omega)\mathbf{P}^H(\omega)$. The separating matrix can be computed as

$$\mathbf{B}(\omega,k) = \mathbf{Q}^H(\omega)\tag{1.97}$$

Hence, to perform separation at any given frequency ω, one needs to compute the eigenvalue decomposition of a normalized cross-correlation matrix.

In practice, most BSS algorithms that use temporal correlation identify the separating system by solving the following nonlinear system of m^2 entries of $\mathbf{B}(\omega,k)=[\mathbf{b}_1\ \mathbf{b}_2\ \cdots\ \mathbf{b}_m]^T$ when $m=n$:

$$\mathbf{b}_i^H E\{\mathbf{x}(\omega,k)\mathbf{x}^H(\omega,k)\}\mathbf{b}_j = \delta_{ij},\quad 1\le i\le j\le m\tag{1.98}$$

$$\mathbf{b}_i^H E\{\mathbf{x}(\omega,k)\mathbf{x}^H(\omega,k+l)\}\mathbf{b}_j = 0,\quad 1\le i< j\le m\tag{1.99}$$

Procedures that solve this type of task are called *joint diagonalization* methods as they search for a matrix whose rows are the eigenvectors of at least two different structured data matrices. Alternatively, a two-stage approach similar to contrast optimization can be used, whereby the first prewhitening stage yields a prewhitened sequence $\mathbf{v}(\omega,k)$ whose eigenvectors are the orthogonal separating matrix through

$$\mathbf{B}^H(\omega,k)E\{\mathbf{v}(\omega,k)\mathbf{v}^H(\omega,k+l)\}\mathbf{B}(\omega,k) = \mathbf{I}\tag{1.100}$$

Solutions based on joint decorrelation involve estimates of the second-order statistics of the data; however, these techniques require a specific constraint on the temporal structure of the sources. In cases where the temporal correlation properties of the sources at a particular frequency bin are similar for a chosen lag value l, the performance of these methods degrade. For this reason, the use of several lag values l is generally recommended to increase the chance of good system identifiability [35]. An application of this approach in frequency-domain BSS is described in [36].

BSS algorithms using nonstationarity. Since speech signals are statistically nonstationary, separation methods based on second-order statistics that exploit the time-varying nature of speech can be developed. In this case, the source covariance matrix at frequency bin ω can be assumed to be

$$\mathbf{R}_s(\omega, k) = \mathbf{\Lambda}_s(\omega, k) = \text{diag}\{\sigma_1^2(\omega, k), \sigma_2^2(\omega, k), \cdots, \sigma_m^2(\omega, k)\} \tag{1.101}$$

In order for separation to occur, the following identifiability conditions must be satisfied:

$$\frac{\sigma_i^2(\omega, k)}{\sigma_j^2(\omega, k)} \neq \text{constant}, \quad 1 \leq i, j \leq m, \quad i \neq j \tag{1.102}$$

The above condition implies that the correlation statistics of each source must vary somewhat independently of each other over a chosen observation interval. Let the covariance matrix $\mathbf{R}_x(\omega, k)$ of measured signals at time k be

$$\begin{aligned} \mathbf{R}_x(\omega, k) &= E\{\mathbf{x}(\omega, k)\mathbf{x}^H(\omega, k)\} \\ &= \mathbf{A}(\omega)E\{\mathbf{s}(\omega, k)\mathbf{s}^H(\omega, k)\}\mathbf{A}^H(\omega) \\ &= \mathbf{A}(\omega)\mathbf{\Lambda}_s(\omega, k)\mathbf{A}^H(\omega) \end{aligned} \tag{1.103}$$

By dividing the total observed signal into different blocks, the estimates $\mathbf{R}_x(\omega, k)$ are formed at different times k_i $1 \leq i \leq K$. Due to the nonstationarity of the sources, the various $\mathbf{R}_x(\omega, k_i)$ will be different from each other. For sufficiently large K, we have sufficient number of matrices to allow the identification of the mixing matrix $\mathbf{A}(\omega)$, allowing us to estimate its inverse. Typically, a separating matrix $\mathbf{B}(\omega)$ is designed to jointly diagonalize a set of K different covariance matrices $\mathbf{R}_x(\omega, k_i)$ $1 \leq i \leq K$, *e.g.* by minimizing the off-diagonal terms of the matrix $\mathbf{B}(\omega)[\mathbf{R}_x(\omega, k)]\mathbf{B}^H(\omega)$ for all K matrices. An application of this technique for frequency-domain convolutive BSS is described in [37].

1.6.2 Permutation and Scaling Ambiguities in Frequency-Domain BSS

Due to the decoupled nature of the solutions across different frequency bins obtained in frequency-domain BSS, there is no guarantee that the separated signals obtained by such procedures will have the same permutation and scaling properties for different ω. Scaling ambiguities can result in unequal scaling of the spectral components before reconstruction, and permutation ambiguities can result in the spectral "mixing" of sources upon reconstruction. Thus, some method for solving permutation and scaling ambiguities is generally necessary to obtain good performance from frequency-domain BSS methods.

Scaling ambiguities are most often resolved by some form of normalization of each separation matrix at each frequency bin. Source permutation, however, is a much more challenging problem and has received considerable attention by several researchers. Most methods for resolving frequency-dependent permutation fall into one of three categories:

1. those that exploit specific signal properties of the DFT;
2. those that exploit specific properties of speech; and
3. those that exploit specific geometric properties of the sensor array, such as directions of arrival (DOAs).

All three classes of methods require additional information about the measurement setup or the signals being separated. Once permutation and scaling ambiguities are resolved, the final time-domain source estimates can be reconstructed using overlap-add methods [38].

Spectral continuity based permutation removal. One of the simplest methods for solving permutation ambiguity is zero-padding of the data prior to processing [39]. This processing imposes a smoothness constraint on the resulting spectra. For simple mixing conditions involving low amounts of reverberation, it was thought that the separating matrices at nearby frequencies will have similar permutations. In the presence of complicated mixing conditions such as those generated from real rooms, however, these techniques are not too useful. A related approach was suggested in [37], in which time-domain truncation of the frequency-domain separation filter is used to impose continuity. The main problem with this approach is that, in order to achieve a sufficiently continuous frequency-domain filter solution, the ratio of the lengths of the frequency-domain filter T and the time-domain truncation filter Q should follow $T/Q \approx 8$ (as suggested in [37]). This constraint makes the DFT length T very large, leading to increased computational complexity. For example, if the time-domain filter has $Q = 512$ taps, the corresponding frequency-domain filter should have $T = 4096$ taps. Moreover, extremely long filter lengths result in block processing that violates the stationarity assumption on the block of data under the analysis window (see Fig. 1.4).

Spectral correlation based permutation removal. To deal with more complicated mixing conditions, a method based on spectral correlation was first proposed in [36]. This method is based on the following two properties of speech signals:

1. Speech signals are inherently nonstationary; however, for short-enough time segments, speech signal can be considered to be stationary.
2. Nonstationarity in speech arises mainly due to amplitude modulation.

The first property allows the use of the DFT for transforming the mixture data in the spectral domain, whereas the second property allows for the

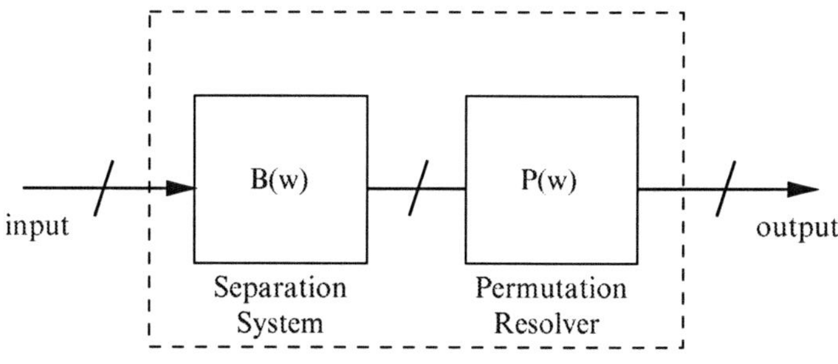

Fig. 1.4. Frequency-domain blind source separation with permutation solver.

grouping of components arising from the same source based on amplitude correlations. For spectral components from two different sources, the amplitude correlations will be zero at all frequencies, *i.e.*

$$\rho_{(|y_i|,|y_j|)} = \frac{(\mu_{(|y_i||y_j|)} - \mu_{|y_i|}\mu_{|y_j|})}{(\sigma_{|y_i|}\sigma_{|y_j|})} = 0, \quad \forall i \neq j \tag{1.104}$$

where μ denotes the mean and σ denotes the variance of the absolute value of the components. Due to their similar amplitude modulation, the spectral components arising from the same sources will have nonzero correlation values.

Beamforming based permutation removal. Connections between blind beamforming [40] and BSS were first indicated in [41]. In this work, directional vectors associated with the sources were obtained by exploiting the spatial independence of the sources. The beamformer response for a narrowband source of frequency ω impinging of the array with a directional vector $\mathbf{d}(\omega, \theta)$ is given as

$$r(\omega, \theta) = \mathbf{b}^H \mathbf{d}(\omega, \theta) \tag{1.105}$$

where $\mathbf{b} = [b_1 \; b_2 \; \cdots \; b_n]^T$ are the complex weights associated with the array sensors (note that for broadband sources, $\mathbf{b}(\omega)$ is a more appropriate notation), and θ is the source direction. The direction vector $\mathbf{d}(\omega, \theta)$ is generally given as

$$\mathbf{d}(\omega, \theta) = [\exp(j\omega\tau_0(\theta)) \; \cdots \; \exp(j\omega\tau_{m-1}(\theta))]^T \tag{1.106}$$

where $\tau_l(\theta) = ld\sin(\theta)/c$ is the time delay associated with the lth sensor with respect to the first sensor, n is the number of sensors in the array, d is the array element separation, and c is the speed of sound. Signals with significant frequency content (*e.g.* audio signals) received at a sensor array will typically have specific directional vectors associated with each of their frequency components.

In beamforming methods, a single source is often of interest. The goal of convolutive BSS is the extraction of all measured sources within the signal mixtures. Thus, m beamformers are generally needed. Each of the m different beamformers will have its own directional pattern. Under a direct-path model, each beamformer forms nulls in the directions of all jammer signals (*i.e.* other sources) and extracts a single target signal. Hence, if one assumes specific knowledge of the sensor array geometry, a comparison of directional responses associated with each frequency in each of the beamformers can be used to resolve permutation ambiguities. This technique was first proposed in [42] and has since been enhanced by several others [43, 44].

There are certain problems, however, associated with the use of this approach. One problem is the ambiguity resulting from sensor spacing which due to the broadband nature of speech is frequency-dependent. An inadequate sensor spacing gives rise to *spatial-aliasing* [45] which manifests itself as the presence of grating lobes in the directional response of the beamformer. In other words, a particular source appears to be coming from different directions to the beamformer. In order to avoid spatial-aliasing, the sensor spacing should be selected such that $d \leq \lambda/2$, where λ is the wavelength of the sound signal. Higher signal frequencies need smaller sensor spacing, and lower signal frequencies need a larger spacing between the sensors to avoid aliasing. Another significant problem is the computational complexity of solving the permutation as the number of sources and sensors increases. Algorithms designed to resolve a two-source ambiguity may not scale well to three, four, or more source mixtures.

Other researchers have suggested the merger of beamforming with BSS algorithms in order to avoid post processing during reconstruction of the sources in the time-domain. If the DOA of each of the sources is known, BSS/ICA algorithms might be initialized/constrained such that the directional response is consistent across all of the frequency bins. One such algorithm of this type was first proposed in [46]. In situations where DOA information is not available *a priori*, procedures for estimating the DOAs as part of the BSS/ICA algorithms have been developed [44, 46].

1.7 Numerical Evaluations

In this section, we explore the abilities of several convolutive BSS algorithms using signals measured from a controlled laboratory environment. Such experiments allow us to carefully evaluate the capabilities of each method in a repeatable fashion while altering the difficulty of the separation task in specific ways.

1.7.1 Laboratory Setup

Data for evaluating the separation methods was generated in an acoustically-isolated laboratory environment with up to three loudspeakers playing recordings of talkers (one female and two male) as the sources. The sources were located 127 cm away from three omnidirectional microphones and were spaced at angles of $-30°$, $0°$, and $27.5°$ from the angle of incidence of the microphone array. Two room reverberation conditions were created in the room corresponding to reverberation times of 300 ms and 425 ms, respectively. Figure 1.5 shows a photograph of the laboratory setup for the 300 ms reverberation experiment with three sources and a uniform linear array. All measurements were made using 7 seconds of data per channel and a 48 kHz sampling rate and were downsampled to an 8 kHz sampling rate for processing. Figures 1.6 and 1.7 show the impulse responses of the loudspeaker/ microphone paths for reverberation times of 300 ms and 425 ms, respectively, as calculated using pseudo-random noise sequences. For purposes of correlating these plots with the photograph in Fig. 1.5, the microphones are labeled as 1, 2, and 3 from right to left in the photograph, and the loudspeakers are labeled as 6, 7, and 8 from right to left in the photograph with directions of arrival of $27.5°$, $0°$, and $-30°$, respectively.

Each algorithm was applied to this measured microphone data for two- and three-source signal mixtures, whereby the 0^0 source was omitted for the two-source mixture. After separation, least-squares methods were used to estimate the contributions of the source recordings to each of the recorded mixtures as well as the output signals from each algorithm. By calculating power ratios from these least-squares estimates, we can compute the average

Fig. 1.5. Laboratory measurement environment used for numerical evaluations.

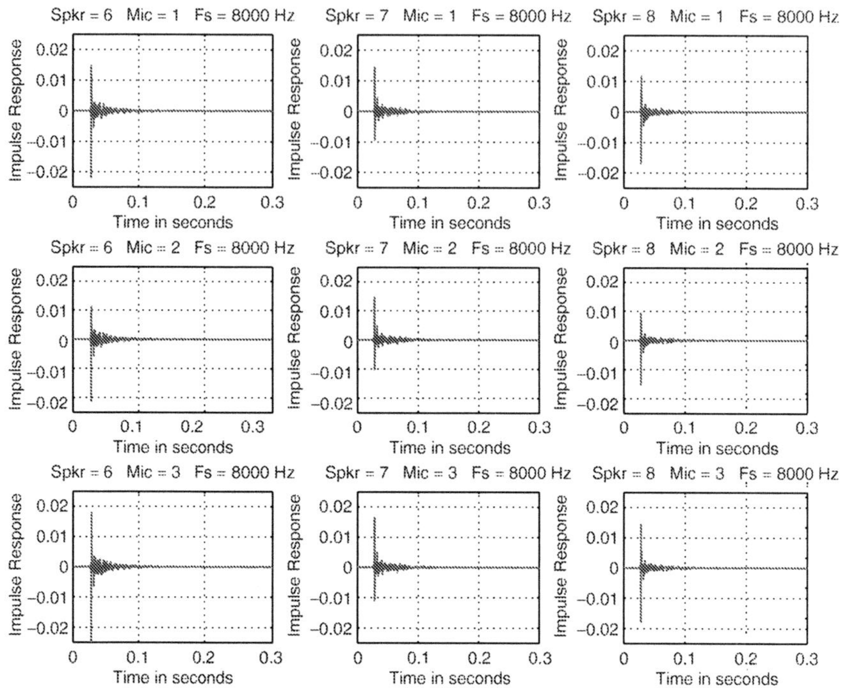

Fig. 1.6. Impulse responses measured from the uniform linear array, $RT = 300\,\text{ms}$ condition.

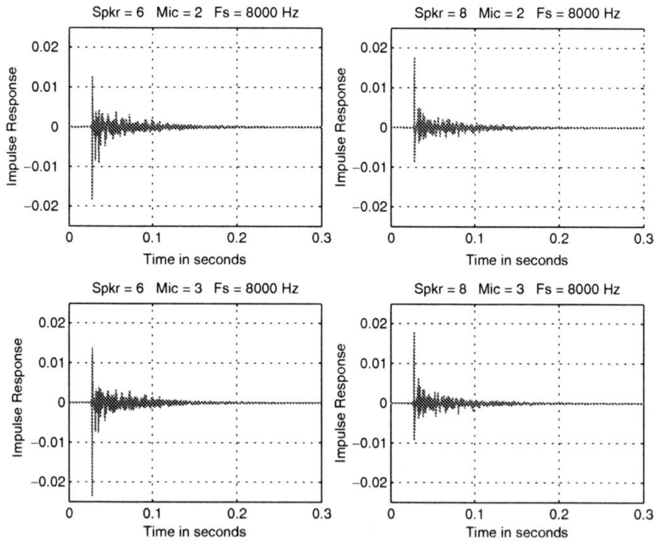

Fig. 1.7. Impulse responses measured from the uniform linear array, $RT = 425\,\text{ms}$ condition.

improvement in signal-to-interference-plus-noise ratio (SINR) for each algorithm in each data case.

1.7.2 Effect of Array Configuration and Room Reverberation

In our first experiment, we evaluate a large class of algorithms to understand performance effects due to room reverberation and the microphone array configuration. Two arrangements of microphones were used: (1) a nearly-uniform linear array with 4 cm spacing, and (2) an equilateral triangular array arranged in a vertical plane with 8 cm spacing. The room reverberation time was adjusted by placing or removing foam tiles from the walls of the room. We evaluated the following algorithms:

- STFICA algorithms, with sequential and symmetric orthogonalization with one stage (for the RT = 300 ms condition) or two stages (for the RT = 425 ms condition) of least squares prewhitening with lengths of $M = 400$ taps per filter [47, 48]. Each separation system used $L = 300$ taps per input–output filter channel.
- Parra's decorrelation-based method with the following choices: Number of diagonalized matrices $K = 5$, number of data blocks averaged $N = 10$, FFT size $L = 1024$, $Q = 400$-tap time-domain filters per input–output channel, and a 1000-iteration limit [37].
- Parra's decorrelation-based method with beamforming initialization using the above parameter settings [46].
- Bin-wise natural gradient frequency-domain (NGFD) method with "center-spike" initialization, $\mu = 0.09$, $L = 1024$, with 100 iterations.
- Bin-wise natural gradient frequency-domain (NGFD) method with beamforming initialization using the above parameter settings.
- Natural gradient time-domain (NGTD) method using causal FIR filters and "center-spike" initialization, a step size schedule of $\mu = 0.0005$ for 200 data passes followed by $\mu = 0.0001$ for a single data pass followed by $\mu = 0.00001$ for a single data pass, $L = 512$ [23].

These algorithms were selected due to their widespread use as tools within modified approaches [43, 44]. Figure 1.8 shows the performance of the various algorithms on data collected using two- and three-microphone uniform arrays in terms of average improvement in SINR. We consider each data case separately.

Case 1: $m = 2$, uniform linear array, RT = 300 ms: The best performing algorithm in this case is the NGFD algorithm with beamforming initialization, in which the average SINR improvement is 13.8 dB across the two outputs. The STFICA1 and STFICA2 algorithms achieve 11.8 dB and 9.8 dB average SINR improvements, respectively. These latter two algorithms, however, do not require knowledge of the directions of arrival of the sources

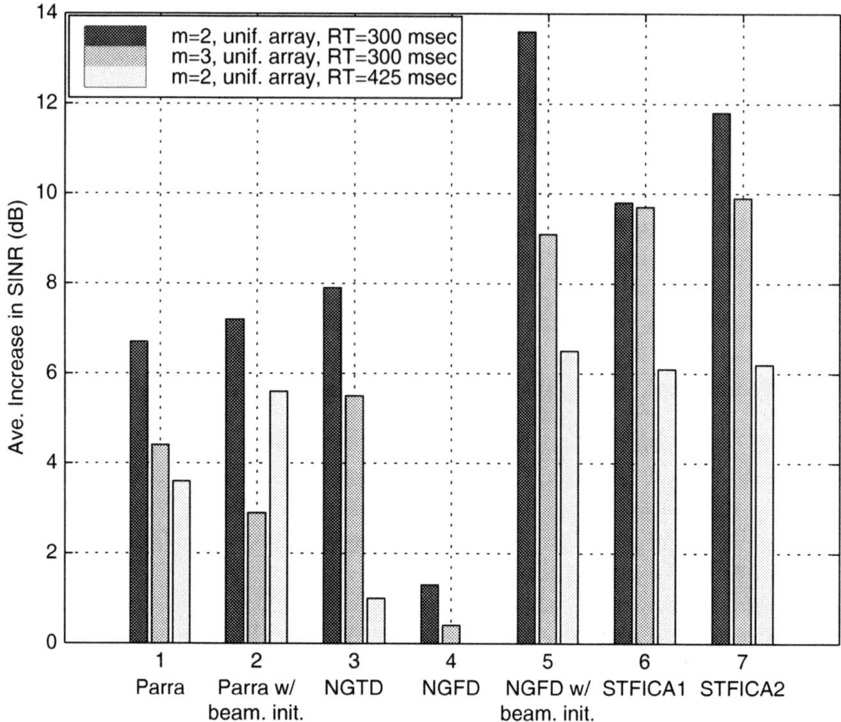

Fig. 1.8. Performance of various separation methods under two reverberation conditions (RT = 300 ms and 425 ms) for a uniform linear microphone array, and m = two and three source case.

for their initialization, whereas the NGFD algorithm obtains only a 1.3 dB average SINR improvement without this information.

Case 2: $m = 3$*, uniform linear array, RT = 300 ms:* In the three-source case, the proposed STFICA algorithms perform the best, achieving 9.7 dB and 9.9 dB of average SINR improvement. The best-performing algorithms in this comparison employed geometric knowledge of the source and sensor positions and obtained a 9.1 dB average SINR improvement in the best case (NGFD).

Case 3: $m = 2$*, uniform linear array, RT = 425 ms:* In this more-reverberant mixing environment, the best-performing algorithm was the NGFD algorithm with beamforming knowledge, in which a 6.5 dB improvement in SINR was obtained. The proposed STFICA algorithms gave nearly the same performance at 6.1 dB and 6.2 dB, respectively, and the latter methods do not require knowledge of sensor positions to work well.

In these experiments, the best of the methods based on non-Gaussian statistics generally outperformed those methods based on second-order decorrelation alone. It should be noted that both STFICA procedures were easy to

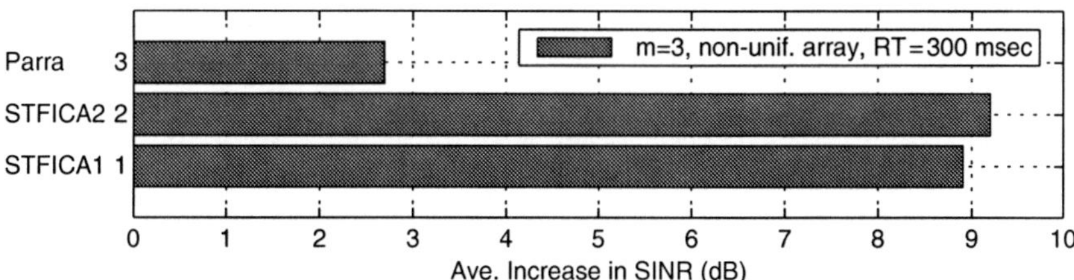

Fig. 1.9. Performance of various separation methods under an RT = 300 ms reverberation condition for a nonuniform linear microphone array, m = three source case.

evaluate in these experiments as they need only to be run once. The other procedures are gradient-based methods that may require significant hand-tuning of step size parameters, careful selection of the number of data passes, and a careful initialization strategy in order to obtain good separation performance.

Figure 1.9 shows the performance of three algorithms in a three-source mixture case for a triangular microphone array. In this case, direction of arrivals are not as easily calculated, so only algorithms with generic (*e.g.* "center-spike") initializations have been evaluated. As can be seen, the proposed STFICA algorithms perform well in this situation despite having an unknown array geometry, unknown source positions, and generic initialization strategies.

1.7.3 Effect of Initial Conditions

We now present numerical evaluations to illustrate the separate effects that beamforming initialization and data prewhitening have on the behaviors of one class of CBSS algorithms. In order to minimize any performance effects due to choice of separation criterion, we focus on the natural gradient algorithms presented in [20] and [23] that attempt to minimize the mutual information of the extracted signals using frequency-domain and time-domain system structures, respectively. For comparison, we show the performance of two other algorithms on this data: one employing decorrelation with geometric beamforming constraints [46], and one using contrast-based optimization with prewhitening [47, 48]. These latter algorithms incorporate either beamforming or prewhitening within their structures and are not claimed to work without such processing. All experiments were performed under the RT = 300 ms mixing conditions. For the normalized natural gradient algorithm in the frequency-domain [25], the parameters chosen were $L = 512$ and $\mu = 0.35$, and 200 passes of the algorithm through the data have been used to adapt the filter. For the natural gradient time-domain algorithm [23], we used $L = 512$ and a step size schedule of $\mu = .0009$ for 150 data passes followed by $\mu = .0001$ for a single data pass followed by $\mu = .00001$ for a single data pass. The data nonlinearity used in each algorithm was $f(y) = y/|y|$, where y in this case

Table 1.3. Improvement in average SINR [dB]; RT = 300 ms.

Algorithm	TWO-SOURCE CASE			THREE- SOURCE CASE		
	Center-Spike	w/Beam-forming	w/Prewhi-tening	Center-Spike	w/Beam-forming	w/Prewhi-tening
SNGFD [25]	0.25	13.56	1.52	3.33	12.55	4.55
NGTD [23]	12.63	13.60	15.34	10.89	17.07	16.80
Parra-GBSSII [46]	–	7.95	–	–	5.42	–
STFICA-Symm [47, 48]	–	–	11.23	–	–	12.66

corresponds to the ith frequency bin output or the ith time domain filter output, respectively.

Table 1.3 shows the SINR improvements obtained by the various algorithms for the various processing strategies on the two-source mixture data. As can be seen, the frequency-domain natural gradient method does not perform well either with center-spike initialization or with data prewhitening. With beamforming initialization, the algorithm achieves good performance on this data that closely matches the time-domain natural gradient algorithm. The latter algorithm's performance is quite good for center-spike initialization on this data, but improvements of 1.0 dB and 2.7 dB are obtained with beamforming initialization and data prewhitening, respectively. Shown for comparison are the behaviors of the decorrelation-based method in [46] as well as the contrast-based method with prewhitening in [25]. As can be seen, the time-domain natural gradient method outperforms both of these competing methods when using the same knowledge of the environment or data preprocessing.

Also shown in Table 1.3 are the SINR improvements obtained by the various processing strategies on the three-source mixture data. Similar performance relationships as in the two-source data case are observed in this case. The time-domain natural gradient algorithm can separate the source mixtures with any of the three strategies employed. Figure 1.10 shows the combined impulse responses at convergence for the natural gradient time-domain algorithm with beamforming initialization when applied to this data, indicating that separation has occurred. It should be noted that prewhitening-based processing strategies do not require knowledge of the source-sensor array geometry.

1.7.4 Effect of Signal Power Mismatch

We now explore the performances of several convolutive BSS algorithms under the realistic scenario in which one source is much weaker in the signal mixtures as compared to the other sources. The algorithms considered include fixed-point fast-ICA algorithms for convolutive mixing [48], a second

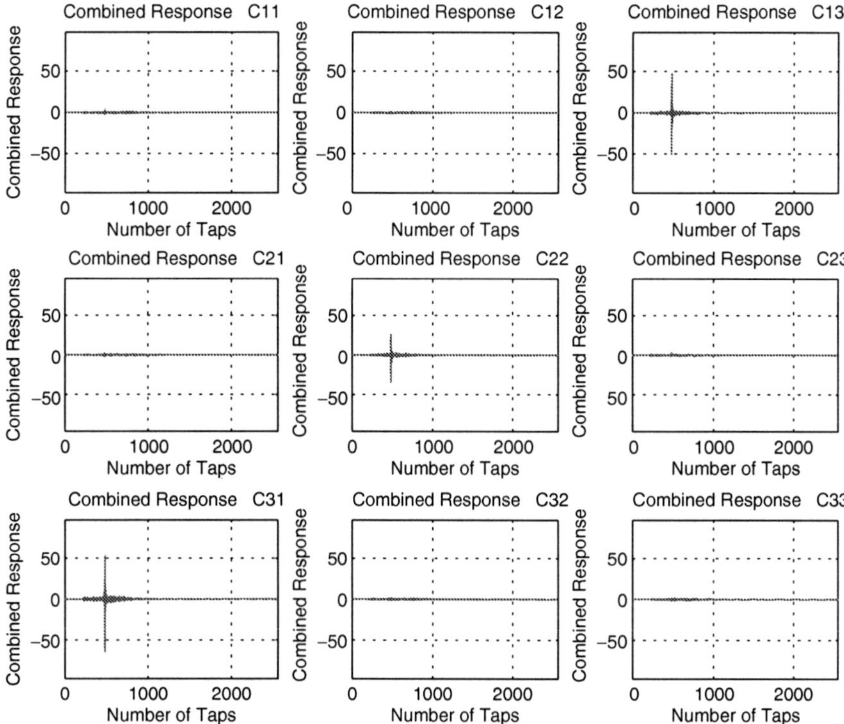

Fig. 1.10. Combined impulse response of the time-domain truncated natural gradient with beamforming initialization, three-source case.

order statistics based frequency-domain joint decorrelation algorithm and its beamforming-constrained version [37, 46] and the natural gradient algorithms in [23, 25].

Data for these evaluations was generated according to a reverberation time of RT = 300 ms; however, the direction of arrivals of the sources was $-35°$, $0°$, and $27°$, respectively, from the array normal. To balance any effects of speech segment power differences, recordings in the two-source case were made using 14 seconds of data per channel, in which the speech segments were switched so that everay talker is played through each loudspeaker at different times. For the three-source mixtures, 21 seconds of data per channel was generated using a similar repeating approach.

For each algorithm, the parameter choices were as follows. For the STFICA algorithms, the least squares prewhitening filter had $M = 400$ taps per filter channel, and each separation system used $L = 300$ taps per input–output channel. For the normalized natural gradient algorithm in the frequency-domain [25], the parameters chosen were $L = 512$ and $\mu = 0.35$, and 200 passes of the algorithm through the data were made. For the natural gradient time-domain algorithm [23], a filter length of $L = 512$ taps was chosen, and a step size schedule of $\mu = .001$ for 200 data passes followed by $\mu = .0001$ for a single data pass followed by $\mu = .00001$ for a single data pass was used. The data nonlinearity used in the natural gradient algorithms was $f(y) = y/|y|$, where y in this case corresponds to the ith frequency bin output or the ith time domain filter output, respectively. For Parra's algorithm [37], with two sources and 14 seconds of data, the parameters chosen were {number of diagonalized matrices K } = 5, {number of data blocks averages $N = 21$}, {FFT size L = 1024}, $Q = 400$-tap time domain filters per input–output channel, and a limit of 1000 iterations. Parra's decorrelation-based method with beamforming constraints (equation (16) in [46]) for two sources employed $K = 5, N = 43, L = 512$, and a 1000 iteration limit. With three sources, the parameters in all of the aforementioned algorithms were kept the same except for [37] and [46], in which the value of N was changed to 32 and 65, respectively.

Shown in Fig. 1.11 are the SIR improvements obtained by the various algorithms for the various processing strategies on the two-source mixture data. We plot the output SIR computed between the outputs of the separation filters and the original speech sources versus the input SIR computed between the outputs of the sensors and the original speech sources. From Fig. 1.11,

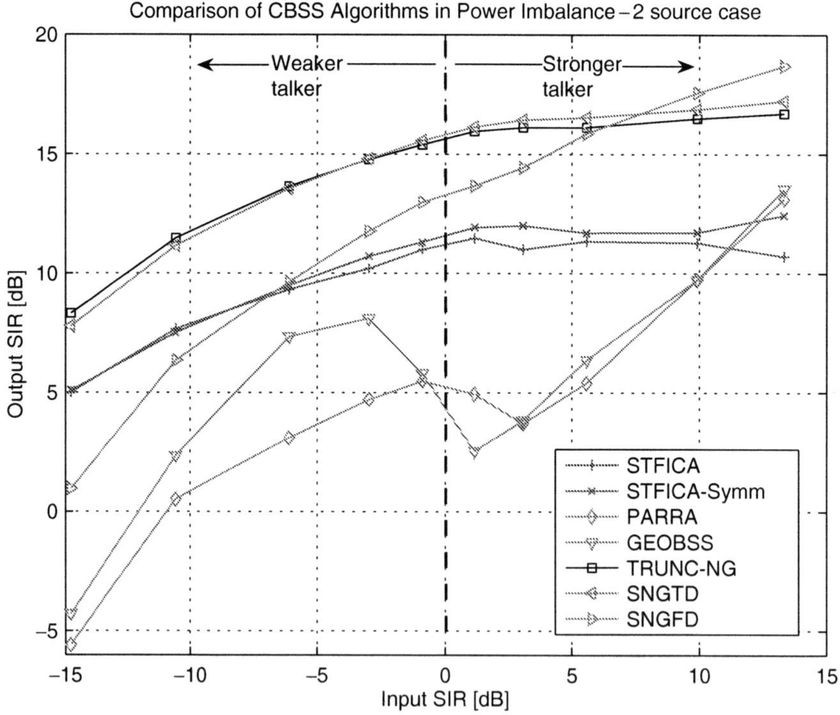

Fig. 1.11. Output SIR vs. input SIR.

we find the interesting result that the *weaker source* in the signal mixture is enhanced to a greater degree as compared to the stronger source within the mixture. This characteristic is most evident in the natural gradient time-domain methods and in the spatio-temporal FastICA methods. We also see that the frequency-domain algorithms provide very little separation under a high signal power imbalance in the mixtures, due to the fact that the strongest talker is generally extracted twice by these methods in such situations.

For three-source mixtures in which one of the sources is much weaker than the other two, Fig. 1.12 shows the output SIR vs. the input SIR for the weakest source for the various algorithms tested. Again, the time-domain natural gradient methods perform best here, obtaining about a 21 dB improvement in SIR for an input SIR of −14.5 . For an input SIR of −2 dB, the SIR gain was almost 16 dB. The STFICA algorithms performed about 3 dB worse than the time-domain natural gradient methods, and the overall performance of the frequency-domain decorrelation-based methods was between 10 dB and 15 dB worse than that of the natural gradient time-domain methods.

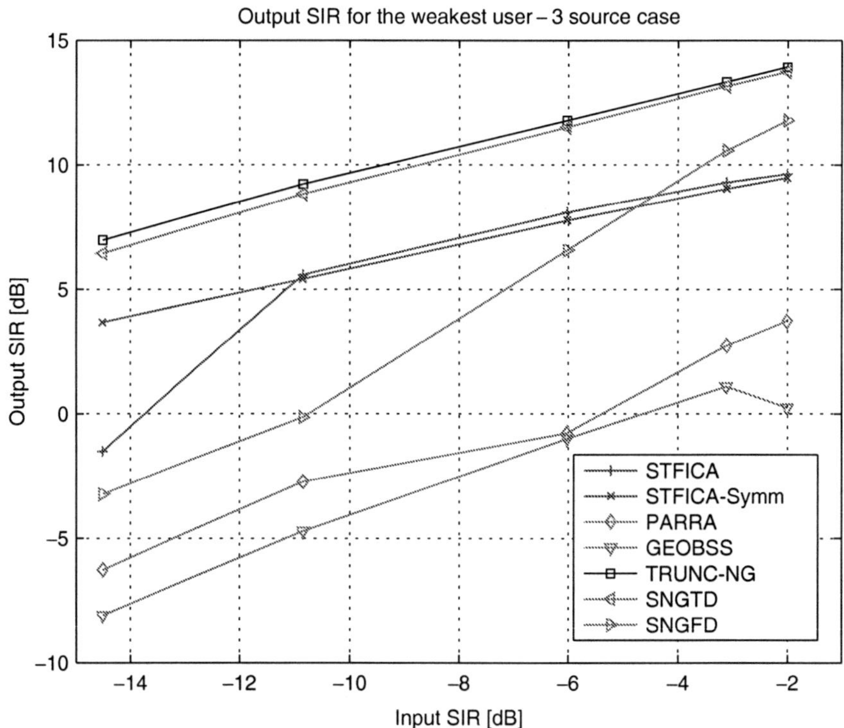

Fig. 1.12. Output SIR vs. input SIR for the weakest source.

1.8 Summary and Conclusions

In this chapter, we have provided a general overview of convolutive blind source separation methods for multi-microphone speech separation. The theoretical motivations for and structural relationship between several classes of separation methods has been described, and novel techniques for scaled natural gradient blind source separation and time-domain extensions of the well-known FastICA algorithm for contrast-based BSS have been provided. Extensive numerical experiments have elucidated the performance relationships between the various methods, indicating that convolutive BSS algorithms can separate speech signals without knowledge of sensor geometries and in the presence of signal-level mismatch in the measured mixtures.

While convolutive blind source separation shows much promise for multi-microphone speech enhancement, several challenges remain. Perhaps the most significant is the fact that most real-world speech enhancement tasks involve measurements of more sources than sensors and significant background noise, and thus careful attention should be placed on how to deal with the subsequent lack of signal diversity and the inherent nonlinear nature of the separation task in such cases. To this end, single-sensor speech enhancement is a natural complement to the multi-microphone techniques described in this chapter, and combinations of single-channel signal enhancement with multi-microphone separation methods are worthy of further study.

References

1. Y. Sato, "A method for self recovering equalization," *IEEE Trans. Communications*, vol. 23, no. 6, pp. 679–682, June 1975.
2. D. N. Godard, "Self-recovering equalization and carrier tracking in two-dimensional data communication systems." *IEEE Trans. Communications*, vol. 28, no. 11, pp. 1867–1875, Nov. 1980.
3. A. Benveniste, M. Goursat, and G. Ruget, "Robust identification of a non-minimum phase system – Blind adjustment of a linear equalizer in data communications," *IEEE Trans. Automatic Control*, vol. 25, no. 3, pp. 385–399, 1980.
4. O. Shalvi and E. Weinstein, "New criteria for blind deconvolution of nonminimum phase systems (channels)," *IEEE Trans. Inform. Theory*, vol. 36, no. 2, pp. 312–321, Mar. 1990.
5. R. A. Wiggins, "Minimum entropy deconvolution," *Geoexploration*, vol. 16, no. 1–2, pp. 21–35, 1978.
6. H. B. Barlow, "The coding of sensory messages," in *Current Problems in Animal Behaviour*, W. H. Thorpe and O. L. Zangwill, Eds. Cambridge University Press, 1960, pp. 331–360.
7. T. M. Cover and J. A. Thomas, *Elements of Information Theory*. Wiley Series in Telecommunications, 1991.
8. R. Linsker, "Local synaptic learning rules suffice to maximize mutual information in a linear-network," *Neural Computation*, vol. 4, no. 5, pp. 691–702, Sept. 1992.

9. ——, "A local learning rule that enables information maximization for arbitrary input distributions," *Neural Computation*, vol. 9, no. 8, pp. 1661–1665, Nov. 1997.
10. J.-P. Nadal and N. Para, "Non-linear neurons in the low noise limit: A factorial code maximizes information transfer," *Network*, vol. 4, pp. 565–581, 1994.
11. A. Bell and T. Sejnowski, "An information-maximization approach to blind separation and blind deconvolution," *Neural Computation*, vol. 7, pp. 1129–1159, 1995.
12. L. Tong, Y. Inouye, and R.-W. Liu, "Waveform-preserving blind estimation of multiple independent sources," *IEEE Trans. Signal Processing*, vol. 41, no. 7, pp. 2461–2470, July 1993.
13. L. R. Rabiner and R. W. Schafer, *Digital Processing of Speech Signals.* Prentice-Hall Signal Processing Series, 1993.
14. H. Buchner, R. Aichner, and W. Kellermann, "Blind source separation for convolutive mixtures exploiting nongaussianity, nonwhiteness and nonstationarity," in *Conf. Rec. IEEE intl. Workshop on Acoustic Echo and Noise Control (IWAENC)*, Kyoto, Japan, Sept. 2003, pp. 275–278.
15. W. B. Davenport, Jr., "A study of speech probability distributions," Tech. Rep. 148, Research Laboratory of Electronics, Massachusetts Inst. of Technology, Cambridge, MA, Aug. 1950.
16. S. C. Douglas, "Blind signal separation and blind deconvolution," in *Handbook of Neural Network Signal Processing*, Y.-H. Hu and J.-N. Hwang, Eds. New York: Wiley, 2001, ch. 7.
17. T. W. Lee, M. Girolami, A. J. Bell, and T. J. Sejnowski, "A unifying information-theoretic framework for independent component analysis," *Computers Mathematics Appl.*, vol. 39, no. 11, pp. 1–21, 2000.
18. J.-F. Cardoso, "Blind signal separation: Statistical principles," *Proc. of the IEEE*, vol. 86, no. 10, pp. 2009–2025, Oct. 1998.
19. S. Amari, T. P. Chen, and A. Cichocki, "Stability analysis of learning algorithms for blind source separation," *Neural Networks*, vol. 10, no. 8, pp. 1345–1351, Nov. 1997.
20. S. Amari, A. Cichocki, and H. Yang, "A new learning algorithm for blind signal separation," in *Advances in Neural Information Processing Systems*, vol. 8, pp. 757–763, 1996.
21. S. Amari, "Natural gradient works efficiently in learning," *Neural Computation*, vol. 10, no. 2, pp. 251–276, Feb. 1998.
22. S. Amari, S. C. Douglas, A. Chichocki, and H. H. Yang, "Multichannel blind deconvolution and equalization using the natural gradient," in *Proc. IEEE Workshop Signal Proc. Adv. Wireless Comm.*, Paris, France, 1997, pp. 101–104.
23. S. C. Douglas, H. Sawada, and S. Makino, "Natural gradient multichannel blind deconvolution and speech separation using causal FIR filters," *IEEE Trans. Speech Audio Processing*, vol. 13, no. 1, pp. 92–104, Jan. 2005.
24. L. X. Yuan, W. W. Wang, and J. A. Chambers, "Variable step-size sign natural gradient algorithm for sequential blind source separation," *IEEE Signal Processing Lett.*, vol. 12, no. 8, pp. 589–592, Aug. 2005.
25. S. C. Douglas and M. Gupta, "Scaled natural gradient algorithms for instantaneous and convolutive blind source separation," *IEEE Int. Conf. Acoust., Speech, Signal Processing*, vol. 2, pp. 637–640, Apr. 2007.

26. N. Delfosse and P. Loubaton, "Adaptive blind separation of independent sources – A deflation approach," *Signal Processing*, vol. 45, no. 1, pp. 59–83, July 1995.
27. G. H. Golub and C. F. V. Loan, *Matrix Computations*, 3rd ed. The John Hopkins University Press, 1996.
28. A. Hyvarinen, J. Karhunen, and E. Oja, *Independent Component Analysis.* Wiley Series on Adaptive and Learning Systems, 2001.
29. A. Hyvarinen and E. Oja, "A fast fixed-point algorithm for independent component analysis," *Neural Computation*, vol. 9, no. 7, pp. 1483–1492, Oct. 1997.
30. P. P. Vaidyanathan, *Multirate Systems and Filter Banks.* Englewood Cliffs, NJ: Prentice-Hall, 1992.
31. S. C. Douglas, "An adaptive constraint method for paraunitary filter banks with applications to spatio-temporal subspace tracking," *Accepted for publication in EURASIP J. Applied Signal Processing.*
32. ——, "The singular value manifold and numerical stabilization of algorithms with orthogonality constraints," in *Proc. IEEE Workshop Sensor Array*, Boston, MA, July 2006.
33. C. W. Therrien, *Discrete Random Signals and Statistical Signal Processing.* Prentice Hall, 1992.
34. L. Molgedey and H. G. Schuster, "Separation of a mixture of independent signals using time-delayed correlations," *Physical Review Letters*, vol. 72, no. 23, pp. 3634–3637, June 1994.
35. A. Belouchrani, K. AbedMeraim, J. F. Cardoso, and E. Moulines, "A blind source separation technique using second-order statistics," *IEEE Trans. Signal Processing*, vol. 45, no. 2, pp. 434–444, Feb. 1997.
36. N. Murata, S. Ikeda, and A. Ziehe, "An approach to blind source separation based on temporal structure of speech signals," *Neurocomputing*, vol. 41, pp. 1–24, Oct. 2001.
37. L. Parra and C. Spence, "Convolutive blind separation of non-stationary sources," *IEEE Trans. Speech Audio Processing*, vol. 8, no. 3, pp. 320–327, May 2000.
38. S. Haykin, *Adaptive Filter Theory*, 4th ed. Prentice Hall, 2001.
39. P. Smaragdis, "Blind separation of convolved mixtures in the frequency domain," *Neurocomputing*, vol. 22, no. 1–3, pp. 21–34, Nov. 1998.
40. B. D. V. Veen and K. M. Buckley, "Beamforming: A versatile approach to spatial filtering," *IEEE ASSP. Mag.*, pp. 4–24, Apr. 1988.
41. J. F. Cardoso and A. Soloumiac, "Blind beamforming for non-Gaussian signals," *IEEE Trans. Signal Processing*, vol. 140, no. 6, pp. 362–370, Dec. 1993.
42. S. kurita, H. Saruwatari, S. Kajita, K. Takeda, and F. Itakura, "Evaluation of blind signal separation method using directivity pattern under reverberant conditions," in *Proc. IEEE ICASSP*, vol. 5, Istanbul, Turkey, June 5–9, 2000, pp. 3140–3143.
43. H. Sawada, R. Mukai, S. Araki, and S. Makino, "A robust and precise method for solving the permutation problem of frequency-domain blind source separation," *IEEE Trans. Speech Audio Processing*, vol. 12, no. 5, pp. 530–538, Sept. 2004.
44. H. Saruwatari, T. Kawamura, T. Nishikawa, A. Lee, and K. Shikano, "Blind source separation based on a fast-convergence algorithm combining ICA and beamforming," *IEEE Trans. Audio Speech Language Processing*, vol. 14, no. 2, pp. 666–678, Mar. 2006.

45. D. H. Johnson and D. E. Dudgeon, *Array Signal Processing: Concepts and Techniques.* Prentice Hall, 1993.
46. L. C. Parra and C. V. Alvino, "Geometric source separation: Merging convolutive source separation with geometric beamforming," *IEEE Trans. Speech Audio Processing*, vol. 10, no. 6, pp. 352–362, Sept. 2002.
47. S. C. Douglas, H. Sawada, and S. Makino, "A spatio-temporal FastICA algorithm for separating convolutive mixtures," *IEEE Int. Conf. Acoust., Speech, Signal Processing*, vol. 5, Mar. 2005, pp. 165–168.
48. S. C. Douglas, M. Gupta, H. Sawada, and S. Makino, "Spatio-temporal FastICA algorithms for the blind separation of convolutive mixtures," *IEEE Trans. Audio Speech Language Processing,* vol. 15, no. 5, pp. 1511–1520, July 2007.

2 Frequency-Domain Blind Source Separation

Hiroshi Sawada, Shoko Araki, and Shoji Makino

NTT Communication Science Laboratories, NTT Corporation
2-4 Hikaridai, Seika-cho, Soraku-gun, Kyoto 619-0237, Japan
E-mail: {sawada, shoko, maki}@cslab.kecl.ntt.co.jp

Abstract. This chapter explains the frequency-domain approach to the blind source separation of acoustic signals mixed in a real room environment. With the application of short-time Fourier transforms, convolutive mixtures in the time domain can be approximated as multiple instantaneous mixtures in the frequency domain. So, separation is performed in each frequency bin with a simple instantaneous separation matrix. We employ complex-valued independent component analysis (ICA) to calculate the separation matrix. Then, the permutation ambiguity of the ICA solutions should be aligned so that the separated signals are constructed properly in the time domain. We estimate the time difference of arrival (TDOA) of a source at microphones from the ICA solutions. The frequency-dependent TDOA estimations are then clustered in order to align the permutation ambiguities. We also consider the use of time–frequency masking for a case where the separation by linear filters is insufficient when the sources outnumber the microphones. Experimental results are shown for a simple 3-source 3-microphone case, and also for a rather complicated case with many background interference signals.

2.1 Introduction

The technique for estimating individual source components from their mixtures at multiple sensors is known as blind source separation (BSS). The estimation is performed blindly, i.e., without possessing information about the mixing situation or the sources, such as the source location and its active time periods. If the sources are to be separated blindly, they should have some distinct characteristics, such as non-Gaussianity, nonstationarity, or nonwhiteness. Independent component analysis (ICA), which is sometimes regarded as synonymous with BSS, relies on non-Gaussianity. Many textbooks have been published on BSS and ICA [1–4].

One well-recognized BSS application is the separation of audio sources that have been mixed and then captured by multiple microphones in a real-room environment (e.g., solving a cocktail party problem). The difficulty of this problem lies in the fact that the mixing system is not simply instantaneous but convolutive, with delay and reflections. Such a mixing situation is generally modeled with the impulse responses from the sound sources to the microphones. In a practical room situation, such impulse responses can have thousands of taps even with an 8 kHz sampling rate, and this makes the

S. Makino et al. (eds.), Blind Speech Separation, 47–78.

convolutive problem difficult to solve. Various efforts have been devoted to the separation of convolutive mixtures. They can be classified into two major approaches: time-domain BSS [5–12] and frequency-domain BSS [13–27].

With time-domain BSS, a cost function is defined for time-domain signals, and optimized with convolutive separation filters. The approach is theoretically sound and attains a good separation once the optimization has converged, since there is no approximation and the cost function correctly evaluates such criteria as non-Gaussianity or nonstationarity. However, the optimization with convolutive separation filters is not as simple as BSS/ICA for instantaneous mixtures, and generally computationally expensive. Some techniques for reducing the computational complexity [11, 12] have recently been proposed.

With frequency-domain BSS, time-domain mixed signals observed at the microphones are converted into frequency-domain time-series signals by a short-time Fourier transform (STFT). If we use a sufficiently long frame for STFT, the convolutive mixture can be approximated with multiple instantaneous mixtures, each of which is defined for each frequency bin. Thus, we can employ any instantaneous ICA/BSS algorithm to separate the mixtures. However, the permutation ambiguity of an ICA solution becomes a serious problem. The ambiguities should be aligned properly so that the separated frequency components that originate from the same source are grouped together. The problem is well-known as the permutation problem [21, 22].

This chapter presents the basic scheme for frequency-domain BSS, and then explains in detail two important processing components, ICA and permutation alignment. Regarding ICA, we describe a three-step procedure in Sect. 2.4, which consists of whitening [3], FastICA [3], and maximum likelihood (ML) estimation [28]. This structure is intended for computational efficiency and a good separation. Another important point is that we need to model source signals with a complex-valued probability density function. We introduce one that can be applied to both FastICA and the ML estimation, and then derive both algorithms in detail for the complex-valued case.

As regards permutation alignment, we review various approaches to the permutation problem in Sect. 2.5. Then, we focus on a method based on estimating the time difference of arrival (TDOA) [29, 30] of a source signal at multiple microphones. TDOAs are estimated frequency by frequency from ICA solutions and then the frequency-dependent estimations are clustered for individual sources. The permutation ambiguities are aligned so that members of the same cluster have the same index. This method is very simple, yet effective as long as the reverberation of a room is not so severe. In addition, the method can identify close and distant sources from the variances of the TDOA clusters.

We performed experiments with two setups, as described in Sect. 2.7. Setup A had a basic configuration with three microphones and three sources. Setup B had an advanced configuration with some sources close to the microphones and six background sources far away. In setup B, we had four microphones, which was fewer than the total number of sources. Even if the total number of sources is greater than the number of microphones, ICA with a square separation matrix produces separated signals that correspond to the close sources [24]. However, a perfect separation with linear filters was impossible and the separated signal had some interference residuals. To suppress these residuals, we applied time–frequency (T–F) masking to the separated signal [24, 31, 32], as explained in Sect. 2.6. T–F masking is a widely used method for separating speeches in underdetermined situations with a limited number of microphones [33–36], as discussed elsewhere in this book.

2.2 Problem Formulation

Let $s_1, \ldots, s_N$ be N source signals. The observation x_{jk} at microphone j that originates only from source k is described by the convolutive model

$$x_{jk}(t) = \sum_{l=0}^{P} h_{jk}(l \cdot t_s)\, s_k(t - l \cdot t_s)\,, \tag{2.1}$$

where t represents discrete time (a multiple of $t_s = 1/f_s$ with f_s being the sampling rate), and h_{jk} is the impulse responses from source k to microphone j modeled with $P+1$ samples. We consider a situation where the N sources are simultaneously active. Hence, the observation x_j at microphone j is modeled by the convolutive mixture model

$$x_j(t) = \sum_{k=1}^{N} x_{jk}(t) = \sum_{k=1}^{N} \sum_{l=0}^{P} h_{jk}(l \cdot t_s)\, s_k(t - l \cdot t_s)\,. \tag{2.2}$$

Figure 2.1 shows the signal notations. Now, the goal of the BSS task is to obtain separated signals $y_1, \ldots, y_N$, each of which corresponds to each of the source signals $s_1, \ldots, s_N$. The task should be performed only with M observed mixtures $x_1, \ldots, x_M$, and without information on the sources s_k and the impulse responses h_{jk}. The number of sources N may also be unknown. As a basic setup, we assume that the number of sources N is no more than

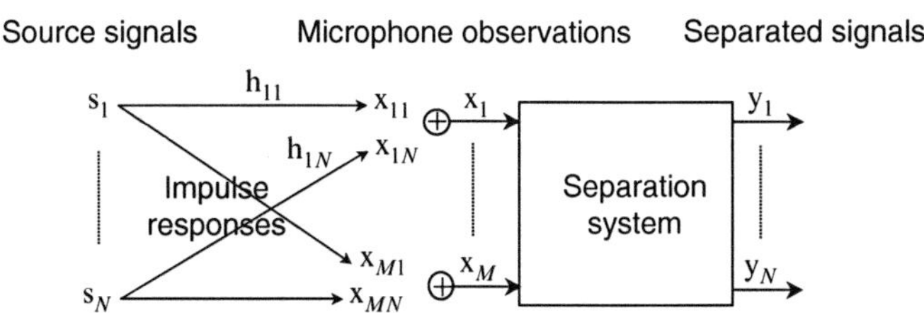

Fig. 2.1. Signal notations.

the number of the microphones M, i.e., $N \leq M$. However, in this chapter, we also consider a case with a larger number $N > M$ of sources. Separating all the sources in such a case is referred to as underdetermined BSS, which is a rather difficult problem and is discussed in other chapters of this book. In this chapter, we try to separate the mixtures into at most M separated signals, and identify which of them contain sources close to the microphones.

An ultimate goal of the BSS task is to make the separated signal y_i close to the corresponding target source s_i, with eliminating all the effects of the impulse responses h_{ji}. This corresponds to blind separation and deconvolution. Even blind deconvolution of a single source (no separation needed) is a challenging task especially with colored sources such as speech [37, 38]. Thus, let us here set an easier goal, where the separated signal y_i should be close the target source component x_{Ji} observed at a selected microphone J. This goal, namely adjusting the separated signal to the microphone observation, has been proposed in many BSS studies [7, 10, 16, 39].

The separation performance is evaluated in terms of signal-to-interference ratio (SIR) improvement and signal-to-distortion ratio (SDR). A larger number represents a better result for both criteria. To calculate these numbers, we need the individual source observations x_{jk} defined in (2.1), which are not available in the BSS procedure. The SIR improvement for output i is calculated by $\mathsf{OutputSIR}_i - \mathsf{InputSIR}_i$. These two types of SIRs are defined by the power ratio between the components related to the target sources and interference sources, at a specific microphone J and at the output i:

$$\mathsf{InputSIR}_i = 10 \log_{10} \frac{\sum_t |\mathrm{x}_{Ji}(t)|^2}{\sum_t |\sum_{k \neq i} \mathrm{x}_{Jk}(t)|^2} \quad \text{(dB)},$$

$$\mathsf{OutputSIR}_i = 10 \log_{10} \frac{\sum_t |\mathrm{y}_{ii}(t)|^2}{\sum_t |\sum_{k \neq i} \mathrm{y}_{ik}(t)|^2} \quad \text{(dB)},$$

where y_{ik} is the component of s_k that appears at output y_i. The signal y_{ik} is calculated by applying the same separation operation to the individual source observations $\mathrm{x}_{1k}, \ldots, \mathrm{x}_{Mk}$ instead of the mixtures $\mathrm{x}_1, \ldots, \mathrm{x}_M$. Such signals are decomposed components of the separated signal: $\mathrm{y}_i(t) = \sum_{k=1}^{N} \mathrm{y}_{ik}(t)$.

The SDR for output i is defined by the power ratio between the individual source observation x_{Ji} at a microphone J and the distortion in y_{ii}:

$$\mathsf{SDR}_i = 10 \log_{10} \frac{\sum_t |\alpha_i \, \mathrm{x}_{Ji}(t - \delta_i)|^2}{\sum_t |\mathrm{y}_{ii}(t) - \alpha_i \, \mathrm{x}_{Ji}(t - \delta_i)|^2} \quad \text{(dB)}.$$

The distortion is defined in the denominator and is minimized by adjusting scalars δ_i and α_i for time and amplitude differences. The optimal δ_i is obtained by maximizing the cross correlation

$$\delta_i = \mathrm{argmax}_\delta \sum_t \mathrm{y}_{ii}(t) \, \mathrm{x}_{Ji}(t - \delta).$$

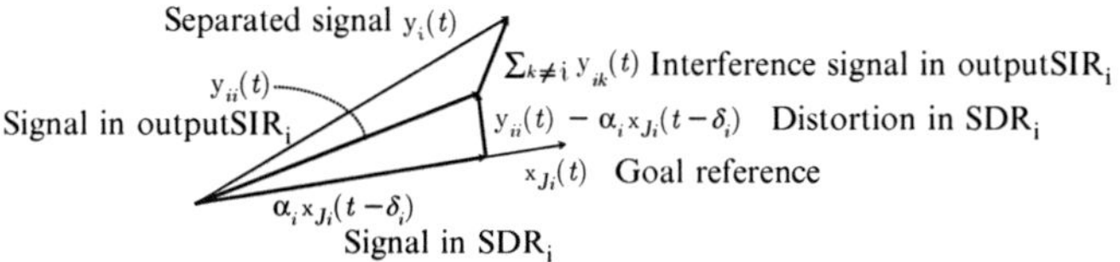

Fig. 2.2. Graphical interpretation of SIR and SDR definitions.

Or, if the time difference between x_J and y_i is known based on the operations of the separation system, this information can just be used for δ_i. In either case, the optimal α_i is then calculated by a least-mean-square estimator

$$\alpha_i = \frac{\sum_t \mathrm{y}_{ii}(t)\, \mathrm{x}_{Ji}(t-\delta_i)}{\sum_t |\mathrm{x}_{Ji}(t-\delta_i)|^2} .$$

Figure 2.2 shows a graphical interpretation of $\mathsf{OutputSIR}_i$ and SDR_i.

2.3 Overview of Frequency-Domain BSS

This section presents an overview of the frequency-domain BSS approach that we consider in this chapter. Figure 2.3 shows the system structure.

STFT

First, each of the time-domain microphone observations $\mathrm{x}_j(t)$ is converted into frequency-domain time-series signals $x_j(n,f)$ by a short-time Fourier transform (STFT) with an L-sample frame and its S-sample shift:

$$x_j(n,f) \leftarrow \sum_t \mathrm{x}_j(t)\, \mathrm{win_a}(t - nSt_s)\, e^{-\imath 2\pi f t}, \tag{2.3}$$

for all discrete frequencies $f \in \{0, \frac{1}{L}f_s, \ldots, \frac{L-1}{L}f_s\}$, and for frame index n. The analysis window $\mathrm{win_a}(t)$ is defined as being nonzero only in the L-sample interval $[-\frac{L}{2}t_s, (\frac{L}{2}-1)t_s]$ and tapers smoothly to zero at each end of the interval, such as a Hanning window $\mathrm{win_a}(t) = \frac{1}{2}(1 + \cos\frac{2\pi t}{Lt_s})$.

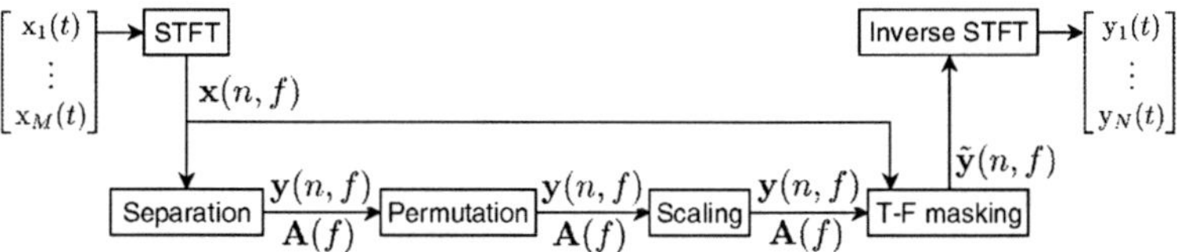

Fig. 2.3. System structure of frequency-domain BSS.

If the frame size L is long enough to cover the main part of the impulse responses h_{jk}, the convolutive model (2.1) can be approximated as an instantaneous model at each frequency:

$$x_{jk}(n, f) = h_{jk}(f) s_k(n, f), \tag{2.4}$$

where $h_{jk}(f)$ is the frequency response from source k to microphone j, and $s_k(n, f)$ is a frequency-domain time-series signal of $\mathrm{s}_k(t)$ obtained by the same operation as (2.3). Consequently, the convolutive mixture model (2.2) can be treated as an instantaneous mixture model:

$$x_j(n, f) = \sum_{k=1}^{N} x_{jk}(n, f) = \sum_{k=1}^{N} h_{jk}(f) s_k(n, f) . \tag{2.5}$$

We also use a vector notation

$$\mathbf{x}(n, f) = \sum_{k=1}^{N} \mathbf{h}_k(f) s_k(n, f), \tag{2.6}$$

where $\mathbf{h}_k = [h_{1k}, \ldots, h_{Mk}]^\mathsf{T}$ is the vector of frequency responses from source s_k to all microphones, and $\mathbf{x} = [x_1, \ldots, x_M]^\mathsf{T}$ is the observation vector.

The following operations, Separation, Permutation, Scaling, and T–F masking, are performed in the frequency domain. We can limit the set of frequencies $\mathcal{F}$ to perform the operations by

$$\mathcal{F} = \left\{0, \frac{1}{L} f_s, \ldots, \frac{1}{2} f_s\right\} \tag{2.7}$$

due to the relationship of complex conjugate:

$$x_j(n, \tfrac{m}{L} f_s) = x_j^*(n, \tfrac{L-m}{L} f_s), \quad m = 1, \ldots, \tfrac{L}{2} - 1 . \tag{2.8}$$

Separation

In the separation stage, the vector of frequency-domain mixtures $\mathbf{x} = [x_1, \ldots, x_M]^\mathsf{T}$ is linearly transformed into the vector of separated frequency components $\mathbf{y} = [y_1, \ldots, y_N]^\mathsf{T}$ by

$$\mathbf{y}(n, f) = \mathbf{W}(f) \mathbf{x}(n, f), \quad f \in \mathcal{F}, \tag{2.9}$$

where $\mathbf{W}$ is an $N \times M$ separation matrix. We employ complex-valued instantaneous ICA to calculate the separation matrix $\mathbf{W}$. Many ICA algorithms have been reported [1–4]. Section 2.4 describes the detailed procedure for the complex-valued ICA used in our implementation and experiments.

Equation (2.9) already provides us with separated frequency components $y_1, \ldots, y_N$. If ICA works well, we expect $y_1, \ldots, y_N$ to be close to the original source frequency components $s_1, \ldots, s_N$. However, the correspondence is up to the scaling and permutation ambiguities that an ICA solution inherently has. Even if we permute the elements of $\mathbf{y} = [y_1, \ldots, y_N]^\mathsf{T}$ or multiply an element by a constant, it is still an ICA solution. In a vector notation,

$$\mathbf{y}(n, f) \leftarrow \mathbf{\Lambda}(f)\, \mathbf{P}(f)\, \mathbf{y}(n, f) \tag{2.10}$$

is also a solution for any permutation $\mathbf{P}(f)$ and diagonal $\mathbf{\Lambda}(f)$ matrices.

To align such ambiguities, it is advantageous to calculate vectors $\mathbf{a}_i = [a_{1i}, \ldots, a_{Mi}]^\mathsf{T}$, $i = 1, \ldots, N$, which we call basis vectors, and to represent the mixture vector $\mathbf{x}$ by a linear combination of the basis vectors:

$$\mathbf{x}(n, f) = \sum_{i=1}^{N} \mathbf{a}_i(f) y_i(n, f) = \mathbf{A}(f)\, \mathbf{y}(n, f)\,, \tag{2.11}$$

where $\mathbf{A} = [\mathbf{a}_1, \cdots, \mathbf{a}_N]$ is a matrix whose columns are the basis vectors. If the separation matrix $\mathbf{W}$ has an inverse, matrix $\mathbf{A}$ is given simply by its inverse, i.e., $\mathbf{A} = \mathbf{W}^{-1}$. Otherwise it is calculated as a least-mean-square estimator [40]

$$\mathbf{A} = \mathrm{E}\{\mathbf{x}\mathbf{y}^\mathsf{H}\}(\mathrm{E}\{\mathbf{y}\mathbf{y}^\mathsf{H}\})^{-1}\,, \tag{2.12}$$

which minimizes $\mathrm{E}\{||\mathbf{x} - \mathbf{A}\mathbf{y}||^2\}$. The equality in (2.11) may not hold exactly in this case.

Equation (2.11) describes a separation of the mixture vector $\mathbf{x}$ into $\mathbf{a}_1 y_1, \ldots, \mathbf{a}_N y_N$. Together with the mixture model (2.6), we have

$$\sum_{i=1}^{N} \mathbf{a}_i(f) y_i(n, f) = \sum_{k=1}^{N} \mathbf{h}_k(f) s_k(n, f)\,.$$

And if ICA works well, we further have

$$\mathbf{a}_i(f) y_i(n, f) \approx \mathbf{h}_k(f) s_k(n, f) \quad i = 1, \ldots, N\,, \tag{2.13}$$

with some correspondences between i and k, which are unknown at this stage because of the permutation ambiguity.

Permutation

The permutation stage determines the permutation matrix $\mathbf{P}(f)$ in (2.10) for each frequency $f \in \mathcal{F}$ so that separated frequency components $y_i(n, f)$ from the same source have the same index i for all $f \in \mathcal{F}$. Such a problem is well-known as the permutation problem, and much effort has been devoted to solving this problem. Section 2.5 reviews various attempts, and then presents one method in detail.

After the permutation matrix $\mathbf{P}$ is determined, the vector of separated components $\mathbf{y}$ and the matrix of basis vectors $\mathbf{A}$ is updated as

$$\mathbf{y}(n,f) \leftarrow \mathbf{P}(f)\,\mathbf{y}(n,f)\,, \quad \forall n, \tag{2.14}$$
$$\mathbf{A}(f) \leftarrow \mathbf{A}(f)\,[\mathbf{P}(f)]^{\mathsf{T}}\,, \tag{2.15}$$

for each frequency $f \in \mathcal{F}$. Equation (2.11) is not changed by the update because a permutation matrix is an orthogonal matrix $\mathbf{P}^{\mathsf{T}}\mathbf{P} = \mathbf{I}$.

Scaling

The scaling stage determines the diagonal matrix $\mathbf{\Lambda}(f)$ in (2.10) for each frequency $f \in \mathcal{F}$. As formulated in Sect. 2.2, our goal is to bring the time-domain separated signal $\mathrm{y}_i(t)$ close to the target source component $\mathrm{x}_{Ji}(t)$ observed at a selected microphone J [7, 10, 16, 39]. This condition can be interpreted in the frequency domain such that $y_i(n,f)$ should be close to $x_{Ji}(n,f)$.

Here we assume that the permutation ambiguities are properly aligned. Then, (2.13) can be written as

$$\mathbf{a}_i(f)\,y_i(n,f) \;\approx\; \mathbf{h}_i(f)\,s_i(n,f)\,, \quad i = 1,\ldots,N\,. \tag{2.16}$$

Considering the J-th element $a_{Ji}(f)\,y_i(n,f) \;\approx\; h_{Ji}(f)\,s_i(n,f)$ and the definition $x_{Ji}(n,f) = h_{Ji}(f)\,s_i(n,f)$ in (2.4), we understand that the operation

$$y_i(n,f) \leftarrow a_{Ji}(f)\,y_i(n,f)$$

makes $y_i(n,f)$ close to $x_{Ji}(n,f)$. For the vector notation (2.10), we have a diagonal matrix

$$\mathbf{\Lambda}(f) = \mathrm{diag}[a_{J1}(f),\ldots,a_{JN}(f)]$$

that solves the scaling problem.

T–F masking

This stage performs time–frequency (T–F) masking with the intention of further improving the SIR. T–F masking is applied to each element of $\mathbf{y} = [y_1,\ldots,y_N]^{\mathsf{T}}$ by

$$\tilde{y}_i(n,f) = \mathcal{M}_i(n,f)\,y_i(n,f), \quad i = 1,\ldots,N\,, \tag{2.17}$$

where $0 \le \mathcal{M}_i(n,f) \le 1$ is a mask specified for each time–frequency slot (n,f). We discuss how to specify these masks in Sect. 2.6. The resultant elements are represented in a vector form $\tilde{\mathbf{y}} = [\tilde{y}_1,\ldots,\tilde{y}_N]^{\mathsf{T}}$ in Fig. 2.3.

The effectiveness of T–F masking depends on the situation, and thus we consider it optional (setting $\mathcal{M} = 1$ results in no change). SIR improvement is generally attained with some sacrifice as regards the SDR, since the operation (2.17) is not linear filtering and introduces some distortion. If we have non-negligible background interference sources, as with setup B in Fig. 2.11, T–F masking is effective in reducing the residuals caused by the limitation of linear filtering (explained in Sect. 2.6), even though some distortions are introduced.

Inverse STFT

At the end of the flow, time-domain output signals $\mathrm{y}_i(t)$ are obtained by the inverse operation of the STFT:

$$\mathrm{y}_i(t) = \sum_n \mathrm{win_s}(t-nSt_s) \sum_{f\in\{0,\,\frac{1}{L}f_s,\,\dots,\,\frac{L-1}{L}f_s\}} \tilde{y}_i(n,f)\, e^{\imath 2\pi f t},$$

where $\mathrm{win_s}(t)$ is a synthesis window defined as nonzero only in the L-sample interval $[-\frac{L}{2}t_s, (\frac{L}{2}-1)t_s]$. The summation over the frame index n is with those that satisfy $-\frac{L}{2}t_s \le t-nSt_s \le (\frac{L}{2}-1)t_s$. To realize a perfect reconstruction, the analysis and synthesis windows should satisfy the condition

$$\sum_n \mathrm{win_s}(t-nSt_s)\,\mathrm{win_a}(t-nSt_s) = 1$$

for any time t. The condition is satisfied by pairing a Hanning window for $\mathrm{win_a}(t)$ and a rectangular window for $\mathrm{win_s}(t)$. However, a synthesis window that tapers smoothly to zero at each end is also preferred in terms of mitigating the edge effect. In such a case, the condition is satisfied, for instance, by using a square-root version of a Hanning window for both $\mathrm{win_a}(t)$ and $\mathrm{win_s}(t)$.

2.4 Complex-Valued ICA

This section presents a complex-valued ICA procedure to separate the mixtures $\mathbf{x}(n,f)$ in the frequency domain. For a simpler notation, here let us omit the frequency index f of the separation formula (2.9):

$$\mathbf{y}(n) = \mathbf{W}\mathbf{x}(n)\,, \tag{2.18}$$

where $\mathbf{x} = [x_1, \dots, x_M]^\mathsf{T}$ is the vector of mixtures, $\mathbf{y} = [y_1, \dots, y_N]^\mathsf{T}$ is the vector of separated signals, and $\mathbf{W}$ is an $N \times M$ separation matrix.

The procedure presented here consists of the following three steps:

1. Whitening and dimension reduction (if necessary): $\mathbf{z}(n) = \mathbf{V}\mathbf{x}(n)$
2. FastICA, separation with a unitary matrix: $\mathbf{y}(n) = \mathbf{U}\mathbf{z}(n)$
3. Maximum likelihood estimation: $\mathbf{y}(n) = \mathbf{W}\mathbf{x}(n)$

This three-step structure provides computational efficiency (step 2) and better separation performance (step 3). The first step only considers the second order statistics (variances) of variables. Thus, the operation is linear and can be done quickly. The second and the third steps consider the density function of speech sources, such as that defined by (2.19), which is usually different from a Gaussian density. Consequently, the operation involves an optimization with nonlinear functions, which needs iterative updates and thus is time consuming. The second step (FastICA) limits the solution space to a unitary matrix. This makes the algorithm fast to converge by employing Newton's method without calculating the matrix inversion. The third step has no constraint for the separation matrix $\mathbf{W}$. In this sense, it might improve the solution $\mathbf{W} = \mathbf{UV}$ obtained by the first and second steps. In the following subsections, we first assume the probability density function of source signals, and then explain each step of the procedure.

2.4.1 Source Density Function

An ICA algorithm generally assumes the source signal model with a probability density function. Many density functions can be considered for modeling a speech source in the frequency domain. In this chapter, we employ the following density function:

$$p(y_i) = \frac{1}{B} \exp\left(-\frac{\sqrt{|y_i|^2 + \alpha}}{b}\right), \tag{2.19}$$

where $b > 0$ specifies the variance, a small nonnegative parameter $\alpha \geq 0$ controls the smoothness around the origin $y_i = 0$, and B should be defined so that the integral of $p(y_i)$ becomes unity: $\int p(y)dy = 1$. Among these parameters, B is unimportant in the ICA algorithm, and does not have to be specified. Also, b is relatively unimportant since it specifies the variance and only affects the scale of the separated signal y_i, which will be aligned afterwards anyway in the frequency-domain BSS scheme.

Figure 2.4 shows some density functions for a complex-valued variable, that follow (2.19) and the Gaussian distribution. The variance is normalized to 1 for all the cases. We see that (2.19) provides a sharper peaked distribution (the probability of the variable being zero is higher) than the Gaussian distribution. Such a distribution models a speech signal in the frequency domain very well. Regarding the parameter α, a smaller value gives a more sharply peaked distribution. However, it should be nonzero if the second order derivative of $\log p(y_i)$ is used in the ICA algorithm. Otherwise, it can be set to zero to make the density function simpler: $p(y_i) \propto \exp(-\frac{|y_i|}{b})$, i.e., a Laplace distribution.

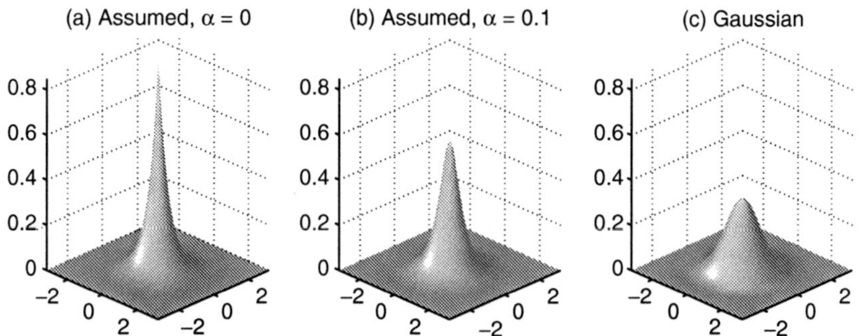

Fig. 2.4. Probability density functions of complex variables. (a) and (b) are those assumed by (2.19).

2.4.2 Whitening

This subsection presents the first step of the procedure. Let us first define the correlation matrix of vectors $\mathbf{x}$ and $\mathbf{y}$ as

$$\mathbf{R_{xy}} = \mathrm{E}\{\mathbf{x}\mathbf{y}^{\mathsf{H}}\},$$

where $\mathbf{y}^{\mathsf{H}}$ is the conjugate transpose of $\mathbf{y}$, and $\mathrm{E}\{\cdot\}$ is the expectation operator. The mathematical definition of the expectation operator requires the density functions of the variables. However, we usually approximate the operation by calculating the sample mean, because the density functions cannot usually be obtained. Thus, the correlation matrix is practically calculated with the sample mean as

$$\mathbf{R_{xy}} = \frac{1}{K}\sum_{n=1}^{K} \mathbf{x}(n)\mathbf{y}^{\mathsf{H}}(n),$$

where K is the number of samples.

The whitening operation is performed with an $M \times M$ matrix $\mathbf{V}$ by

$$\mathbf{z}(n) = \mathbf{V}\mathbf{x}(n)$$

such that the correlation matrix of the output vector $\mathbf{z} = [z_1, \ldots, z_M]$ becomes an identity matrix $\mathbf{R_{zz}} = \mathbf{I}$. Because $\mathbf{R_{zz}} = \mathbf{V}\mathbf{R_{xx}}\mathbf{V}^{\mathsf{H}}$, the whitening matrix $\mathbf{V}$ is simply given by

$$\mathbf{V} = \mathbf{D}^{-1/2}\mathbf{E}^{\mathsf{H}}, \tag{2.20}$$

if we have an eigenvalue decomposition $\mathbf{R_{xx}} = \mathbf{E}\mathbf{D}\mathbf{E}^{\mathsf{H}}$.

If we know the number of source signals N, it may be a good idea to reduce the dimension of the output vector $\mathbf{z}$ from M to N. This reduces the computational complexity of the ICA algorithm that follows, and may stabilize the algorithm for a robust convergence by reducing the search space. However, there is a possibility that a better solution resides in the discarded space. So, the effectiveness of the dimension reduction depends on the situation. In reducing the dimension, the source signal components should be kept while the noise components should be reduced. For this purpose, let us here sort the eigenvalues $d_1 \geq \cdots \geq d_M$ in the diagonal matrix $\mathbf{D} = \text{diag}(d_1, \ldots, d_M)$, and also sort the corresponding eigenvectors in the matrix $\mathbf{E} = [\mathbf{e}_1, \ldots, \mathbf{e}_M]$. Then, dimension reduction is simply performed by keeping the first N elements of the output vector $\mathbf{z} = [z_1, \ldots, z_N, \ldots, z_M]^\mathsf{T}$. If the dimension reduction is not performed, simply let $N = M$ for the uniformity of the following descriptions.

The whitening operation is not intended for source separation, but simply for making the elements $z_1, \ldots, z_N$ uncorrelated and unit-variance. As regards frequency-domain BSS, this kind of power normalization (making unit-variance) is very effective for a rapid and robust convergence to an ICA solution with a predefined step-size for iterative updates. Without such power normalization, the convergence would not be uniform from frequency to frequency, since audio sources are generally highly colored and thus there is a large variation in the total power from frequency to frequency. Also, the whitening operation constitutes prerequisite preprocessing for the FastICA algorithm in the next step.

2.4.3 FastICA

The second step employs the FastICA algorithm [3] to produce a vector of separated signals $\mathbf{y} = [y_1, \ldots, y_N]$ with a unitary transformation $\mathbf{U}$ applied to the whitened vector $\mathbf{z}$:

$$\mathbf{y}(n) = \mathbf{U}\mathbf{z}(n) \,. \tag{2.21}$$

The core part of the algorithm is designed to extract one separated signal y_i with a unit-norm vector $\mathbf{u}_i$ by

$$y_i(n) = \mathbf{u}_i^\mathsf{H}\mathbf{z}(n) \,.$$

Thus, let us first explain the one-unit algorithm and then extend it to extract all the N elements $\mathbf{y} = [y_1, \ldots, y_N]^\mathsf{T}$ with $\mathbf{U} = [\mathbf{u}_1, \ldots, \mathbf{u}_N]^\mathsf{H}$.

The one-unit FastICA algorithm can be formulated as follows. Consider a function $G(y_i) = -\log p(y_i)$ with a probability density function $p(y_i)$. The expected value $\text{E}\{G(y_i)\}$ can be an approximation of the entropy of y_i, if the density function p provides a good approximation of the true distribution of y_i. The criterion is to minimize the expected value $\text{E}\{G(y_i)\}$ under the unit-norm constraint $||\mathbf{u}_i|| = 1$ for the vector $\mathbf{u}_i$. This can be formulated by introducing a Lagrange multiplier λ:

$$\mathcal{L}(\mathbf{u}_i, \lambda) = \text{E}\{G(y_i)\} - \lambda(||\mathbf{u}_i||^2 - 1) \,, \quad y_i = \mathbf{u}_i^\mathsf{H}\mathbf{z} \,.$$

The partial derivatives of $\mathcal{L}(\mathbf{u}_i, \lambda)$ with respect to $\mathbf{u}_i^*$ and λ are

$$\frac{\partial \mathcal{L}(\mathbf{u}_i, \lambda)}{\partial \mathbf{u}_i^*} = \mathrm{E}\{g(y_i)\,\mathbf{z}\} - \lambda \mathbf{u}_i = 0\,, \quad g(y_i) = \frac{\partial G(y_i)}{\partial y_i}\,, \tag{2.22}$$

$$\frac{\partial \mathcal{L}(\mathbf{u}_i, \lambda)}{\partial \lambda} = -||\mathbf{u}_i||^2 + 1 = 0\,, \tag{2.23}$$

respectively. Now the minimization problem with the constraint is reduced to finding solutions of these simultaneous equations (2.22) and (2.23). By employing Newton's method for root finding, and with several mathematical manipulations, we obtain the main update rule for FastICA:

$$\mathbf{u}_i \leftarrow \mathrm{E}\{g'(y_i)\}\mathbf{u}_i - \mathrm{E}\{g(y_i)\,\mathbf{z}\}, \quad y_i = \mathbf{u}_i^{\mathsf{H}}\mathbf{z}, \tag{2.24}$$

where $g(y_i)$ and $g'(y_i)$ are the first and second derivatives of $G(y_i)$. After every update, the vector $\mathbf{u}_i$ should be normalized to unit-norm

$$\mathbf{u}_i \leftarrow \frac{\mathbf{u}_i}{||\mathbf{u}_i||} \tag{2.25}$$

to satisfy the constraint. In summary, the FastICA algorithm for extracting a single element y_i iterates the two updates (2.24) and (2.25) until convergence.

Regarding the functions $g(y_i)$ and $g'(y_i)$, if we assume (2.19) for the density function $p(y_i)$ with the variance parameter $b = 1$, we have

$$G(y_i) = -\log p(y_i) = \sqrt{|y_i|^2 + \alpha} + \log B\,,$$

and the first and second derivatives used in the update (2.24) are given by

$$g(y_i) = \frac{\partial G(y_i)}{\partial y_i} = \frac{y_i^*}{2\sqrt{|y_i|^2 + \alpha}}\,, \tag{2.26}$$

$$g'(y_i) = \frac{\partial g(y)}{\partial y_i^*} = \frac{1}{2\sqrt{|y_i|^2 + \alpha}}\left[1 - \frac{1}{2}\frac{|y_i|^2}{|y_i|^2 + \alpha}\right]. \tag{2.27}$$

Now let us extend the algorithm to extract all the elements $\mathbf{y} = [y_1, \ldots, y_N]^{\mathsf{T}}$ with a unitary matrix $\mathbf{U} = [\mathbf{u}_1, \ldots, \mathbf{u}_N]^{\mathsf{H}}$ by (2.21). The parallel applications of the conjugate-transposed version

$$\mathbf{u}_i^{\mathsf{H}} \leftarrow \mathrm{E}\{g'(y_i)\}\mathbf{u}_i^{\mathsf{H}} - \mathrm{E}\{g(y_i)^*\,\mathbf{z}^{\mathsf{H}}\}$$

of the update (2.24) for all $i = 1, \ldots, N$ can be described by

$$\mathbf{U} \leftarrow \mathbf{\Gamma}\mathbf{U} - \mathrm{E}\{\mathbf{\Phi}(\mathbf{y})\mathbf{z}^{\mathsf{H}}\}\,, \tag{2.28}$$

with

$$\mathbf{\Gamma} = \mathrm{diag}(\mathrm{E}\{g'(y_1)\}, \ldots, \mathrm{E}\{g'(y_N)\})\,, \quad \mathbf{\Phi}(\mathbf{y}) = [g(y_1)^*, \ldots, g(y_N)^*]^{\mathsf{T}}\,.$$

Here we assume that $g'(y_i)$ is a real valued function that is not affected by the conjugate operator $\cdot^*$. After each update by (2.28), the matrix $\mathbf{U}$ should be adjusted to a unitary matrix. This can be done by

$$\mathbf{U} \leftarrow (\mathbf{U}\mathbf{U}^{\mathsf{H}})^{-1/2}\mathbf{U}\,. \tag{2.29}$$

For a matrix $\mathbf{R}$, the inverse square root $\mathbf{R}^{-1/2}$ is calculated by $\mathbf{R}^{-1/2} = \mathbf{E}\mathbf{D}^{-1/2}\mathbf{E}^{\mathsf{H}}$ with the eigenvalue decomposition of the matrix $\mathbf{R} = \mathbf{E}\mathbf{D}\mathbf{E}^{\mathsf{H}}$. In summary, the FastICA algorithm for multiple elements $\mathbf{y} = [y_1, \ldots, y_N]^{\mathsf{T}}$ iterates the two updates (2.28) and (2.29) until convergence.

FastICA is a very efficient algorithm and generally converges to a good solution with several iterations, even from an initial solution that is simply an identity matrix $\mathbf{U} = \mathbf{I}$. The efficiency comes from the fact that the input $\mathbf{z}$ is whitened and the solution $\mathbf{U}$ is limited to a unitary matrix. However, the solution may be improved by another ICA algorithm that does not have such a unitary constraint. One reason is that the output $\mathbf{y}$ of FastICA is whitened $\mathrm{E}\{\mathbf{y}\mathbf{y}^{\mathsf{H}}\} = \mathbf{I}$ and therefore uncorrelated, whereas the original sources $s_1, \ldots, s_N$ are not always completely uncorrelated with a limited number of samples.

2.4.4 Maximum Likelihood Estimation

The third step solves the ICA equation in a general form

$$\mathbf{y}(n) = \mathbf{W}\mathbf{x}(n), \tag{2.30}$$

where $\mathbf{W}$ has no constraint. Here we perform maximum likelihood estimation for $\mathbf{W}$, which is basically equivalent to the information-maximization approach [41], as shown in [28]. The likelihood function with the observation vector samples $\mathbf{x}(1), \ldots, \mathbf{x}(K)$ is given by

$$\mathcal{L}(\mathbf{W}|\mathbf{x}) = \prod_{n=1}^{K} p(\mathbf{x}(n)|\mathbf{W})\,. \tag{2.31}$$

A solution for the separation matrix $\mathbf{W}$ is obtained by maximizing the likelihood function.

Since the vectors $\mathbf{x}$ and $\mathbf{y}$ are linearly related with the matrix $\mathbf{W}$ in (2.30), their density functions can be described as [41, 42]

$$p(\mathbf{y}) = \frac{1}{|\det \mathbf{W}|} p(\mathbf{x}) \;\; \Leftrightarrow \;\; p(\mathbf{x}) = |\det \mathbf{W}|\, p(\mathbf{y})\,.$$

Also, the separated signals y_i in the vector are assumed to be independent of each other

$$p(\mathbf{y}) = \prod_{i=1}^{N} p(y_i)\,.$$

Putting these equations into the likelihood function (2.31), and then taking its logarithm, we have the log likelihood function

$$\log \mathcal{L} = K \log |\det \mathbf{W}| + \sum_{n=1}^{K} \sum_{i=1}^{N} \log p(y_i(n)) .$$

Instead of maximizing (2.31), let us maximize this function. For notational simplicity, let us divide $\log \mathcal{L}$ by the number of samples K and replace the sample mean with the expectation operator. Then, the cost function to be maximized is described by

$$\mathcal{J} = \log |\det \mathbf{W}| + \sum_{i=1}^{N} \mathrm{E}\{\log p(y_i)\} . \tag{2.32}$$

The maximization is performed by iterative updates with a small step-size parameter η

$$\mathbf{W} \leftarrow \mathbf{W} + \eta \frac{\partial \mathcal{J}}{\partial \mathbf{W}^*} ,$$

where the gradient is given by

$$\frac{\partial \mathcal{J}}{\partial \mathbf{W}^*} = (\mathbf{W}^{\mathsf{H}})^{-1} - \mathrm{E}\{\mathbf{\Phi}(\mathbf{y})\mathbf{x}^{\mathsf{H}}\} \tag{2.33}$$

with

$$\mathbf{\Phi}(\mathbf{y}) = [\Phi(y_1), \ldots, \Phi(y_N)]^{\mathsf{T}}, \quad \Phi(y_i) = -\frac{\partial \log p(y_i)}{\partial y_i^*} .$$

Assuming (2.19) for the density function $p(y_i)$, we have

$$\Phi(y_i) = \frac{y_i}{2b\sqrt{|y_i|^2 + \alpha}} . \tag{2.34}$$

We notice that $\Phi(y_i)$ with $b = 1$ is the complex conjugate of $g(y_i)$ in (2.26) used in FastICA. However, let us set $b = 1/2$ and $\alpha = 0$ for (2.34) to make the function simpler

$$\Phi(y_i) = \frac{y_i}{|y_i|} ,$$

which is the sign of the complex number y_i. We can set $\alpha = 0$ because the second order derivative of $\log p(y_i)$ is not used here in maximizing the cost function (2.32).

The drawback of the gradient (2.33) is that it involves matrix inversion, which is computationally demanding. Thus, the natural gradient [4, 43]

$$\frac{\partial \mathcal{J}}{\partial \mathbf{W}^*} \mathbf{W}^{\mathsf{H}} \mathbf{W} = \left[\mathbf{I} - \mathrm{E}\{\mathbf{\Phi}(\mathbf{y})\mathbf{y}^{\mathsf{H}}\}\right] \mathbf{W}$$

is commonly used instead. The natural gradient has many attractive properties including the following. It does not involve matrix inversion. Moreover, it has the equivariance property [44] meaning that the update does not depend on the condition of the mixing system. In summary, the separation matrix is iteratively optimized by

$$\mathbf{W} \leftarrow \mathbf{W} + \eta \left[\mathbf{I} - \mathrm{E}\{\boldsymbol{\Phi}(\mathbf{y})\mathbf{y}^{\mathsf{H}}\} \right] \mathbf{W}\,. \tag{2.35}$$

We can start with a good initial solution $\mathbf{W} = \mathbf{UV}$ obtained by the first and the second steps.

2.5 Permutation Alignment

This section discusses how to specify the permutation matrix $\mathbf{P}(f)$ in (2.14) and (2.15). We first review various methods and approaches to the permutation problem. Then, we describe a method that exploits the source location diversity by estimating TDOAs, as a simple and intuitive one. We also discuss how to handle permutations, such as $[\,1\;2\;3\,] \rightarrow [\,3\;1\;2\,]$, in the alignment procedure.

2.5.1 Various Approaches

We classify various approaches into four strategies. The first strategy basically makes the separation matrices $\mathbf{W}$ smooth in the frequency domain. This can be realized simply by windowing the separation filters in the time domain [9, 13–15], or averaging the separation matrices among adjacent frequencies [13]. However, this operation makes the separation matrix $\mathbf{W}$ different from the ICA solution (2.9), and generally degrades the separation performance. A possible way to solve this problem is to interleave the ICA update, e.g., (2.35), and this smoothing operation until convergence.

The second strategy estimates information related to the source locations, such as direction of arrival (DOA) or time difference of arrival (TDOA). The beamforming approach [20–22] analyzes the directivity patterns formed by the separation matrices $\mathbf{W}$ to identify the DOA of each source (see e.g., [22] for illustrative examples). However, analysis of the directivity patterns is practically possible only for a two-source case, and becomes intractable when there are more sources. A separation matrix $\mathbf{W}$ and basis vectors $\mathbf{A} = [\mathbf{a}_1, \ldots, \mathbf{a}_N]$ represent the same information because they are directly related by the inverse or the least-mean-square estimator (2.12). And utilizing basis vectors is more general and straightforward since they represent the mixing situation explicitly. Therefore, a better way is to estimate DOAs and/or attenuation ratios [22, 23] from basis vectors, or to represent such information in frequency-normalized basis vectors [24]. Then we cluster such estimations to align the permutation ambiguities. The next subsection presents a related method based on TDOA estimation. These methods work well as long as the reverberation is not severe in the mixing system and thus the location-related information can be distinguished for each source.

The third and fourth strategies exploit the dependence of separated signals across frequencies. The advantage of these strategies is that they are less affected by a mixing system in a bad condition than the second strategy, such as severe reverberations or closely located sources. The third strategy evaluates the dependence of already separated bin-wise signals y_i. The correlation coefficient of envelopes $|y_i|$ has commonly been used [16, 22] for the dependence evaluation. Since the inter-frequency dependence of envelopes is clearly exhibited only among a small set of frequencies (see e.g., [22] for illustrative examples), the drawback of using envelopes is that a misalignment at a frequency may lead to a complete misalignment beyond that frequency. Recently in [45], we propose a new measure that shows the inter-frequency dependence more clearly among all frequencies. Consequently, the measurements form clusters for individual sources, and thus permutations can be aligned more robustly.

The fourth strategy incorporates the inter-frequency dependence evaluation into an ICA/BSS criterion and algorithm. Thus the algorithm is extended from a simple instantaneous one. In [17], the signal envelopes are decorrelated for both bin-wise signal separation and permutation alignment. A frequency-independent time-frame-dependent scale factor is introduced in the density function of bin-wise separated signals to model the source signal activity in [18]. Recent work [25–27] introduces a multivariate density function that models the separated signals of all frequencies at a time, instead of assuming a density function for each frequency. By assuming an appropriate distribution, such as a spherical distribution, the update rule of the ICA algorithm can be kept simple.

2.5.2 Method Based on TDOA Estimations

Of the various approaches reviewed above, here we describe a method that employs the second strategy, which estimates the time difference of arrivals (TDOAs) from basis vectors and then clusters the estimations for permutation alignment. Figure 2.5 models the mixing system with the time delay τ_{jk} from source k to microphone j. If we have M microphones, we can define $\frac{1}{2}M(M-1)$ TDOAs for a source, for each pair of microphones. However, they are somewhat redundant, for example as $\tau_{1k} - \tau_{2k} = (\tau_{3k} - \tau_{2k}) + (\tau_{1k} - \tau_{3k})$. Thus, let us consider M TDOAs

$$r_{jk} = \tau_{jk} - \tau_{Jk}\,, \quad j = 1, \ldots, M \tag{2.36}$$

for a source k, by specifying a reference microphone J. Let us omit the J dependence from the r_{jk} notation for simplicity.

Frequency dependent TDOA estimations We estimate TDOAs based on analyzing the elements $a_{ji}(f)$ of the basis vectors $\mathbf{a}_i(f)$ obtained in (2.11).

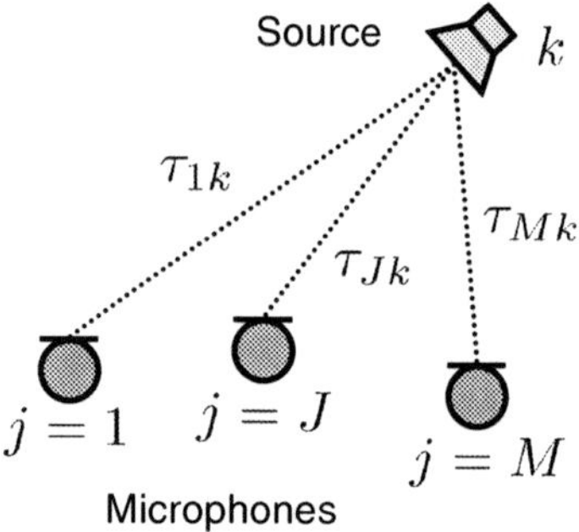

Fig. 2.5. Modeling mixing system with time delays.

Thus, they are frequency dependent. Let $\hat{r}_{ji}(f)$ be an estimation of r_{jk} at frequency f with the corresponding basis vector elements $a_{ji}(f)$ and $a_{Ji}(f)$. We use different subscripts i and k for the source index, because the permutation ambiguities are not aligned at this stage. The frequency-dependent estimation can be calculated by

$$\hat{r}_{ji}(f) = -\frac{\arg\left[a_{ji}(f)/a_{Ji}(f)\right]}{2\pi f} \,. \tag{2.37}$$

We explain the reason below. With the time delay model in Fig. 2.5, the frequency response $h_{jk}(f)$ that appears in (2.4) can be approximated as

$$h_{jk}(f) \approx e^{-\imath 2\pi f \tau_{jk}} \,. \tag{2.38}$$

And with (2.13) we have

$$\frac{a_{ji}(f)}{a_{Ji}(f)} = \frac{a_{ji}y_i}{a_{Ji}y_i} \approx \frac{h_{jk}s_k}{h_{Jk}s_k} = \frac{h_{jk}(f)}{h_{Jk}(f)} \approx e^{-\imath 2\pi f(\tau_{jk}-\tau_{Jk})} \,. \tag{2.39}$$

Therefore, taking the argument gives the formula (2.37) for estimating (2.36).

Valid frequency range We need to pay attention to the frequency range where (2.37) gives a valid estimation. First, frequency f should not be zero because of the division. A small value for f also results in an unstable estimation. Secondly, the phase wrapping problem should be considered. The argument $\arg(\cdot)$ of a complex number $\cdot$ is generally in the range $-\pi < \arg(\cdot) \leq \pi$. If $-2\pi f(\tau_{jk} - \tau_{Jk})$ in (2.39) is outside this range, information regarding TDOA cannot properly be extracted. The estimation (2.37) is guaranteed to be effective only for a frequency range

$$\mathcal{F}_L = \{f : -\pi < -2\pi f(\tau_{jk} - \tau_{Jk}) \leq \pi, \forall j,k\} \cap \mathcal{F} \,.$$

Let v be the propagation velocity of the signals, and $d_{\max}$ be the maximum distance between the reference microphone J and any other microphone.

Then the TDOAs are bounded by

$$\max_{j,k} |\tau_{jk} - \tau_{Jk}| \le d_{\max}/v\,.$$

Therefore, the valid frequency range is given by

$$\mathcal{F}_L = \left\{ f : 0 < f < \frac{v}{2\,d_{\max}} \right\} \cap \mathcal{F}\,.$$

In this chapter, we consider only cases where the valid range covers the set $\mathcal{F}$. For higher frequencies, an advanced method [46] is applicable that takes the phase wrapping possibility into consideration.

Clustering Let us define a vector of frequency dependent TDOA estimations (2.37)

$$\hat{\mathbf{r}}_i(f) = [\hat{r}_{1i}(f), \ldots, \hat{r}_{Mi}(f)]^\mathsf{T}\,. \tag{2.40}$$

If ICA works well and the following TDOA estimations are mostly good, such a TDOA estimation vector is close to one of the true TDOA vectors $\mathbf{r}_k = [r_{1k}, \ldots, r_{Mk}]^\mathsf{T}$, $k = 1, \ldots, N$ for the sources. As a result, the TDOA estimation vectors $\hat{\mathbf{r}}_i(f)$ for all $i = 1, \ldots, N$ and $f \in \mathcal{F}_L$ forms N clusters, each of which corresponds to a source signal. Figure 2.6 shows some examples of TDOA estimations and their clustering results.

A variety of criteria and algorithms can be employed for clustering. Here we simply minimize the average squared distances between the centroid $\mathbf{c}_k$ and cluster members $\hat{\mathbf{r}}$ for each cluster k:

$$\mathcal{J}_k = \frac{1}{|\mathcal{F}_L|} \sum_{f \in \mathcal{F}_L} ||\hat{\mathbf{r}}_i(f) - \mathbf{c}_k||^2 \,\Big|_{i=\Pi_f(k)}, \tag{2.41}$$

where $|\mathcal{F}_L|$ is the number of elements (cardinality) of the set, and Π_f is a permutation $\{1, \ldots, N\} \to \{1, \ldots, N\}$ that maps each of the source indices k to each of the ICA output indices i. The minimization can be performed efficiently by the following algorithm, which is similar to k-means clustering [47]. The algorithm iterates the following two updates until convergence:

$$\Pi_f \leftarrow \operatorname{argmin}_\Pi \sum_{k=1}^{N} ||\hat{\mathbf{r}}_{\Pi(k)}(f) - \mathbf{c}_k||^2, \;\; \forall f \in \mathcal{F}_L, \tag{2.42}$$

$$\mathbf{c}_k \leftarrow \frac{1}{|\mathcal{F}_L|} \sum_{f \in \mathcal{F}_L} \hat{\mathbf{r}}_i(f) \,\Big|_{i=\Pi_f(k)}, \;\; \forall k\,. \tag{2.43}$$

The first update (2.42) optimizes the permutation Π_f for each frequency with the current centroids $\mathbf{c}_1, \ldots, \mathbf{c}_N$. The second update (2.43) calculates the centroid $\mathbf{c}_k$ for each source with the current permutations. In Fig. 2.6,

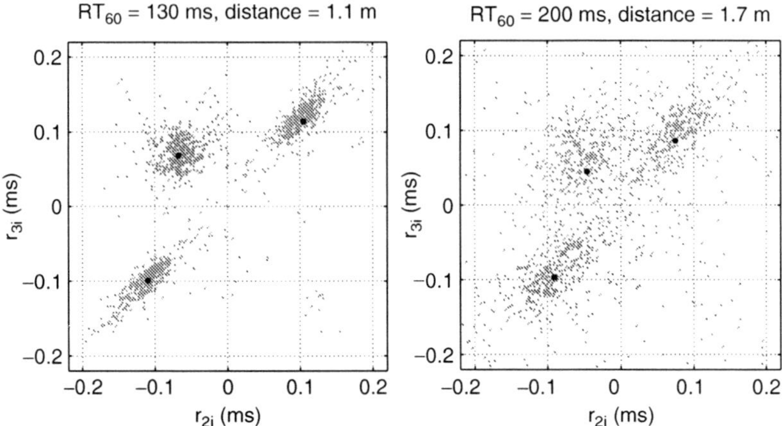

Fig. 2.6. TDOA vectors $\hat{\mathbf{r}}_i(f) = [\hat{r}_{1i}(f), \ldots, \hat{r}_{Mi}(f)]$ estimated for all output indices $i = 1, \ldots, N$ and frequency bins $f \in \mathcal{F}_L$ in setup A, whose condition is described in Fig. 2.9. The first microphone ($J = 1$) was set as a reference. Consequently, $\hat{r}_{1i}(f)$ is zero for all i and f, so is omitted from the plots. Two combinations of reverberation time and source distance are demonstrated in this figure. In the closer-source less-reverberant condition (left hand plot), three clusters of TDOA estimations are clearly formed. In the other case (right hand plot), the TDOA estimations are more scattered, but they still form clusters and their centroids (plotted in circles) are identified.

the centroid of each cluster is depicted by a large black dot. Such centroids can be regarded as estimations for the true TDOA vectors.

We observe in Fig. 2.6 that the closer-source less-reverberant condition (left hand plot) results in denser clusters, meaning that the TDOA estimations are more accurate than those of the other condition (right hand plot). The accuracy of the estimations depends on how well the approximation (2.38) for the frequency response holds. Since the approximation ignores all the reverberations in the mixing system, the estimation accuracy is in inverse proportion to the reverberation. Under the same reverberant condition, a source closer to the microphones results in a more accurate estimation, because the direct-path component of the impulse responses are more distinct.

Identifying close sources The above relationship between cluster density and source distance can be used to identify close sources. We performed experiments in setup B shown in Fig. 2.11, where we had some sources close (0.8 m) to the microphones and six background sources. In this setup, we used just four microphones, and this was insufficient to separate all the sources with linear filters (2.30). So, our goal was to separate the mixtures into four signals (the same as the number of microphones), and then to identify the separated signals that corresponded to the close sources.

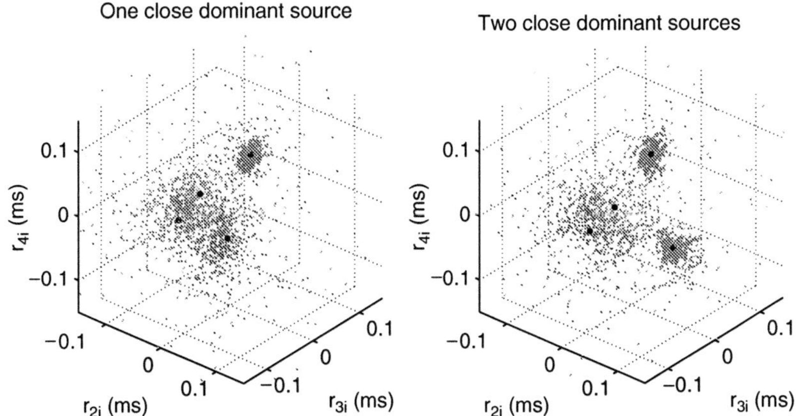

Fig. 2.7. TDOA vectors $\hat{\mathbf{r}}_i(f) = [\hat{r}_{1i}(f), \ldots, \hat{r}_{Mi}(f)]$ estimated for all output indices $i = 1, \ldots, N$ and frequency bins $f \in \mathcal{F}_L$ in setup B, whose condition is described in Fig. 2.11. Since the first microphone ($J = 1$) was set as a reference, $\hat{r}_{1i}(f)$ is zero for all i and f and omitted from the plots. In the experiment for the left hand plot, we used only one close source a120 together with all six background sources. As regards the right hand plot, we added one more close source b120, which resulted in eight sources in total. In both plots, the dense clusters correspond to TDOA estimations for the close sources. Cluster centroids are plotted with circles.

Figure 2.7 shows TDOA estimations obtained in the experiments. As explained in the caption, close sources are related to dense clusters. The density of a cluster is represented by the average squared distance $\mathcal{J}_k$ (2.41). Thus, if the average squared distances are sorted in ascending order

$$\mathcal{J}_1 \leq \cdots \leq \mathcal{J}_N \,, \tag{2.44}$$

the separated signals corresponding to the close sources would have a small source index k.

2.5.3 Handling Permutations

In the clustering algorithm presented in the previous subsection, a permutation Π_f is optimized for each frequency f in (2.42). The criterion is to minimize the sum of the squared distances $||\hat{\mathbf{r}}_i(f) - \mathbf{c}_k||^2$ with $i = \Pi_f(k)$ over all source indices $k = 1, \ldots, N$. Although an exhaustive search with all possible permutations provides the best solution, it is computationally demanding and not realistic for a large source number N. Instead, a simple heuristic can be considered such that the mapping $i = \Pi_f(k)$ related to the smallest distance is decided immediately and the corresponding row and column are eliminated in the next step.

Table 2.1 shows some examples of squared distances. If we apply the above heuristic to the data on the left side, a mapping $\Pi_f(1) = 3$ is first decided

Table 2.1. Examples of the squared distances $||\hat{\mathbf{r}}_i(f) - \mathbf{c}_k||^2$ between centroids and TDOA estimation vectors. Scaled at $\times 10^{-7}$. The bold typeface represents the permutation obtained by the simple heuristic described in the main text.

At frequency f of 922 Hz

	$\hat{\mathbf{r}}_1(f)$	$\hat{\mathbf{r}}_2(f)$	$\hat{\mathbf{r}}_3(f)$
$\mathbf{c}_1$	0.135	0.237	**0.019**
$\mathbf{c}_2$	0.620	**0.035**	0.119
$\mathbf{c}_3$	**0.063**	0.760	0.397

At frequency f of 445 Hz

	$\hat{\mathbf{r}}_1(f)$	$\hat{\mathbf{r}}_2(f)$	$\hat{\mathbf{r}}_3(f)$
$\mathbf{c}_1$	0.218	**0.014**	0.035
$\mathbf{c}_2$	**0.013**	0.312	0.123
$\mathbf{c}_3$	1.005	0.249	**0.516**

because 0.019 is the smallest in the matrix. Then, the first row and the third column are eliminated, and the smallest value 0.035 is selected next. In this way, a permutation $\Pi_f : [1\ 2\ 3] \to [3\ 2\ 1]$ is obtained, which is the best solution among all possible permutations. However, for the data of another frequency on the right side, the situation is not so straightforward. The above heuristic provides a permutation $\Pi_f : [1\ 2\ 3] \to [2\ 1\ 3]$, which results in a larger distance sum than the best permutation $\Pi_f : [1\ 2\ 3] \to [3\ 1\ 2]$. But the heuristic still works well, if we can ignore the third cluster $\mathbf{c}_3$, for example for the reason that only the first two sources are of importance.

The clustering information represented by Π_f is then used to align the permutation ambiguity of separated frequency components $y_1, \ldots, y_N$ and basis vectors $\mathbf{a}_1, \ldots, \mathbf{a}_N$ frequency by frequency. In (2.14) and (2.15), we describe the alignment procedure with a permutation matrix $\mathbf{P}(f)$ instead of using a permutation Π_f. These two different notations represent the same permutation if we describe the permutation matrix as

$$\mathbf{P}(f) = \begin{bmatrix} \mathbf{e}_{\Pi_f(1)} \\ \vdots \\ \mathbf{e}_{\Pi_f(N)} \end{bmatrix},$$

where $\mathbf{e}_i$ is a row vector in which the i-th element is 1 and all the other elements are 0. Thus, the two sets of operations

$$\begin{aligned} &\mathbf{y} \leftarrow \mathbf{P}\mathbf{y}, \quad \mathbf{A} \leftarrow \mathbf{A}\mathbf{P}^\mathsf{T}, \\ &\mathbf{y} \leftarrow [y_{\Pi(1)}, \ldots, y_{\Pi(N)}]^\mathsf{T}, \ \mathbf{A} \leftarrow [\mathbf{a}_{\Pi(1)}, \ldots, \mathbf{a}_{\Pi(N)}] \end{aligned}$$

have the same effect for $\mathbf{y} = [y_1, \ldots, y_N]^\mathsf{T}$ and $\mathbf{A} = [\mathbf{a}_1, \ldots, \mathbf{a}_N]$. For instance, if a permutation $\Pi : [1\ 2\ 3] \to [3\ 1\ 2]$ is given, then the corresponding permutation matrix becomes

$$\mathbf{P} = \begin{bmatrix} 0 & 0 & 1 \\ 1 & 0 & 0 \\ 0 & 1 & 0 \end{bmatrix},$$

and the above two sets of operations equivalently perform

$$\begin{bmatrix} y_1 \\ y_2 \\ y_3 \end{bmatrix} \leftarrow \begin{bmatrix} y_3 \\ y_1 \\ y_2 \end{bmatrix}, \quad [\mathbf{a}_1, \mathbf{a}_2, \mathbf{a}_3] \leftarrow [\mathbf{a}_3, \mathbf{a}_1, \mathbf{a}_2] \, .$$

2.6 Time–Frequency Masking

In this section, we first discuss the motivation behind using time–frequency (T–F) masking, and then present a method for specifying a mask $\mathcal{M}_i(n, f)$ in (2.17) for each output i and each time frequency slot (n, f).

Limitation of linear filtering The use of T–F masking is motivated by the limitation of linear filtering (2.9). Let us focus on the i-th row of the filtering. If we denote $\mathbf{W} = [\mathbf{w}_1, \ldots, \mathbf{w}_N]^{\mathsf{H}}$ and take the mixing model (2.6) into account, we have

$$y_i(n) = \mathbf{w}_i^{\mathsf{H}}\mathbf{x}(n) = \mathbf{w}_i^{\mathsf{H}}\mathbf{h}_i s_i(n) + \sum_{k \neq i} \mathbf{w}_i^{\mathsf{H}}\mathbf{h}_k s_k(n) \, , \tag{2.45}$$

where we omit the dependence of frequency f for simplicity. If the number of microphones is sufficient, i.e., $N \leq M$, there exists $\mathbf{w}_i$ that satisfies $\mathbf{w}_i^{\mathsf{H}}\mathbf{h}_k = 0$ for all $k \neq i$ and makes the second term zero. In contrast, if we have more sources than microphones $N > M$, there is a set $\mathcal{K} \subseteq \{1, \ldots, i{-}1, i{+}1, \ldots, N\}$ such that $\mathbf{w}_i^{\mathsf{H}}\mathbf{h}_k \neq 0, \; \forall k \in \mathcal{K}$. And $y_i(n)$ contains unwanted residuals $\sum_{k\in\mathcal{K}} \mathbf{w}_i^{\mathsf{H}}\mathbf{h}_k s_k(n)$. The purpose of T–F masking is to obtain another version of output $\tilde{y}_i(n)$ that contains less power of the residuals $\sum_{k\in\mathcal{K}} \mathbf{w}_i^{\mathsf{H}}\mathbf{h}_k s_k(n)$ than $y_i(n)$.

Calculating masks Let us rewrite the formula of T–F masking (2.17) omitting the frequency dependence:

$$\tilde{y}_i(n) = \mathcal{M}_i(n)\, y_i(n), \quad i = 1, \ldots, N \, .$$

The basic idea when designing a mask $\mathcal{M}_i(n)$ is that $y_i(n)$ should be passed through ($\mathcal{M}_i = 1$) if the target source term $\mathbf{w}_i^{\mathsf{H}}\mathbf{h}_i s_i(n)$ is dominant in (2.45) and conversely $y_i(n)$ should be masked out ($\mathcal{M}_i = 0$) if only the residual term is active. We adopt the following idea to distinguish these two situations.

We can say that the target source term is dominant in (2.45) if $s_i(n)$ is dominant in the mixtures $\mathbf{x}(n)$ since the filter vector $\mathbf{w}_i$ is designed to enhance $s_i(n)$. Also from (2.6), if $s_i(n)$ is dominant in $\mathbf{x}(n)$, then $\mathbf{x}(n) \approx \mathbf{h}_i s_i(n)$. If ICA and the following permutation alignment work well, (2.16) is satisfied and $\mathbf{a}_i y_i(n) \approx \mathbf{h}_i s_i(n)$. Therefore, in summary, if $\mathbf{x}(n) \approx \mathbf{a}_i y_i(n)$, we strongly expect the target source term to be dominant in (2.45).

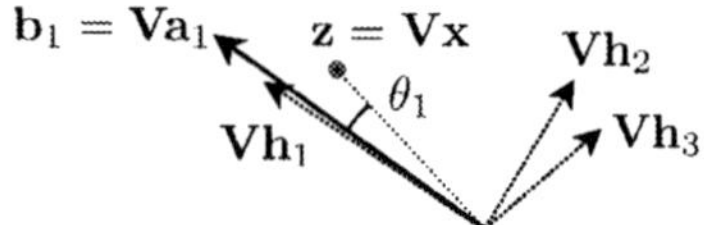

Fig. 2.8. Angle θ_1 calculated in whitened space.

Thus, the mask $\mathcal{M}_i(n)$ can be designed by analyzing the closeness between the mixture $\mathbf{x}(n)$ and the basis vector $\mathbf{a}_i$. In order to normalize the closeness measure, let us transform the vectors $\mathbf{x}(n)$ and $\mathbf{a}_i$ with the whitening matrix $\mathbf{V}$ defined in (2.20):

$$\mathbf{z}(n) = \mathbf{V}\mathbf{x}(n)\,, \quad \mathbf{b}_i = \mathbf{V}\mathbf{a}_i\,.$$

Then the angle

$$\theta_i(n) = \arccos \frac{|\mathbf{b}_i^{\mathsf{H}}\, \mathbf{z}(n)|}{||\mathbf{b}_i|| \cdot ||\mathbf{z}(n)||}$$

between the two vectors is calculated. If $\theta_i(n)$ is close to zero, we estimate that the two vectors $\mathbf{x}(n)$ and $\mathbf{a}_i$ are close and the $s_i(n)$ term is dominant. In contrast, if $\theta_i(n)$ is close to $\pi/2$, we estimate that only the residual term is active. Hence the mask is specified by

$$\mathcal{M}_i(n) = \begin{cases} 1 & \text{if } \theta_i(n) \leq \theta_T, \\ 0 & \text{if } \theta_i(n) > \theta_T, \end{cases} \tag{2.46}$$

where θ_T is a predefined threshold for controlling the trade-off between the suppression of interference and the introduction of distortions. As θ_T becomes smaller, the residuals of interference in $\tilde{y}_i$ decrease but the distortion in y_i increases.

The role of whitening matrix $\mathbf{V}$ is to make the mixing vectors $\mathbf{h}_1, \ldots, \mathbf{h}_N$ orthogonal to each other in the whitened space. If we have enough microphones $N \leq M$, the orthogonality is guaranteed if the N source signals are uncorrelated. This is not the case if the sources outnumber the microphones $N > M$. However, we mainly consider a case where the target source is close to the microphones and thus louder than the other background sources at the microphones. In such a case, the mixing vector $\mathbf{h}_1$ of the target source s_1 tends to form large angles with the other mixing vectors in the whitened space, as shown in Fig. 2.8. Therefore, the angle θ_i between $\mathbf{z} = \mathbf{V}\mathbf{x}$ and $\mathbf{b}_i = \mathbf{V}\mathbf{a}_i$ provides useful information about whether or not s_i is dominant in the mixtures.

2.7 Experiments

Experiments were conducted to examine the effectiveness of the BSS procedure presented in this chapter. We used two experimental setups. Setup A

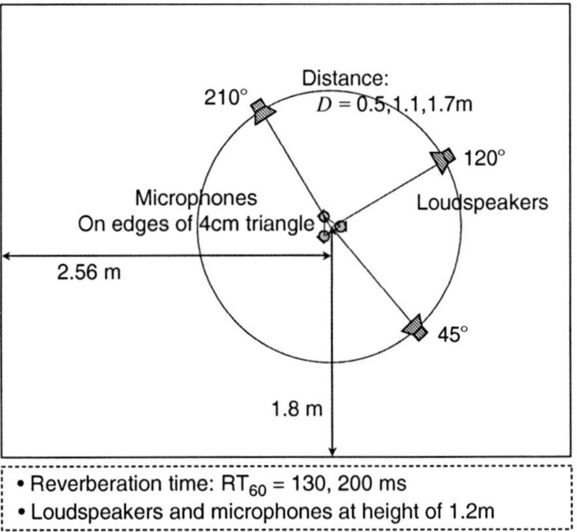

Fig. 2.9. Experimental setup A.

Table 2.2. Common experimental configurations.

Room size	4.45 × 3.55 × 2.5 m
Source signals	Speeches of 3 s, 8 combinations
Sampling rate	$f_s = 8$ kHz, $t_s = 0.125$ ms
STFT frame size	1024 points (128 ms)
STFT frame shift	256 points (32 ms)
Maximum distance between microphones	$d_{\max} = 4$ cm
Parameter for density function (2.19)	$\alpha = 0.1$ for FastICA
Step-size parameter in (2.35)	$\eta = 0.05$

in Fig. 2.9 is considered to be a basic one where we have three sources and three microphones. Setup B in Fig. 2.11 is rather complicated, in that we have some target sources close to the microphones and also six background interference sources. Table 2.2 summarizes the configurations common to both setups. We measured the impulse responses from the loudspeakers to the microphones, and convolved each of them with 3-second speech sources. We then mixed them together to construct each of the microphone observations. The sampling rate was 8 kHz and the maximum distance between the microphones was 4 cm in both setups. Consequently, the phase wrapping problem did not occur.

2.7.1 Basic 3-Source 3-Microphone Case

In setup A shown in Fig. 2.9, we had three microphones arranged in a triangle and three loudspeakers located at the same distance D from the microphones.

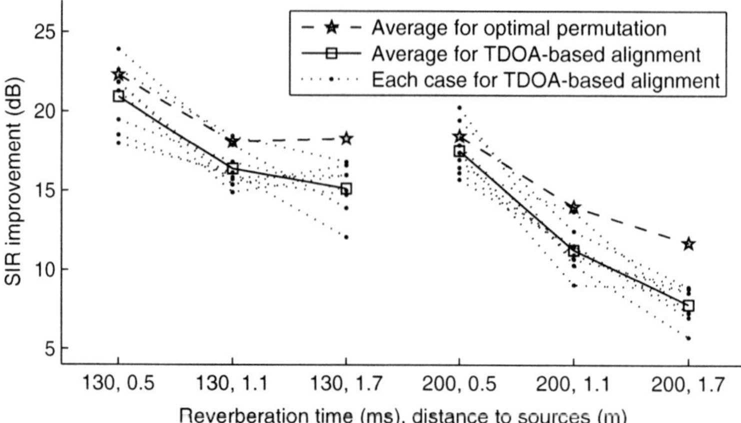

Fig. 2.10. SIR improvement with different reverberation times RT_{60} and distances from the sources to the microphones D in setup A.

We selected three different distances D of 0.5, 1.1, and 1.7 m. Also, we set two different reverberation times $RT_{60} = 130$ and 200 ms by changing the number of cushioned walls in the room. Since the number of microphones was sufficient for the number of sources, time–frequency masking was not employed. The separation performance was evaluated in terms of the SIR improvement averaged over all three outputs.

Figure 2.10 shows the SIR improvements achieved for different reverberation times and distances from the sources to the microphones. For each condition, 8 combinations of 3 speech signals were tested. The permutation ambiguities of the ICA solution were aligned based on the frequency dependent TDOA estimations presented in Sect. 2.5.2. The average SIR improvement is shown by the solid line, and the average SIR improvement with the optimal permutation is depicted with the dashed line. The difference between these two lines indicates the separation degradation caused by the permutation misalignment. The degradation becomes large as the reverberation time and/or the distance increases. But in general the permutation alignment method worked well under such conditions. Even when we employed optimal permutation, the separation performance was affected by the reverberation time and the distance. This is mainly because the instantaneous mixture model (2.5) tends to be less accurate if the STFT frame, 128 ms in this case, cannot cover the whole effect of the impulse responses.

Then, we demonstrate the efficiency of the three-step procedure for ICA explained in Sect. 2.4. We fixed the reverberation time and distance at $RT_{60} = 130$ ms and $D = 1.1$ m. Table 2.3 shows the results of various procedures. Whitening was employed in all cases. The combination of FastICA and maximum likelihood (ML) estimation, i.e., the three-step procedure, provides good separations with an efficient computational cost. The ML estimation

Table 2.3. SIR improvement and computational time with various ICA procedures in setup A. The averages for 8 combinations of 3 speech signals. ML stands for maximum likelihood. To focus on the effect of the ICA procedures, the permutations were aligned with optimal solutions. The computational time was only for the ICA procedure. The BSS program was coded in Matlab and run on an AMD 2.4 GHz Athlon 64 processor. The reverberation time and distance were $\mathrm{RT}_{60} = 130$ ms and $D = 1.1$ m, respectively.

	FastICA+ML	FastICA	ML (50 ite.)	ML (200 ite.)
SIR improvement	18.1 dB	17.1 dB	13.3 dB	18.4 dB
Computational time	3.1 s	1.1 s	2.1 s	7.8 s

update (2.35) was iterated 50 times in the combination case. With FastICA alone, the computational time was small but there was slightly less SIR improvement. With ML estimation alone, when we started with the whitening matrix as an initial solution, 50 times iterations of the update (2.35) were insufficient for us to obtain good results. We needed many iterations (200 in this case) to obtain a good separation, which results in a large computational time.

2.7.2 With Many Background Interference Sources

Then, let us consider a more complicated situation. In setup B shown in Fig. 2.11, we have four microphones arranged 3-dimensionally, and three sources close to the microphones. Some of them served as target sources for each simulation. We also have six background sources far from the microphones, which were all active and treated as interference sources in every simulation. Since four microphones were insufficient to separate all the sources with linear filtering, only the target sources close to the microphones were separated. For example with a single target, the first output should contain the separated signal corresponding to the target source, and the remaining three outputs should contain the interference sources in a mixed manner. Close target sources were successfully identified in all cases by sorting the TDOA clusters according to their variances (2.44). When permutations were decided based on the squared distances as shown in Table 2.1, more priority was given to a cluster with a smaller index. We employed time–frequency masking to suppress the background interference further.

Table 2.4 shows the SIR improvement and SDR evaluated at each output corresponding to the target source. The performance seems to depend on the target position, as that of c120 is apparently worse than the others. This is because the background interference sources came from similar directions. As regards the effect of T–F masking, SIR were improved further with the application, but SDR became worse. By changing the threshold θ_T in (2.46), we clearly observed the trade-off between SIR improvement and SDR.

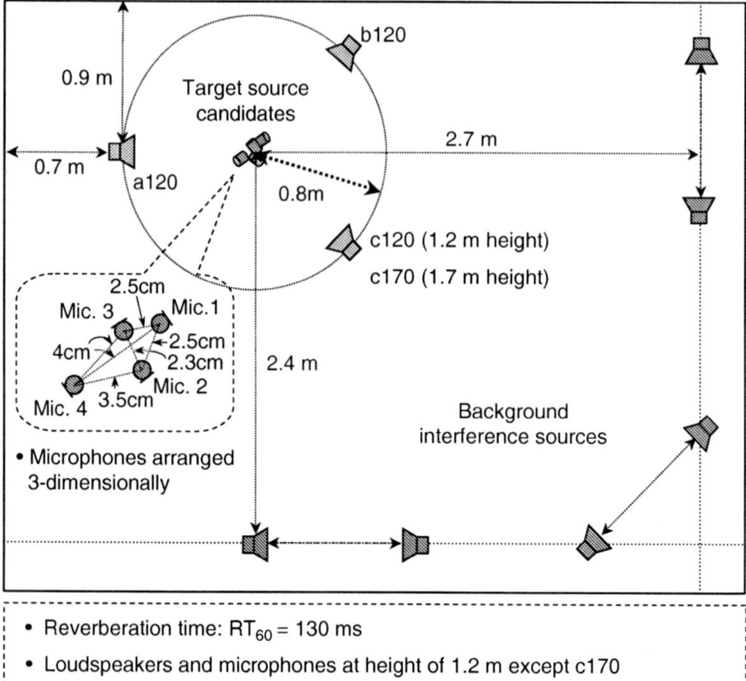

Fig. 2.11. Experimental setup B.

Table 2.4. SIR improvement (above) and SDR (below) evaluated in setup B. The averages for 8 combinations of many speech signals. All six background interference sources were active in every situation. The columns a120, b120, c120 and c170 show the results for single target cases where only the corresponding target source was active. The column a120, b120 shows the results with the two target sources active.

SIR improvement (dB)

Target position	a120	b120	c120	c170	a120, b120	
Without T–F masking	8.2	8.9	7.5	8.4	10.6	11.4
With T–F masking, $\theta_T = 0.4\pi$	9.2	10.0	8.8	9.4	11.6	12.4
With T–F masking, $\theta_T = 0.35\pi$	9.6	10.6	9.5	9.8	12.1	12.9
With T–F masking, $\theta_T = 0.3\pi$	10.0	11.0	10.0	10.2	12.7	13.5

SDR (dB)

Target position	a120	b120	c120	c170	a120, b120	
Without T–F masking	10.8	12.7	8.5	8.6	12.0	12.8
With T–F masking, $\theta_T = 0.4\pi$	9.6	11.1	7.6	7.5	10.6	11.1
With T–F masking, $\theta_T = 0.35\pi$	8.6	9.8	6.9	6.8	9.5	9.7
With T–F masking, $\theta_T = 0.3\pi$	7.2	8.3	5.9	5.8	8.2	8.4

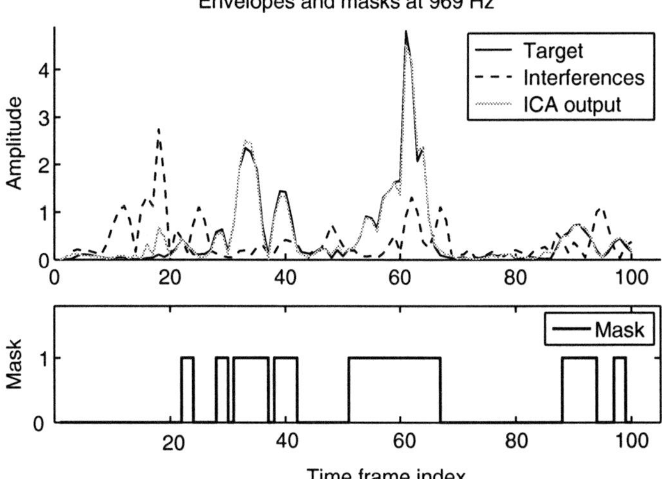

Fig. 2.12. The envelopes at 969 Hz of a target signal, the total sum of all interferences, and the ICA output corresponding to the target signal (above). The masks calculated for each time frame (below).

Figure 2.12 shows an example where T–F masking works well. Although the ICA output was close to the target component at many time frames, there were some interference components in the ICA output at some frames (e.g., from 15 to 20). This shows the limitation of ICA as a spatial filter. The lower plot shows masks calculated for each frame by (2.46) with $\theta_T = 0.4\pi$. By using these masks, the BSS system eliminated the interference components at frames 15 to 20, and improved the SIR at the output.

2.8 Conclusion

This chapter presented a comprehensive description of the frequency-domain approach to the blind separation of convolutive mixtures. The chapter focused on the complex-valued ICA procedure and the permutation alignment procedure based on the estimation of TDOAs from basis vectors. The use of T–F masking was also considered as a way of suppressing interference residuals caused by the limitations of linear filtering. Experimental results for real room situations demonstrated the effectiveness of the described procedures.

References

1. T. W. Lee, *Independent Component Analysis – Theory and Applications.* Kluwer Academic Publishers, 1998.
2. S. Haykin, Ed., *Unsupervised Adaptive Filtering (Volume I: Blind Source Separation).* John Wiley & Sons, 2000.

3. A. Hyvärinen, J. Karhunen, and E. Oja, *Independent Component Analysis*. John Wiley & Sons, 2001.
4. A. Cichocki and S. Amari, *Adaptive Blind Signal and Image Processing*. John Wiley & Sons, 2002.
5. S. Amari, S. Douglas, A. Cichocki, and H. Yang, "Multichannel blind deconvolution and equalization using the natural gradient," in *Proc. IEEE Workshop on Signal Processing Advances in Wireless Communications*, Apr. 1997, pp. 101–104.
6. M. Kawamoto, K. Matsuoka, and N. Ohnishi, "A method of blind separation for convolved non-stationary signals," *Neurocomputing*, vol. 22, pp. 157–171, 1998.
7. K. Matsuoka and S. Nakashima, "Minimal distortion principle for blind source separation," in *Proc. ICA 2001*, Dec. 2001, pp. 722–727.
8. S. C. Douglas and X. Sun, "Convolutive blind separation of speech mixtures using the natural gradient," *Speech Communication*, vol. 39, pp. 65–78, 2003.
9. H. Buchner, R. Aichner, and W. Kellermann, "Blind source separation for convolutive mixtures: A unified treatment," in *Audio Signal Processing for Next-Generation Multimedia Communication Systems*, Y. Huang and J. Benesty, Eds. Kluwer Academic Publishers, Feb. 2004, pp. 255–293.
10. T. Takatani, T. Nishikawa, H. Saruwatari, and K. Shikano, "High-fidelity blind separation of acoustic signals using SIMO-model-based independent component analysis," *IEICE Trans. Fundamentals*, vol. E87-A, no. 8, pp. 2063–2072, Aug. 2004.
11. S. C. Douglas, H. Sawada, and S. Makino, "A spatio-temporal FastICA algorithm for separating convolutive mixtures," in *Proc. ICASSP 2005*, vol. V, Mar. 2005, pp. 165–168.
12. R. Aichner, H. Buchner, F. Yan, and W. Kellermann, "A real-time blind source separation scheme and its application to reverberant and noisy acoustic environments," *Signal Process.*, vol. 86, no. 6, pp. 1260–1277, 2006.
13. P. Smaragdis, "Blind separation of convolved mixtures in the frequency domain," *Neurocomputing*, vol. 22, pp. 21–34, 1998.
14. L. Parra and C. Spence, "Convolutive blind separation of non-stationary sources," *IEEE Trans. Speech Audio Processing*, vol. 8, no. 3, pp. 320–327, May 2000.
15. L. Schobben and W. Sommen, "A frequency domain blind signal separation method based on decorrelation," *IEEE Trans. Signal Processing*, vol. 50, no. 8, pp. 1855–1865, Aug. 2002.
16. N. Murata, S. Ikeda, and A. Ziehe, "An approach to blind source separation based on temporal structure of speech signals," *Neurocomputing*, vol. 41, no. 1–4, pp. 1–24, Oct. 2001.
17. J. Anemüller and B. Kollmeier, "Amplitude modulation decorrelation for convolutive blind source separation," in *Proc. ICA 2000*, June 2000, pp. 215–220.
18. N. Mitianoudis and M. Davies, "Audio source separation of convolutive mixtures," *IEEE Trans. Speech and Audio Processing*, vol. 11, no. 5, pp. 489–497, Sept. 2003.
19. F. Asano, S. Ikeda, M. Ogawa, H. Asoh, and N. Kitawaki, "Combined approach of array processing and independent component analysis for blind separation of acoustic signals," *IEEE Trans. Speech Audio Processing*, vol. 11, no. 3, pp. 204–215, May 2003.

20. H. Saruwatari, S. Kurita, K. Takeda, F. Itakura, T. Nishikawa, and K. Shikano, "Blind source separation combining independent component analysis and beamforming," *EURASIP Journal on Applied Signal Processing*, vol. 2003, no. 11, pp. 1135–1146, Nov. 2003.
21. M. Z. Ikram and D. R. Morgan, "Permutation inconsistency in blind speech separation: Investigation and solutions," *IEEE Trans. Speech Audio Processing*, vol. 13, no. 1, pp. 1–13, Jan. 2005.
22. H. Sawada, R. Mukai, S. Araki, and S. Makino, "A robust and precise method for solving the permutation problem of frequency-domain blind source separation," *IEEE Trans. Speech Audio Processing*, vol. 12, no. 5, pp. 530–538, Sept. 2004.
23. R. Mukai, H. Sawada, S. Araki, and S. Makino, "Frequency-domain blind source separation of many speech signals using near-field and far-field models," *EURASIP Journal on Applied Signal Processing*, vol. 2006, pp. Article ID 83 683, 13 pages, 2006.
24. H. Sawada, S. Araki, R. Mukai, and S. Makino, "Blind extraction of dominant target sources using ICA and time–frequency masking," *IEEE Trans. Audio, Speech and Language Processing*, pp. 2165–2173, Nov. 2006.
25. A. Hiroe, "Solution of permutation problem in frequency domain ICA using multivariate probability density functions," in *Proc. ICA 2006 (LNCS 3889)*. Springer, Mar. 2006, pp. 601–608.
26. T. Kim, H. T. Attias, S.-Y. Lee, and T.-W. Lee, "Blind source separation exploiting higher-order frequency dependencies," *IEEE Trans. Audio, Speech and Language Processing*, pp. 70–79, Jan. 2007.
27. I. Lee, T. Kim, and T.-W. Lee, "Complex FastIVA: A robust maximum likelihood approach of MICA for convolutive BSS," in *Proc. ICA 2006 (LNCS 3889)*. Springer, Mar. 2006, pp. 625–632.
28. J.-F. Cardoso, "Infomax and maximum likelihood for blind source separation," *IEEE Signal Processing Letters*, vol. 4, no. 4, pp. 112–114, Apr. 1997.
29. J. H. DiBiase, H. F. Silverman, and M. S. Brandstein, "Robust localization in reverberant rooms," in *Microphone Arrays*, M. Brandstein and D. Ward, Eds. Springer, 2001, pp. 157–180.
30. J. Chen, Y. Huang, and J. Benesty, "Time delay estimation," in *Audio Signal Processing*, Y. Huang and J. Benesty, Eds. Kluwer Academic Publishers, 2004, pp. 197–227.
31. N. Roman and D. Wang, "Binaural sound segregation for multisource reverberant environments," in *Proc. ICASSP 2004*, vol. II, May 2004, pp. 373–376.
32. D. Kolossa and R. Orglmeister, "Nonlinear postprocessing for blind speech separation," in *Proc. ICA 2004 (LNCS 3195)*, Sept. 2004, pp. 832–839.
33. M. Aoki, M. Okamoto, S. Aoki, H. Matsui, T. Sakurai, and Y. Kaneda, "Sound source segregation based on estimating incident angle of each frequency component of input signals acquired by multiple microphones," *Acoustical Science and Technology*, vol. 22, no. 2, pp. 149–157, 2001.
34. S. Rickard, R. Balan, and J. Rosca, "Real-time time–frequency based blind source separation," in *Proc. ICA2001*, Dec. 2001, pp. 651–656.
35. Ö. Yılmaz and S. Rickard, "Blind separation of speech mixtures via time–frequency masking," *IEEE Trans. Signal Processing*, vol. 52, no. 7, pp. 1830–1847, July 2004.

36. D. Wang, "On ideal binary mask as the computational goal of auditory scene analysis," in *Speech Separation by Humans and Machines*, P. Divenyi, Ed. Kluwer Academic Publishers, 2004, pp. 181–197.
37. T. Nakatani, K. Kinoshita, and M. Miyoshi, "Harmonicity-based blind dereverberation for single-channel speech signals," *IEEE Trans. Audio, Speech and Language Processing*, vol. 15, no. 1, pp. 80–95, Jan. 2007.
38. M. Delcroix, T. Hikichi, and M. Miyoshi, "Precise dereverberation using multichannel linear prediction," *IEEE Trans. Audio, Speech and Language Processing*, vol. 15, no. 2, pp. 430–440, Feb. 2007.
39. J.-F. Cardoso, "Multidimensional independent component analysis," in *Proc. ICASSP 1998*, vol. 4, May 1998, pp. 1941–1944.
40. T. Kailath, A. H. Sayed, and B. Hassibi, *Linear Estimation*. Prentice Hall, 2000.
41. A. Bell and T. Sejnowski, "An information-maximization approach to blind separation and blind deconvolution," *Neural Computation*, vol. 7, no. 6, pp. 1129–1159, 1995.
42. A. Papoulis and S. U. Pillai, *Probability, Random Variables and Stochastic Processes*. McGraw-Hill, 2002.
43. S. Amari, A. Cichocki, and H. H. Yang, "A new learning algorithm for blind signal separation," in *Advances in Neural Information Processing Systems*, vol. 8. The MIT Press, 1996, pp. 757–763.
44. J.-F. Cardoso and B. H. Laheld, "Equivariant adaptive source separation," *IEEE Trans. Signal Processing*, vol. 44, no. 12, pp. 3017–3030, Dec. 1996.
45. H. Sawada, S. Araki, and S. Makino, "Measuring dependence of bin-wise separated signals for permutation alignment in frequency-domain BSS," in *Proc. ISCAS 2007*, May 2007, pp. 3247–3250.
46. H. Sawada, S. Araki, R. Mukai, and S. Makino, "Solving the permutation problem of frequency-domain BSS when spatial aliasing occurs with wide sensor spacing," in *Proc. ICASSP 2006*, vol. V, May 2006, pp. 77–80.
47. R. O. Duda, P. E. Hart, and D. G. Stork, *Pattern Classification*, 2nd ed. Wiley Interscience, 2000.

3 Blind Source Separation using Space–Time Independent Component Analysis

Mike Davies[1], Maria Jafari[2], Samer Abdallah[2], and Emmanuel Vincent[3], and Mark Plumbley[2]

[1] IDCoM & Joint Research Institute for Signal and Image Processing, Edinburgh University, Edinburgh, UK
E-mail: mike.davies@ed.ac.uk

[2] Centre for Digital Music, Department of Electronic Engineering, Queen Mary University of London, UK
E-mail: mark.plumbley, maria.jafari, samer.abdallah@elec.qmul.ac.uk

[3] IRISA-INRIA, Campus Universitaire de Beaulieu, 35042 Rennes cedex, France
E-mail: emmanuel.vincent@irisa.fr

Abstract. We consider the problem of convolutive blind source separation (BSS). This is usually tackled through either multichannel blind deconvolution (MCBD) or using frequency-domain independent component analysis (FD-ICA). Here, instead of using a fixed time or frequency basis to solve the convolutive blind source separation problem we propose learning an adaptive spatial–temporal transform directly from the speech mixture. Most of the learnt space–time basis vectors exhibit properties suggesting that they represent the components of individual sources as they are observed at the microphones. Source separation can then be performed by projection onto the appropriate group of basis vectors. We go on to show that both MCBD and FD-ICA techniques can be considered as particular forms of this general separation method with certain constraints. While our space–time approach involves considerable additional computation it is also enlightening as to the nature of the problem and has the potential for performance benefits in terms of separation and de-noising.

3.1 Introduction

The convolutive blind audio source separation problem arises when an array of microphones records mixtures of sound sources. The problem is typically addressed in either the time domain or the frequency domain. In the time domain this is generally tackled using a multichannel blind deconvolution (MCBD) algorithm, e.g. [1–3], which is an extension of classical independent component analysis (ICA). However these time domain techniques assume that the source signals are temporally independent which can lead to over-whitening of the outputs. There is also the computational burden of dealing with long acoustic impulse response functions (e.g. of the order of 4000 taps for 16 kHz sampling), which has led the bulk of research in audio BSS to explore solutions within the frequency domain, e.g. [4]. This entails performing instantaneous ICA independently at each frequency bin, and can result

S. Makino et al. (eds.), Blind Speech Separation, 79–99.

in simpler, computationally more efficient algorithms. However frequency domain approaches have the serious drawback of introducing a nontrivial permutation problem.

In this chapter we consider a full space–time ICA (ST-ICA) decomposition of the data that also provides source separation. This was originally proposed in [5] and was subsequently explored further in [6]. No assumptions are made as to the number of microphones, or even the type of mixing (e.g. instantaneous or convolutive) in the underlying model. However we will see that these characteristics are implicitly encoded in the recovered dictionary elements.

The ST-ICA basis vectors provide an efficient space–time representation of the sources. We will see that the basis vectors learned have a strong time–frequency structure, but unlike the FD-ICA representation, tend to be temporally localized in the high-frequencies. Separation is achieved by grouping together appropriate basis vectors that identify independent spatial–temporal subspaces. The individual sources can then be retrieved by projecting the data onto these subspaces. Interestingly this achieves separation without explicitly estimating the room impulse responses or their inverses.

In Sect. 3.2 we introduce the space–time representation, and relate this to the MCBD and FD-ICA methods. We then propose the ST-ICA scheme and explain how all three approaches can be viewed as special cases of multidimensional independent component analysis (MICA) within the space–time context. However the ST-ICA representation is clearly distinct from either the FD-ICA or MCBD solutions and provides a sparser representation of the data.

In a similar vein, we show that the component grouping problem in ST-ICA is essentially equivalent to the permutation problem of FD-ICA. However we will see that the localized nature of the high frequency ST-ICA basis vectors make the grouping more robust than in the FD-ICA approach. Here we demonstrate, that, similar to FD-ICA, we can construct an unsupervised clustering method based on direction of arrival information. To clarify these ideas further we examine the ST-ICA vectors learnt from a two-source, two-microphone system and compare the separation performance with an existing FD-ICA technique.

We begin by briefly reviewing the structure of the convolutive BSS problem and explain how MCBD, FD-ICA and ST-ICA are related. Computer simulations are carried out to compare the performance of FD-ICA and ST-ICA. The chapter is concluded with a discussion of various practical issues including computational complexity and data requirements.

3.2 Review of Convolutive Blind Source Separation

We are interested in the problem of convolutive linear mixing that occurs when an array of microphones records a mixture of sound sources in a room with delays and echoes. This scenario is often modeled as a causal finite impulse response (FIR) mixture, given by

$$x_q(n) = \sum_{p=1}^{P} \sum_{l=0}^{L-1} a_{qp}(l) s_p(n-l), \quad q = 1, \ldots, Q \tag{3.1}$$

where $x_q(n)$ is the signal recorded at the qth microphone at time n, $s_p(n)$ is the pth source signal, $a_{qp}(l)$ denotes the impulse response from source p to sensor q, and L is the maximum length of all impulse responses. We will always assume that the number of sources, P, is less than or equal to the number of microphones, Q.

The aim of convolutive blind source separation is to recover the original source signals $s_p(n)$ given only the mixtures $x_q(n)$. This is usually achieved by estimating a matrix of unmixing filters, $w_{pq}(m)$ giving:

$$y_p(n) = \sum_{q=1}^{Q} \sum_{m=0}^{M-1} w_{pq}(m) x_q(n-m) \quad p = 1, \ldots, P \tag{3.2}$$

where $y_p(n)$ is an estimate of one of the original sources and M the length of the unmixing filters which are assumed to be sufficiently long to be able to approximately invert (3.1). In practice, the room impulse response functions may not be well represented by stationary FIR filters and there may also be additional background noise. For simplicity we will not consider these problems here.

Unfortunately, due to the convolutive nature of the mixing operator, convolutive BSS is only solvable up to unknown *permutation* and *filtering* operations [3]. The filtering ambiguity is equivalent to the scaling ambiguity in the standard instantaneous ICA and can therefore introduce significant spectral distortion. However, if we can estimate the sources as heard at the microphones we have elegantly sidestepped the issue since the source signals at the microphones are uniquely defined [7].

Typically the blind speech separation problem has been approached from one of two directions. The first is to attempt to solve the problem using MCBD by assuming spatial–temporal independence. The second approach maps the system into the (short time) frequency domain and therefore assumes independence across frequency. In the following sections we will briefly compare these two approaches and put them into a unifying framework.

3.2.1 Multi-Channel Blind Deconvolution

In MCBD algorithms, such as [1], the inverse filter operation $w_{pq}(m)$ is directly estimated in the time domain so that the output signals are both spatially and temporally independent, as typically measured via a nonlinear decorrelation. Thus the separating solution satisfies:

$$E\{\phi(y_i(t)) y_j(t-m)\} = \delta_{ij}\delta_m \tag{3.3}$$

where δ is the Kronecker delta function and ϕ is some nonlinear activation function.

Application of the MCBD framework to speech separation has a number of difficulties. Firstly it requires that the source signals have a non-Gaussian sample distribution for their time samples. While speech signals are reasonably super-Gaussian, other audio sources, such as music, cannot be so clearly classified. In some cases the marginal distributions in the time domain can look very close to Gaussian. Secondly, as can be seen from (3.3), MCBD attempts to output signals that are both component-wise independent and temporally independent. This can introduce a great deal of spectral distortion as it typically over-whitens the speech signals [3]. While mapping back to the observation domain removes this ambiguity, this involves estimating both the mixing and unmixing filters and is a difficult task to perform in the time domain.

A final criticism of using MCBD techniques for speech signals is that typical room impulse responses can be quite long (e.g. of the order of 4000 taps for 16 kHz sampling) even in medium sized rooms. This can make working in the time domain quite a burden. One solution here is to perform the filtering operations in the frequency domain (e.g. [8]). Having moved partially towards a frequency domain solution it is natural to ask whether we can solve the whole problem within the frequency domain [9]. This technique is described next.

3.2.2 Frequency Domain ICA (FD-ICA)

The use of the frequency, or more accurately time–frequency (TF) domain for BSS has been motivated by two factors. Firstly, in the frequency domain, convolutions can be replaced by multiplications, yielding reduced computational complexity. Secondly, it has been observed that, when mapping signals such as speech and audio into the TF domain, the statistics of the sources tend to be much sparser and hence more non-Gaussian. This, in turn, means that FD-ICA algorithms are likely to achieve better performance [10].

The computational benefits can be seen by transforming (3.2) into the frequency domain:

$$\mathbf{y}(\omega) = \mathbf{W}(\omega)\mathbf{x}(\omega) \tag{3.4}$$

where $\mathbf{y}(\omega)$ and $\mathbf{x}(\omega)$ are respectively the recovered source estimates and observed data at frequency ω, and $\mathbf{W}(\omega)$ is the transfer function of the unmixing filter matrix also at frequency ω. Thus we can, in principle, replace convolution by multiplication. In practice, we still need to gather statistics about the mixtures $x_q(n)$. We therefore divide the input sequences up into frames of length N and use the short-time Fourier transform (STFT), $X_q(f, t)$, of the signal, $x_q(t)$, observed at the qth microphone:

$$X_q(f,t) = \sum_{m=0}^{N-1} x_q(m+t)\gamma(m)e^{-j2\pi mf/N} \tag{3.5}$$

where $\gamma(m)$ is some windowing function (e.g. Hamming), t is the STFT block index, and $f \in \{0,\dots,N-1\}$, denotes the normalized frequency. To keep the quantity of data down, $X_q(f,t)$ is also usually down-sampled, typically by a factor of $N/2$. Alternatively, other subband transforms may also be used, e.g. [11].

Once the time–frequency domain has been chosen appropriately the unmixing model can be approximated by:

$$\mathbf{Y}(f,t) = \mathbf{W}(f)\mathbf{X}(f,t) \tag{3.6}$$

where $\mathbf{Y}(f,t)$ and $\mathbf{X}(f,t)$ are respectively the vector of recovered source estimates in the STFT domain, and the vector of microphone observations, again in the STFT domain. As above, $\mathbf{W}(f)$ denotes the frequency dependent separating matrix that is to be estimated. The convolutive BSS problem is thus transformed into multiple complex valued ICA problems in the time–frequency domain, with a suitable ICA algorithm (e.g. [12–14]) used to estimate $\mathbf{W}(f)$ separately in each frequency bin.

Once we have estimated $\mathbf{W}(f)$ it is possible to estimate the image of the pth source at the qth microphone as [7]:

$$X_{qp}(f,t) = \sum_{j=1}^{Q} \mathbf{A}_{q,i}(f)\mathbf{W}_{i,j}(f)\mathbf{x}_j(f,t) \tag{3.7}$$

where $\mathbf{A}(f) = \mathbf{W}^{-1}(f)$ is the estimate of the mixing operator at frequency f.

To transform back into the time domain we need to apply the inverse STFT to $X_{qp}(f,t)$. However this requires us to know the ordering of the sources within each of the frequency bands. Since ICA has an inherent permutation ambiguity we arrive at the well-known *permutation problem*: somehow we need to group components across the frequency bands.

Solving the Permutation Problem Various methods have been proposed for solving the permutation problem in FD-ICA. These are primarily based upon either higher level dependencies or additional geometric information about the microphone array.

For a fully blind approach it is necessary to identify additional dependency structure across the frequency bins. Ikeda and Murata [15] proposed that the amplitude envelopes of the different frequency bins should be correlated, and that the correct permutations across frequency can be determined by matching the amplitude envelopes. Davies [16] and Lee et al. [17] proposed a time–frequency probabilistic source model which similarly couples the frequency bins using higher level dependencies.

An alternative *semi-blind* technique that has proved very effective in FD-ICA is to use prior knowledge of the geometry of the microphone array, and apply ideas from beamforming [18]. This allows us to interpret the unmixing matrix at a given frequency bin as a set of narrowband beamformers. By assuming that the room impulse responses are dominated by the direct path, we can then match the permutations based on the *direction of arrival* (DOA) of the different sources. Kurita et al. first used this technique to address the permutation problem [19, 20]. However, to ensure the direction of arrival calculation is unique, the inter-microphone spacing must satisfy $d < \lambda_{\min}/2 = c/(2f_{\max})$, otherwise a *phase ambiguity* occurs (also known as *spatial aliasing*) and there are multiple possible solutions. To avoid this problem the beamforming permutation solution can only be applied unambiguously to the frequencies $f < f_{\max}$ [21]. For example, with $f_{\max} = 8\,\text{kHz}$ and $c = 340\,\text{m/s}$ we get $d \leq (340/16000)\text{m} \approx 2.1\,\text{cm}$ [22].

While the MCBD and FD-ICA approaches appear to be very different, they can, in fact, be viewed as two constrained cases within a more general framework. This framework, in turn, will lead us to propose an alternative, ST-ICA method. We will furthermore see that there is an equivalence between solving the permutation problem in FD-ICA and the grouping of components in ST-ICA.

3.2.3 A Unifying Framework

Following [23], let us assume that we can replace the filtering operations in (3.1) by circulant convolutions. Then the mixing model can be written in a form that closely resembles a standard $QN \times PN$-dimensional ICA problem:

$$\begin{pmatrix} \mathbf{x}(1) \\ \vdots \\ \\ \vdots \\ \mathbf{x}(N) \end{pmatrix} = \begin{pmatrix} A(0) & 0 & & A(L-1) \cdots & A(1) \\ \vdots & \ddots & & \ddots & \vdots \\ & & & & A(L-1) \\ A(L-1) & & & & \\ \vdots & & & \ddots & 0 \\ 0 & \cdots A(L-1) & & \cdots & A(0) \end{pmatrix} \begin{pmatrix} \mathbf{s}(1) \\ \vdots \\ \\ \vdots \\ \mathbf{s}(N) \end{pmatrix} \tag{3.8}$$

where $\mathbf{x}(k) = [x_1(k), \ldots, x_Q(k)]^T$, $\mathbf{s}(k) = [s_1(k), \ldots, s_P(k)]^T$ and each $A(k)$ is a $Q \times P$ matrix whose elements are the filter coefficients $a_{qp}(k)$. Note that to do this we have constructed space–time vectors of our observed signals. The idea of forming space–time vectors is graphically illustrated in Fig. 3.1. We can write (3.8) more conveniently as:

$$\tilde{\mathbf{x}} = \tilde{\mathbf{A}}\tilde{\mathbf{s}} \tag{3.9}$$

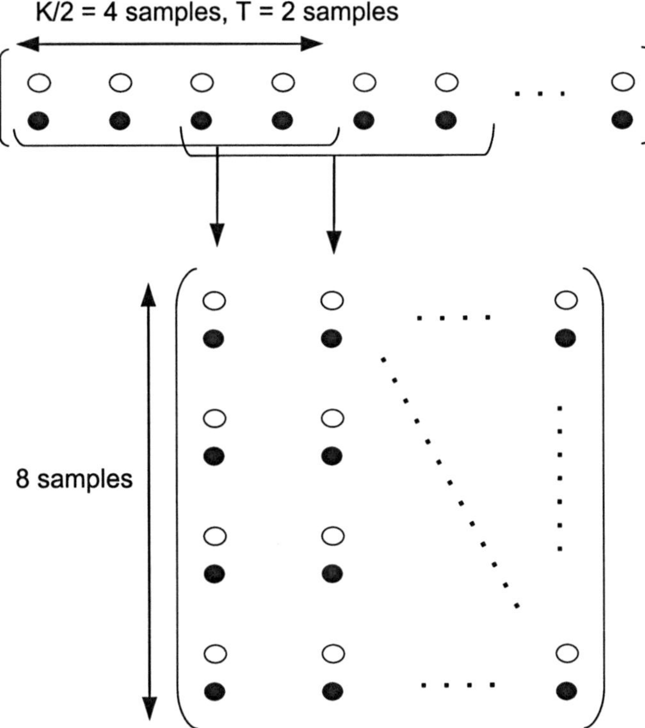

Fig. 3.1. An illustration of the reshaping of the sensor vector into a space–time form. In this illustration, we have $K/2 = 4$ sample pairs per frame, with an overlap of $T = 2$ samples.

where we will use a tilde to indicate the space–time form. Furthermore, because $\tilde{\mathbf{A}}$ is block circulant, when $P = Q$ there generally exists a unique inverse operator $\tilde{\mathbf{W}}$ that is also block circulant.

One of the key benefits of the space–time form is that it makes explicit the number of space and time degrees of freedom available to us in our problem. This is also the basis for the popular space–time adaptive processing (STAP) in radar [24].

We can now consider the differences between MCBD and FD-ICA within this framework. It was previously shown in [23] that MCBD is equivalent to the application of a standard ICA algorithm to (3.8) but with the constraint that the mixing matrix (or equivalently the unmixing matrix) is constrained to be block circulant.

To consider FD-ICA we note that the block circulant matrix is block diagonalizable by a block discrete Fourier transform (DFT) matrix, $\tilde{\mathcal{F}}$. That is, we can write:

$$\tilde{\mathbf{A}} = \tilde{\mathcal{F}}^H \tilde{\mathbf{A}}^f \tilde{\mathcal{F}} \tag{3.10}$$

where the space–time matrix $\tilde{\mathbf{A}}^f$ has the form:

$$\tilde{\mathbf{A}}^f = \begin{bmatrix} A^f(0) & 0 & \cdots & 0 \\ 0 & A^f(1) & & \\ \vdots & & & \vdots \\ 0 & & \cdots & A^f(N-1) \end{bmatrix} \tag{3.11}$$

where each $A^f(k)$ is a $Q \times P$ complex matrix associated with the kth frequency bin of the DFT.

Following the same line of argument as [23] FD-ICA is equivalent to the application of a standard ICA algorithm to:

$$\tilde{\mathbf{x}} = \tilde{\mathcal{F}}^H \tilde{\mathbf{A}}^f \tilde{\mathcal{F}} \tilde{\mathbf{s}} \tag{3.12}$$

but with the constraint that the mixing matrix (or equivalently the unmixing matrix) is block diagonal. We can therefore see that both MCBD and FD-ICA are estimating space–time adaptive filters, but with different constraints.

Note also that in each case the pth source lies in an N dimensional subspace within the full $Q \times N$-dimensional observation space. In the MCBD case, this subspace is spanned by the N (circularly) shifted versions of the mixing filter $[a_{1p}(0), \ldots, a_{qp}(0), \ldots, a_{qp}(L-1), 0, \ldots]^T$. In the case of FD-ICA the pth source again lies in an N-dimensional subspace which when transformed into the frequency domain is spanned by a single basis vector from each $A^f(k)$, $k = 1 \ldots N$.

Thus, in the space–time representation, both MCBD and FD-ICA identify independent subspaces that contain each source. That is, we have an instance of MICA [25]. This is also sometimes referred to as independent subspace analysis (ISA). In MCBD the grouping of the basis functions is trivial as they are constrained to be shifted versions of each other. In FD-ICA there is no direct mechanism for grouping the components, resulting in the permutation problem, and additional means need to be found to group the components correctly.

The above arguments are, of course, only technically valid for circulant structures unless we take the limit of $N \to \infty$, at which point circulant matrices become similar to Toeplitz matrices [26]. Note however that the assumption in MCBD that we can invert an FIR filter matrix by an FIR filter matrix is similarly not valid unless $N \to \infty$ for related reasons. For finite N we must accept a drop in performance in MCBD while modifications, such as windowing, are required in FD-ICA to avoid the substantial aliasing distortion that can otherwise be introduced [27].

3.3 Space–Time ICA

We now propose what, on the face of it, appears to be a very different solution to the convolutive source separation problem.

The key difference between FD-ICA and MCBD is that the former first applies a frequency transform to the data before identifying the independent subspaces. Computational savings aside this can provide the benefit of making the data more sparse (more non-Gaussian), which in turn should improve the statistical performance of the BSS estimator [10]. A natural question, therefore is: is there another transform that could make the data even sparser. Furthermore could we learn an optimal sparse representation at the same time as performing source separation?

In fact we can use ICA to learn such a sparsifying transform. However, instead of using it across mixture variables we use it across time samples. In an early application of this method, Bell and Sejnowski [28] found that ICA trained on time frames of monophonic recordings of 'tooth taps' discovered features (basis vectors) exhibiting localized time and phase structure, while those learned by e.g. principal components analysis (PCA) did not. It has also been noted that the infomax algorithm of Bell and Sejnowski [29] is equivalent to the model by Olshausen and Field [30] for learning sparse factorial codebooks.

Studies on monophonic audio signals [31, 32] have reported that the basis vectors learned by ICA from speech signals are typically strongly localized in both time and frequency, yielding a representation that exhibits wavelet-like bases. The resulting representation of the sounds transformed into this learned basis are sparse, i.e. with most coefficients close to zero, and give a representation that is very reminiscent of the response of auditory nerve fibers [32].

We now extend this idea to learn a multichannel and multisource representation that is both sparse and separating. First we construct space–time observation vectors as in (3.8). We can then use any standard ICA algorithm to learn a sparse representation for this data in a similar manner to [30, 31]. However, from Sect. 3.2.3 above, this can also be viewed as a MICA representation. We therefore know that each of the 1-dimensional space–time basis vectors (i.e. the columns of the estimated space–time mixing matrix: $\tilde{\mathbf{A}}$) should lie within one of the independent subspaces associated with an individual source [25]. The subspaces (and hence the sources) can be identified by first applying a standard ICA algorithm to determine the 1-dimensional component vectors. These can then be grouped into sets, using, for example, some measure of higher level dependency [5]. Finally separation can be achieved by projecting the observation data onto one of the independent subspaces.

Notice that, from a MICA perspective, the *component grouping problem* in ST-ICA is directly equivalent to the permutation problem in FD-ICA. Also, as we have formulated it, both the MCBD and FD-ICA structures are potential candidate solutions. However we will see that in practice a different basis structure tends to emerge, suggesting that neither the time domain nor the STFT provide the optimal space within which to work.

3.3.1 Learning the ST-ICA Basis

The learning process for ST-ICA is conceptually very simple. First we construct the QN-dimensional space–time vectors, $\tilde{\mathbf{x}}(k)$, from the observation data. Then we apply a standard ICA algorithm to learn the unmixing matrix $\tilde{\mathbf{W}}$ and hence obtain estimates for the ST-ICA basis vectors, $\tilde{\mathbf{a}}_k$, as the columns of the space–time mixing matrix $\tilde{\mathbf{A}} = \tilde{\mathbf{W}}^{-1}$.

Here we use a natural gradient maximum likelihood (ML) algorithm [12] which takes the form:

$$\Delta\tilde{\mathbf{W}} = \eta(\mathbf{I} - E\{\phi(\tilde{\mathbf{y}})\tilde{\mathbf{y}}^T\})\tilde{\mathbf{W}} \tag{3.13}$$

where η is the learning rate, and $\phi(\tilde{\mathbf{y}}) = -\nabla_{\tilde{\mathbf{y}}} \log p(\tilde{\mathbf{y}})$ is the ML activation function, using $p(\tilde{\mathbf{y}}) = \prod_{p=1}^{P} p(\tilde{y}_p)$ for some fixed prior $p(\tilde{y}_p)$. For $p(\tilde{y}_p)$ we use the generalized exponential prior $p(\tilde{y}_p) \propto \exp(-|\tilde{y}_p|^{\alpha})$ where the exponent α can either be fixed or estimated during the adaptation.

Figure 3.2 shows a learnt space–time basis for a stereo pair of audio signals consisting of two people speaking simultaneously. Each basis vector, $\tilde{\mathbf{a}}_k$, has been de-interleaved into pairs of waveforms, $\mathbf{a}_k^{(1)} = \tilde{\mathbf{a}}_k(2n)$ and $\mathbf{a}_k^{(2)} = \mathbf{a}_k(2n-1)$, $n = 1, \ldots, N$, to show their spatial and temporal structure. The individual waveforms exhibit the same wavelet-like localization in time and frequency, as in the monophonic case. Notice that they neither look like

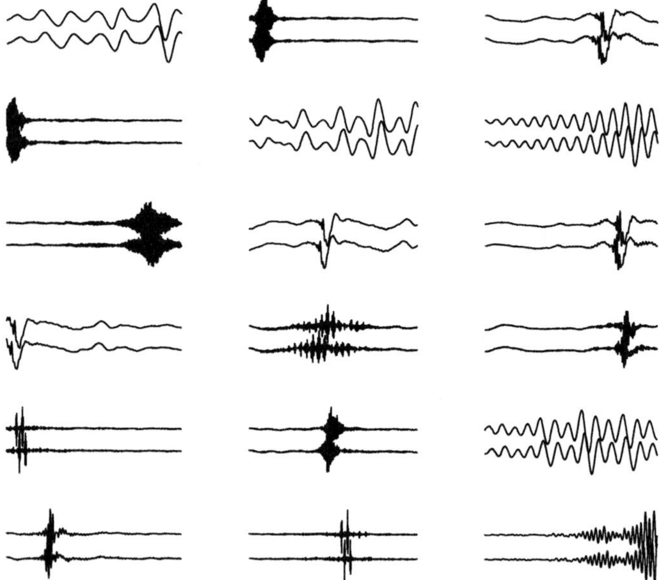

Fig. 3.2. Examples of stereo basis vectors extracted with the ICA algorithm described in Sect. 3.3.1. Note that the basis vectors have be de-interleaved to show their spatial and temporal structure.

the basis vectors for FD-ICA nor for MCBD. Unlike MCBD, in which the de-interleaved basis vectors contain shifted versions of the source-microphone impulse responses, the ST-ICA basis vectors is also signal dependent, showing a strong frequency localization [5]. Yet unlike the STFT vectors used in FD-ICA, which are effectively windowed sinusoids, the learnt ST-ICA vectors become increasingly more localized in time as their center frequency increases. Furthermore, the waveforms display relative amplitude differences and time delays between the two channels. This is because they encode *both* the source structure and the channel characteristics. This also means that ST-ICA can separate the sources without directly learning the mixing filters or their inverses.

3.3.2 Component Grouping in ST-ICA

In the space–time formalism both FD-ICA and ST-ICA generate PN component vectors that need to be grouped into P sets to identify the appropriate independent subspaces. In this respect the two clustering problems are equivalent and in principle any method for solving the permutation problem in FD-ICA is potentially applicable to cluster the components in ST-ICA. The main differences in the two problems are: (1) in FD-ICA a source is constrained to have a component from each frequency bin, while there is no such constraint on the index of the components associated with a given source in ST-ICA; and (2) the components in FD-ICA are by construction narrowband, while the components in ST-ICA tend to have a wider bandwidth, particularly at higher frequencies.

As discussed in Sect. 3.2.2 there are various ways to tackle the permutation problem, most notably by estimating higher level dependencies between 1-dimensional components or by using additional geometric information about the microphone array. Both these methods can equally be used for component grouping in ST-ICA.

In [5] Abdallah and Plumbley showed that higher level dependency structure could be used to group the space–time components using $\mathcal{F}$-correlations. Projecting onto the resulting subspaces then produced an effective separation of the individual sources. In [6] it was shown that components could equally be grouped using information about the microphone array. Furthermore, since ST-ICA components tend to have a wider bandwidth it was found that the phase ambiguity problem was avoided.

3.3.3 Clustering using Source DOAs

We briefly give details of the DOA based clustering scheme that was first introduced in [6]. First, recall that when an ST-ICA basis vector for a given source is de-interleaved it is representative of the source as observed at the individual microphones. That is: ST-ICA has also encoded the room impulse response functions into the basis vectors. If, as we did with FD-ICA in

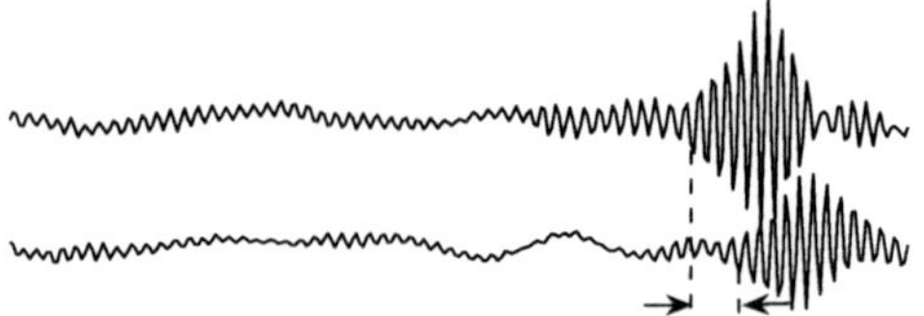

Fig. 3.3. A single space–time component vector de-interleaved into a two-microphone vector pair. The relative delay between the microphones is clearly visible.

Sect. 3.2.2 above, we assume that the room impulse responses are dominated by the direct path, then we should expect to see roughly the same waveform at each microphone, except for a relative delay associated with the source DOA.

An example of a pair of waveforms for a single ST-ICA basis vector is shown in Fig. 3.3. Examination of the figure shows that the bottom waveform closely resembles the top waveform apart from a temporal shift. Note also that since the vectors are strongly localized (i.e. wideband) there is no phase ambiguity and we can explicitly estimate the relative delay.

There are many time delay estimation techniques in use with microphone arrays. Here we use the well known generalized cross-correlation with phase transform (GCC-PHAT) algorithm [33] which is appropriate for finding the relative delay between a pair of signals.

We first de-interleave the basis vector, $\tilde{\mathbf{a}}_k$, into its two constituent waveforms, $\mathbf{a}_k^{(1)}$ and $\mathbf{a}_k^{(2)}$. We can then estimate the relative delay, τ_k, between the two waveforms by finding the maximum of the GCC-PHAT function:

$$R_k(\tau) = \int_{-\infty}^{\infty} \frac{A_k^{(1)}(\omega) A_k^{(2)}(\omega)^*}{|A_k^{(1)}(\omega) A_k^{(2)}(\omega)^*|} e^{j\omega\tau} d\omega \tag{3.14}$$

where $A_k^{(1)}(\omega)$ and $A_k^{(2)}(\omega)$ are the Fourier transforms of $\mathbf{a}_k^{(1)}$ and $\mathbf{a}_k^{(2)}$ respectively. Because GCC-PHAT considers all frequencies together and because the waveforms tend to be relatively wideband we have typically observed that the function $R_k(\tau)$ exhibits a single sharp peak at the lag corresponding to the time delay between the two signals. Thus the phase ambiguity has been avoided.

Once we have the relative delays τ_k for each basis vector, k, we can construct the histogram of the delays and use the K-means clustering algorithm to group the basis vectors into index sets γ_p, $p = 1, \ldots, P$ that have similar relative delays. We can then identify the independent subspace associated with the pth source as: $E_p = \mathrm{span}\{\tilde{\mathbf{a}}_k, k \in \gamma_p\}$.

3.3.4 Separation and Reconstruction

The final stage of our blind signal separation scheme is the separation and reconstruction which is achieved by projecting the observed data onto a source subspace. As this gives the estimated source as observed at the microphones the filter ambiguity is circumvented.

The projection operator, P_{E_p}, that projects $\tilde{\mathbf{x}}$ onto E_p is given by:

$$P_{E_p} := \tilde{\mathbf{A}}\mathbf{\Lambda}_p\tilde{\mathbf{W}} \tag{3.15}$$

where $\mathbf{\Lambda}_p$ is the diagonal matrix whose kth diagonal element is given as:

$$\mathbf{\Lambda}_p(k,k) = \begin{cases} 1 & \text{if } k \in \gamma_p \\ 0 & \text{otherwise} \end{cases} \tag{3.16}$$

We can then obtain the estimate for the contribution of the pth source to the space–time vector $\tilde{\mathbf{x}}$ as:

$$\tilde{\mathbf{x}}_p = P_{E_p}\tilde{\mathbf{x}} \tag{3.17}$$

To retrieve the time domain signals it is finally necessary to de-interleave the space–time vector $\tilde{\mathbf{x}}_p$. Applying this scheme to contiguous nonoverlapping space–time blocks (i.e. critically down-sampling) will provide a unique solution for the estimated sources.

If, on the other hand, we use a sliding space–time window we will be N-times oversampling the data (akin to a sliding DFT [27]). Thus there will be N different source estimates. An effective solution is to take the average over these estimates. We can think of this as a multichannel version of cycle-spinning [34] and averages out any block artefacts that occur in critically sampled transforms. This also induces a separation operator for the pth source takes the form of spatial–temporal *time invariant* linear filter.

3.4 An Experiment with Stereo Mixtures

We now provide an illustrative comparison between the FD-ICA method and the proposed ST-ICA scheme in the context of blind audio source separation with some results from stereo observations of two male speech sources mixed in a synthesized reverberant room. The speech sources were sampled at 16 kHz and had a duration of 1 minute each. To allow us to control the room Reverberation Time (RT) and the Input Signal-to-Noise Ratio (ISNR), and to provide us with an absolute ground truth, the speech sources were mixed using simulated room impulse responses, determined by the image technique using McGovern's RIR Matlab function [35]. The positions of the microphones and the loudspeakers are illustrated in Fig. 3.4. Six different mixing conditions were obtained by varying RT between 20 ms (320 samples), 80 ms

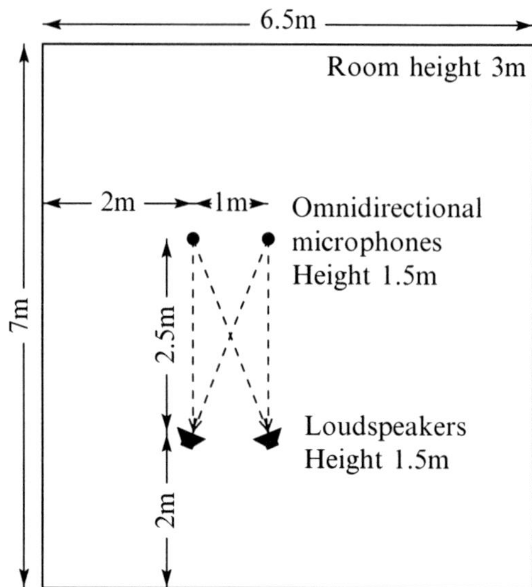

Fig. 3.4. Experimental setup for simulated speech recordings. The reverberation times were set to either 20 ms, 80 ms or 320 ms.

(1300 samples) and 320 ms (5100 samples), and adding white noise to the mixture with ISNRs of 40 dB and 20 dB.

We chose the frame lengths separately for each algorithm, but kept them fixed for all the reverberation times tested. We used the FD-ICA algorithm with the MuSIC-based permutation alignment algorithm described by Mitianoudis and Davies [36], setting the STFT frame size to 2048 samples, which was previously found to be the minimum appropriate for this algorithm at a 16 kHz sampling rate [6, 7].

For the proposed ST-ICA algorithm we used a frame size of 512 samples. This was limited by both computational considerations and statistical ones. That is, for a given data length, we need to restrict the dimension of the space–time vector if we are going to avoid over-fitting in the estimation of the ST-ICA basis, $\tilde{\mathbf{A}}$. Note this is 4 times smaller than the frame size we are using in FD-ICA.

Excerpts of the original mixture and source signals and of the estimated source signals are available for listening on our demo web page[1]. The upper plot in Fig. 3.5 shows an example of the time-delay estimates obtained with GCC-PHAT, for the different ST-ICA basis vectors. The histogram of the estimated time-delays is shown in the lower plot of Fig. 3.5. The figure shows that the directions of the two sources (corresponding to relative delays of approximately ±9 samples) are correctly identified. Also most of the basis functions are associated with one or other of the directions of arrival.

[1] `http://www.elec.qmul.ac.uk/people/mariaj/asb_demo/`

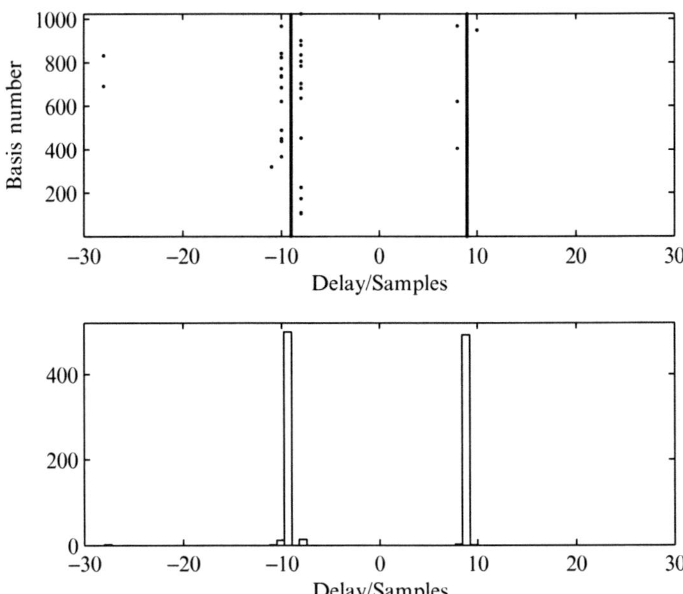

Fig. 3.5. Plot of the time delays estimated for all basis vectors (upper plot), and its histogram (lower plot).

3.4.1 Evaluation

We evaluated the performance of the methods using the Signal-to-Distortion Ratio (SDR), Signal-to-Interference Ratio (SIR), Signal-to-Noise Ratio (SNR) and Signal-to-Artefacts Ratio (SAR) as defined in [37]. SDR measures the difference between an estimated source and a target source allowing for possible linear filtering between the estimated and target source: we allowed for time-invariant filtering of filter length 1024 samples when calculating SDR. SIR, SNR and SAR provide a more detailed diagnosis of the performance by distinguishing between the elements of the total distortion which are due to unwanted interfering sources (SIR), remaining mixing noise (SNR) and other artefacts (SAR).

These SDR, SIR, SNR and SAR criteria are defined in [37] on a per-source basis. To gain a single figure for all sources, we averaged the criteria across all microphones and all sources. The results are presented in Table 3.1.

We see that with short reverberation times (RT = 20 ms, i.e. 320 samples at 16 kHz) our proposed method outperforms FD-ICA by more than 8 dB SDR in relatively clean conditions (ISNR = 40 dB) and by about 2 dB SDR in more noisy conditions (ISNR = 20 dB). In terms of SIR the improvements are 15 dB and 7 dB respectively. As the reverberation time increases, the performance of the proposed method degrades faster than FD-ICA. This is, of course, to be expected as we were constrained to use a much smaller (by

Table 3.1. Objective performance of FD-ICA and ST-ICA with default frame sizes on simulated speech recordings. All values are expressed in decibels (dB). Bold numbers indicate the best SDR for each mixing condition. See text for comments.

Mixing conditions	ISNR	40 dB			20 dB		
	RT	20 ms	80 ms	320 ms	20 ms	80 ms	320 ms
FD-ICA	SDR	7.0	**11.2**	**6.3**	6.2	6.5	**4.2**
	SIR	10.4	16.1	9.1	12.3	14.0	9.1
	SNR	19.1	19.9	28.9	26.7	10.7	25.8
	SAR	11.1	14.2	10.3	7.7	11.4	7.0
ST-ICA	SDR	**15.4**	7.7	1.3	**8.3**	**6.8**	-4.2
	SIR	25.7	16.3	8.9	19.7	17.8	7.4
	SNR	20.2	28.0	22.9	12.5	26.3	16.9
	SAR	18.2	9.8	4.2	12.6	7.5	-2.1

a factor of 4) frame size for the ST-ICA algorithm. Indeed in the high noise, long reverberation time case the ST-ICA failed to separate the sources. While supervised clustering based on the known source locations improved things slightly the resulting SDR was still negative (-0.6 dB).

3.5 Future Work

We have shown that ST-ICA can be used to learn a space–time transform that provides both a sparse representation and a mechanism for source separation. We now briefly discuss potential areas for further work. In particular, we consider algorithmic improvements that may help reduce the high computational load and large data requirements that ST-ICA currently has. We also discuss the possibilities of noise reduction in ST-ICA and the use of local ST-ICA bases.

3.5.1 Computation and Data Requirements

One computational saving we could make would be to use a "Newton-like" fixed point ICA algorithm [14]. Such techniques tend to converge much more rapidly than gradient based approaches. Furthermore, the prewhitening step that these methods require can be efficiently done using a block DFT or block Discrete Cosine Transform (DCT), since for large frame sizes both these transforms provide an approximate diagonalization of the covariance matrix for wide sense stationary processes [38]. This has the added advantage that the imposed the Toeplitz structure implicit in the stationarity assumption reduces the number of free parameters that we need to estimate and should therefore reduce the variance of the estimator. It might also be possible to exploit similar structure in the higher order stage of the algorithm.

Other improvements to the statistical reliability of the estimates could be achieved by further restricting the degree of adaption in learning the space–time representation $\tilde{\mathbf{A}}$. Of course, MCBD and FD-ICA are both particular cases of constrained adaption. However, there are also a number of other possibilities. For example, using a restricted "beamspace" has been a popular approach in beamforming [39] and STAP [24] and we believe that similar ideas could be applied here. Restricting the search space in this way is also likely to reduce the computation load.

For example, one immediate insight that we can glean from the ST-ICA representation presented here is that the "preferred" time–frequency space for speech separation does not appear linear in frequency. Thus, instead of an STFT-type transform something more akin to an auditory filterbank appears to have emerged with a Q-factor that grows with frequency (c.f. [32]). Whether we should use a fixed auditory filterbank as is popular in Computational Auditory Scene Analysis (CASA) [40], or whether one could devise a partially adaptive technique that optimized this time–frequency structure based on the signal are open questions. However we suspect that something should be possible using, for example, fast adaptive time–frequency representations [41, Sect. 9.4].

3.5.2 Noise Reduction

Although not explored in this present work, the signal adapted nature of ST-ICA suggests that we should be able to do much more effective noise reduction in this domain. There are a couple of approaches that could be adopted here. To begin with, we are not constrained to have fixed dimensional subspaces. While in the current work we have assumed that all sources are full band and therefore live in a full N dimensional subspace, in practice, the sources may reside in lower dimensional subspaces. In a high SNR scenario this could be identified using a principal component analysis (PCA) preprocessing step. However it may also be possible in the component grouping stage to identify an extra independent 'noise' subspace, in addition to subspaces associated with the individual sources. The dimensions of the various subspaces will depend upon which basis vectors are grouped together. It might also be possible to make 'soft' allocation of certain basis vectors, by grouping each vector in none, one or more than one source subspace. Finally, since the ST-ICA representation is adapted to maximize sparsity it is a good candidate for additional noise reduction within the source subspace via sparse code shrinkage [42].

3.5.3 Local ST-ICA Bases

One other final area that we feel is ripe for further exploration is the possibility of generating local short time ST-ICA representations. Over short time intervals speech signals can be considered as quasi-stationary and will often have a highly restricted frequency support. This in turn means that the signal subspace will be greatly reduced. Furthermore it opens up the possibility of separating out more sources than sensors in a similar manner to the degenerate unmixing estimation technique (DUET) [43] since not all sources will simultaneously occupy the full frequency range.

Indeed, while this chapter has presented ST-ICA as a space–time generalization of conventional convolutive blind source separation, we can also think of it as a spatial or multichannel extension of Casey and Westner's single channel source separation [44], where local ICA representations of a monophonic signal were used to separate the signal into independent sources. One would naturally expect that the inclusion of additional spatial information here should further aid separation.

3.6 Conclusions

In this chapter we have presented the convolutive blind source separation problem in a space–time framework. Both MCBD and FD-ICA were seen to be special restrictive cases of working in this domain. We then considered how to learn a general unconstrained ST-ICA representation and saw how this representation was adapted to both the structure of the source signals and the mixing filters. Specifically the learnt basis vectors tended to be localized in both time and frequency and were unlike either the equivalent MCBD or FD-ICA representations. Indeed, one of the potential benefits of the ST-ICA framework is that the resulting representation is better adapted to the source signals. That is, the ST-ICA transform generates a sparser (more super-Gaussian) representation of the source signals than either the time or frequency domain representations.

The ST-ICA representation also provides us directly with a means of source separation since in the space-time framework the individual sources reside in subspaces within the full vector space. Therefore, in this domain, the problem of blind source separation is one of MICA and we need to decompose the space into independent subspaces. Projecting onto one of these subspaces then separates out the associated source. We also saw an equivalence between the component grouping problem in ST-ICA and the permutation problem in FD-ICA. Indeed we have seen that similar techniques can be applied to either.

Of course, there is still much to do. In particular, if ST-ICA is to be competitive with the state-of-the-art frequency domain ICA techniques, the two issues of how to reduce the high computational load and how to obtain good statistical estimation from limited data will need to be addressed. Nevertheless, we believe that this general framework represents a promising avenue for future research in audio source separation.

Acknowledgments

This research was partly funded through EPSRC research grants GR/R54620 and GR/S85900. MD also acknowledges support for his position from the Scottish Funding Council and for their support of the Joint Research Institute with Heriot-Watt University, as part of the Edinburgh Research Partnership.

References

1. S.-I. Amari, S. Douglas, A. Cichocki, and H. Yang, "Multichannel blind deconvolution and equalization using the natural gradient," in *Proc. IEEE Workshop on Signal Processing Advances in Wireless Communications*, Apr. 1997, pp. 101–104.
2. K. Torkkola, "Blind separation of convolved sources based on information maximization," in *Proc. of the IEEE Workshop on Neural Networks for Signal Processing (NNSP)*, 1996, pp. 423–432.
3. S. C. Douglas and X. Sun, "Convolutive blind separation of speech mixtures using the natural gradient," *Speech Communication*, vol. 39, pp. 65–78, 2003.
4. S. Makino, H. Sawada, R. Mukai, and S. Araki, "Blind source separation of convolutive mixtures of speech in frequency domain," *IEICE Trans. Fundamentals* vol. E88, pp. 1640–1655, 2005.
5. S. Adballah and M. Plumbley, "Application of geometric dependency analysis to the separation of convolved mixtures," in *Proc. ICA 2004 (LNCS 3195)*, Sept. 2004, pp. 540–547.
6. M. G. Jafari, S. A. Abdallah, M. D. Plumbley, and M. E. Davies, "Sparse coding for convolutive blind audio separation," in *Proc. ICA 2006 (LNCS 3889)*, Mar. 2006, pp. 132–139.
7. N. Mitianoudis and M. Davies, "Audio source separation of convolutive mixtures," *IEEE Trans. Audio and Speech Processing*, vol. 11, pp. 489–497, 2003.
8. R. H. Lambert, *Multichannel Blind Deconvolution: FIR Matrix Algebra and Separation of Multipath Mixtures*, Ph.D. dissertation, University of Southern California, Los Angeles, CA., 1996.
9. P. Smaragdis, "Blind separation of convolved mixtures in the frequency domain," *Neurocomputing*, vol. 22, pp. 21–34, 1998.
10. J.F. Cardoso, "Blind signal separation: statistical principles," *Proceedings of the IEEE*, vol. 86, no. 10, pp. 2009–2025, 1998.
11. S. Araki, S. Makino, R. Aichner, T. Nishikawa, and H. Saruwatari, "Subband based blind source separation with appropriate processing for each frequency band," in *Proc. ICA 2003*, Apr. 2003, pp. 499–504.
12. J.-F. Cardoso and B. Laheld, "Equivariant adaptive source separation," *IEEE Trans. Signal Processing*, vol. 44, pp. 3017–3030, 1996.
13. S. Amari and A. Cichocki, "Adaptive blind signal processing – neural network approaches," *Proceedings of the IEEE*, vol. 86, no. 10, pp. 2026–2048, 1998.
14. A. Hyvärinen, "Fast and robust fixed-point algorithm for independent component analysis," *IEEE Trans. Neural Networks*, vol. 10, no. 3, pp. 626–634, 1999.
15. S. Ikeda and N. Murata, "A method of ICA in time–frequency domain," in *Proc. ICA99*, Jan. 1999, pp. 365–371.

16. M. E. Davies, "Audio source separation," in *Mathematics of Signal Processing V*, Eds. J. G. McWhirter and I. K. Proudler, Oxford University Press, 2002, pp. 57–68.
17. I. Lee, T. Kim, and T.-W. Lee, "Complex FastIVA: A robust maximum likelihood approach of MICA for convolutive BSS", in *Proc. ICA 2006 (LNCS 3889)*, Mar. 2006, pp. 625–632.
18. S. Araki, S. Makino, R. Mukai, Y. Hinamoto, T. Nishikawa, and H. Saruwatari, "Equivalence between frequency domain blind source separation and frequency domain adaptive beamforming," in *Proc. ICASSP 2002*, May 2002, pp. 1785–1788.
19. S. Kurita, H. Saruwatari, S. Kajita K. Takeda, and F. Itakura, "Evaluation of blind signal separation method using directivity pattern under reverberant conditions," in *Proc. ICASSP 2000*, June 2000, pp. 3140–3143.
20. H. Saruwatari, S. Kurita, and K. Takeda, "Blind source separation combining frequency-domain ICA and beamforming," in *Proc. ICASSP 2001*, May 2001, pp. 2733–2736.
21. M. Z. Ikram and D. R. Morgan, "A beamforming approach to permutation alignment for multichannel frequency-domain blind speech separation," in *Proc. ICASSP 2002*, May 2002, pp. 881–884.
22. H. Sawada, R. Mukai, S. Araki, and S. Makino, "A robust and precise method for solving the permutation problem of frequency-domain blind source separation," *IEEE Trans. Speech and Audio Processing*, vol. 12, pp. 530–538, 2004.
23. S. C. Douglas and S. Haykin, "Relationships between blind deconvolution and blind source separation," in *Unsupervised Adaptive Filtering, vol. 2*, Ed. S. Haykin, John Wiley & Sons, 2002, pp. 113–145.
24. W. L. Melvin, "A STAP overview," *IEEE A&E Systems Magazine*, vol. 19, no. 1, pp. 19–35, 2004.
25. J.-F. Cardoso, "Multidimensional independent component analysis," in *Proc. ICASSP'98*, May 1998, pp. 1941–1944.
26. R. M. Gray, "On the asymptotic eigenvalue distribution of Toeplitz matrices," *IEEE Trans. Info. Theory*, vol. IT-18, no. 6, pp. 725–730, 1972.
27. J. J. Shynk, "Frequency-domain and multirate adaptive filtering," *IEEE Signal Processing Magazine*, vol. 9, no. 1, pp. 14–37, Jan. 1992.
28. A. Bell and T. Sejnowski, "Learning the higher-order structure of a natural sound," *Network: Computation in Neural Systems* vol. 7, pp. 261–266, 1996.
29. A. J. Bell and T. J. Sejnowski, "An information maximization approach to blind separation and blind deconvolution," *Neural Computation*, vol. 7, no. 6, pp. 1129–1159, 1995.
30. B. A. Olshausen and D. J. Field, "Emergence of simple-cell receptive-filed properties by learning a sparse code of natural images," *Nature*, vol. 381, pp. 607–609, 1996.
31. S. A. Abdallah and M. D. Plumbley, "If edges are the independent components of natural images, what are the independent components of natural sounds?," in *Proc. ICA 2001*, Dec. 2001, pp. 534–539.
32. M. S. Lewicki, "Efficient coding of natural sounds," *Nature Neuroscience* vol. 5, no. 4, pp. 356–363, 2002.
33. C. Knapp and G. Carter, "The generalized correlation method for estimation of time delay," *IEEE Trans. Acoustic, Speech and Signal Processing*, vol. 24, pp. 320–327, 1976.

34. R. R. Coifman and D. L. Donoho, "Translation-invariant de-noising," in *Wavelets and Statistics*, vol. 103, *Springer Lecture Notes in Statistics*, Eds. A. Antoniadis and G. Oppenheim, Springer-Verlag: New York, 1995, pp. 126–150.
35. S. McGovern, "A model for room acoustics," Available at http://2pi.us/rir.html (2003).
36. N. Mitianoudis and M. Davies, "Permutation alignment for frequency domain ICA using subspace beamforming methods," in *Proc. ICA 2004 (LNCS 3195)*, Sept. 2004, pp. 669–676.
37. E. Vincent, R. Gribonval, and C. Févotte, "Performance measurement in blind audio source separation," *IEEE Trans. Audio, Speech and Language Processing* vol. 14, no. 4, pp. 1462–1469, 2006.
38. M. Unser, "On the approximation of the discrete Karhunen–Loeve transform for stationary processes," *Signal Processing*, vol. 7, pp 231–249, 1984.
39. B. D. Van Veen and K. M. Buckley, "Beamforming: a versatile approach to spatial filtering," *IEEE ASSP Magazine*, vol. 5, no. 2, pp. 2–24, Apr. 1988.
40. G. J. Brown, *Computational Auditory Scene Analysis: a Representational Approach*, Ph.D. dissertation, Computer Science Dept., Sheffield Univ, 1992.
41. S. Mallat, *A Wavelet Tour of Signal Processing*, Academic Press, 1999.
42. A. Hyvärinen, "Sparse code shrinkage: Denoising of nongaussian data by maximum likelihood estimation," *Neural Computation*, vol. 11, no. 7, pp. 1739–1768, 1999.
43. A. Jourjine, S. Rickard, and Ö. Yilmaz, "Blind separation of disjoint orthogonal signals: demixing n sources from 2 mixtures," in *Proc. ICASSP 2000*, June 2000, pp. 2985–2988.
44. M. Casey and A. Westner, "Separation of mixed audio sources by independent subspace analysis," in *Proc. ICMC 2000*, Aug. 2000, pp. 154–161.

4 TRINICON-based Blind System Identification with Application to Multiple-Source Localization and Separation

Herbert Buchner[1]*, Robert Aichner[2], and Walter Kellermann[2]

[1] Deutsche Telekom Laboratories
Technical University Berlin
Ernst-Reuter-Platz 7
D-10587 Berlin, Germany
E-mail: hb@buchner-net.com

[2] Multimedia Communications and Signal Processing
University of Erlangen-Nuremberg
Cauerstr. 7
D-91058 Erlangen, Germany
E-mail: {aichner, wk}@LNT.de

Abstract. This contribution treats blind system identification approaches and how they can be used to localize multiple sources in environments where multipath propagation cannot be neglected, e.g., acoustic sources in reverberant environments. Based on TRINICON, a general framework for broadband adaptive MIMO signal processing, we first derive a versatile blind MIMO system identification method. For this purpose, the basics of TRINICON will be reviewed to the extent needed for this application, and some new algorithmic aspects will be emphasized. The generic approach then allows us to study various illustrative relations to other algorithms and applications. In particular, it is shown that the optimization criteria used for blind system identification allow a generalization of the well-known Adaptive Eigenvalue Decomposition (AED) algorithm for source localization: Instead of one source as with AED, several sources can be localized simultaneously. Performance evaluation in realistic scenarios will show that this method compares favourably with other state-of-the-art methods for source localization.

4.1 Introduction

4.1.1 Overview

The area of broadband signal aquisition by sensor arrays in multipath or convolutive environments can be divided into two general tasks: the acquisition of clean source signals, and the analysis of the scene, e.g., in order to extract

*This work was mainly performed while the first author was with Multimedia Communications and Signal Processing, University of Erlangen-Nuremberg.

S. Makino et al. (eds.), Blind Speech Separation, 101–147.

the source positions or the reverberation time of the environment. A challenging and important example for such environments are 'natural' acoustic human/machine interfaces which use multiple microphones to support sound signal acquisition so that the users should be untethered and mobile. Due to the distance between the sources and the sensors, the sensor signal processing generally has to cope with two basic problems, namely the presence of additive noise and interferers, and the disturbing effect of reflections and scattering of the desired source signals in the recordings. Intuitively, if all propagation paths from the desired and interfering sources to all sensors were known exactly we would in principle be able to ideally solve all the above-mentioned tasks and associated problems. However, since both the original source signals and the propagation signals are generally unknown in practice, a blind estimation of the propagation paths, i.e., a *blind system identification* (BSI) of the multiple-input multiple-output (MIMO) system is desirable in order to analyze the scene with the given sensor signals.

This chapter consists of two parts. In the first part, consisting of Sects. 4.2 and 4.3, a general treatment of BSI for MIMO systems is presented, based on TRINICON, a previously introduced versatile framework for broadband adaptive MIMO signal processing [1–4], which is especially well suited for speech and audio signals. We also show a practically important relation between BSI and blind source separation (BSS) for convolutive mixtures. In addition to the inherent broadband structure necessary for a proper system identification, the top-down, i.e., *deductive* approach of the TRINICON framework also allows us to present both relations to already known and new efficient algorithms. Section 4.2 follows the ideas outlined in [5, 6]. Some of these ideas were also developed independently in [7] in a slightly different way.

An important and particularly illustrative application of broadband MIMO BSI considered in the second part of this chapter, Sects. 4.4 to 4.6, is the acoustic localization of multiple simultaneously active sources in reverberant environments. A popular method to the estimation of the position of an acoustic source in a room is to apply a two-stage approach, consisting of the estimation of *time differences of arrival* (TDOAs) between microphone pairs, followed by the (possibly multidimensional) determination of the position by a purely geometrical calculation. In contrast to another method, based on a farfield assumption and the estimation of *directions of arrival* (DOAs), the TDOA-based method also allows an accurate localization of sources in the nearfield. For the signal processing part of these methods, there are already some popular and conceptually simple approaches in the literature both for a single source, such as the generalized cross-correlation method with its numerous variants [8, 9], and for multiple sources, such as the subspace methods, known as, e.g., MUSIC [10] or ESPRIT [11] and their variants, e.g., [9]. However, most of the source localization methods were originally designed only for freefield propagation and/or narrowband applications so that none of the above-mentioned approaches takes multipath propagation and dispersion such as room reverberation in acoustic scenarios into account.

Each of the unmodeled reverberation paths causes an additional peak in the correlation function like an additional source which in turn causes ambiguities in these methods [12–14]. A considerable advantage of the BSI-based source localization method over the conventional correlation-based methods is that due to the explicit multipath model the reverberation does no longer act as a disturbance to the position estimates so that the above-mentioned ambiguity is inherently solved by this method. So far the literature on efficient algorithmic solutions for blind adaptive system identification and their application to source localization has mainly focused on single-input multiple-output (SIMO) systems, i.e., for a single source [15, 16]. As we will see in this chapter, the broadband MIMO solution based on TRINICON results in a general multidimensional localization scheme for multiple sources in reverberant environments. The TRINICON-based TDOA estimation for multiple sources was first demonstrated in [6]. Due to the system identification, this approach is also suitable for an accurate localization in the nearfield of the sources. Moreover, a further differentiating and practically important feature of the TRINICON-based approach [6] is that its signal-separating property also inherently resolves a fundamental spatial ambiguity. This ambiguity generally arises in the *multidimensional* case of any multiple-source localization task where the multiple TDOAs/DOAs corresponding to the multiple sources in each dimension must be assigned to the corresponding multiple TDOAs/DOAs of the same sources for the other dimensions. In [17] it was demonstrated that this assignment is made possible due to the inherent blind source separation ability of this approach, i.e., because of the availability of the separated signals due to the relation between BSI and BSS as mentioned above.

4.1.2 Blind Adaptive MIMO Filtering Tasks and Matrix Formulation

Blind signal processing on convolutive mixtures of unknown time series is desirable for several application domains, a prominent example being the so-called cocktail party problem in acoustics, where we want to recover the speech signals of multiple speakers who are simultaneously talking in a real room. The room may be very reverberant due to reflections on the walls, i.e., the original source signals $s_q(n)$, $q = 1, \ldots, Q$ are filtered by a linear multiple-input and multiple-output (MIMO) system before they are picked up by the sensors yielding the sensor signals $x_p(n)$, $p = 1, \ldots, P$. In this chapter, we describe this MIMO mixing system by length-M finite impulse response (FIR) filters, i.e.,

$$x_p(n) = \sum_{q=1}^{Q} \sum_{\kappa=0}^{M-1} h_{qp,\kappa} s_q(n - \kappa), \tag{4.1}$$

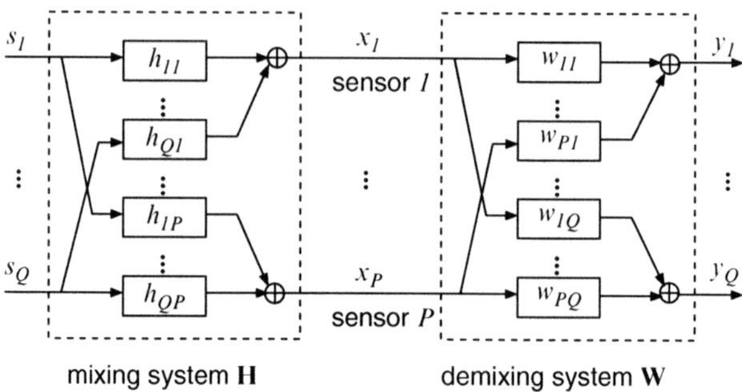

Fig. 4.1. Setup for blind MIMO signal processing.

where $h_{qp,\kappa}$, $\kappa = 0, \ldots, M-1$ denote the coefficients of the FIR filter model from the q-th source signal $s_q(n)$ to the p-th sensor signal $x_p(n)$ according to Fig. 4.1. Moreover, we assume throughout this chapter that the number Q of sources is less or equal to the number P of sensors. These cases $Q \leq P$ are of particular interest in the context of blind system identification as detailed below, and they are commonly known as *overdetermined* and *determined*, respectively. Note that in general, the sources $s_q(n)$ may or may not be all simultaneously active at a particular instant of time.

Obviously, since only the sensor signals, i.e., the output signals of the mixing system are accessible by the blind signal processing, *any* type of linear blind adaptive MIMO signal processing may be described by the serial structure shown in Fig. 4.1. Thus, according to a certain optimization criterion, we are interested in finding a corresponding demixing system by the blind adaptive signal processing whose output signals $y_q(n)$ are described by

$$y_q(n) = \sum_{p=1}^{P} \sum_{\kappa=0}^{L-1} w_{pq,\kappa} x_p(n-\kappa). \tag{4.2}$$

The parameter L denotes the FIR filter length of the demixing filters $w_{pq,\kappa}$.

Depending on the chosen coefficient optimization criterion, we distinguish two general classes of blind signal processing problems[1]:

- **"Direct blind adaptive filtering problems":** This class summarizes here blind system identification (BSI) and blind source separation (BSS)/blind interference cancellation for convolutive mixtures.
  In the BSS approach, we want to determine a MIMO FIR demixing filter which separates the signals *up to an – in general arbitrary – filtering and permutation* by forcing the output signals to be mutually independent.

[1]Note that in supervised adaptive filtering we may distinguish the analogous general classes of problems. In this case we classify system identification and interference cancellation after [18] as the "direct supervised adaptive filtering problems", whereas inverse modeling and linear prediction after [18] may be classified as the "inverse supervised adaptive filtering problems".

Traditionally, and perhaps somewhat misleadingly, BSS has often been considered to be an inverse problem in the literature, e.g., in [19, 20]. In another interpretation, BSS may be considered as a set of *blind beamformers* [21, 22] under certain restricting conditions, most notably the fulfilment of the spatial sampling theorem by the microphone array. Moreover, under the farfield assumption, the DOAs may be extracted from the corresponding array patterns, which in turn may be calculated from the BSS filter coefficients.

In this chapter we will see that more generally, a properly designed and configured broadband BSS system actually performs blind MIMO system identification (which is independent of the spatial sampling theorem). The general broadband approach shown in this chapter thus allows us to unify the BSS and BSI concepts and provides various algorithmic synergy effects and new applications. This general class of direct blind adaptive signal processing problems is the main focus of this chapter.

- **"Inverse blind adaptive filtering problems":** This class stands here for multichannel blind deconvolution (MCBD)[2] w.r.t. the mixing system $\mathbf{H}$.

  Here, in addition to the separation, we want to recover the original signals *up to an arbitrary* (frequency-independent) *scaling, possibly a time shift, and a permutation*, i.e., in the acoustic applications we want to dereverberate the signals. In terms of the MIMO system description, for this task, effectively, an inversion of (long and usually nonminimum phase) room impulse responses is necessary. However, using the multiple-input/output inverse theorem (MINT) [23] any MIMO system $\mathbf{H}$ can exactly be inverted if P, Q, and L are suitably chosen, and if h_{qp} $\forall$ $p \in \{1, \dots, P\}$ do not have common zeros in the z-plane. Therefore, in principle, there is a general solution to the MCBD problem by using multiple sensors. Obviously, to realize MCBD two different fundamental approaches are conceivable. One approach is to first perform blind MIMO system identification as mentioned above, followed by a (MINT-based) inversion of the estimated mixing system, e.g., [25, 26]. The other, theoretically equivalent but in practice often more reliable approach is to perform directly a blind estimation of the actual inverse of the MIMO mixing system, e.g., [3, 27–29]. As we may expect, in any case there are various relations between the general classes of direct and inverse blind adaptive filtering problems, i.e., BSI and MCBD, and the corresponding algorithms. As a side aspect, this chapter also tries to highlight some of these relations.

---

[2] Later in Sect. 4.3.2 we will see that in practical systems for the blind deconvolution tasks it is important to take the spectral characteristics of the source signals into account. The method of multichannel blind partial deconvolution (MCBPD), introduced in Sect. 4.3.2 to address this issue also belongs to the class of inverse blind adaptive filtering problems.

Matrix Formulation. To analyze and to formulate the above-mentioned blind adaptive MIMO filtering problems compactly, we introduce the following matrix formulation of the overall system consisting of the mixing and demixing systems. Moreover, this matrix formulation is also used directly in the TRINICON framework described later in Sect. 4.3 in order to blindly estimate the adaptive demixing filter coefficients.

As a compact formulation of the mixing filter coefficients $h_{qp,\kappa}$, $\kappa = 0,\ldots,M-1$ and the demixing filter coefficients $w_{pq,\kappa}$, $\kappa = 0,\ldots,L-1$, $p=1,\ldots,P$, $q=1,\ldots,Q$, we form the $QM \times P$ mixing coefficient matrix

$$\check{\mathbf{H}} = \begin{bmatrix} \mathbf{h}_{11} & \cdots & \mathbf{h}_{1P} \\ \vdots & \ddots & \vdots \\ \mathbf{h}_{Q1} & \cdots & \mathbf{h}_{QP} \end{bmatrix} \tag{4.3}$$

and the $PL \times Q$ demixing coefficient matrix

$$\check{\mathbf{W}} = \begin{bmatrix} \mathbf{w}_{11} & \cdots & \mathbf{w}_{1Q} \\ \vdots & \ddots & \vdots \\ \mathbf{w}_{P1} & \cdots & \mathbf{w}_{PQ} \end{bmatrix}, \tag{4.4}$$

respectively, where

$$\mathbf{h}_{qp} = \left[h_{qp,0},\ldots,h_{qp,M-1}\right]^{\mathrm{T}}, \tag{4.5}$$

$$\mathbf{w}_{pq} = \left[w_{pq,0},\ldots,w_{pq,L-1}\right]^{\mathrm{T}} \tag{4.6}$$

denote the coefficient vectors of the FIR subfilters of the MIMO systems, and superscript ^{T} denotes transposition of a vector or a matrix. The downwards pointing hat symbol on top of $\mathbf{H}$ and $\mathbf{W}$ in (4.3) and (4.4) serves to distinguish these *condensed* matrices from the corresponding larger matrix structures as introduced below in (4.10) for the case of the mixing system. The rigorous distinction between these different matrix structures is also an essential aspect of the general TRINICON framework as shown later.

Analogously, the coefficients $c_{qr,\kappa}$, $q = 1,\ldots,Q$, $r = 1,\ldots,Q$, $\kappa = 0,\ldots,M+L-2$ of the overall system of length $M+L-1$ from the sources to the adaptive filter outputs are combined into the $Q(M+L-1)\times Q$ matrix

$$\check{\mathbf{C}} = \begin{bmatrix} \mathbf{c}_{11} & \cdots & \mathbf{c}_{1Q} \\ \vdots & \ddots & \vdots \\ \mathbf{c}_{Q1} & \cdots & \mathbf{c}_{QQ} \end{bmatrix}, \tag{4.7}$$

where

$$\mathbf{c}_{qr} = \left[c_{qr,0},\ldots,c_{qr,M+L-2}\right]^{\mathrm{T}}. \tag{4.8}$$

All these subfilter coefficients $c_{qr,\kappa}$ are obtained by convolving the mixing filter coefficients with the demixing filter coefficients. In general, a convolution of two such finite-length sequences can also be written as a matrix-vector product so that the coefficient vector for the model from the q-th source to the r-th output reads here

$$\mathbf{c}_{qr} = \sum_{p=1}^{P} \mathbf{H}_{qp,[L]} \mathbf{w}_{pr}. \tag{4.9}$$

The so-called *convolution* or *Sylvester matrix* $\mathbf{H}_{qp,[L]}$ of size $M + L - 1 \times L$ in this equation exhibits a special structure, containing the M filter taps in each column,

$$\mathbf{H}_{qp,[L]} = \begin{bmatrix} h_{qp,0} & 0 & \cdots & 0 \\ h_{qp,1} & h_{qp,0} & \ddots & \vdots \\ \vdots & h_{qp,1} & \ddots & 0 \\ h_{qp,M-1} & \vdots & \ddots & h_{qp,0} \\ 0 & h_{qp,M-1} & \ddots & h_{qp,1} \\ \vdots & & \ddots & \vdots \\ 0 & \cdots & 0 & h_{qp,M-1} \end{bmatrix}. \tag{4.10}$$

The additional third index in brackets denotes the width of the Sylvester matrix which has to correspond to the length of the column vector $\mathbf{w}_{pr}$ in (4.9) so that the matrix-vector product is equivalent to a linear convolution. The brackets serve to emphasize this fact and to clearly distinguish the meaning of this index from the meaning of the third index of the individual elements of the matrices, e.g., in (4.10).

We may now express the overall system matrix $\check{\mathbf{C}}$ after (4.7) compactly using this Sylvester matrix formulation to finally obtain

$$\check{\mathbf{C}} = \mathbf{H}_{[L]} \check{\mathbf{W}}, \tag{4.11}$$

where $\mathbf{H}_{[L]}$ denotes the $Q(M + L - 1) \times PL$ MIMO block Sylvester-matrix combining all channels,

$$\mathbf{H}_{[L]} = \begin{bmatrix} \mathbf{H}_{11,[L]} & \cdots & \mathbf{H}_{1P,[L]} \\ \vdots & \ddots & \vdots \\ \mathbf{H}_{Q1,[L]} & \cdots & \mathbf{H}_{QP,[L]} \end{bmatrix}. \tag{4.12}$$

Based on this matrix formulation, we are now in a position to formulate the blind adaptive MIMO filtering tasks compactly, and to discuss the corresponding ideal solutions, regardless of how the adaptation is actually performed in practice (note that this also means that the results are valid for both blind and supervised adaptation). The blind adaptation of the coefficients towards these ideal solutions will be treated later in Sect. 4.3.

Condition for Ideal Solution of Inverse Adaptive Filtering Problems (*Ideal Inversion Solution*). As mentioned above, the aim of the inverse adaptive filtering problem is to recover the original signals $s_q(n)$, $q = 1, \ldots, Q$, as shown in Fig. 4.1, up to an arbitrary frequency-independent scaling, time shift, and possibly a permutation of the demixing filter outputs. Disregarding the potential permutation among the output signals[3], this condition may be expressed in terms of an *ideal overall system matrix*

$$\check{\mathbf{C}}_{\text{ideal,inv}} = \text{Bdiag}\left\{[0,\ldots,0,1,0,\ldots,0]^{\mathrm{T}},\ldots,[0,\ldots,0,1,0,\ldots,0]^{\mathrm{T}}\right\}\boldsymbol{\Lambda}_{\alpha}, \tag{4.13}$$

where the Bdiag$\{\cdot\}$ operator describes a block-diagonal matrix containing the listed vectors on the main diagonal. Here, these target vectors represent pure delays. The diagonal matrix $\boldsymbol{\Lambda}_{\alpha} = \text{Diag}\left\{[\alpha_1,\ldots,\alpha_P]^{\mathrm{T}}\right\}$ accounts for the scaling ambiguity. The *condition for the ideal inversion solution* thus reads

$$\mathbf{H}_{[L]}\check{\mathbf{W}} = \check{\mathbf{C}}_{\text{ideal,inv}}. \tag{4.14}$$

This system of equations may generally be solved exactly or approximately by the Moore–Penrose pseudoinverse, denoted by $\cdot^{+}$, so that

$$\begin{aligned}\check{\mathbf{W}}_{\text{LS,inv}} &= \mathbf{H}_{[L]}^{+}\check{\mathbf{C}}_{\text{ideal,inv}}\\ &= \left[\mathbf{H}_{[L]}^{\mathrm{T}}\mathbf{H}_{[L]}\right]^{-1}\mathbf{H}_{[L]}^{\mathrm{T}}\check{\mathbf{C}}_{\text{ideal,inv}}.\end{aligned} \tag{4.15}$$

Note that this expression corresponds to the least-squares (LS) solution

$$\check{\mathbf{W}}_{\text{LS,inv}} = \arg\min_{\check{\mathbf{W}}}\|\mathbf{H}_{[L]}\check{\mathbf{W}} - \check{\mathbf{C}}_{\text{ideal,inv}}\|^2. \tag{4.16}$$

It can be shown that under certain practically realizable conditions this solution becomes the ideal inversion solution, i.e., the pseudoinverse in (4.15) turns into the true matrix inverse,

$$\check{\mathbf{W}}_{\text{ideal,inv}} = \mathbf{H}_{[L]}^{-1}\check{\mathbf{C}}_{\text{ideal,inv}}. \tag{4.17}$$

This method is known as the Multiple-input/output INverse Theorem (MINT) [23] and is applicable even for mixing systems with nonminimum

[3] It could formally be described by an additional permutation matrix in the ideal solution. However, since in many practical cases this ambiguity may easily be resolved (e.g., by a correlation analysis), we renounced on this formal treatment for clarity.

phase. The basic requirement for $\mathbf{H}_{[L]}$ in order to be invertible is that it is of full rank. This assumption can be interpreted such that the FIR acoustic impulse responses contained in $\mathbf{H}_{[L]}$ do not possess any common zeros in the z-domain, which usually holds in practice for a sufficient number of sensors [23]. Another requirement for invertibility of $\mathbf{H}_{[L]}$ is that the number of its rows equals the number of its columns, i.e., $Q(M+L-1) = PL$ according to the dimensions noted above (4.12). From this condition, we immediately obtain the *optimum filter length for inversion* [24]:

$$L_{\text{opt,inv}} = \frac{Q}{P-Q}(M-1). \tag{4.18}$$

An important conclusion of this consideration is that the MIMO mixing system can be inverted exactly even with a finite-length MIMO demixing system, as long as $P > Q$, i.e., the number of sensors is greater than the number of sources. Note that P, Q, M must be such that $L_{\text{opt,inv}}$ is an integer number in order to allow the matrix inversion in (4.17). Otherwise, we have to resort to the LS approximation (4.15) with $L_{\text{opt,inv}} = \lceil Q(M-1)/(P-Q) \rceil$.

Conditions for Ideal Solution of Signal Separation Problems (*Ideal Separation Solution*). The goal of any separation algorithm, such as BSS or conventional beamforming, is to eliminate the crosstalk between the different sources $s_q(n)$, $q = 1, \ldots, Q$, as shown in Fig. 4.1, in the output signals $y_q(n)$, $q = 1, \ldots, Q$ of the demixing system. Disregarding again a potential permutation among the output signals as above, this condition may be expressed in terms of the overall system matrix $\check{\mathbf{C}}$ as

$$\check{\mathbf{C}} - \text{bdiag}\left\{\check{\mathbf{C}}\right\} = \text{boff}\left\{\check{\mathbf{C}}\right\} = \mathbf{0}. \tag{4.19}$$

Here, the operator bdiag$\{\cdot\}$ applied to a block matrix consisting of several submatrices or vectors sets all submatrices or vectors on the off-diagonals to zero. Analogously, the boff$\{\cdot\}$ operation sets all submatrices or vectors on the diagonal to zero.

With the overall system matrix (4.11), the condition for the ideal separation is expressed as

$$\text{boff}\left\{\mathbf{H}_{[L]}\check{\mathbf{W}}\right\} = \mathbf{0}. \tag{4.20}$$

This relation for the ideal solution of the *direct blind adaptive filtering problems* is the analogous expression to the relation (4.14) for the ideal solution of the inverse blind adaptive filtering problems.

As we will see in the next section, the relation (4.20) allows us to

- derive an explicit expression of the ideal separation solution analogously to (4.17);
- establish a link between BSS and BSI;

- establish the conditions for BSI;
- derive the optimum separating FIR filter length $L_{\mathrm{opt,sep}}$ analogously to (4.18) for which the ideal separation solution (4.19) can be achieved.

If we are only interested in separation with certain other constraints to the output signals (e.g., minimal signal distortion between sensor signals and output signals), but not in system identification, we may impose further explicit conditions to the block-diagonal elements of $\mathbf{H}_{[L]}\check{\mathbf{W}}$ in addition to the condition (4.20) on the block-offdiagonals. For instance, the so-called *minimum distortion principle* after [30] may in fact be regarded as such an additional condition. However, since this is not within the scope of system identification we will not discuss these conditions further in this chapter.

4.2 Blind MIMO System Identification and Relation to Blind Source Separation

Traditionally, blind source separation (BSS) has often been considered as an inverse problem. In this section we show that the theoretically ideal convolutive (blind) source separation solution corresponds to blind MIMO system identification. By choosing an appropriate filter length we show that for broadband algorithms the well-known filtering ambiguity can be avoided. Ambiguities in instantaneous BSS algorithms are scaling and permutation [19]. In narrowband convolutive BSS these ambiguities occur independently in each frequency bin so that arbitrary scaling becomes arbitrary filtering, as mentioned above. For additional measures to solve the internal permutation problem appearing independently in each frequency bin, see, e.g., [31] and for the arbitrary filtering, e.g., [30]. On the other hand, broadband time-domain BSS approaches are known to avoid the bin-wise permutation ambiguity. However, traditionally, multichannel blind deconvolution (MCBD) algorithms are often used in the literature [20, 30], which have the drawback of whitening the output signals when applied to acoustic scenarios. Repair measures for this problem have been proposed in [30] (minimum distortion principle) and in [20] (linear prediction). In the following we consider the ideal broadband solution of mere MIMO separation approaches and relate it to the known blind system identification approach based on single-input multiple-output (SIMO) models [15, 25, 26]. This section follows the ideas outlined in [5, 6]. Some of these ideas were also developed independently in [7] in a slightly different way.

This section discusses the ideal separation condition $\mathrm{boff}\left\{\mathbf{H}_{[L]}\check{\mathbf{W}}\right\} = \mathbf{0}$ illustrated in Fig. 4.2 for the case $Q = P = 3$. Since in this equation we impose explicit constraints only on the block-offdiagonal elements of $\check{\mathbf{C}}$, this is equivalent to establishing a set of homogeneous systems of linear equations

$$\mathbf{H}_{(:\backslash q):,[L]}\check{\mathbf{W}}_{:q} = \mathbf{0}, \quad q = 1, \ldots, Q \tag{4.21}$$

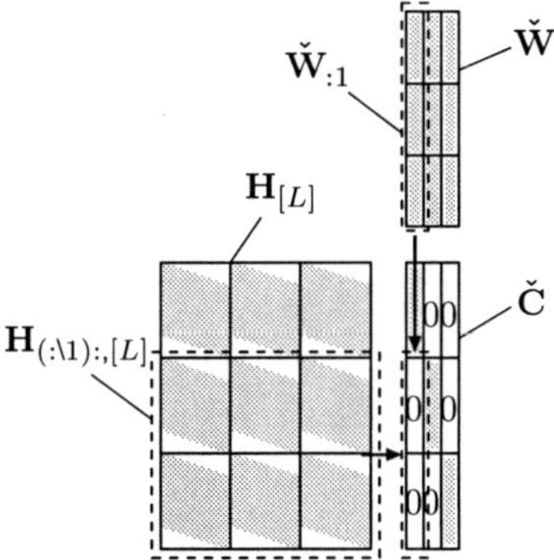

Fig. 4.2. Overall system $\check{\mathbf{C}}$ for the ideal separation, illustrated for $P = Q = 3$.

to be solved. Each of these systems of equations results from the constraints on one column of $\check{\mathbf{C}}$, as illustrated in Fig. 4.2 for the first column. The notation in the indices in (4.21) indicates that for the q-th column $\check{\mathbf{W}}_{:q}$ of the demixing filter matrix $\check{\mathbf{W}}$, we form a submatrix $\mathbf{H}_{(:\backslash q):,[L]}$ of $\mathbf{H}_{[L]}$ by striking out the q-th row $\mathbf{H}_{q:,[L]}$ of Sylvester-submatrices of the original matrix $\mathbf{H}_{[L]}$.

For homogeneous systems of linear equations such as (4.21) it is known that nontrivial solutions $\check{\mathbf{W}}_{:q} \not\equiv \mathbf{0}$ are indeed obtained if the rank of $\mathbf{H}_{(:\backslash q):,[L]}$ is smaller than the number of elements of $\check{\mathbf{W}}_{:q}$. Later in this section, we will also derive an expression of the optimum separation filter length $L_{\text{opt,sep}}$ for an arbitrary number of sensors and sources analogously to the optimum inversion filter length $L_{\text{opt,inv}}$ in (4.18). This derivation will be based on this observation.

In the following subsections, we first discuss the solution of (4.21) for the case $P = Q = 2$, and then generalize the results to more than two sources and sensors.

4.2.1 Square Case for Two Sources and Two Sensors

For the case $Q = P = 2$, the set of homogeneous linear systems of equations (4.21) reads

$$\mathbf{H}_{11,[L]}\mathbf{w}_{12} + \mathbf{H}_{12,[L]}\mathbf{w}_{22} = \mathbf{0}, \tag{4.22a}$$
$$\mathbf{H}_{21,[L]}\mathbf{w}_{11} + \mathbf{H}_{22,[L]}\mathbf{w}_{21} = \mathbf{0}. \tag{4.22b}$$

Since the matrix-vector products in these equations represent convolutions of FIR filters they can equivalently be written as a multiplication in the z-domain:

$$H_{11}(z)W_{12}(z) + H_{12}(z)W_{22}(z) = 0, \tag{4.23a}$$
$$H_{21}(z)W_{11}(z) + H_{22}(z)W_{21}(z) = 0. \tag{4.23b}$$

Due to the FIR filter structure the z-domain representations can be expressed by the zeros $z_{0H_{qp},\nu}$, $z_{0W_{pq},\mu}$ and the gains $A_{H_{qp}}$, $A_{H_{pq}}$ of the filters $H_{qp}(z)$ and $W_{pq}(z)$, respectively:

$$\begin{aligned} A_{H_{11}} \prod_{\nu=1}^{M-1} (z - z_{0H_{11},\nu}) A_{W_{12}} \prod_{\mu=1}^{L-1} (z - z_{0W_{12},\mu}) \\ = -A_{H_{12}} \prod_{\nu=1}^{M-1} (z - z_{0H_{12},\nu}) A_{W_{22}} \prod_{\mu=1}^{L-1} (z - z_{0W_{22},\mu}), \end{aligned} \tag{4.24a}$$

$$\begin{aligned} A_{H_{21}} \prod_{\nu=1}^{M-1} (z - z_{0H_{21},\nu}) A_{W_{11}} \prod_{\mu=1}^{L-1} (z - z_{0W_{11},\mu}) \\ = -A_{H_{22}} \prod_{\nu=1}^{M-1} (z - z_{0H_{22},\nu}) A_{W_{21}} \prod_{\mu=1}^{L-1} (z - z_{0W_{21},\mu}). \end{aligned} \tag{4.24b}$$

Analogously to the case of MINT [23] described in the previous section, we assume that the impulse responses contained in $\mathbf{H}_{(:\backslash q):,[L]}$, i.e., $H_{11}(z)$ and $H_{12}(z)$ in (4.24a) do not share common zeros. In the same way, we assume that $H_{21}(z)$ and $H_{22}(z)$ in (4.24b) do not share common zeros. If no common zeros exist and if we choose the *optimum filter length for the case* $Q = P = 2$ as $L_{\mathrm{opt,sep}} = M$, then the equality in (4.24a) can only hold if the zeros of the demixing filters are chosen as $z_{0W_{12},\mu} = z_{0H_{12},\mu}$ and $z_{0W_{22},\mu} = z_{0H_{11},\mu}$ for $\mu = 1, \ldots, M-1$. Analogously, the equality in (4.24b) can only hold if $z_{0W_{11},\mu} = z_{0H_{22},\mu}$ and $z_{0W_{21},\mu} = z_{0H_{21},\mu}$ for $\mu = 1, \ldots, M-1$. Additionally, to fulfill the equality, the gains of the demixing filters in (4.24a) have to be chosen as $A_{W_{22}} = \alpha_2 A_{H_{11}}$ and $A_{W_{12}} = -\alpha_2 A_{H_{12}}$, where α_2 is an arbitrary scalar constant. Thus, the demixing filters are only determined up to a scalar factor α_2. Analogously, for the equality (4.24b) the gains of the demixing filters are given as $A_{W_{11}} = \alpha_1 A_{H_{22}}$ and $A_{W_{21}} = -\alpha_1 A_{H_{21}}$ with the scalar constant α_1.

In summary, this leads to the ideal separating filter matrix $\check{\mathbf{W}}_{\mathrm{ideal,sep}}$ given in the time domain as

$$\check{\mathbf{W}}_{\mathrm{ideal,sep}} = \begin{bmatrix} \alpha_1 \mathbf{h}_{22} & -\alpha_2 \mathbf{h}_{12} \\ -\alpha_1 \mathbf{h}_{21} & \alpha_2 \mathbf{h}_{11} \end{bmatrix} = \begin{bmatrix} \mathbf{h}_{22} & -\mathbf{h}_{12} \\ -\mathbf{h}_{21} & \mathbf{h}_{11} \end{bmatrix} \begin{bmatrix} \alpha_1 & 0 \\ 0 & \alpha_2 \end{bmatrix}, \tag{4.25}$$

where due to the scaling ambiguity each column is multiplied by an unknown scalar α_q.

From (4.25) we see that under the conditions put on the zeros of the mixing system in the z-domain, and for $L = L_{\mathrm{opt,sep}}$, this *ideal separation solution corresponds to a MIMO system identification up to an arbitrary scalar constant.* Thus, a suitable algorithm which is able to perform *broadband* BSS *under these conditions* can be used for blind MIMO system identification (if the source signals provide sufficient spectral and temporal support for exciting the mixing system). In Sect. 4.3, we present such a suitable algorithmic framework for this task. Moreover, as we will see in the following subsection, this approach may be seen as a generalization of the state-of-the-art method for the blind identification of SIMO systems.

Finally, since we did not impose an explicit constraint on the block-diagonal elements of the overall system $\check{\mathbf{C}}$ in the original separation condition (4.20), we are now interested in the resulting overall system in the case of the ideal separating solution (4.25). By inserting this solution into (4.11), we readily obtain

$$\begin{aligned}\check{\mathbf{C}}_{\text{ideal,sep}} &= \mathbf{H}_{[M]}\check{\mathbf{W}}_{\text{ideal,sep}} \\ &= \begin{bmatrix} \mathbf{H}_{11,[M]} & \mathbf{H}_{12,[M]} \\ \mathbf{H}_{21,[M]} & \mathbf{H}_{22,[M]} \end{bmatrix} \begin{bmatrix} \alpha_1 \mathbf{h}_{22} & -\alpha_2 \mathbf{h}_{12} \\ -\alpha_1 \mathbf{h}_{21} & \alpha_2 \mathbf{h}_{11} \end{bmatrix} \\ &= \begin{bmatrix} \alpha_1 \left(\mathbf{H}_{11,[M]}\mathbf{h}_{22} - \mathbf{H}_{12,[M]}\mathbf{h}_{21}\right) & \alpha_2 \left(\mathbf{H}_{12,[M]}\mathbf{h}_{11} - \mathbf{H}_{11,[M]}\mathbf{h}_{12}\right) \\ \alpha_1 \left(\mathbf{H}_{12,[M]}\mathbf{h}_{22} - \mathbf{H}_{22,[M]}\mathbf{h}_{21}\right) & \alpha_2 \left(\mathbf{H}_{22,[M]}\mathbf{h}_{11} - \mathbf{H}_{21,[M]}\mathbf{h}_{12}\right) \end{bmatrix} \\ &= \begin{bmatrix} \alpha_1 \left(\mathbf{H}_{11,[M]}\mathbf{h}_{22} - \mathbf{H}_{12,[M]}\mathbf{h}_{21}\right) & \mathbf{0} \\ \mathbf{0} & \alpha_2 \left(\mathbf{H}_{22,[M]}\mathbf{h}_{11} - \mathbf{H}_{21,[M]}\mathbf{h}_{12}\right) \end{bmatrix}, \end{aligned} \tag{4.26}$$

where in the last line the commutativity of the convolution has been exploited so that the crosstalk between the channels is cancelled out perfectly. The output signals of the overall system are filtered (but not arbitrarily filtered) versions of the original source signals.

4.2.2 Relation to SIMO System Identification

BSS algorithms aiming at the ideal solution (4.25) can be interpreted as a generalization of the popular class of blind SIMO system identification approaches, e.g., [25, 26, 32], as illustrated in Fig. 4.3 (a). The main reason for the popularity of this SIMO approach is that it can be implemented as a relatively simple least-squares error minimization. From Fig. 4.3 (a) and for $e(n) = 0$ it follows for sufficient excitation $s(n)$ that

$$h_1(n) * w_1(n) = -h_2(n) * w_2(n). \tag{4.27}$$

This can be expressed in the z-domain as $H_1(z)W_1(z) = -H_2(z)W_2(z)$. Comparing this error cancelling condition with the ideal separation conditions (4.23a)/(4.23b), we immediately see that the SIMO-based approach indeed corresponds exactly to one of the separation conditions, and for deriving the ideal solution, we may apply exactly the same reasoning as in the MIMO case above. Thus, assuming that $H_1(z)$ and $H_2(z)$ have no common zeros, the equality of (4.27) can only hold if the filter length is chosen again as $L = M$. Then, this leads to the ideal cancellation filters $W_1(z) = \alpha H_2(z)$ and $W_2(z) = -\alpha H_1(z)$ which can be determined up to an arbitrary scaling by the factor α as in the MIMO case. For $L > M$ the scaling ambiguity would

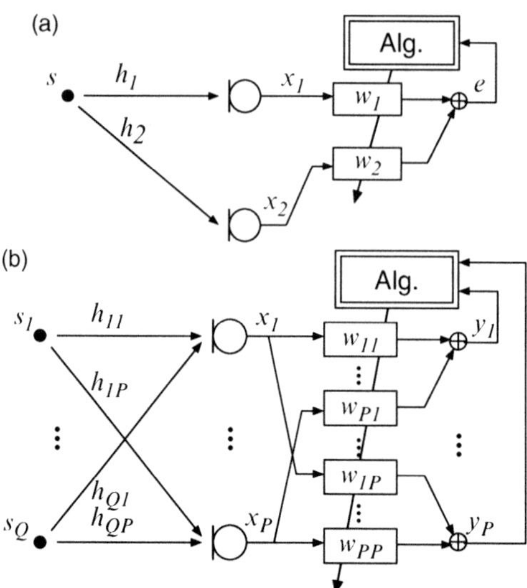

Fig. 4.3. Blind system identification based on (a) SIMO and (b) MIMO models.

result in arbitrary *filtering*. For the SIMO case this scaling ambiguity was derived similarly in [26].

Note that the SIMO case may also be interpreted as a special 2×2 MIMO case according to Fig. 4.3 (b) with the specialization being that one of the sources is always identical to zero so that the BSS output corresponding to this (virtual) source also must always be identical to zero, whereas the other BSS output signal is not of interest in this case. This leads again to the cancellation condition (4.27), and illustrates that the relation between broadband BSS and SIMO-based BSI will also hold from an algorithmic point of view, i.e., known adaptive solutions for SIMO BSI may also be derived as special cases of the algorithmic framework for the MIMO case.

Adaptive algorithms performing the error minimization mentioned above for the SIMO structure have been proposed in the context of blind deconvolution, e.g., in [25, 26] and for blind identification used for passive source localization, e.g., in [15, 16]. It is also known as the *adaptive eigenvalue decomposition* (AED) algorithm. This name comes from the fact that in the SIMO case, the homogeneous system of equations (4.21) may be reformulated into an analogous signal-dependent homogeneous system of equations containing the sensor-signal correlation matrix instead of the mixing filter matrix. The solution vector (in the SIMO case the matrix $\check{\mathbf{W}}$ reduces to a vector) of the homogeneous system can then be interpreted as the *eigenvector corresponding to the zero-valued (or smallest) eigenvalue* of the sensor correlation matrix. In [16, 25] this SIMO approach, i.e., the single-source case, was also generalized to more than two microphone channels.

4.2.3 Ideal Separation Solution in the General Square Case for More than Two Sources and Sensors

The factorized formulation of the ideal separation solution for the case $Q = P = 2$ in the second part of (4.25) suggests that it may be expressed more generally by the adjoint of the matrix $\check{\mathbf{H}}$ where the entries $\mathbf{h}_{qp}$ are treated like scalar values. We formalize this operation and call it the *block-adjoint operator* $\mathrm{badj}_P\{\cdot\}$, where the index P denotes the number of submatrices in each row or column of the argument. Using the block-adjoint operator the general form of (4.25) for an arbitrary number $Q = P$ of sensors or sources reads [5]

$$\check{\mathbf{W}}_{\mathrm{ideal,sep}} = \mathrm{badj}_P\left\{\check{\mathbf{H}}\right\}\ \boldsymbol{\Lambda}_\alpha, \tag{4.28}$$

where the diagonal matrix $\boldsymbol{\Lambda}_\alpha = \mathrm{Diag}\left\{[\alpha_1, \ldots, \alpha_P]^{\mathrm{T}}\right\}$ again describes the scaling ambiguity. Note that the size of $\mathrm{badj}_P\left\{\check{\mathbf{H}}\right\}$ is determined for $P, Q > 2$ by the internal convolutions of the FIR filters contained in $\check{\mathbf{H}}$. We may easily verify that the resulting size after the convolutions is $[P(P-1)(M-1)+1]\times P$.

To verify that the approach (4.28) is indeed the ideal separating solution for $P, Q \geq 2$, we may calculate the overall system matrix as in the 2×2 case above. Extending the well-known property of the conventional adjoint of a square matrix $\mathbf{A}$

$$\mathbf{A}\mathrm{adj}\left\{\mathbf{A}\right\} = \det\left\{\mathbf{A}\right\} \cdot \mathbf{I} = \mathrm{Diag}\left\{\det\left\{\mathbf{A}\right\}, \ldots, \det\left\{\mathbf{A}\right\}\right\} \tag{4.29}$$

to the analogous formulation of the block-adjoint it may be shown that

$$\begin{aligned} \check{\mathbf{C}}_{\mathrm{ideal,sep}} &= \mathbf{H}_{[L]}\mathrm{badj}_P\left\{\check{\mathbf{H}}\right\}\ \boldsymbol{\Lambda}_\alpha \\ &= \mathrm{Bdiag}\left\{\mathrm{bdet}_P\left\{\check{\mathbf{H}}\right\}, \ldots, \mathrm{bdet}_P\left\{\check{\mathbf{H}}\right\}\right\}\ \boldsymbol{\Lambda}_\alpha, \end{aligned} \tag{4.30}$$

where the operator $\mathrm{bdet}_P\left\{\check{\mathbf{H}}\right\}$ denotes a *block-determinant operator* on the mixing system. Similarly to the block-adjoint, the block-determinant operator generalizes the conventional determinant operator so that we work with submatrices as its entries rather than scalar values. Thus, in contrast to the conventional determinant the block-determinant $\mathrm{bdet}_P\left\{\check{\mathbf{H}}\right\}$ is still a matrix and can be interpreted as a MIMO system with FIR filters of length $P(M-1)+1$ due to the P internal convolutions. The submatrices on the block-diagonal in (4.26) represent this operation for the 2×2 case.

4.2.4 Ideal Separation Solution and Optimum Separating Filter Length for an Arbitrary Number of Sources and Sensors

As mentioned above, for homogeneous systems of linear equations such as the ideal separation conditions (4.21) it is known that nontrivial solutions $\check{\mathbf{W}}_{:q} \neq \mathbf{0}$ are obtained if the rank of $\mathbf{H}_{(:\backslash q):,[L]}$ is smaller than the number of elements of $\check{\mathbf{W}}_{:q}$. Additionally, as in the case of MINT [23] described in the previous section, we assume that the impulse responses contained in $\mathbf{H}_{(:\backslash q):,[L]}$ do not share common zeros in the z-domain so that $\mathbf{H}_{(:\backslash q):,[L]}$ is assumed to have full row rank. Thus, combining these conditions leads to the requirement that the matrix $\mathbf{H}_{(:\backslash q):,[L]}$ is *wide*, i.e., the number PL of its columns must be greater than the number $(Q-1)(M+L-1)$ of its rows to obtain nontrivial solutions, i.e., $PL > (Q-1)(M+L-1)$. Solving this inequality for L yields the lower bound for the separating filter length as

$$L_{\text{sep}} > \frac{Q-1}{P-Q+1}(M-1). \tag{4.31}$$

The difference between the number of columns of $\mathbf{H}_{(:\backslash q):,[L]}$ and the number of rows further specifies the dimension of the space of possible nontrivial solutions $\check{\mathbf{W}}_{:q}$, i.e., the number of linearly independent solutions spanning the solution space. Obviously, due to the bound derived above, the best choice we can make to narrow down the solutions is a one-dimensional solution space, i.e., $PL = (Q-1)(M+L-1)+1$. Solving now this *equality* for L and choosing the integer value to be strictly larger than the above bound finally results in the *optimum separating filter length* as

$$L_{\text{opt,sep}} = \frac{(Q-1)(M-1)+1}{P-Q+1}. \tag{4.32}$$

Note that narrowing down the solution space to a one-dimensional space by this choice of filter length precisely means that in this case the *filtering ambiguity of BSS reduces to an arbitrary scaling.* These considerations show that this is possible even for an arbitrary number P of sensors and an arbitrary number Q of sources, where $P \geq Q$. However, the parameters P, Q, M must be such that $L_{\text{opt,sep}}$ is an integer number in order to allow the ideal separation solution. Otherwise, we have to resort to approximations by choosing, e.g., $L_{\text{opt,sep}} = \lceil [(Q-1)(M-1)+1]/(P-Q+1) \rceil$.

To actually obtain the ideal separation solution $\check{\mathbf{W}}_{\text{ideal,sep}}$ with (4.32) for the general, i.e., not necessarily square case $P \geq Q$, we may not straightforwardly apply the block-adjoint and block-determinant operators introduced in the previous subsection. We therefore consider again the original set of homogeneous systems of linear equations (4.21). For the choice $L = L_{\text{opt,sep}}$, we may easily augment the matrix $\mathbf{H}_{(:\backslash q):,[L]}$ to a square matrix $\tilde{\mathbf{H}}_{(:\backslash q):,[L]}$ by adding one row of zeros on both sides of (4.21). The corresponding augmented set of linear systems of equations

$$\tilde{\mathbf{H}}_{(:\backslash q):,[L]} \check{\mathbf{W}}_{:q} = \mathbf{0}, \quad q = 1, \ldots, Q \tag{4.33}$$

is equivalent to the original set (4.21). However, we may now express the *general solution vector* $\check{\mathbf{W}}_{:q}$ *of (4.21) for the q-th column of* $\check{\mathbf{W}}$ as *the eigenvector corresponding to the zero-valued eigenvalue of the augmented matrix* $\tilde{\mathbf{H}}_{(:\backslash q):,[L]}$.

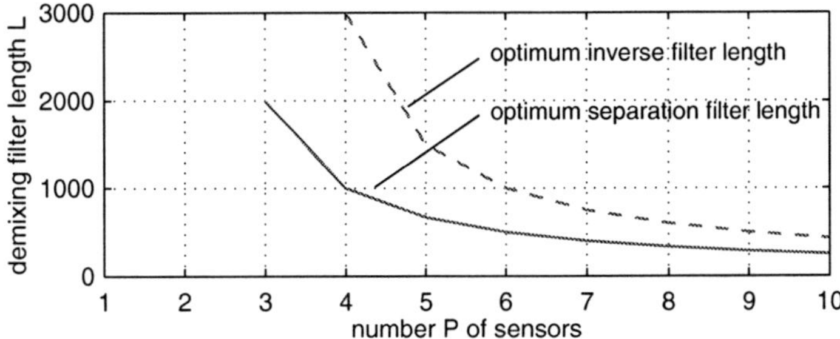

Fig. 4.4. Comparison of the optimum filter lengths for $M = 1000$ and $Q = 3$.

The general equation (4.32) for the optimum separation filter length is the expression which is analogous to the optimum inverse filter length considered earlier in (4.18). Comparing these two equations, we can verify that in contrast to the inversion, which requires $P > Q$ for the ideal solution using FIR filters, the ideal separation condition is already possible for $P = Q$. Moreover, for the special case $P = Q = 2$, the general expression (4.32) also confirms the choice $L_{\mathrm{opt,BSS}} = M$ as already obtained in Sect. 4.2.1. Figure 4.4 illustrates the different optimum filter lengths by an example.

In practice it is obviously difficult to choose the optimum filter length in the blind applications precisely since the length M of the mixing system is generally unknown. Moreover, in many applications we do not require a complete identification of all reflections within the mixing system but only of the dominant ones (e.g., in the localization application considered in later in this chapter). Fortunately, this is in line with the above-mentioned requirement to avoid an overestimation of the filter length in order to narrow down the solution space, i.e., to prevent the filtering ambiguity. Thus, in any case, the choice $L \leq L_{\mathrm{opt,sep}}$ is preferable in practice.

4.2.5 General Scheme for Blind System Identification

In the previous Sects. 4.2.1 and 4.2.2 we have explicitly shown the relation between the ideal separation solution and the mixing system for the two-sensor cases. These considerations did also result in a link to the well-known SIMO-based system identification method (note that for BSI with more than two sensors, one simple approach is to apply several of these schemes in parallel), and also to a generalization of this method to the MIMO case with two simultaneously active sources. In the case of more than two sources we may not directly pick the estimated mixing system coefficients from the separation solution $\check{\mathbf{W}}$. The previous Sects. 4.2.3 and 4.2.4 generalized the considerations on the two-sensor cases for the *separation* task. In this section, we now outline the generalization of the two-sensor cases in Sects. 4.2.1 and 4.2.2 for the *identification* task. The considerations so far suggest the following generic *two-step BSI scheme for an arbitrary number of sources* (where $P \geq Q$):

(1) Based on the available sensor signals, perform a properly designed broadband BSS (see Sect. 4.3) resulting in an estimate of the demixing system matrix.
(2) Analogously to the relation (4.21) between the mixing and demixing systems, and the associated considerations in Sect. 4.2.4 for the separation task, determine an *estimate of the mixing system matrix* using the estimated demixing system from the first step.

In general, to perform step (2) for more than two sources, some further considerations are required. First, an equivalent reformulation of the homogeneous system of equations (4.33) is necessary so that now the *demixing system matrix* instead of the mixing system matrix is formulated *as a blockwise Sylvester matrix*. Note that this corresponds to a *block-transposition* (which we denote here by superscript $\cdot^{\mathrm{bT}}$) of (4.21), i.e.,

$$\left(\mathbf{W}^{\mathrm{bT}}\right)_{(:\backslash q):,[M]} \left(\check{\mathbf{H}}^{\mathrm{bT}}\right)_{:q} = \mathbf{0}, \quad q = 1, \ldots, Q. \tag{4.34}$$

The block-transposition is an extension of the conventional matrix transposition. It means that we keep the original form of the channel-wise submatrices but we may change the order of the mixing and demixing subfilters by exploiting the commutativity of the convolutions similarly as in (4.26). Note that the commutativity property does not hold for the MIMO system matrices as a whole, i.e., $\mathbf{W}_{(:\backslash q):,[M]}$ and $\check{\mathbf{H}}_{:q}$, so that they have to be block-transposed to change their order.

Similarly to Sect. 4.2.4, we may then calculate the corresponding estimate of the mixing system in terms of eigenvectors using the complementary form (4.34) of the homogeneous system of equations. Based on this system of equations, we can devise various powerful strategies for BSI in the general MIMO case.

4.2.6 Summary

We have defined and analyzed the signal separation and deconvolution problems using clear conditions for the involved linear mixing and demixing systems. For both problems, the ideal demixing filter coefficients have been derived. Thereby, signal separation was classified as a direct blind adaptive filtering problem, which is in contrast to deconvolution as an inverse adaptive filtering problem. Under certain conditions, such as a suitable filter length for the demixing system and sufficient excitation by the source signals, blind MIMO system identification can be achieved by blind signal separation. In this case, the solutions are unique up to a scaling factor (and a possible permutation of the output channels). From this uniqueness and the correspondence between BSS and BSI we can draw the conclusions that (1) arbitrary filtering may be prevented with broadband approaches, (2) the known whitening problem is avoided, and (3) the BSS framework also allows for several new applications, such as simultaneous localization of multiple sources, as shown later in this chapter.

4.3 TRINICON – A General Framework for Adaptive MIMO Signal Processing and Application to the Blind Adaptation Problems

For the blind estimation of the coefficients corresponding to the desired solutions discussed in the previous section, we have to consider and to exploit the properties of the excitation signals, such as their nonstationarity, their spectral characteristics, and their probability densities.

In the existing literature, the BSS problem has mostly been addressed for instantaneous mixtures or narrowband approaches in the frequency domain which adapt the coefficients independently in each DFT bin, e.g., [19, 33, 34]. On the other hand, in the case of MCBD, many approaches either aim at whitening the output signals as they are based on an i.i.d. model of the source signals (e.g., [27, 28]), which is undesirable for speech and audio signals which should not be whitened, or are rather heuristically motivated, e.g., [29].

The aim of this section is to present an overview of the algorithmic part of broadband blind adaptive MIMO filtering based on TRINICON ('TRIple-N Independent component analysis for CONvolutive mixtures'), a generic concept for adaptive MIMO filtering which takes all the above-mentioned signal properties (nonwhiteness, nonstationarity, and non–Gaussianity) into account, and allows a unified treatment of broadband BSS as needed for a proper BSI, and MCBD algorithms applicable to speech and audio signals in real acoustic environments [1–4]. This framework generally uses multivariate stochastic signal models in the cost function to describe the temporal structure of the source signals. This versatile approach provides a powerful cost function for both, BSS/BSI and MCBD, and, for the latter, also leads to improved solutions for speech dereverberation.

As in the previous sections, we will again mainly focus on the direct blind adaptive filtering problems, such as BSS. In [4], a direct relation between the BSS adaptation mechanism and the ideal separation solution (4.20) was established. Moreover, although both time-domain and equivalent broadband frequency-domain formulations of TRINICON have been developed with the corresponding multivariate models in both the time domain and the frequency domain [2, 4], we consider in this chapter mainly the time-domain formulation. We discuss here only gradient-based coefficient updates for clarity of presentation. The algorithmic TRINICON framework is directly based on the matrix notation developed above.

Throughout this section, we regard the standard BSS model where the number Q of *maximum simultaneously active source signals* $s_q(n)$ is equal to the number of sensor signals $x_p(n)$, i.e., $Q = P$. However, it should be noted that in contrast to other BSS algorithms we do not assume prior knowledge about the exact number of active sources. Thus, even if the algorithms will be derived for $Q = P$, the number of simultaneously active sources may change throughout the application of the BSS algorithm and only the condition $Q \leq P$ has to be fulfilled.

4.3.1 Cost Function and Gradient-Based Coefficient Optimization

Matrix Notation for Convolutive Mixtures. To introduce an algorithm for broadband processing of convolutive mixtures, we first need to formulate the convolution of the FIR demixing system of length L in the following matrix form [4]:

$$\mathbf{y}^{\mathrm{T}}(n) = \mathbf{x}^{\mathrm{T}}(n)\mathbf{W}, \tag{4.35}$$

where n denotes the time index, and

$$\mathbf{x}^{\mathrm{T}}(n) = [\mathbf{x}_1^{\mathrm{T}}(n), \ldots, \mathbf{x}_P^{\mathrm{T}}(n)], \tag{4.36}$$

$$\mathbf{y}^{\mathrm{T}}(n) = [\mathbf{y}_1^{\mathrm{T}}(n), \ldots, \mathbf{y}_P^{\mathrm{T}}(n)], \tag{4.37}$$

$$\mathbf{W} = \begin{bmatrix} \mathbf{W}_{11} & \cdots & \mathbf{W}_{1P} \\ \vdots & \ddots & \vdots \\ \mathbf{W}_{P1} & \cdots & \mathbf{W}_{PP} \end{bmatrix}, \tag{4.38}$$

$$\mathbf{x}_p^{\mathrm{T}}(n) = [x_p(n), \ldots, x_p(n - 2L + 1)], \tag{4.39}$$

$$\mathbf{y}_q^{\mathrm{T}}(n) = [y_q(n), \ldots, y_q(n - D + 1)] \tag{4.40}$$

$$= \sum_{p=1}^{P} \mathbf{x}_p^{\mathrm{T}}(n)\mathbf{W}_{pq}. \tag{4.41}$$

The parameter D in (4.40), $1 \leq D < L$, denotes the number of lags taken into account to exploit the nonwhiteness of the source signals as shown below. $\mathbf{W}_{pq}$, $p = 1, \ldots, P$, $q = 1, \ldots, P$ denote $2L \times D$ Sylvester matrices that contain all coefficients of the respective filters:

$$\mathbf{W}_{pq} = \begin{bmatrix} w_{pq,0} & 0 & \cdots & 0 \\ w_{pq,1} & w_{pq,0} & \ddots & \vdots \\ \vdots & w_{pq,1} & \ddots & 0 \\ w_{pq,L-1} & \vdots & \ddots & w_{pq,0} \\ 0 & w_{pq,L-1} & \ddots & w_{pq,1} \\ \vdots & & \ddots & \vdots \\ 0 & \cdots & 0 & w_{pq,L-1} \\ 0 & \cdots & 0 & 0 \\ \vdots & & \vdots & \vdots \\ 0 & \cdots & 0 & 0 \end{bmatrix}. \tag{4.42}$$

Note that for $D = 1$, (4.35) simplifies to the well-known vector formulation of a convolution, as it is used extensively in the literature on supervised adaptive filtering, e.g., [18].

Optimization Criterion. Various approaches exist to blindly estimate the demixing matrix $\mathbf{W}$ for the above-mentioned tasks by utilizing the following source signal properties [19] which we all combine into an efficient and versatile algorithmic framework [1–3]:

(i) Non-Gaussianity is exploited by using higher-order statistics for independent component analysis (ICA). ICA approaches can be divided into several classes. Although they all lead to similar update rules, the minimization of the mutual information (MMI) among the output channels can be regarded as the most general approach for BSS [19]. To obtain an estimator not only allowing spatial separation but also temporal separation for MCBD, we use the Kullback–Leibler divergence (KLD) [35] between a certain *desired* joint pdf (essentially representing a hypothesized stochastic source model) and the joint pdf of the actually estimated output signals. The desired pdf is factorized w.r.t. the different sources (for BSS) and possibly also w.r.t. certain temporal dependencies (for MCBD) as shown below. The KLD is guaranteed to be positive [35], which is a necessary condition for a useful cost function.

(ii) Nonwhiteness is exploited by simultaneous minimization of output cross-relations over multiple time-lags. We therefore consider multivariate pdfs, i.e., 'densities including D time-lags'.

(iii) Nonstationarity is exploited by simultaneous minimization of output cross-relations at different time-instants. We assume ergodicity within blocks of length N so that the ensemble average is replaced by time averages over these blocks.

Based on the KLD, we now define the following general cost function taking into account all three fundamental signal properties (i)–(iii):

$$\mathcal{J}(m, \mathbf{W}) = -\sum_{i=0}^{\infty} \beta(i, m) \frac{1}{N} \sum_{j=iL}^{iL+N-1} \left\{ \log(\hat{p}_{s,PD}(\mathbf{y}(j))) - \log(\hat{p}_{y,PD}(\mathbf{y}(j))) \right\}, \tag{4.43}$$

where $\hat{p}_{s,PD}(\cdot)$ and $\hat{p}_{y,PD}(\cdot)$ are the assumed or estimated PD-variate source model (i.e., desired) pdf and output pdf, respectively. The index m denotes the block time index for a block of N output samples shifted by L samples relatively to the previous block. Furthermore, D is the memory length, i.e., the number of time-lags to model the nonwhiteness of the P signals as above. β is a window function with finite support that is normalized so that $\sum_{i=0}^{m} \beta(i, m) = 1$, allowing for online, offline, and block-online algorithms [2, 36].

Gradient-Based Coefficient Update. In this chapter we concentrate on iterative gradient-based block-online coefficient updates which can be written in the general form

$$\check{\mathbf{W}}^0(m) := \check{\mathbf{W}}(m-1), \tag{4.44a}$$

$$\check{\mathbf{W}}^\ell(m) = \check{\mathbf{W}}^{\ell-1}(m) - \mu \Delta \check{\mathbf{W}}^\ell(m), \quad \ell = 1, \ldots, \ell_{\max}, \tag{4.44b}$$

$$\check{\mathbf{W}}(m) := \check{\mathbf{W}}^{\ell_{\max}}(m), \tag{4.44c}$$

where μ is a step-size parameter, and the superscript index ℓ denotes an iteration parameter to allow for multiple iterations ($\ell = 1, \ldots, \ell_{\max}$) within each block m. The $LP \times P$ coefficient matrix $\check{\mathbf{W}}$ (defined in (4.4)) to be optimized is smaller than the $2LP \times DP$ Sylvester matrix $\mathbf{W}$ used above for the formulation of the cost function, and it contains only the nonredundant elements of $\mathbf{W}$.

The simplest case of the above procedure (4.44a)–(4.44c) is the gradient descent update, which is defined by

$$\Delta \check{\mathbf{W}}^\ell(m) = \nabla_{\check{\mathbf{W}}} \mathcal{J}(m, \mathbf{W})|_{\check{\mathbf{W}} = \check{\mathbf{W}}^\ell(m)}. \tag{4.45}$$

Obviously, when calculating this gradient explicitly, we are confronted with the problem of the different coefficient matrix formulations $\mathbf{W}$ and $\check{\mathbf{W}}$ in the cost function and in the optimization procedure, respectively. This is a direct consequence of taking into account the nonwhiteness signal property, as mentioned above, and – although it may seem less obvious at this point – it leads to an important building block whose actual implementation is fundamental to the properties of the resulting algorithm, the so-called *Sylvester constraint* ($\mathcal{SC}$) on the coefficient update [2, 4]. Using the Sylvester constraint operator the gradient descent update (4.45) can be rewritten as

$$\Delta \check{\mathbf{W}}^\ell(m) = \mathcal{SC}\left\{ \nabla_{\mathbf{W}} \mathcal{J}(m, \mathbf{W}) \right\}|_{\mathbf{W} = \mathbf{W}^\ell(m)}. \tag{4.46}$$

Depending on the particular realization of $(\mathcal{SC})$, we are able to select both, well known and also novel improved adaptation algorithms [36]. As discussed in [36] there are two particularly simple and popular realizations of $(\mathcal{SC})$ leading to two different classes of algorithms:

(1) Computing only the *first column* of each channel of the update matrix to obtain the new coefficient matrix $\check{\mathbf{W}}$. This method is denoted as $(\mathcal{SC}_{\mathrm{C}})$.
(2) Computing only the *L-th row* of each channel of the update matrix to obtain the new coefficient matrix $\check{\mathbf{W}}$. This method is denoted as $(\mathcal{SC}_{\mathrm{R}})$.

It can be shown that in both cases the update process is significantly simplified [36]. However, in general, both choices require some trade-off in the algorithm performance. While $\mathcal{SC}_{\mathrm{C}}$ may provide a potentially more robust convergence behavior, it will not work for arbitrary source positions (e.g., in the case of two sources, they are required to be located in different half-planes w.r.t. the orientation of the microphone array), which is in contrast to the more versatile $\mathcal{SC}_{\mathrm{R}}$ [36]. Note that the choice of $\mathcal{SC}$ also determines the appropriate coefficient initialization [36].

Next, in this chapter, we derive a novel *generic* Sylvester constraint to further formalize and clarify this concept.

Let $W_{kj}^{KJ} = [\mathbf{W}]_{kj}^{KJ}$ denote the kj-th component of the *Sylvester matrix* after (4.42) for the KJ-th channel corresponding to the KJ-th submatrix in (4.38). According to [2, 4], the gradient of $\mathcal{J}$ w.r.t. these components is transformed by a certain choice of $(\mathcal{SC})$ to the gradient w.r.t. the components $\check{W}_m^{MN} = \left[\check{\mathbf{W}}\right]_m^{MN}$ of the *condensed matrix* as used above in (4.45). This can be expressed concisely by applying the chain rule for matrix derivatives in the following form:

$$
\begin{aligned}
\frac{\partial \mathcal{J}}{\partial \check{W}_m^{MN}} &= \sum_{k,j,K,J} \frac{\partial \mathcal{J}}{\partial W_{kj}^{KJ}} \frac{\partial W_{kj}^{KJ}}{\partial \check{W}_m^{MN}} \\
&= \sum_{k,j,K,J} \frac{\partial \mathcal{J}}{\partial W_{kj}^{KJ}} \delta_{KM} \delta_{JN} \delta_{k,(m+j-1)} \\
&= \sum_{k,j} \frac{\partial \mathcal{J}}{\partial W_{kj}^{MN}} \delta_{k,(m+j-1)},
\end{aligned} \tag{4.47}
$$

where

$$
\delta_{ij} = \begin{cases} 1 & \text{for } i = j \\ 0 & \text{for } i \neq j \end{cases} \tag{4.48}
$$

denotes the Kronecker symbol. Hence, we have the simple linear relation

$$
[\nabla_{\check{\mathbf{W}}} \mathcal{J}]_m^{MN} = \sum_{k,j} [\nabla_{\mathbf{W}} \mathcal{J}]_{kj}^{MN} \, \delta_{k,(m+j-1)} \tag{4.49}
$$

between the MN-th submatrices of $\nabla_{\mathbf{W}}\mathcal{J}$ and $\nabla_{\check{\mathbf{W}}}\mathcal{J}$. If we consider now for illustration of (4.49) the individual elements of $\nabla_{\check{\mathbf{W}}}\mathcal{J}$ for one channel in more detail, i.e.,

$$
\begin{aligned}
[\nabla_{\check{\mathbf{W}}}\mathcal{J}]_1^{MN} &= \sum_j [\nabla_{\mathbf{W}}\mathcal{J}]_{jj}^{MN} \\
[\nabla_{\check{\mathbf{W}}}\mathcal{J}]_2^{MN} &= \sum_{k,j} [\nabla_{\mathbf{W}}\mathcal{J}]_{kj}^{MN}\,\delta_{k,(j+1)} = \sum_j [\nabla_{\mathbf{W}}\mathcal{J}]_{j+1,j}^{MN} \\
&\vdots \\
[\nabla_{\check{\mathbf{W}}}\mathcal{J}]_L^{MN} &= \sum_{k,j} [\nabla_{\mathbf{W}}\mathcal{J}]_{kj}^{MN}\,\delta_{k,(j+L-1)} = \sum_j [\nabla_{\mathbf{W}}\mathcal{J}]_{j+L-1,j}^{MN},
\end{aligned}
$$

we can readily see that the generic Sylvester constraint corresponds – up to the constant D denoting the width of the submatrices – to a *channel-wise arithmetic averaging* of elements according to Fig. 4.5.

Note that the previously introduced approaches, classified by the choice $(\mathcal{SC}_\mathrm{C})$ or $(\mathcal{SC}_\mathrm{R})$ as mentioned above, thus correspond to certain approximations by neglecting some of the elements within this averaging process, as illustrated in Fig. 4.6.

Natural Gradient-Based Coefficient Update. It can be shown (after a somewhat tedious but straightforward derivation) that by taking the *natural gradient* [19] of $\mathcal{J}(m)$ with respect to the demixing filter matrix $\mathbf{W}(m)$ [4],

$$
\Delta\check{\mathbf{W}} \propto \mathcal{SC}\left\{\mathbf{W}\mathbf{W}^\mathrm{T}\frac{\partial\mathcal{J}}{\partial\mathbf{W}}\right\}, \tag{4.50}
$$

we obtain the following generic TRINICON-based update rule:

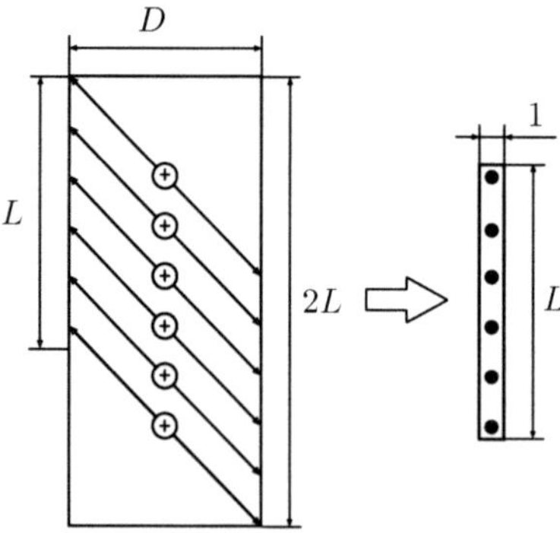

Fig. 4.5. Illustration of the generic Sylvester constraint $(\mathcal{SC})$ for one channel, i.e., the MN-th submatrix $[\nabla_{\check{\mathbf{W}}}\mathcal{J}]^{MN}$ of $\nabla_{\check{\mathbf{W}}}\mathcal{J}$.

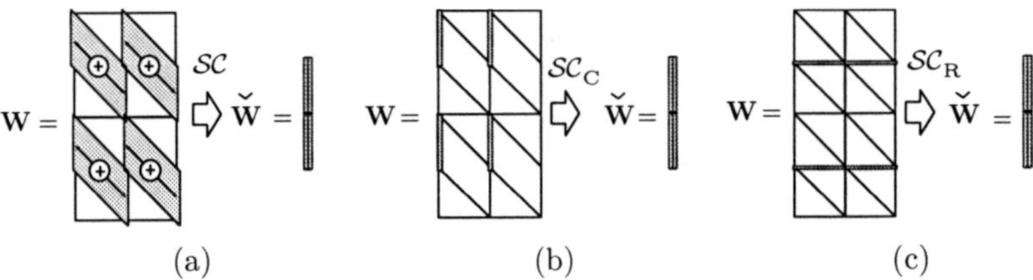

Fig. 4.6. Illustration of two efficient approximations of (a) the generic Sylvester constraint $\mathcal{SC}$; (b) the column Sylvester constraint $\mathcal{SC}_{\mathrm{C}}$; and (c) the row Sylvester constraint $\mathcal{SC}_{\mathrm{R}}$.

$$\check{\mathbf{W}}(m) = \check{\mathbf{W}}(m-1) - \mu \Delta \check{\mathbf{W}}(m), \tag{4.51a}$$

$$\Delta \check{\mathbf{W}}(m) = \frac{1}{N} \sum_{i=0}^{\infty} \beta(i,m)\, \mathcal{SC} \left\{ \sum_{j=iL}^{iL+N-1} \mathbf{W}(i)\mathbf{y}(j) \right. \\ \left. \cdot \left[\boldsymbol{\Phi}_{s,PD}^{\mathrm{T}}(\mathbf{y}(j)) - \boldsymbol{\Phi}_{y,PD}^{\mathrm{T}}(\mathbf{y}(j)) \right] \right\}, \tag{4.51b}$$

with the *desired* score function

$$\boldsymbol{\Phi}_{s,PD}(\mathbf{y}(j)) = -\frac{\partial \log \hat{p}_{s,PD}(\mathbf{y}(j))}{\partial \mathbf{y}(j)} \tag{4.51c}$$

resulting from the hypothesized source model, and the actual score function

$$\boldsymbol{\Phi}_{y,PD}(\mathbf{y}(j)) = -\frac{\partial \log \hat{p}_{y,PD}(\mathbf{y}(j))}{\partial \mathbf{y}(j)}. \tag{4.51d}$$

The hypothesized source model $\hat{p}_{s,PD}(\cdot)$ in (4.51c) is chosen according to the class of signal processing problem to be solved. For instance, a factorization of $\hat{p}_{s,PD}(\cdot)$ among the sources yields BSS, i.e.,

$$\hat{p}_{s,PD}(\mathbf{y}(j)) \overset{\text{(BSS)}}{=} \prod_{q=1}^{P} \hat{p}_{y_q,D}(\mathbf{y}_q(j)), \tag{4.52}$$

while a complete factorization leads to the traditional MCBD approach,

$$\hat{p}_{s,PD}(\mathbf{y}(j)) \overset{\text{(MCBD)}}{=} \prod_{q=1}^{P} \prod_{d=1}^{D} \hat{p}_{y_q,1}(\mathbf{y}_q(j-d)). \tag{4.53}$$

4.3.2 Special Cases and Illustration in the Time Domain

Besides the various options to design the Sylvester constraint, there are many further interesting known and novel practical approximations within the framework. To begin with, we first consider algorithms based on second-order statistics (SOS) as they are particularly illustrative.

Realizations Based on Second-Order Statistics. Here, the source models are simplified to sequences of multivariate Gaussian functions described by $PD \times PD$ correlation matrices $\mathbf{R}_{..}$ within the length-N signal blocks. This leads to the coefficient update [3]

$$\begin{aligned}\Delta\check{\mathbf{W}}(m) &= \sum_{i=0}^{\infty}\beta(i,m)\,\mathcal{SC}\left\{\mathbf{W}(i)\hat{\mathbf{R}}_{\mathbf{yy}}\left[\hat{\mathbf{R}}_{\mathbf{ss}}^{-1}-\hat{\mathbf{R}}_{\mathbf{yy}}^{-1}\right]\right\}\\ &= \sum_{i=0}^{\infty}\beta(i,m)\,\mathcal{SC}\left\{\mathbf{W}(i)\left[\hat{\mathbf{R}}_{\mathbf{yy}}-\hat{\mathbf{R}}_{\mathbf{ss}}\right]\hat{\mathbf{R}}_{\mathbf{ss}}^{-1}\right\}. \qquad (4.54)\end{aligned}$$

Generic SOS-based BSS. The BSS variant of the generic SOS natural gradient update (4.54) follows immediately by setting

$$\hat{\mathbf{R}}_{\mathbf{ss}}(i) = \operatorname{bdiag}\hat{\mathbf{R}}_{\mathbf{yy}}(i). \qquad (4.55)$$

The update (4.54) together with (4.55) was originally obtained independently in [4] as a generalization of the cost function of [37]:

$$\mathcal{J}_{\mathrm{SOS}}(m) = \sum_{i=0}^{\infty}\beta(i,m)\left\{\log\det\hat{\mathbf{R}}_{\mathbf{ss}}(i) - \log\det\hat{\mathbf{R}}_{\mathbf{yy}}(i)\right\}. \qquad (4.56)$$

In Fig. 4.7 the mechanism of (4.54) based on the model (4.55) is illustrated. By minimizing $\mathcal{J}_{\mathrm{SOS}}(m)$, all cross-correlations for D time-lags are reduced and will ideally vanish, while the auto-correlations are untouched to preserve the structure of the individual signals. This class of algorithms leads to very robust practical solutions even for a large number of filter taps due to an inherent normalization by the auto-correlation matrices, reflected by the inverse in (4.54) of $\operatorname{bdiag}\hat{\mathbf{R}}_{\mathbf{yy}}$. Note that there are also various efficient approximations of this broadband algorithm, e.g, [36, 38, 39], with a reduced

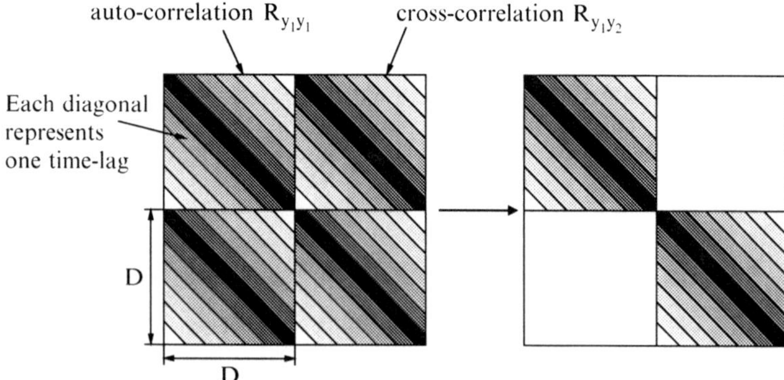

Fig. 4.7. Illustration of SOS-based BSS.

computational complexity allowing already real-time operation on a regular PC platform. These efficient implementations also form a powerful basis for blind system identification and for simultaneous localization of multiple acoustic sources, as shown later in this chapter. Moreover, a close link has been established [2, 4] to various popular frequency-domain algorithms, as we discuss in more detail in Sect. 4.3.3.

Inverse blind adaptive filtering problems. To illustrate that TRINICON also offers a powerful framework for *inverse* blind adaptive filtering problems in the same way as it does for the *direct* blind adaptive filtering problems, we give a brief overview of some of the most important ideas in this context in the following two paragraphs, based on [3].

MCBD based on SOS. Traditionally, ICA-based MCBD algorithms assume i.i.d. source models, e.g., [27, 28]. In the SOS case, this corresponds to a complete whitening of the output signals by not only applying a joint de-cross-correlation, but also a joint de-auto-correlation, i.e., $\hat{\mathbf{R}}_{\mathbf{ss}} = \operatorname{diag} \hat{\mathbf{R}}_{\mathbf{yy}}$ over multiple time-instants, as illustrated in Fig. 4.9 (b).

MCBPD based on SOS. Signal sources which are non i.i.d. should not become i.i.d. at the output of the blind adaptive filtering stage. Therefore, their statistical dependencies should be preserved. In other words, the adaptation algorithm has to distinguish between the statistical dependencies within the source signals, and the statistical dependencies introduced by the mixing system $\check{\mathbf{H}}$. We denote the corresponding generalization of the traditional MCBD technique as *MultiChannel Blind Partial Deconvolution* (MCBPD) [3]. Equations (4.51b)–(4.51d) inherently contain a statistical source model (signal properties (i)–(iii) in Sect. 4.3.1), expressed by the multivariate densities, and thus provide all necessary requirements for the MCBPD approach.

A typical example for MCBPD applications is speech dereverberation, which is especially important for distant-talking automatic speech recognition (ASR), as there is a very strong need for speech dereverberation without introducing artifacts to the signals. In this application, MCBPD allows to distinguish between the actual speech production system, i.e., the vocal tract, and the reverberant room (Fig. 4.8). Ideally, only the influence of the room

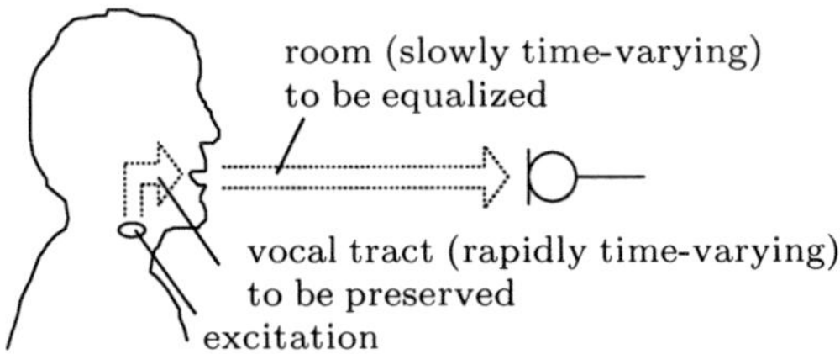

Fig. 4.8. Illustration of speech dereverberation as an MCBPD application.

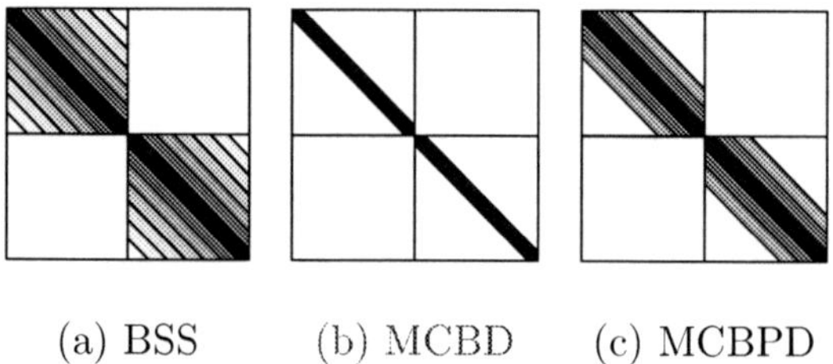

Fig. 4.9. Desired correlation matrices $\hat{\mathbf{R}}_{\mathbf{ss}}$ for BSS, MCBD, and MCBPD with TRINICON in the SOS case.

acoustics should be minimized. In the *SOS case*, the auto-correlation structure of the speech signals can be taken into account, as shown in Fig. 4.9 (c). While the room acoustics influences all off-diagonals, the effect of the vocal tract is concentrated in the first few off-diagonals around the main diagonal. These first off-diagonals of $\hat{\mathbf{R}}_{\mathbf{yy}}$ are now taken over into $\hat{\mathbf{R}}_{\mathbf{ss}}$, as shown in Fig. 4.9 (c). Alternatively, the structure in Fig. 4.9 (c) may be approximated by small submatrices making its handling somewhat more efficient. Note that there is a close link to linear prediction techniques which gives guidelines for the number of lags to be preserved.

Realizations Based on Higher-Order Statistics. The general HOS approach (4.51b)–(4.51d) provides the possibility to take into account all available information on the statistical properties of the desired source signals. This provides an increased flexibility and improved performance of BSS relative to the SOS case. Moreover, the more accurate modeling of the desired source signals yields also an improved MCBPD.

To apply the general approach in a real-world scenario, appropriate multivariate score functions (4.51c) and (4.51d) have to be determined. Fortunately, there is an efficient solution to this problem by assuming so-called spherically invariant random processes (SIRPs) [40–42]. The general form of correlated SIRPs of D-th order is given with a properly chosen function $f_D(\cdot)$ by

$$\hat{p}_D(\mathbf{y}_p(j)) = \frac{1}{\sqrt{\pi^D \det(\mathbf{R}_{\mathbf{y}_p\mathbf{y}_p}(i))}} f_D\left(\mathbf{y}_p^{\mathrm{T}}(j)\mathbf{R}_{\mathbf{y}_p\mathbf{y}_p}^{-1}(i)\mathbf{y}_p(j)\right) \tag{4.57}$$

for the p-th channel, where $\mathbf{R}_{\mathbf{y}_p\mathbf{y}_p}$ denotes the corresponding auto-correlation matrix with the corresponding number of lags. These models are representative for a wide class of stochastic processes. Speech signals in particular can very accurately be represented by SIRPs [42]. A great advantage arising from the SIRP model is that multivariate pdfs can be derived analytically from the corresponding univariate pdf together with the (lagged) correlation matrices. The function $f_D(\cdot)$ can thus be calculated from the well-known univariate models for speech, e.g., the Laplacian density. Using the chain rule, the corresponding score function (4.51c) can be derived from (4.57), as shown in [1,2] in more detail.

The calculation of the other score function (4.51d) becomes particularly simple in most practical realizations by transforming the output pdf $\hat{p}_{y,PD}(\cdot)$ into the corresponding multivariate input signal pdf using $\mathbf{W}$, which is considered as a mapping matrix of a linear transformation (see [1,2] for the general broadband case where $\mathbf{W}$ exhibits a blockwise-Sylvester structure). The derivative of the input signal pdf vanishes as it is independent of the demixing system.

Note that the multivariate Gaussian pdf is a special case of a SIRP and thus, the above described SOS-based algorithms represent special cases of the corresponding algorithms based on SIRPs [1,2]. As in the SOS case, by transforming the model into the DFT domain, various links to novel and existing popular frequency-domain algorithms can be established [2], as we discuss in more detail in Sect. 4.3.3.

4.3.3 On Frequency-Domain Realizations

For convolutive mixtures, the classical approach of frequency-domain BSS appears to be an attractive alternative where all techniques originally developed for instantaneous BSS are typically applied independently in each frequency bin, e.g., [19]. Unfortunately, this traditional narrowband approach exhibits several limitations as identified in, e.g., [43–45]. In particular, the permutation problem, which is inherent in BSS, may then also appear independently in each frequency bin so that extra repair measures have to be taken to address this *internal* permutation. Problems caused by circular convolution effects due to the narrowband approximation are reported in, e.g., [44].

In [2] it is shown how the equations of the TRINICON framework can be transformed into the frequency domain in a rigorous way (i.e., without any approximations) in order to avoid the above-mentioned problems. As in the case of the time-domain algorithms, the resulting generic DFT-domain BSS may serve both as a unifying framework for existing algorithms, and also as a guideline for developing new improved algorithms by certain suitable *selective* approximations as shown in, e.g., [2] or [38]. Figure 4.10 gives an overview on the most important classes of DFT-domain BSS algorithms known so far (various more special cases may be developed in the future). A very important observation from this framework using multivariate pdfs is that, in general, all frequency components are linked together so that the internal permutation problem is avoided (the following elements are reflected in Fig. 4.10 by the different approximations of the generic SIRP-based BSS):

1. Constraint matrices appearing in the generic frequency-domain formulation (see, e.g., [2]) describe the inter-frequency correlation between DFT components.

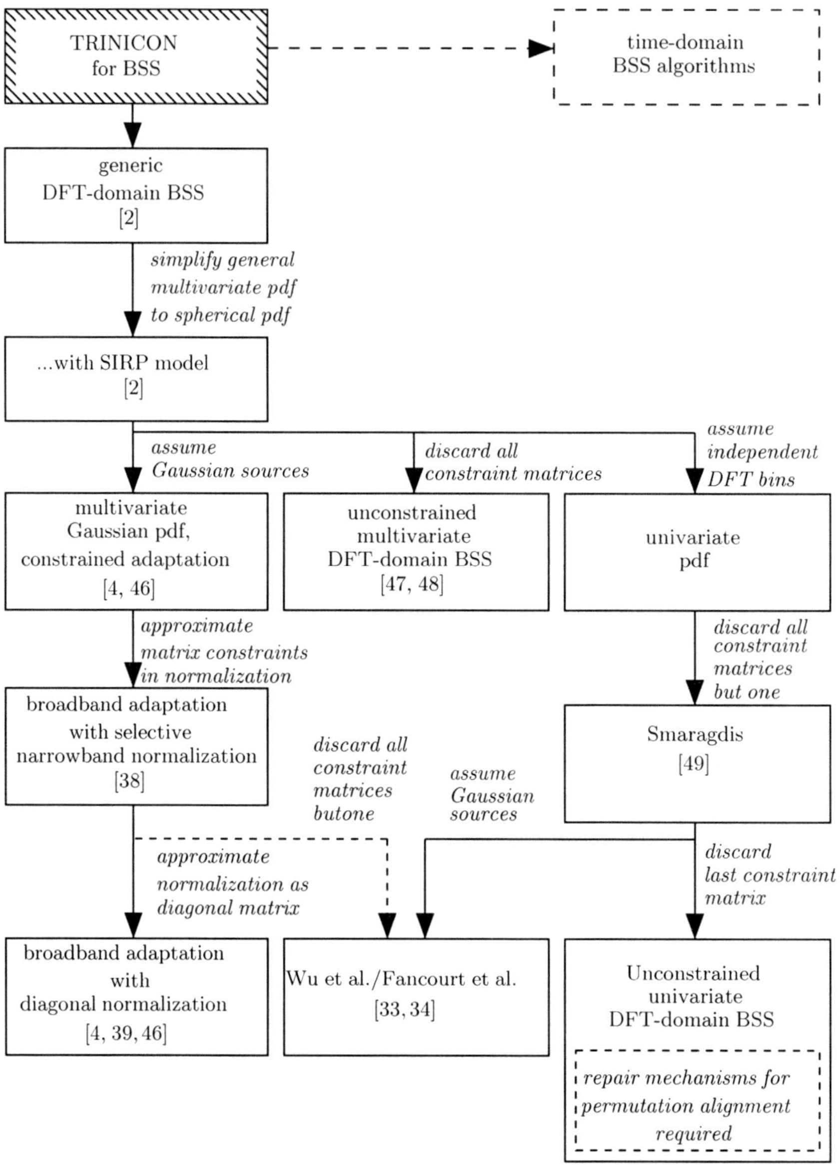

Fig. 4.10. Overview of BSS algorithms in the DFT domain.

2. The multivariate score function, derived from the multivariate pdf is a broadband score function. As an example, for SIRPs the argument of the multivariate score function (which is a nonlinear function in the higher-order case) is $\mathbf{y}_p^{\mathrm{T}}(j)\mathbf{R}_{\mathbf{y}_p\mathbf{y}_p}^{-1}(i)\mathbf{y}_p(j)$ according to (4.57). Even for the simple case $\mathbf{R}_{\mathbf{y}_p\mathbf{y}_p}^{-1}(i) = \mathbf{I}$ where we have $\mathbf{y}_p^{\mathrm{T}}(j)\mathbf{y}_p(j) = \|\mathbf{y}_p(j)\|^2$, i.e., the

quadratic norm, and – due to the Parseval theorem – the same in the frequency domain, i.e., the quadratic norm over all DFT components, we immediately see that all DFT-bins are taken into account simultaneously so that the internal permutation problem is avoided. Note that the traditional narrowband approach (with the internal permutation problem) would result as a special case if we assumed all DFT components to be statistically independent from each other (which is of course not the case for real-world broadband signals such as speech and audio signals). In contrast to this independence approximation the dependencies among all frequency components (including higher-order dependencies) are inherently taken into account in TRINICON in an optimal way. Actually, in the traditional narrowband approach, the additionally required repair mechanisms for permutation alignment try to exploit such inter-frequency dependencies.

From the viewpoint of the blind system identification the *broadband algorithms with constraint matrices* (i.e., the algorithms represented in the first column of Fig. 4.10) are of particular interest. Among these algorihms, the system described in [38] has turned out to be very efficient in this context and for multiple source localization as described later in this chapter. A pseudo-code of this algorithm is also included in [38].

Another important consideration for the practical implementation of BSI is the proper choice of the Sylvester constraint. Since the *column constraint* $\mathcal{SC}_{\mathrm{C}}$ is not suited for arbitrary source positions, it is generally *not appropriate* for BSI and the source localization application discussed next in this chapter. Thus, in the implementations discussed below the *row constraint* $\mathcal{SC}_{\mathrm{R}}$ is used.

4.4 Acoustic Source Localization: An Overview of the Approaches and TRINICON as a Generic Source Localization Scheme

The precision of acoustic source localization is determined by several factors. Critical parameters of the acoustic environment itself are the Signal-to-Noise Ratio (SNR, additive distortion) and the reverberation time (T_{60}) or the power ratio between the direct path and the reverberation of the room (Signal-to-Reverberation Ratio, SRR). Further important conditions are determined by the sources (number of sources, spatial diversity, velocity of the motion, signal statistics) and the sensors (number of sensors, array geometry, temporal sampling rate).

In the literature (e.g., [9]) existing approaches for acoustic source localization are often roughly divided into three categories:

(1) Maximization of the output power of steered beamformers (SRP, *steered response power*)

(2) Approaches based on high-resolution spectral estimation (also called *subspace* approaches)
(3) Approaches based on the estimation of *time differences of arrival* (TDOA) as an intermediate step

All of these methods may essentially be seen as two-step methods consisting of a certain signal processing stage, based on the available sensor signals, and a certain mapping from the signal processing output to the geometrical source position(s) by taking into account the sensor array geometry.

The first category of source localization approaches is based on the variation of the spatial alignment of a beamformer and results in systematic scanning of the acoustic environment so that basically this method may provide a very accurate localization even for multiple sources. Note that in this method, both of the above-mentioned two steps, i.e., the signal processing, and the geometric mapping step are included in the scanning process. In principle, the number of microphones is easily scalable in this concept. It also allows a relatively high robustness to additive interferences [9]. However, known disadvantages of this technique are that due to the (fixed) beamformer design one normally has to assume an ideal freefield propagation of the acoustic signals, and the search process necessarily becomes computationally very demanding if the desired spatial resolution increases.

The second category includes a class of algorithms which can be considered as an advancement and a systematization of the first category. The corresponding approach is based on the $P \times P$ correlation matrix of the P microphone signals (without time lags) and it allows also inherently a simultaneous localization of multiple active sources. If the number Q of sources is less or equal to the number P of sensors then it can be shown [50] that the eigenvectors for the Q largest eigenvalues of the correlation matrix span a certain subspace ('signal subspace'). This subspace corresponds to the one resulting from the direction vectors of the sources. From these direction vectors, we can extract the corresponding directions of arrival (DOA) in a separate mapping step. The remaining $P - Q$ eigenvectors, i.e., the eigenvectors corresponding to the $P - Q$ smallest eigenvalues of the correlation matrix, constitute the subspace of the background noise which is orthogonal to the signal subspace. This concept forms the basis for several well-known algorithms proposed in the literature, e.g., MUSIC [10] and ESPRIT [11]. Unfortunately, however, these algorithms were originally developed only for narrowband signals and therefore they are not immediately applicable to acoustic broadband signals, such as speech. One of the problems is that, in general, each frequency component yields a different signal subspace. Moreover, just as above in category (1), the room reverberation is not modeled by this method. Therefore, numerous modifications of these algorithms have been proposed in the literature. In order to solve the problem due to the narrowband assumption, [51] proposes an introduction of a *focussing* to a certain center frequency so that only a single signal subspace is obtained which ideally contains the complete information on the source positions. Unfortunately, in practice this is often problematic due to robustness issues and due to the necessity of a good initial guess which is difficult to obtain. Therefore, so far this approach has not been widely used for audio signals [9]. In addition, if there is multipath propagation due to spatial reflections, then these reflected signal components act on the microphones as additional correlated sources. In order to solve the narrowband problem, and thus the focussing problem in an optimal way, a new approach has been proposed in [52] which takes into account the underlying physics of wave propagation. There, the sound field for freefield propagation is decomposed into eigenfunctions. Thereby even the scattering on the microphone array itself can be efficiently taken into account, and by using a circular or spherical array, a full 360 degrees field-of-view is possible.

Finally, according to the above discussions in this section and in Sect. 4.2, broadband BSS may be also regarded as a generalized subspace approach based on the block-diagonalization of the signal correlation matrix in the case of second-order statistics. Due to the systematic incorporation of time lags into the correlation matrix in contrast to the instantaneous correlation matrix used in the conventional subspace methods, we are now able to take into account the room reverberation. Thus, as illustrated in Fig. 4.11, the blind broadband adaptive MIMO filtering approach generalizes and unifies both the traditional subspace methods, and the SIMO-based BSI. Note that both of these traditional methods are based on the calculation of the eigenvector(s) corresponding to the smallest eigenvalue(s) of the (lagged or instantaneous) sensor correlation matrix (see also the note at the end of Sect. 4.2.2 for the case of SIMO-based BSI).

Category (3) is by far the most widely used. As in the previous category, we split here the determination of the source position using multiple microphone signals into two separate steps. In contrast to the first two categories, the first step is here the *explicit* estimation of the temporal signal delays between different pairs of microphones (*time difference of arrival*, TDOA). The second step constitutes the calculation of the position in the three-dimensional space or in the two-dimensional plane using these estimates. Under the assumption that the relative microphone positions are known a-priori, the problem of source localization from a given set of TDOAs can be reduced to a purely geometrical problem.

For the explicit determination of TDOAs many different techniques have been proposed in the literature [9, 16]. Of particular interest is the fact that there are some more recent and powerful TDOA estimation techniques based directly on the blind system identification methods presented earlier in this chapter. Thus, since these techniques inherently take the room reflections into account, they promise a high robustness even in real reverberant environments.

Another advantage of the TDOA-based method is that it also allows for an accurate localization of sources in the *nearfield*.

In summary, the TDOA-based method using a blind system identification technique may be considered as the most general and versatile source localization approach. However, most of the BSI techniques known so far from the literature are based on SIMO systems, as shown in Sect. 4.2.2, i.e., the localization systems based on these techniques are only suitable for one source. The MIMO BSI technique described earlier in this chapter thus also generalizes this type of TDOA-based method to allow a *simultaneous localization of multiple sources in reverberant environments* even in the nearfield. In the following we therefore consider this two-step TDOA-based approach in more detail. To begin with, the geometrical considerations in Sect. 4.5.1 concentrate on the second step for the actual localization, particularly on the necessary number of different TDOA measurements and the array geometry. In contrast, for the first step, i.e., the TDOA estimation, the acoustic conditions of the room play a very important role. For that, we also consider the popular *generalized cross-correlation* (GCC) method as a reference in Sect. 4.5.3.

Figure 4.11 summarizes the above considerations and illustrates that the TRINICON-based MIMO BSI scheme may be considered as a *generic source localization approach.* Note also that in principle, broadband BSS can be applied to all three of the above-mentioned categories of acoustic source localization approaches since broadband BSS may be considered as

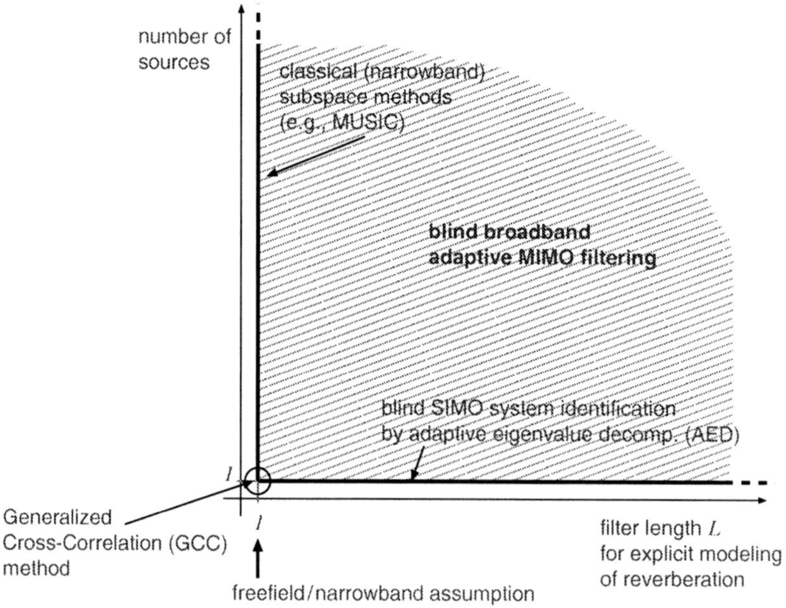

Fig. 4.11. Relations between various localization approaches.

(1) multiple blind beamformers, (2) a generalized subspace approach based on the block-diagonalization of the signal correlation matrix, and (3) a method for MIMO-based BSI, as discussed above.

4.5 Acoustic Source Localization Based on Time-Differences of Arrival

4.5.1 Basic Geometric Considerations

The obtained information on the source position $\hat{\mathbf{r}}_\mathrm{s}$ from an estimated TDOA $\hat{\tau}_{ij}$ between the signals of microphones i and j can be expressed as

$$c\hat{\tau}_{ij} = \|\hat{\mathbf{r}}_\mathrm{s} - \mathbf{r}_i\| - \|\hat{\mathbf{r}}_\mathrm{s} - \mathbf{r}_j\| \,, \tag{4.58}$$

where c denotes the velocity of sound and the vectors $\mathbf{r}_i$ and $\mathbf{r}_j$ denote the three-dimensional (or two-dimensional) positions of microphones i and j, respectively.

In three-dimensional space such an equation describes a hyperboloïd, as exemplarily shown in Fig. 4.12 (a).

From information on another time difference one obtains a second hyperboloïd. If the two pairs of microphones are placed on one straight line, the points fulfilling both conditions are describing a circle. To provide information on the position of the source on the circle, the TDOA of a third microphone pair has to be taken into account. To avoid linear dependencies, this microphone pair must not be placed on the same straight line as the former two. For example, if we want to restrict ourselves for practical reasons to a two-dimensional microphone array in the xy-plane, the third microphone pair will

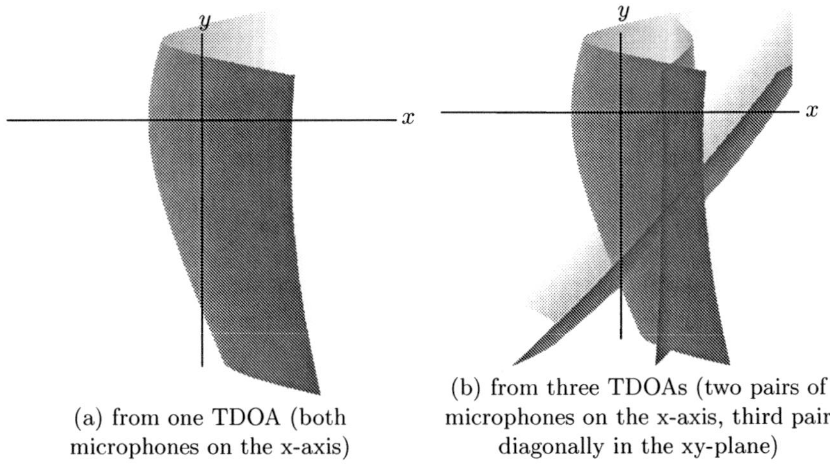

(a) from one TDOA (both microphones on the x-axis)

(b) from three TDOAs (two pairs of microphones on the x-axis, third pair diagonally in the xy-plane)

Fig. 4.12. Determination of the possible source positions in the room.

be in the same plane. This results in two unique intersection points, one with positive, and one with negative z-coordinate, as shown in Fig. 4.12 (b). In many scenarios one can already uniquely determine the position with such a setup since either the negative or positive z-coordinates may be excluded (e.g., if the array is mounted on a wall).

4.5.2 Microphone Array Geometry and Grid of Potential Source Positions

Equation (4.58) represents a nonlinear set of equations. Considering this set of equations, we readily see that the geometry of the microphone array has a major influence on the calculation of the positions. In the following, we exemplarily study this influence for two different arrays for the two-dimensional case where two TDOA estimates are needed (see Sect. 4.5.1).

The setup in Fig. 4.13 uses the signal from the center microphone simultaneously for two equations by estimating the TDOA between microphones 1 and 2, and between 2 and 3, respectively. This has two advantages. It not only reduces the necessary number of microphones, but the calculation of the source positions, based on the nonlinear set of equations (4.58) will become much easier as there are fewer geometrical parameters if we set the origin on the position of the center microphone. Note that in general, the solution of this set of equations is not trivial and in some cases there is no closed-form solution. Thus, for more complicated array geometries, we have to resort to numerical and/or approximate solutions, rather than exact closed-form solutions. There is a rich literature on this problem which also includes the consideration of overdetermined sets of equations, i.e., taking into account more TDOA estimates in order to further improve the robustness to measurement errors. Important distinctions between these methods include likelihood-based [53–56] together with iterative optimization (e.g., using the Newton–Raphson [57] or Gauss–Newton [58] methods) versus least-squares and linear approximation versus closed-form algorithms [59–65].

On the other hand, there may also be a major disadvantage with the microphone setup after Fig. 4.13, depending on the chosen TDOA estimation method, as we will discuss in the following.

The TDOAs, estimated by the methods discussed in Sect. 4.5.3 are ordinarily represented by *integer* numbers, corresponding to discrete sampling instants along the time axis. Therefore the estimates of the potential

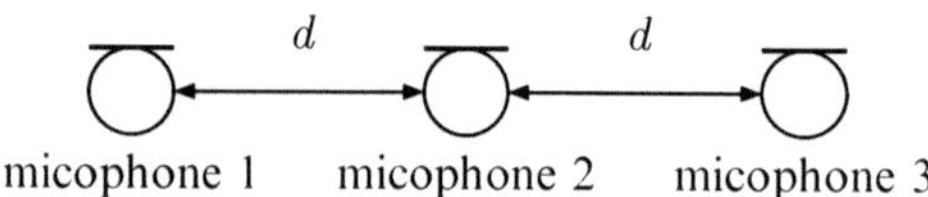

Fig. 4.13. Microphone array 1.

source positions are restricted to a grid of *discrete* positions. The density of this grid depends on the sampling rate and on the positions of the microphones. Note that in principle this density is *independent* of the chosen TDOA estimation method for a given array geometry.

Figure 4.15 shows the possible positions that can be obtained using the microphone array after Fig. 4.13 with a spacing of $d = 16$ cm and a temporal sampling rate of $f_s = 48$ kHz. As we can see, the spatial resolution decreases with increasing distance (*range*) between the source and the microphone array.

To obtain a better resolution with a fixed sampling rate and integer TDOAs, we can either place more pairs of microphones in the room that are closer to the respective source positions, or we change the geometric parameters of the setup in Fig. 4.13. A first possibility is to increase the distance d, i.e., the *spatial diversity* between the microphones. For $d = 50$ cm the resolution is already dramatically improved, as can be seen in Fig. 4.16, and in the case of, e.g., $d = 200$ cm, we would obtain a very dense grid, and a much wider coverage. This is due to the fact that the maximum time difference of arrival is increased ($\tau_{\max} = \frac{cd}{f_s}$). With increasing $\tau_{\max}$ the number of potential TDOA values for each microphone pair is also increased, and thus the number of potential positions is increased significantly. The full potential of this method can be exploited with *TDOA estimators based on blind system identification* (Sect. 4.5.3). With other TDOA estimators, one drawback of this method is that the precision of the TDOA estimation itself may be affected due to spatial aliasing if the microphone spacing is too large. This ambiguity problem typically occurs with narrowband implementations, i.e., the signal processing is carried out independently for each frequency bin in that implementations. The GCC is often implemented in this way. Therefore, an alternative for that case would be to use the modified setup after Fig. 4.14. Instead of increasing the spacing within the microphone pairs, we only increase the distance between the individual pairs as this affects the geometrical calculation. The TDOA estimation is only affected by the spacing d. Note, however, that a small spacing generally increases the error variance, i.e., small TDOA deviations will have a larger influence on the final position estimate. Therefore, in any case a broadband implementation of the blind system identification is recommended for accurate source localization. Moreover, to further improve the spatial resolution of the localizer at a low computational cost, *fractional* delays can be obtained with the BSI-based method by performing a sinc interpolation [66] on the filters of the unmixing

Fig. 4.14. Microphone array 2.

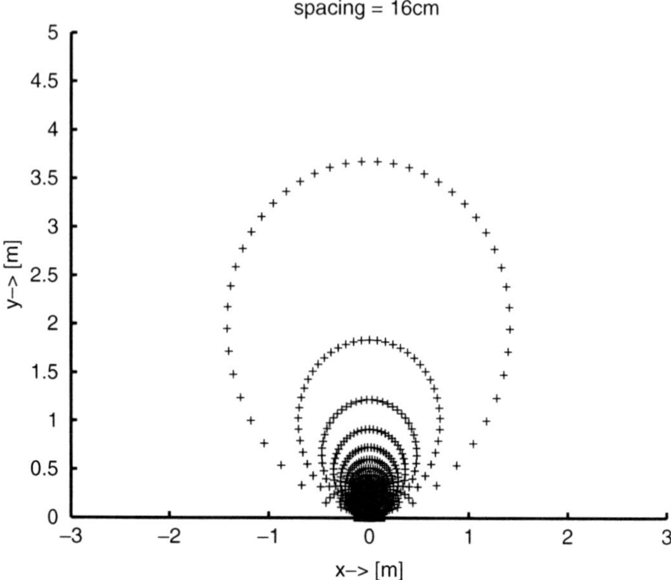

Fig. 4.15. Potential source positions using three equidistant microphones with a spacing of 16 cm at a sampling rate of 48 kHz.

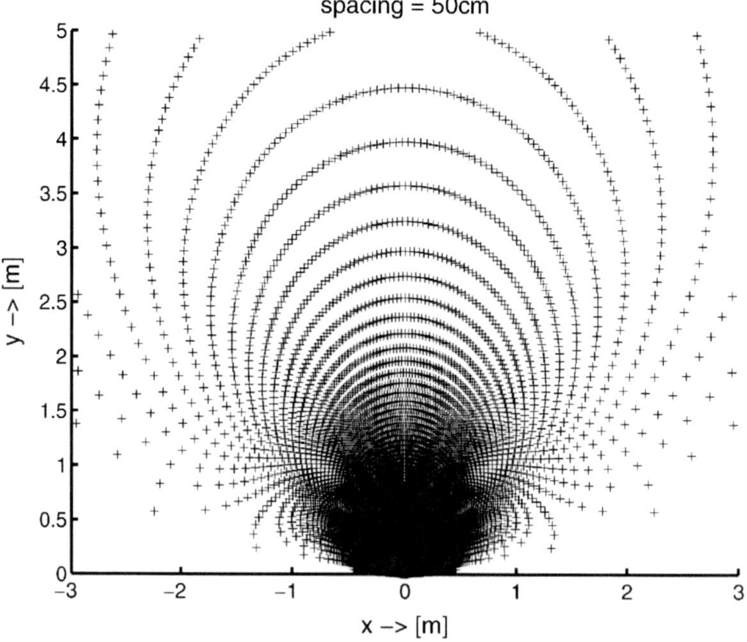

Fig. 4.16. Potential source positions using three equidistant microphones with a spacing of 50 cm at a sampling rate of 48 kHz.

system $\check{\mathbf{W}}$ before performing the effective TDOA estimations, given in (4.60) and (4.61a)/(4.61b), without further increasing the sampling rate for the BSS operations.

4.5.3 Estimation of Time Differences of Arrival

The key for an effective localization with the two-step approach is an accurate and robust TDOA estimator. In the following we examine different possibilities for TDOA estimation. Thereby, as in the previous section, we concentrate here on the use of microphone pairs for clarity, although it has recently become possible to simultaneously take into account multiple microphone signals [16].

The Method of the Generalized Cross-Correlation. The method of the generalized cross-correlation (GCC) [8] is based on the ideal free-field propagation model $x_i(t) = \alpha_i s(t - \tau_i) + b_i(t)$, where b_i is an additive noise signal on the i-th microphone, and α_i denotes an attenuation factor. Due to its simplicity the method is nevertheless often applied in reverberant environments so that to date it is still the most widely used method for single-source localization.

The basic principle of this technique consists of the maximization of the inverse Fourier transformation of a weighted cross-power spectral density, i.e.,

$$\hat{\tau}_{ij} = \arg\max_{\tau_{ij}} \mathcal{F}^{-1}\left\{\Phi(f) S_{x_i x_j}(f)\right\}, \tag{4.59}$$

where $\Phi(f)$ is the weighting function, and $S_{x_i x_j}(f)$ denotes the estimated cross-power spectral density between x_i and x_j.

The numerous variants of the GCC method [8, 67] differ mainly in their weighting functions and in the estimation procedure for $S_{x_i x_j}(f)$. The classical cross-correlation method (CC) uses $\Phi(f) = 1$. The weighting function $\Phi(f) = \frac{1}{|S_{x_i x_j}(f)|}$ yields the *Phase Transform* (PHAT) technique in which a behavior independent of the spectral support is achieved by the normalization by the magnitude of the power spectral density [8].

Other improvements concentrate on the pre-filtering of the input signal, such as, e.g., the cepstral processing in [68].

However, these methods suffer from the fact that the underlying signal model of the GCC does not reflect the conditions in real acoustic environments. This becomes particularly obvious in very reverberant environments. It can be shown that here the robustness can break down abruptly with increasing reverberation time T_{60} [12, 13] as will also be confirmed by our experimental evaluation in Sect. 4.6.

The Blind System Identification Method. To treat this reverberation problem, a completely different approach was presented for single-source localization in [15] which is based on blind adaptive filtering using the adaptive eigenvalue decomposition algorithm. According to Sect. 4.2.2, the AED algorithm adapts itself directly to the impulse responses h_1 and h_2, i.e., the SIMO model between a source s and the microphones. Therefore, this approach is inherently based on the realistic convolutive propagation model. Note that the scaling ambiguity in blind system identification is uncritical for the TDOA estimation (as can be easily seen by (4.60) below).

To perform the adaptation, a wide range of adaptation algorithms, such as the Least-Mean-Squares (LMS) algorithm [18] (in modified form [15]), realized in the time domain, or in efficient frequency-domain realization [18] can be used.

Based on the estimated filter coefficients, the TDOA can then be calculated after each coefficient update according to

$$\begin{aligned}\hat{\tau} &= \arg\max_n |\hat{h}_{2,n}| - \arg\max_n |\hat{h}_{1,n}| \\ &= \arg\max_n |w_{1,n}| - \arg\max_n |w_{2,n}|. \end{aligned} \tag{4.60}$$

Note that here, as above in the case of GCC, we consider only one microphone pair. However, there are generalizations, both of GCC [16] and AED [16] for more than two sensors, in order to further increase the robustness by spatial redundancy.

Motivated by the high accuracy of the above-mentioned adaptive SIMO filter approach for localizing only one source, the more general approach of blind adaptive MIMO filtering for simultaneous localization of *multiple simultaneously active sources* was proposed in [6], based on the considerations discussed in Sect. 4.2 of this chapter. Thereby, the objective was to maintain the realistic convolutive propagation model for the localization, as in the case of AED. As with AED, we may calculate the Q TDOAs for the Q sources from the FIR filters w_{pq} once they are estimated by a TRINICON-based broadband adaptation algorithm, such as [38], as discussed in Sect. 4.3.3. Thereby we make the reasonable assumption that the sources are mutually uncorrelated. The extraction of the multiple TDOAs from the estimated MIMO filter coefficients is based on the relationship between the broadband BSS framework and the AED, as discussed in Sect. 4.2.2. For instance, in the case of two simultaneously active sources, (4.22a) is the corresponding equation to estimate the TDOA of source 1, while (4.22b) gives the TDOA of source 2. Moreover, since the coefficient initialization in the case of Sylvester constraint ($\mathcal{SC}_{\mathrm{R}}$), described in [6], also corresponds to the one recommended for the AED in [15], we can expect similar steady-state performances due to this close link. This is verified in Sect. 4.6. From these findings, we can express the TDOA estimates immediately in the same way as in (4.60) as

$$\hat{\tau}_1 = \arg\max_n |w_{12,n}| - \arg\max_n |w_{22,n}|, \tag{4.61a}$$

$$\hat{\tau}_2 = \arg\max_n |w_{11,n}| - \arg\max_n |w_{21,n}|. \tag{4.61b}$$

4.6 Simultaneous Localization of Multiple Sound Sources in Reverberant Environments Using Blind Adaptive MIMO System Identification

The audio data used for the evaluation have been recorded at a sampling rate of 48 kHz in a TV studio with a reverberation time of $T_{60} \approx 700$ ms. These data are made available as part of an audio-visual database [70]. This database also includes reference data of the speaker positions measured using infrared sensors. From the reference positions reference TDOAs are calculated by geometric considerations. environment. From the database, we chose two scenes in the same environment with one fixed and one moving source, respectively. Those are used separately for the SIMO-based approaches, and a superposition (Fig. 4.17) is used for the MIMO-based approach. The distance between the two microphones was 16 cm. For the adaptation algorithms, the filter lengths were chosen to 1024. (Obviously, this length is shorter than $L_{\mathrm{opt,sep}}$, given the above-mentioned reverberation time. However, for the localization application in the given scenario, it turned out to be sufficient in order to capture the dominant reflections. Moreover, as discussed in Sect. 4.2.4, the disturbing filtering ambiguity is still avoided in this case.) The block length for the GCC (using a phase-transform (PHAT) weighting rule [8]) has been set to 1024. GCC and AED have been complemented by a signal power-based voice-activity detector. Figures 4.18 (a) and (b) show the reference and estimated TDOAs for the fixed and the moving speakers, respectively. In these first experiments, only one speaker was active (also in

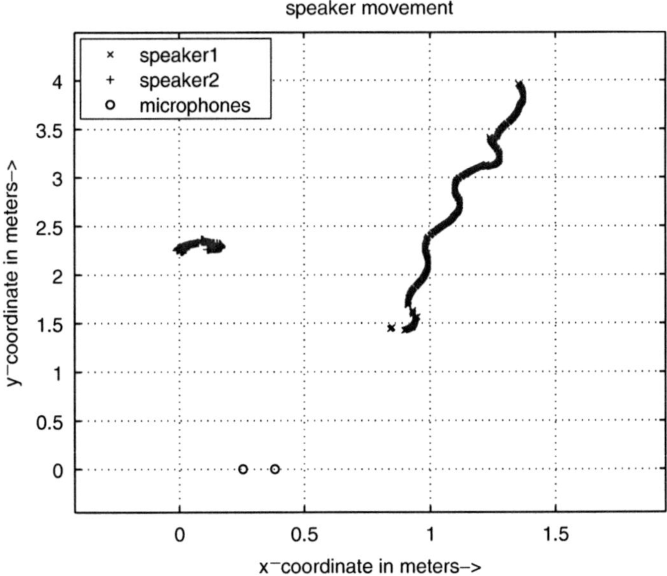

Fig. 4.17. Scenario used for the simulations.

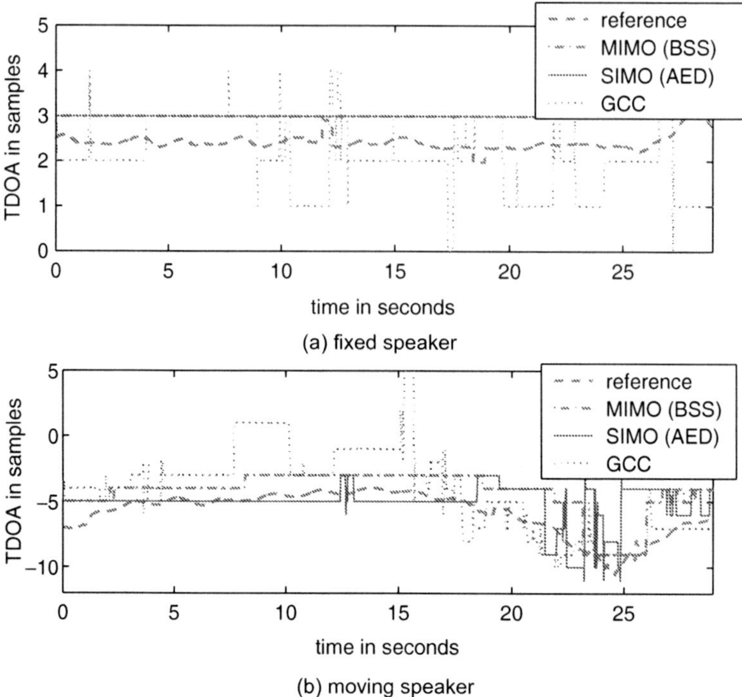

Fig. 4.18. TDOA estimation for one source.

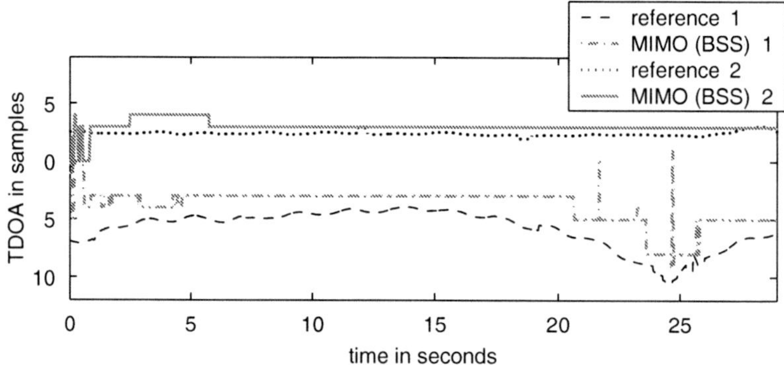

Fig. 4.19. Simultaneous TDOA estimation for two sources.

case of the MIMO-based approach). Subplot (a) confirms that both of the blind adaptation algorithms lead to the same accurate TDOA estimates in this static case, as expected from the considerations in Sect. 4.5. Note that the TDOA estimates can only attain integer values.

In Fig. 4.19 we consider the simultaneous estimation of two TDOAs by the proposed MIMO approach. Due to the scenario in Fig. 4.17 the two

TDOAs exhibit different signs. The estimates deviate only slightly from the corresponding results of the MIMO-based approach in Figs. 4.18 (a) and (b) during some very short time intervals. This may be explained by the different speech activity of the two sources which is typical and inevitable for realistic situations. However, the short peaks in Fig. 4.19 may be easily removed by appropriate postprocessing. Further experimental results, which also illustrate the robustness of the multiple TDOA estimates with respect to background noise and shadowing effects caused by objects placed between the sensors, may be found in [69].

4.7 Conclusions

In this chapter we have shown the relation between convolutive broadband BSS and blind MIMO system identification. From this we can draw the conclusions that (1) for a suitable choice of the filter length arbitrary filtering is prevented with broadband approaches, (2) the known whitening problem is avoided, and (3) the BSS framework also allows for several new applications, such as the simultaneous localization of multiple sources. Based on the relationship between MIMO BSI and broadband BSS we were also able to further clarify the relations between various source localization approaches. As a side aspect, also some relations and similarities to the inverse problems, such as blind dereverberation of speech signals have been illuminated. For all of these applications it became obvious that a proper broadband adaptation of the filter coefficients is desirable and in some cases even absolutely necessary. In many ways the TRINICON framework turned out to be a useful tool to solve the associated problems. On the algorithmic side, we have derived a generic Sylvester constraint which unifies previous algorithmic results and may serve as a guideline for the development of new efficient algorithms.

References

1. H. Buchner, R. Aichner, and W. Kellermann, "Blind source separation for convolutive mixtures exploiting nongaussianity, nonwhiteness, and nonstationarity," in *Proc. Int. Workshop Acoustic Echo and Noise Control (IWAENC)*, Kyoto, Japan, pp. 223–226, Sept. 2003.
2. H. Buchner, R. Aichner, and W. Kellermann, "Blind source separation for convolutive mixtures: A unified treatment," in Y. Huang and J. Benesty (eds.), *Audio Signal Processing for Next-Generation Multimedia Communication Systems*, Kluwer Academic Publishers, Boston, pp. 255–293, Feb. 2004.
3. H. Buchner, R. Aichner, and W. Kellermann, "TRINICON: A versatile framework for multichannel blind signal processing," in *Proc. IEEE Int. Conf. Acoustics, Speech, Signal Processing (ICASSP)*, Montreal, Canada, vol. 3, pp. 889–892, May 2004.
4. H. Buchner, R. Aichner, and W. Kellermann, "A generalization of blind source separation algorithms for convolutive mixtures based on second-order statistics," *IEEE Trans. Speech Audio Process.*, vol. 13, no. 1, pp. 120–134, Jan. 2005.

5. H. Buchner, R. Aichner, and W. Kellermann, "Relation between blind system identification and convolutive blind source separation," in *Proc. Joint Workshop Hands-Free Speech Communication and Microphone Arrays (HSCMA)*, Piscataway, NJ, USA, Mar. 2005 (additional presentation slides with more results downloadable from the web site `www.LNT.de/lms/`).
6. H. Buchner, R. Aichner, J. Stenglein, H. Teutsch, and W. Kellermann, "Simultaneous localization of multiple sound sources using blind adaptive MIMO filtering," in *Proc. IEEE Int. Conf. Acoustics, Speech, Signal Processing (ICASSP)*, Philadelphia, PA, USA, Mar. 2005.
7. M. Hofbauer, *Optimal Linear Separation and Deconvolution of Acoustical Convolutive Mixtures*, Dissertation, Hartung-Gorre Verlag, Konstanz, May 2005.
8. C.H. Knapp and G.C. Carter, "The generalized correlation method for estimation of time delay," *IEEE Trans. Acoust., Speech, Signal Processing*, vol. ASSP-24, pp. 320–327, Aug. 1976.
9. M.S. Brandstein and D.B. Ward, *Microphone Arrays: Signal Processing Techniques and Applications*, Springer, Berlin, 2001.
10. R.O. Schmidt, "Multiple emitter location and signal parameter estimation," *IEEE Trans. Antennas Propagation,* vol. AP-34, no. 3, pp. 276–280, Mar. 1986.
11. R. Roy and T. Kailath, "ESPRIT – estimation of signal parameters via rotational invariance techniques," *IEEE Trans. Acoust., Speech, Signal Processing,* vol. 37. no. 7, pp. 984–995, July 1989.
12. B. Champagne, S. Bedard, and A. Stéphenne,"Performance of time-delay estimation in the presence of room reverberation," *IEEE Trans. Speech Audio Process.,* vol. 4, pp. 148–152, Mar. 1996.
13. J.P. Ianniello, "Time delay estimation via cross-correlation in the presence of large estimation errors," *IEEE Trans. Acoust., Speech, Signal Processing,* vol. ASSP-30, no. 6, pp. 998–1003, Dec. 1982.
14. J. Scheuing and B. Yang, "Disambiguation of TDOA estimates in multi-path multi-source environments (DATEMM)," in *Proc. IEEE Int. Conf. Acoust., Speech, Signal Processing (ICASSP)*, Toulouse, France, 2006.
15. J. Benesty, "Adaptive eigenvalue decomposition algorithm for passive acoustic source localization," *J. Acoust. Soc. Am.*, vol. 107, pp. 384–391, Jan. 2000.
16. J. Chen, Y. Huang, and J. Benesty, "Time delay estimation" in Y. Huang and J. Benesty (eds.), *Audio Signal Processing for Next-Generation Multimedia Communication Systems*, Kluwer Academic Publishers, Boston, pp. 197–227, Feb. 2004.
17. A. Lombard, H. Buchner, and W. Kellermann, "Multidimensional localization of multiple sound sources using blind adaptive MIMO system identification," in *Proc. IEEE Int. Conf. Multisensor Fusion and Integration for Intelligent Systems (MFI)*, Heidelberg, Germany, Sept. 2006.
18. S. Haykin, *Adaptive Filter Theory*, 4th ed., Prentice-Hall, Englewood Cliffs, NJ, 2002.
19. A. Hyvärinen, J. Karhunen, and E. Oja, *Independent Component Analysis*, Wiley & Sons, Inc., New York, 2001.
20. S.C. Douglas, "Blind separation of acoustic signals" in M. Brandstein and D. Ward (eds.), *Microphone Arrays: Signal Processing Techniques and Applications*, pp. 355–380, Springer, Berlin, 2001.
21. J.-F. Cardoso and A. Souloumiac, "Blind beamforming for non gaussian signals," *IEE Proceedings-F*, vol. 140, no. 6, pp. 362–370, Dec. 1993.

22. S. Araki et al., "Equivalence between frequency-domain blind source separation and frequency-domain adaptive beamforming," in *Proc. IEEE Int. Conf. Acoustics, Speech, Signal Processing (ICASSP)*, Orlando, FL, USA, pp. 1785–1788, May 2002.
23. M. Miyoshi and Y. Kaneda, "Inverse filtering of room acoustics," *IEEE Trans. Acoust., Speech, Signal Processing*, vol. 36, no. 2, pp. 145–152, Feb. 1988.
24. K. Furuya, "Noise reduction and dereverberation using correlation matrix based on the multiple-input/output inverse-filtering theorem (MINT)," in *Proc. Int. Workshop Hands-Free Speech Communication (HSC)*, Kyoto, Japan, pp. 59–62, Apr. 2001.
25. M.I. Gürelli and C.L. Nikias, "EVAM: An eigenvector-based algorithm for multichannel blind deconvolution of input colored signals," *IEEE Trans. Signal Process.*, vol. 43, no. 1, pp. 134–149, Jan. 1995.
26. K. Furuya and Y. Kaneda, "Two-channel blind deconvolution of nonminimum phase FIR systems," *IEICE Trans. Fundamentals*, vol. E80-A, no. 5, pp. 804–808, May 1997.
27. S. Amari et al.,"Multichannel blind deconvolution and equalization using the natural gradient," in *Proc. IEEE Int. Workshop Signal Processing Advances in Wireless Communications*, pp. 101–107, 1997.
28. S. Choi et al., "Natural gradient learning with a nonholonomic constraint for blind deconvolution of multiple channels," in *Proc. Int. Symp. Independent Component Analysis Blind Source Separation (ICA)*, pp. 371–376, 1999.
29. B.W. Gillespie and L. Atlas, "Strategies for improving audible quality and speech recognition accuracy of reverberant speech," in *Proc. IEEE Int. Conf. Acoustics, Speech, Signal Processing (ICASSP)*, Hongkong, China, Apr. 2003.
30. K. Matsuoka and S. Nakashima, "Minimal distortion principle for blind source separation," in *Proc. Int. Symp. Independent Component Analysis Blind Signal Separation (ICA)*, San Diego, CA, USA, Dec. 2001.
31. H. Sawada, R. Mukai, S. Araki, and S. Makino, "A robust and precise method for solving the permutation problem of frequency-domain blind source separation," *IEEE Trans. Speech Audio Process.*, vol. 12, no. 8, Sept. 2004.
32. H. Liu, G. Xu, and L. Tong, "A deterministic approach to blind identification of multi-channel FIR systems," in *Proc. IEEE Int. Conf. Acoustics, Speech, Signal Processing (ICASSP)*, Adelaide, Australia, Apr. 1994.
33. H.-C. Wu and J.C. Principe,"Simultaneous diagonalization in the frequency domain (SDIF) for source separation," in *Proc. Int. Symp. Independent Component Analysis Blind Signal Separation (ICA)*, pp. 245–250, 1999.
34. C.L. Fancourt and L. Parra, "The coherence function in blind source separation of convolutive mixtures of non-stationary signals," in *Proc. Int. Workshop Neural Networks Signal Processing (NNSP)*, 2001, pp. 303–312.
35. T.M. Cover and J.A. Thomas, *Elements of Information Theory*, Wiley & Sons, New York, 1991.
36. R. Aichner, H. Buchner, F. Yan, and W. Kellermann, "A real-time blind source separation scheme and its application to reverberant and noisy acoustic environments," *Signal Processing*, vol. 86, no. 6, pp.1260–1277, 2006.
37. M. Kawamoto, K. Matsuoka, and N. Ohnishi, "A method of blind separation for convolved non-stationary signals," *Neurocomputing*, vol. 22, pp. 157–171, 1998.

38. R. Aichner, H. Buchner, and W. Kellermann, "Exploiting narrowband efficiency for broadband convolutive blind source separation," *EURASIP Journal on Applied Signal Processing*, vol. 2007, pp. 1–9, Sept. 2006.
39. T. Nishikawa, H. Saruwatari, and K. Shikano, "Comparison of time-domain ICA, frequency-domain ICA and multistage ICA for blind source separation," in *Proc. European Signal Processing Conference (EUSIPCO)*, vol. 2, pp. 15–18, Sept. 2002.
40. K. Yao, "A representation theorem and its applications to spherically-invariant random processes," *IEEE Trans. Inform. Theor.*, vol. 19, no. 5, pp. 600–608, Sept. 1973.
41. J. Goldman, "Detection in the presence of spherically symmetric random vectors," *IEEE Trans. Inform. Theor.*, vol. 22, no. 1, pp. 52–59, Jan. 1976.
42. H. Brehm and W. Stammler, "Description and generation of spherically invariant speech-model signals," *Signal Processing*, vol. 12, pp. 119–141, 1987.
43. S. Araki et al., "The fundamental limitation of frequency-domain blind source separation for convolutive mixtures of speech," *IEEE Trans. Speech Audio Process.*, vol. 11, no. 2, pp. 109–116, Mar. 2003.
44. H. Sawada et al., "Spectral smoothing for frequency-domain blind source separation," in *Proc. Int. Workshop Acoustic Echo and Noise Control (IWAENC)*, Kyoto, Japan, Sept. 2003, pp. 311–314.
45. M.Z. Ikram and D.R. Morgan, "Exploring permutation inconsistency in blind separation of speech signals in a reverberant environment," in *Proc. IEEE Int. Conf. Acoustics, Speech, Signal Processing (ICASSP)*, Istanbul, Turkey, June 2000, vol. 2, pp. 1041–1044.
46. H. Buchner, R. Aichner, and W. Kellermann, "A generalization of a class of blind source separation algorithms for convolutive mixtures," in *Proc. Int. Symp. Independent Component Analysis Blind Signal Separation (ICA)*, Nara, Japan, Apr. 2003.
47. T. Kim, T. Eltoft, and T.-W. Lee, "Independent vector analysis: an extension of ICA to multivariate components," in *Proc. Int. Conf. Independent Component Analysis Blind Signal Separation (ICA)*, Mar. 2006.
48. A. Hiroe, "Solution of permutation problem in frequency domain ICA using multivariate probability density functions," in *Proc. Int. Conf. Independent Component Analysis Blind Signal Separation (ICA)*, pp. 601–608, Mar. 2006.
49. P. Smaragdis, "Blind separation of convolved mixtures in the frequency domain," *Neurocomputing*, vol. 22, pp. 21–34, July 1998.
50. D.H. Johnson and D.E. Dudgeon, *Array Signal Processing*, Prentice Hall, New Jersey, 1993.
51. H. Wang and M. Kaveh, "Coherent Signal-Subspace Processing for the Detection and Estimation of Angles of Arrival of Multiple Wide-Band Sources," *IEEE Trans. Acoust., Speech, Signal Processing*, vol. ASSP-33, no. 4, pp. 823–831, Aug. 1985.
52. H. Teutsch and W. Kellermann, "Acoustic source detection and localization based on wavefield decomposition using circular microphone arrays," *J. Acoust. Soc. Am.*, vol. 120, no. 5, Nov. 2006.
53. W.R. Hahn and S.A. Tretter, "Optimum processing for delay-vector estimation in passive signal arrays," *IEEE Trans. Inform. Theory*, vol. IT-19, pp. 608–614, May 1973.

54. M. Wax and T. Kailath, "Optimum localization of multiple sources by passive arrays," *IEEE Trans. Acoust., Speech, Signal Processing,* vol. ASSP-31, no. 5, pp. 1210–1218, Oct. 1983.
55. P.E. Stoica and A. Nehorai, "MUSIC, maximum likelihood and Cramer-Rao bound," *IEEE Trans. Acoust., Speech, Signal Processing,* vol. 37, pp. 720–740, May 1989.
56. J.C. Chen, R.E. Hudson, and K. Yao, "Maximum-likelihood source localization and unknown sensor location estimation for wideband signals in the near-field," *IEEE Trans. Signal Process.,* vol. 50, pp. 1843–1854, Aug. 2002.
57. Y. Bard, *Nonlinear Parameter Estimation*, Academic Press, New York, 1974.
58. W.H. Foy, "Position-location solutions by Taylor-series estimation," *IEEE Trans. Aerosp. Electron. Syst.,* vol. AES-12, pp. 187–194, Mar. 1976.
59. R.O. Schmidt, "A new approach to geometry of range difference location," *IEEE Trans. Aerosp. Electron.,* vol. AES-8, pp. 821–835, Nov. 1972.
60. H.C. Schau and A.Z. Robinson, "Passive source localization employing intersecting spherical surfaces from time-of-arrival differences," *IEEE Trans. Acoust., Speech, Signal Processing,* vol. ASSP-35, no. 8, pp. 1223–1225, Aug. 1987.
61. J.O. Smith and J.S. Abel, "Closed-form least-squares source location estimation from range-difference measurements," *IEEE Trans. Acoust., Speech, Signal Processing,* vol. ASSP-35, no. 12, pp. 1661–1669, Dec. 1987.
62. Y.T. Chan and K.C. Ho, "A simple and efficient estimator for hyperbolic location," *IEEE Trans. Signal Process.,* vol. 42, no. 8, pp. 1905–1915, Aug. 1994.
63. Y.T. Chan and K.C. Ho, "An efficient closed-form localization solution from time difference of arrival measurements," in *Proc. IEEE Int. Conf. Acoustics, Speech, Signal Processing (ICASSP),* 1994, vol. 2, pp. 393–396.
64. Y. Huang, J. Benesty, G.W. Elko, and R.M. Mersereau, "Real-time passice source localization: an unbiased linear-correction least-squares approach," *IEEE Trans. Speech Audio Process.,* vol. 9, no. 8, pp. 943–956, Nov. 2001.
65. J.S. Abel and J.O. Smith, "The spherical interpolation method for closed-form passive source localization using range difference measurements," in *Proc. IEEE Int. Conf. Acoustics, Speech, Signal Processing (ICASSP),* vol. 1, pp. 471–474, 1987.
66. T.I. Laakso et al., "Splitting the unit delay," *IEEE Signal Processing Mag.,* vol. 13, pp. 30–60, 1996.
67. M.S. Brandstein and H.F. Silverman, "A robust method for speech signal time-delay estimation in reverberant rooms," in *Proc. IEEE Int. Conf. Acoustics, Speech, Signal Processing (ICASSP)*, Munich, Apr. 1997.
68. A. Stéphenne and B. Champagne, "A new cepstral prefiltering technique for estimating time delay under reverberant conditions," *Signal Processing*, vol. 59, pp. 253–266, 1997.
69. R. Aichner, H. Buchner, S. Wehr, and W. Kellermann, "Robustness of acoustic multiple-source localization in adverse environments," in *Proc. ITG Fachtagung Sprachkommunication*, Kiel, Germany, Apr. 2006.
70. M. Krinidis et al., "An audio-visual database for evaluating person tracking algorithms," in *Proc. IEEE Int. Conf. Acoustics, Speech, Signal Processing (ICASSP)*, Philadelphia, PA, USA, Mar. 2005.

5 SIMO-Model-Based Blind Source Separation – Principle and its Applications

Hiroshi Saruwatari, Tomoya Takatani, and Kiyohiro Shikano

Graduate School of Information Science, Nara Institute of Science and Technology
8916-5 Takayama-cho, Ikoma, Nara, 630-0192, JAPAN
E-mail: sawatari@is.anist.jp

Abstract. In this chapter, we describe a new framework of blind source separation (BSS), i.e., Single-Input Multiple-Output (SIMO)-model-based ICA (SIMO-ICA), and we discuss its applicability to acoustic signal processing. The term "SIMO" represents a specific transmission system in which the input is a single source signal and the outputs are its transmitted signals observed at multiple microphones. The SIMO-ICA consists of multiple ICAs and a fidelity controller, and each ICA runs in parallel under the fidelity control of the entire separation system. In the SIMO-ICA scenario, unknown multiple source signals which are mixed through unknown acoustical transmission channels are detected at the microphones, and these signals can be separated, not into monaural source signals but into SIMO-model-based signals from independent sources as they are at the microphones. Thus, the separated signals of the SIMO-ICA can maintain the spatial qualities of each sound source. This attractive feature of the SIMO-ICA shows the promise of applicability to many high-fidelity acoustic signal processing systems. As a good examples of SIMO-ICA's application, binaural signal separation and blind separation–deconvolution processing are described.

5.1 Introduction

Blind source separation (BSS) is the approach taken to estimate original source signals using only the information of the mixed signals observed in each input channel. This technique is applicable to various fields of signal processing such as digital communications systems, radar antenna systems, and image and acoustic signal processing systems. One promising example in acoustic signal processing is a high-quality hands-free telecommunication systems involving a microphone array [1, 2].

Independent component analysis (ICA) [3] is commonly used in the BSS framework, and various ICA-based methods have been proposed for separation of acoustical sounds which corresponds to the convolutive mixture case [4–9]. However, the existing BSS methods can only separate the mixed sound sources into each *monaural* independent signal. Thus, from the practical point of view, these methods have a serious drawback in that the separated sounds cannot maintain information about the directivity, localization, or spatial qualities of each sound source. Since the above-mentioned information is essential for human hearing, the drawback prevents any BSS methods from being applied to high-fidelity audio technology, e.g., binaural signal processing

S. Makino et al. (eds.), Blind Speech Separation, 149–168.

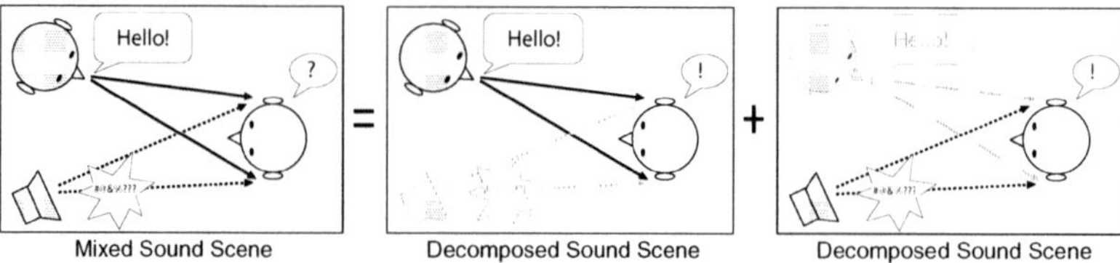

Fig. 5.1. Typical relation in sound scene decomposition processing. Mixed sounds are regarded as the superposition of Single-Input Multiple-Output (SIMO)-model-based acoustic signals from independent sources, and each SIMO component is separated and reproduced in SIMO-ICA framework.

[10], auditory signal processing, or sound reproduction systems [11], which construct an indispensable basis for audio virtual reality technology.

In this chapter, we review a newly proposed blind separation framework for *sound scene decomposition and reconstruction*, in which Single-Input Multiple-Output (SIMO)-model-based acoustic signals are mainly treated with the extended ICA algorithm, SIMO-ICA [12–15]. Here the term "SIMO" represents the specific transmission system in which the input is a single source signal and the outputs are its transmitted signals observed at multiple sensors. In the SIMO-ICA scenario (see Fig. 5.1), unknown multiple source signals which are mixed through unknown acoustical transmission channels are detected at the microphones. Here the mixed sounds are regarded as the superposition of SIMO-model-based acoustic signals from independent sources, and can be separated, not into monaural source signals but into SIMO-model-based signals from independent sources as they are at the microphones. Thus, the separated signals of SIMO-ICA can maintain the spatial qualities of each sound source, i.e., they represent the decomposed sound scenes. Obviously the attractive feature of SIMO-ICA is highly applicable to various audio systems because we can separately control, modify, and reproduce each of sound scenes in this framework.

As the good example of SIMO-ICA's application, first, binaural-sound separation experiments [14] are carried out under a reverberant condition. The experimental results reveal that (a) the signal separation performance of the proposed SIMO-ICA is the same as that of the conventional ICA, and (b) the sound quality of the separated signals in SIMO-ICA is remarkably superior to that in the conventional ICA, particularly for the spatial quality. Secondly, blind separation and deconvolution (BSD) of Multiple-Input Multiple-Output (MIMO) systems with colored inputs [16], are described.

5.2 Mixing Process and Conventional BSS

5.2.1 Mixing Process

In this study, the number of array elements (microphones) is K and the number of multiple sound sources is L. In general, the observed signals in which multiple source signals are mixed linearly are expressed as

$$\boldsymbol{x}(t) = \sum_{n=0}^{N-1} \boldsymbol{a}(n)\boldsymbol{s}(t-n) = \boldsymbol{A}(z)\boldsymbol{s}(t), \tag{5.1}$$

where $\boldsymbol{s}(t)$ is the source signal vector, $\boldsymbol{x}(t)$ is the observed signal vector, $\boldsymbol{a}(n)$ is the mixing filter matrix with the length of N, and $\boldsymbol{A}(z)$ is the z-transform of $\boldsymbol{a}(n)$; these are given as

$$\boldsymbol{s}(t) = [s_1(t), \ldots, s_L(t)]^{\mathrm{T}}, \tag{5.2}$$

$$\boldsymbol{x}(t) = [x_1(t), \ldots, x_K(t)]^{\mathrm{T}}, \tag{5.3}$$

$$\boldsymbol{a}(n) = \begin{bmatrix} a_{11}(n) & \cdots & a_{1L}(n) \\ \vdots & \ddots & \vdots \\ a_{K1}(n) & \cdots & a_{KL}(n) \end{bmatrix}, \tag{5.4}$$

$$\boldsymbol{A}(z) = [A_{ij}(z)]_{ij} = \left[\sum_{n=0}^{N-1} a_{ij}(n) z^{-n} \right]_{ij}, \tag{5.5}$$

where z^{-1} is used as the unit-delay operator, i.e., $z^{-n} \cdot x(t) = x(t-n)$, a_{kl} is the impulse response between the k-th microphone and the l-th sound source, and $[X]_{ij}$ denotes the matrix which includes the element X in the i-th row and the j-th column. Hereafter, we only deal with the case of $K = L$ in this paper.

5.2.2 Conventional ICA-Based BSS Method

As the conventional ICA algorithm, we consider the time-domain ICA (TDICA), in which each element of the separation matrix is represented as an FIR filter. In the TDICA, we optimize the separation matrix by using only the fullband observed signals without subband processing (see Fig. 5.2). The separated signal $\boldsymbol{y}(t)$ is expressed as

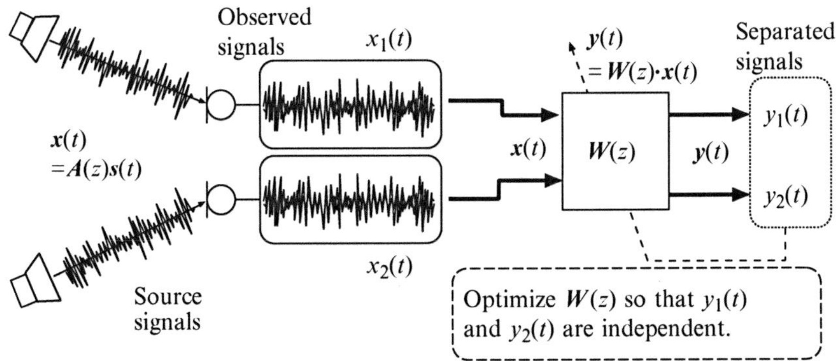

Fig. 5.2. Configuration of conventional TDICA, in which each element of the separation matrix is represented as an FIR filter. The separation filter matrix is optimized so that the output signals become mutually independent.

$$\begin{aligned}\boldsymbol{y}(t) &= [y_1(t), \dots, y_L(t)]^{\mathrm{T}} \\ &= \sum_{n=0}^{D-1} \boldsymbol{w}(n)\boldsymbol{x}(t-n) = \boldsymbol{W}(z)\boldsymbol{x}(t) \\ &= \boldsymbol{W}(z)\boldsymbol{A}(z)\boldsymbol{s}(t), \end{aligned} \tag{5.6}$$

where $\boldsymbol{w}(n)$ is the separation filter matrix, $\boldsymbol{W}(z)$ is the z-transform of $\boldsymbol{w}(n)$, and D is the filter length of $\boldsymbol{w}(n)$. In our study, the separation filter matrix is optimized by minimizing the Kullback–Leibler divergence (KLD) between the joint probability density function (PDF) of $\boldsymbol{y}(t)$ and the product of marginal PDFs of $y_l(t)$. The iterative learning rule is given by [6]

$$\begin{aligned}&\boldsymbol{w}^{[j+1]}(n) \\ &= \boldsymbol{w}^{[j]}(n) - \alpha \sum_{d=0}^{D-1} \left\{ \text{off-diag} \left\langle \boldsymbol{\varphi}(\boldsymbol{y}^{[j]}(t))\boldsymbol{y}^{[j]}(t-n+d)^{\mathrm{T}} \right\rangle_t \right\} \cdot \boldsymbol{w}^{[j]}(d),\end{aligned} \tag{5.7}$$

where α is the step-size parameter, the superscript $[j]$ is used to express the value of the j-th step in the iterations, $\langle \cdot \rangle_t$ denotes the time-averaging operator, and off-diag $\boldsymbol{W}(z)$ is the operation for setting every diagonal element of the matrix $\boldsymbol{W}(z)$ to be zero. Also, we define the nonlinear vector function $\boldsymbol{\varphi}(\cdot)$ as

$$\boldsymbol{\varphi}(\boldsymbol{y}(t)) = [\tanh(y_1(t)), \dots, \tanh(y_L(t))]^{\mathrm{T}}. \tag{5.8}$$

5.2.3 Problems in Conventional ICA

The conventional ICA is basically a means of extracting each of the independent sound sources as a monaural signal (see Fig. 5.3). In addition, the quality

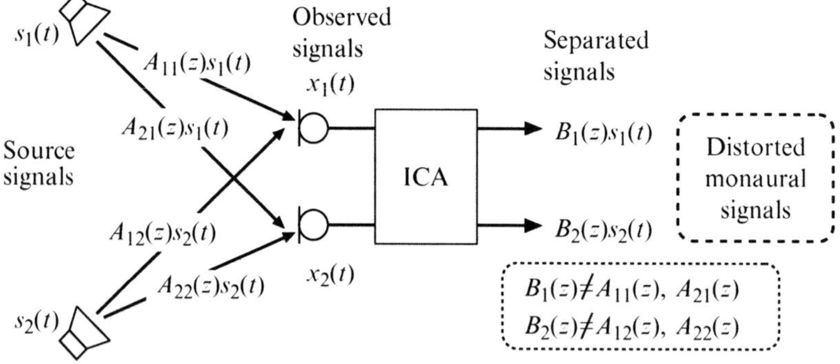

Fig. 5.3. Input and output relations in conventional ICA. Since $B_l(z)$ is possible to be an arbitrary filter ($B_l(z) \neq A_{kl}(z)$), the separated signals include the spectral distortions.

of the separated sound cannot be guaranteed, i.e., the separated signals can possibly include spectral distortions because the modified separated signals which convolved with arbitrary linear filters are still mutually independent. As shown in Fig. 5.3, $y_l(t) = B_l(z)s_l(t)$, where $B_l(z)$ ($\neq A_{kl}(z)$) is an arbitrary filter, is a possible solution obtained from the conventional ICA using (5.7). Therefore, the conventional ICA has a serious drawback in that the separated sounds cannot maintain information about the directivity, localization, or spatial qualities of each sound source. In order to resolve the problem, particularly for the sound quality, Matsuoka and Nakashima have proposed a modified ICA based on the Minimal Distortion Principle [17]. However, this method is valid only for monaural outputs, and the fidelity of the output signals as SIMO-model-based signals cannot be guaranteed.

5.3 SIMO-Model-Based ICA: Algorithm

In order to solve the above-mentioned problems, a new SIMO-model-based ICA algorithm, *SIMO-ICA*, has been recently proposed [13]. SIMO-ICA consists of $(L-1)$ ICA parts and a *fidelity controller*, and each ICA runs in parallel under the fidelity control of the entire separation system (see Fig. 5.4). The separated signals of the l-th ICA ($l = 1, \ldots, L-1$) in SIMO-ICA are defined by

$$\boldsymbol{y}_{(\mathrm{ICA}l)}(t) = \sum_{n=0}^{D-1} \boldsymbol{w}_{(\mathrm{ICA}l)}(n)\boldsymbol{x}(t-n), \tag{5.9}$$

where $\boldsymbol{w}_{(\mathrm{ICA}l)}(n) = [w_{ij}^{(\mathrm{ICA}l)}(n)]_{ij}$ is the separation filter matrix in the l-th ICA. Regarding the fidelity controller, we calculate the following signal vector, in which the all elements are to be mutually independent,

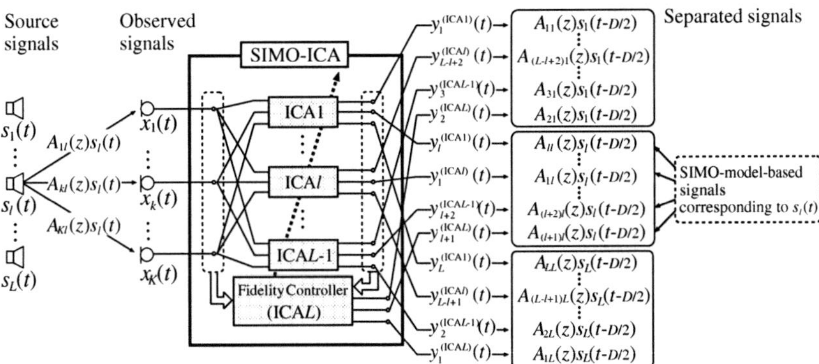

Fig. 5.4. Example of input and output relations in proposed SIMO-ICA, where exclusively-selected permutation matrices $\boldsymbol{P}_l$ are given by (5.21). The SIMO-ICA consists of multiple ICA parts and a fidelity controller, and each ICA runs in parallel under the fidelity control of the entire separation system. In this system, the separated signals maintain their spatial qualities.

$$\boldsymbol{y}_{(\mathrm{ICA}L)}(t) = \boldsymbol{x}(t - D/2) - \sum_{l=1}^{L-1} \boldsymbol{y}_{(\mathrm{ICA}l)}(t). \tag{5.10}$$

Hereafter, we regard $\boldsymbol{y}_{(\mathrm{ICA}L)}(t)$ as an output of a *virtual* "L-th" ICA, and define its *virtual* separation filter as

$$\boldsymbol{w}_{(\mathrm{ICA}L)}(n) = \boldsymbol{I}\delta\left(n - \frac{D}{2}\right) - \sum_{l=1}^{L-1} \boldsymbol{w}_{(\mathrm{ICA}l)}(n), \tag{5.11}$$

where $\boldsymbol{I}$ is the identity matrix, and $\delta(n)$ is a delta function. From (5.11), we can rewrite (5.10) as

$$\boldsymbol{y}_{(\mathrm{ICA}L)}(t) = \sum_{n=0}^{D-1} \boldsymbol{w}_{(\mathrm{ICA}L)}(n)\boldsymbol{x}(t - n). \tag{5.12}$$

The reason why we use the word "*virtual*" here is that the L-th ICA does not have own separation filters unlike the other ICAs, and $\boldsymbol{w}_{(\mathrm{ICA}L)}(n)$ is subject to $\boldsymbol{w}_{(\mathrm{ICA}l)}(n)$ $(l = 1, \ldots, L-1)$.

To explicitly show the meaning of the fidelity controller, we rewrite (5.10) as

$$\sum_{l=1}^{L} \boldsymbol{y}_{(\mathrm{ICA}l)}(t) - \boldsymbol{x}(t - D/2) = 0. \tag{5.13}$$

Equation (5.13) means a constraint to force the sum of all ICAs' output vectors $\sum_{l=1}^{L} \boldsymbol{y}_{(\mathrm{ICA}l)}(t)$ to be the sum of all SIMO components $[\sum_{l=1}^{L} A_{kl}(z)s_l(t - D/2)]_{k1}(= \boldsymbol{x}(t - D/2))$. Using (5.9) and (5.10), we can obtain the appropriate separated signals and maintain their spatial qualities as follows.

Theorem: If the independent sound sources are separated by (5.9), and simultaneously the signals obtained by (5.10) are also mutually independent, then the output signals converge on unique solutions, up to the permutation, as

$$\boldsymbol{y}_{(\mathrm{ICA}l)}(t) = \mathrm{diag}\left[\boldsymbol{A}(z)\boldsymbol{P}_l^{\mathrm{T}}\right]\boldsymbol{P}_l\boldsymbol{s}(t - D/2), \tag{5.14}$$

where $\boldsymbol{P}_l$ $(l = 1, \ldots, L)$ are exclusively-selected permutation matrices which satisfy

$$\sum_{l=1}^{L} \boldsymbol{P}_l = [1]_{ij}. \tag{5.15}$$

Proof of Theorem: The necessity is obvious. The sufficiency is shown below. Let $\boldsymbol{D}_l(z)$ $(l = 1, \ldots, L)$ be arbitrary diagonal polynomial matrices and $\boldsymbol{Q}_l$ be arbitrary permutation matrices. The general expression of the l-th ICA's output is given by

$$\boldsymbol{y}_{(\mathrm{ICA}l)}(t) = \boldsymbol{D}_l(z)\boldsymbol{Q}_l\boldsymbol{s}\left(t - \frac{D}{2}\right). \tag{5.16}$$

If $\boldsymbol{Q}_l$ are not exclusively selected matrices, i.e., $\sum_{l=1}^{L} \boldsymbol{Q}_l \neq [1]_{ij}$, then there exists at least one element of $\sum_{l=1}^{L} \boldsymbol{y}_{(\mathrm{ICA}l)}(t)$ which does not include all of the components of $s_l(t-D/2)$ $(l=1,\ldots,L)$. This obviously makes the left-hand side of (5.13) nonzero because the observed signal vector $\boldsymbol{x}(t-D/2)$ includes all of the components of $s_l(t-D/2)$ in each element. Accordingly, $\boldsymbol{Q}_l$ should be $\boldsymbol{P}_l$ specified by (5.15), and we obtain

$$\boldsymbol{y}_{(\mathrm{ICA}l)}(t) = \boldsymbol{D}_l(z)\boldsymbol{P}_l\boldsymbol{s}\left(t-\frac{D}{2}\right). \tag{5.17}$$

In (5.17) under (5.15), the arbitrary diagonal matrices $\boldsymbol{D}_l(z)$ can be substituted by $\mathrm{diag}[\boldsymbol{B}(z)\boldsymbol{P}_l^{\mathrm{T}}]$, where $\boldsymbol{B}(z) = [B_{ij}(z)]_{ij}$ is a single arbitrary matrix, because all diagonal entries of $\mathrm{diag}[\boldsymbol{B}(z)\boldsymbol{P}_l^{\mathrm{T}}]$ for all l are also exclusive. Thus,

$$\boldsymbol{y}_{(\mathrm{ICA}l)}(t) = \mathrm{diag}\left[\boldsymbol{B}(z)\boldsymbol{P}_l^{\mathrm{T}}\right]\boldsymbol{P}_l\boldsymbol{s}\left(t-\frac{D}{2}\right). \tag{5.18}$$

The substitution of (5.18) into (5.10) leads to the following equation:

$$\begin{aligned}
&\mathrm{diag}\left[\boldsymbol{B}(z)\boldsymbol{P}_L^{\mathrm{T}}\right]\boldsymbol{P}_L\boldsymbol{s}\left(t-\frac{D}{2}\right)\\
&= \boldsymbol{x}\left(t-\frac{D}{2}\right) - \sum_{l=1}^{L-1}\mathrm{diag}\left[\boldsymbol{B}(z)\boldsymbol{P}_l^{\mathrm{T}}\right]\boldsymbol{P}_l\boldsymbol{s}\left(t-\frac{D}{2}\right),
\end{aligned} \tag{5.19}$$

and consequently

$$\begin{aligned}
&\sum_{l=1}^{L}\mathrm{diag}\left[\boldsymbol{B}(z)\boldsymbol{P}_l^{\mathrm{T}}\right]\boldsymbol{P}_l\boldsymbol{s}\left(t-\frac{D}{2}\right) - \boldsymbol{x}\left(t-\frac{D}{2}\right)\\
&= \left[\sum_{l=1}^{L} B_{kl}(z)s_l\left(t-\frac{D}{2}\right)\right]_{k1} - \left[\sum_{l=1}^{L} A_{kl}(z)s_l\left(t-\frac{D}{2}\right)\right]_{k1}\\
&= \left[\sum_{l=1}^{L}\left\{B_{kl}(z)-A_{kl}(z)\right\}s_l\left(t-\frac{D}{2}\right)\right]_{k1}\\
&= [0]_{ij}.
\end{aligned} \tag{5.20}$$

Equation (5.20) is satisfied if and only if $B_{kl}(z) = A_{kl}(z)$ for all k and l. Thus, (5.18) results in (5.14). This completes the Proof of Theorem.

Obviously the solutions given by (5.14) provide necessary and sufficient SIMO components, $A_{kl}(z)S_l(t-D/2)$, for each l-th source. However, the condition (5.15) allows multiple possibilities for the combination of $\boldsymbol{P}_l$. For example, one possibility is shown in Fig. 5.4 and this corresponds to

$$\boldsymbol{P}_l = [\delta_{im(k,l)}]_{ki}, \tag{5.21}$$

where δ_{ij} is Kronecker's delta function, and

$$m(k,l) = \begin{cases} k+l-1 & (k+l-1 \leq L)\\ k+l-1-L & (k+l-1 > L) \end{cases}. \tag{5.22}$$

In this case, (5.14) yields

$$\boldsymbol{y}_{(\mathrm{ICA}l)}(t) = [A_{km(k,l)} s_{m(k,l)}(t - D/2)]_{k1} \quad (l = 1, \ldots, L). \tag{5.23}$$

In order to obtain (5.14), the natural gradient [4] of KLD of (5.12) with respect to $\boldsymbol{w}_{(\mathrm{ICA}l)}(n)$ should be added to the iterative learning rule of the separation filter in the l-th ICA ($l = 1, \ldots, L-1$). Using (5.11), we obtain the partial differentiation of the KLD, $\mathrm{KL}(\boldsymbol{y}_{(\mathrm{ICA}L)}(t))$, with respect to $\boldsymbol{w}_{(\mathrm{ICA}l)}(n)$ ($l = 1, \ldots, L-1$) as

$$\begin{aligned}\frac{\partial \mathrm{KL}(\boldsymbol{y}_{(\mathrm{ICA}L)}(t))}{\partial \boldsymbol{w}_{(\mathrm{ICA}l)}(n)} &= \left[\frac{\partial \mathrm{KL}(\boldsymbol{y}_{(\mathrm{ICA}L)}(t))}{\partial w_{ij}^{(\mathrm{ICA}L)}(n)} \cdot \frac{\partial w_{ij}^{(\mathrm{ICA}L)}(n)}{\partial w_{ij}^{(\mathrm{ICA}l)}(n)}\right]_{ij} \\ &= \left[\frac{\partial \mathrm{KL}(\boldsymbol{y}_{(\mathrm{ICA}L)}(t))}{\partial w_{ij}^{(\mathrm{ICA}L)}(n)} \cdot (-1)\right]_{ij},\end{aligned} \tag{5.24}$$

where $w_{ij}^{(\mathrm{ICA}L)}(n)$ is the element of $\boldsymbol{w}_{(\mathrm{ICA}L)}(n)$. Thus, the natural gradient of (5.24) is given as

$$\begin{aligned}&-\frac{\partial \mathrm{KL}(\boldsymbol{y}_{(\mathrm{ICA}L)}(t))}{\partial \boldsymbol{w}_{(\mathrm{ICA}L)}(n)} \cdot \boldsymbol{W}_{(\mathrm{ICA}L)}(z^{-1})^{\mathrm{T}} \boldsymbol{W}_{(\mathrm{ICA}L)}(z) \\ &= \sum_{d=0}^{D-1} \left\{ \left(\boldsymbol{I}\delta(n-d) - \left\langle \boldsymbol{\varphi}\left(\boldsymbol{x}\left(t - \frac{D}{2}\right)\right.\right.\right.\right. \\ &\quad \left.\left. - \sum_{l=1}^{L-1} \boldsymbol{y}_{(\mathrm{ICA}l)}(t) \right) \cdot \left(\boldsymbol{x}\left(t - n + d - \frac{D}{2}\right)\right. \\ &\quad \left.\left.\left.\left. - \sum_{l=1}^{L-1} \boldsymbol{y}_{(\mathrm{ICA}l)}(t - n + d)^{\mathrm{T}} \right) \right\rangle_t \right\} \\ &\quad \cdot \left(\boldsymbol{I}\delta\left(d - \frac{D}{2}\right) - \sum_{l=1}^{L-1} \boldsymbol{w}_{(\mathrm{ICA}l)}(d) \right),\end{aligned} \tag{5.25}$$

where $\boldsymbol{W}_{(\mathrm{ICA}L)}(z)$ is the z-transform of $\boldsymbol{w}_{(\mathrm{ICA}L)}(n)$. In order to deal with the colored signals, we apply the nonholonomic constraint to (5.25), and then combine the modified (5.25) with an existing ICA algorithm [6]. The new iterative algorithm of the l-th ICA part ($l = 1, \ldots, L-1$) in SIMO-ICA is given as

$$
\begin{aligned}
&\boldsymbol{w}_{(\mathrm{ICA}l)}^{[j+1]}(n) \\
&= \boldsymbol{w}_{(\mathrm{ICA}l)}^{[j]}(n) - \alpha \sum_{d=0}^{D-1} \left[\left\{ \text{off-diag} \left\langle \boldsymbol{\varphi}\left(\boldsymbol{y}_{(\mathrm{ICA}l)}^{[j]}(t) \right) \right. \right. \right. \\
&\quad \left. \left. \boldsymbol{y}_{(\mathrm{ICA}l)}^{[j]}(t-n+d)^{\mathrm{T}} \right\rangle_t \right\} \cdot \boldsymbol{w}_{(\mathrm{ICA}l)}^{[j]}(d) \\
&\quad - \left\{ \text{off-diag} \left\langle \boldsymbol{\varphi}\left(\boldsymbol{x}\left(t - \frac{D}{2}\right) - \sum_{l=1}^{L-1} \boldsymbol{y}_{(\mathrm{ICA}l)}^{[j]}(t) \right) \right. \right. \\
&\quad \left. \left. \cdot \left(\boldsymbol{x}\left(t-n+d-\frac{D}{2}\right) - \sum_{l=1}^{L-1} \boldsymbol{y}_{(\mathrm{ICA}l)}^{[j]}(t-n+d)^{\mathrm{T}} \right) \right\rangle_t \right\} \\
&\quad \left. \cdot \left(\boldsymbol{I}\delta\left(d - \frac{D}{2}\right) - \sum_{l=1}^{L-1} \boldsymbol{w}_{(\mathrm{ICA}l)}^{[j]}(d) \right) \right],
\end{aligned} \tag{5.26}
$$

where α is the step-size parameter. In (5.26), the updating $\boldsymbol{w}_{(\mathrm{ICA}l)}(n)$ for all l should be simultaneously performed in parallel because each iterative equation is associated with the others via $\sum_{l=1}^{L-1} \boldsymbol{y}_{(\mathrm{ICA}l)}^{[j]}(t)$. Also, the initial values of $\boldsymbol{w}_{(\mathrm{ICA}l)}(n)$ for all l should be different.

After the iterations, the separated signals should be classified into SIMO components of each source because the permutation arises. This can be easily achieved by using a cross correlation between time-shifted separated signals, $\max_n \langle y_k^{(l)}(t) y_{k'}^{(l')}(t-n) \rangle_t$, where $l \neq l'$ and $k \neq k'$. The large value of the correlation indicates that $y_k^{(l)}(t)$ and $y_{k'}^{(l')}(t)$ are SIMO components of the same sources.

5.4 Application of SIMO-ICA: Binaural Sound Separation

5.4.1 Conditions for Experiments

We carried out binaural sound separation experiments [14]. In this experiment, speech signals are convolved with impulse responses which is recorded using Head and Torso Simulator (HATS by Brüel & Kjær) under the experimental room as shown in Fig. 5.5. The speech signals are assumed to arrive from two directions, −30° and 45°. The distance between HATS and the loudspeakers is 1.5 m. Two kinds of sentences, spoken by two male and two female speakers, are used as the original speech samples. Using these sentences, we obtain 6 combinations. The sampling frequency is 8 kHz and the length of speech is limited to 3 seconds. The length of $\boldsymbol{w}(n)$ is 512, and the initial values are inverse filters of HRTFs whose directions of

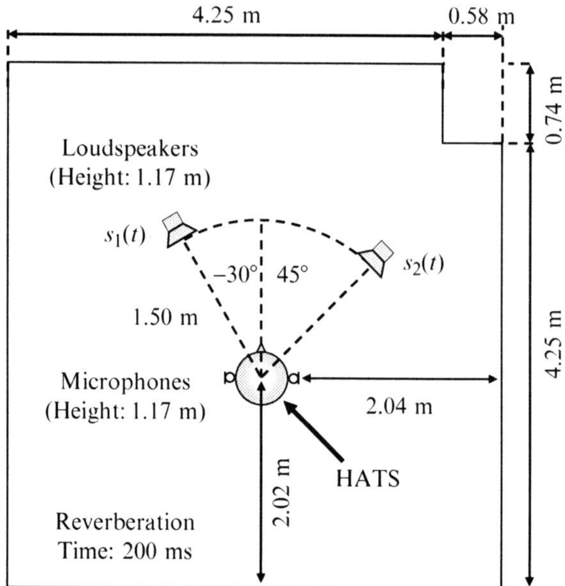

Fig. 5.5. Layout of reverberant room used in binaural-sound separation experiments.

sources are ±60°. The number of iterations in ICA is 5000. Regarding the conventional ICA for comparison, we used the nonholonomic ICA [6]. The step-size parameter α is changed from 1.0×10^{-8} to 2.0×10^{-6} to find the optima.

5.4.2 Results and Discussion

SIMO-model accuracy (SA) is used as an evaluation score. The SA is defined as

$$\mathrm{SA} = \frac{1}{K}\frac{1}{L}\sum_{k=1}^{K}\sum_{l=1}^{L} 10\log_{10}\frac{\| A_{km(k,l)}(z)s_{m(k,l)}(t) \|^2}{\| y_k^{(\mathrm{ICA}l)}(t) - A_{km(k,l)}(z)s_{m(k,l)}(t) \|^2}. \tag{5.27}$$

The SA indicates a degree of the similarity between the separated signals of the ICA and real SIMO-model-based signals. Figure 5.6 shows the results of SA for different speaker combinations. The bars on the right of this figure correspond to the averaged results of each combination. In the averaged scores, the improvement of SA in SIMO-ICA is 9.5 dB compared with the conventional ICA. From these results, it is evident that the separated signals in the SIMO-ICA is obviously superior to that in the conventional ICA-based method. Thus, we can conclude that SIMO-ICA has the potential to decompose the mixed binaural signals into SIMO-model-based signals without loss of information about spatial qualities of each sound source.

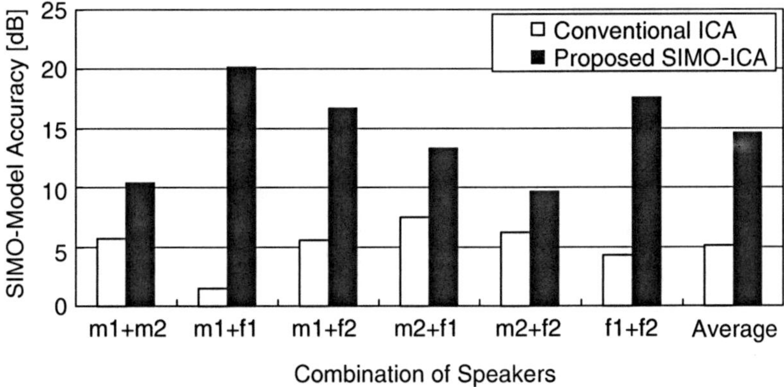

Fig. 5.6. Results of SIMO-model accuracy in separation experiments of binaural sounds recorded using HATS. The labels "m1" and "m2" mean two male speakers, and "f1" and "f2" mean two female speakers.

5.5 Application of SIMO-ICA: BSD of MIMO Systems Driven by Colored Sources

5.5.1 System Overview

In this section, we address the BSD problem for colored signals which are very common in many practical situations. However, this was considered as a tough problem because of the difficulty that how we can apply the deconvolution only to the mixing system without destroying the coloration characteristics of the sources.

Meanwhile, in the previous section, SIMO-ICA was proposed and we showed that SIMO-ICA can separate the mixed colored signals into SIMO-model-based signals at the microphone points. This finding has motivated us to combine our SIMO-ICA and an existing blind deconvolution technique for the SIMO model [18–20]. The configuration of the proposed method [16] is depicted in Fig. 5.7. In the proposed method, the separation–deconvolution problem is resolved into two stages, an SIMO-model-based separation stage and an SIMO-model-based deconvolution stage. Each of the separation–deconvolution problems can be solved efficiently using the following reasonable properties. **(a)** The assumption of the mutual independence among the acoustic sound sources usually holds, and consequently, this can be used in the SIMO-ICA-based separation. **(b)** The temporal-correlation property of the source signals and the nonminimum phase property of the mixing system can be taken into account in the blind multichannel inverse filtering for the SIMO model.

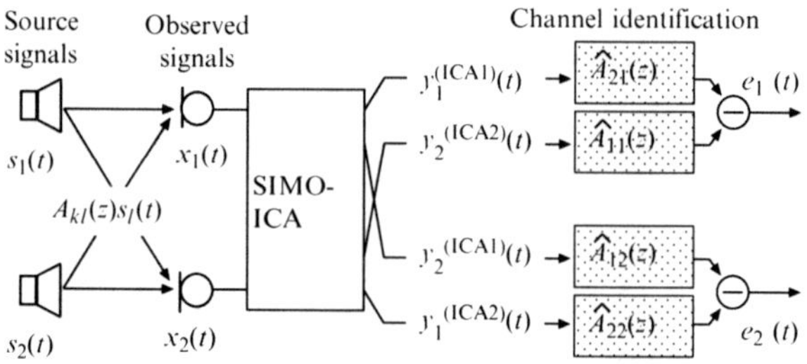

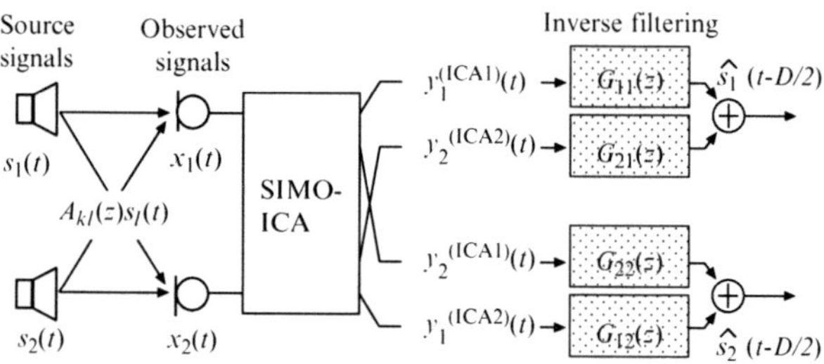

Fig. 5.7. Input and output relations in (a) blind channel identification and (b) multichannel inverse filtering performed in second stage. This corresponds to the specific case of (5.21) with $K = L = 2$.

5.5.2 Conditions for Experiment

In illustration of the BSD method, we carried out a computer-simulation-based experiment. The mixing filter matrix $\boldsymbol{A}(z)$ is taken to be

$$A_{11}(z) = 1 - 0.7z^{-1} - 0.3z^{-2}, \tag{5.28}$$
$$A_{21}(z) = z^{-1} + 0.7z^{-2} + 0.4z^{-3}, \tag{5.29}$$
$$A_{12}(z) = z^{-1} + 0.7z^{-2} + 0.4z^{-3}, \tag{5.30}$$
$$A_{22}(z) = 1 - 0.7z^{-1} - 0.3z^{-2}. \tag{5.31}$$

The impulse responses $a_{ij}(n)$ corresponding to $A_{ij}(z)$ are shown in Fig. 5.8. Two sentences spoken by two male speakers are used as the original speech samples $\boldsymbol{s}(t)$. The sampling frequency is 8 kHz and the length of speech is limited to 7 seconds. The number of iterations in ICA is 15000.

Cepstral distortion (CD) is used as the indication of both separation and deconvolution performances. In this study, we defined the CD as the distance between the spectral envelope of the original source signal $s_l(t - D/2)$ and that of the separated output. The 16th-order Mel-scaled cepstrum based on the smoothed FFT spectrum is used. The CD gives the final BSD score, and will be decreased to zero if the separation–deconvolution processing is performed perfectly.

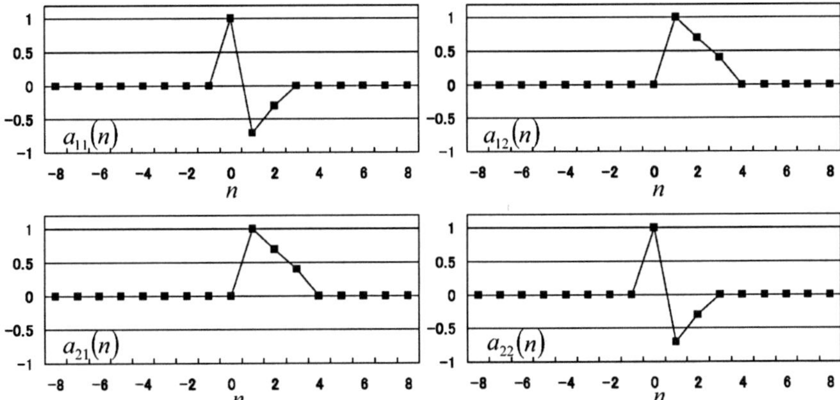

Fig. 5.8. Elements of mixing matrix $\boldsymbol{a}(n)$ used in simulation, which are given by inverse z-transform of (5.28)–(5.31).

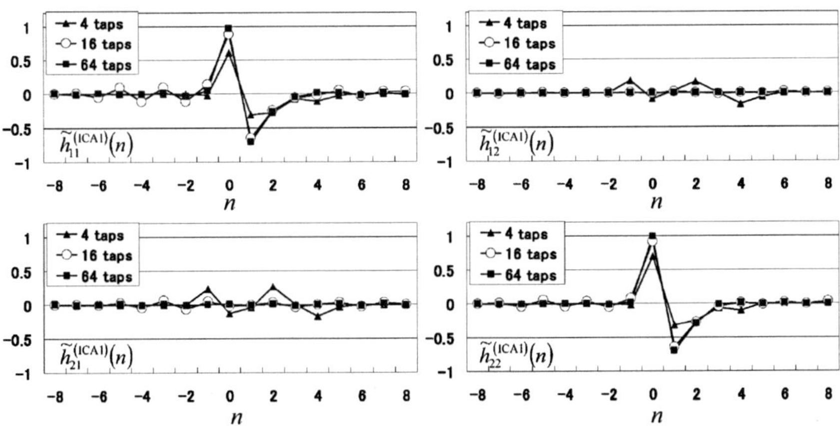

Fig. 5.9. Elements of composite filter matrix $[\tilde{h}_{ij}^{(\mathrm{ICA1})}(n)]_{ij} = z^{D/2}\sum_{d=0}^{D-1}\boldsymbol{w}_{(\mathrm{ICA1})}(d)\boldsymbol{a}(n-d)$ for different filter lengths D. This represents the whole system characteristics of the mixing and separation processes in the first ICA part of the SIMO-ICA.

5.5.3 Results and Discussion

Figures 5.9 and 5.10 show the elements of a *composite* filter matrix of $\boldsymbol{w}_{(\mathrm{ICA}l)}(n)$ and $\boldsymbol{a}(n)$. The composite filter matrix represents the whole system characteristics of the mixing and separation processes in SIMO-ICA; this is defined as

$$\left[\tilde{h}_{ij}^{(\mathrm{ICA1})}(n)\right]_{ij} = z^{D/2}\sum_{d=0}^{D-1}\boldsymbol{w}_{(\mathrm{ICA1})}(d)\boldsymbol{a}(n-d), \tag{5.32}$$

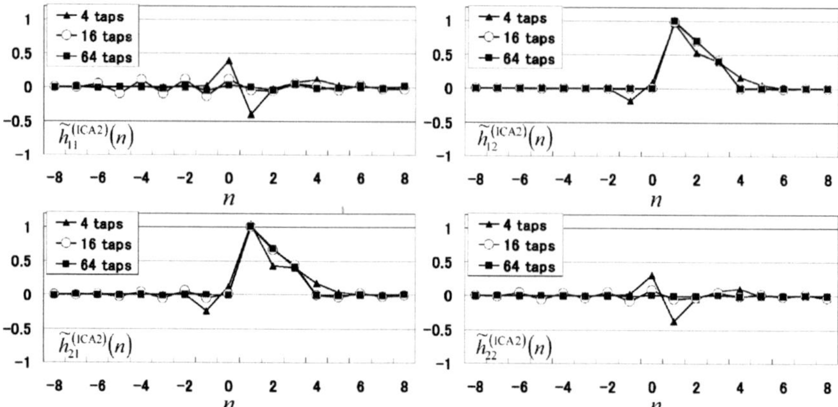

Fig. 5.10. Elements of composite filter matrix $[\tilde{h}_{ij}^{(\mathrm{ICA2})}(n)]_{ij} = z^{D/2}\sum_{d=0}^{D-1} \boldsymbol{w}_{(\mathrm{ICA2})}(d)\boldsymbol{a}(n-d)$ for different filter lengths D. This represents the whole system characteristics of the mixing and separation processes in the second ICA part of the SIMO-ICA.

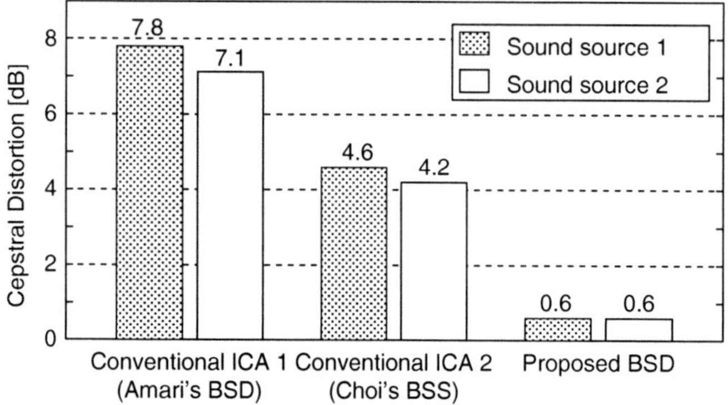

Fig. 5.11. Results of CD in BSD experiments.

$$\left[\tilde{h}_{ij}^{(\mathrm{ICA2})}(n)\right]_{ij} = z^{D/2}\sum_{d=0}^{D-1} \boldsymbol{w}_{(\mathrm{ICA2})}(d)\boldsymbol{a}(n-d), \tag{5.33}$$

where $z^{D/2}$ is used to cancel the time delay of $D/2$ in $\boldsymbol{w}_{(\mathrm{ICA}l)}(n)$ for centering the impulse responses in each of the figures. From these figures, it is confirmed that the composite filter matrix for the first ICA, $[\tilde{h}_{ij}^{(\mathrm{ICA1})}(n)]_{ij}$, becomes $\mathrm{diag}[\boldsymbol{a}(n)]$ as the length of the separation filter, D, is increased to more than the length of the mixing system. Also, the composite filter matrix for the second ICA, $[\tilde{h}_{ij}^{(\mathrm{ICA2})}(n)]_{ij}$, becomes off-$\mathrm{diag}[\boldsymbol{a}(n)]$ as the length of the separation filter is increased. This obviously indicates that each ICA part in the SIMO-ICA can work so as to separately extract each SIMO component.

Figure 5.11 shows the results of CD for different methods, namely, conventional ICA-based BSD (Amari's holonomic ICA [4]), conventional BSS

(Choi's nonholonomic ICA [6]), and the proposed BSD. First, it is evident that the CD of Amari's BSD is obviously high, i.e., the resultant speech is whitened by the decorrelation in the conventional method. Next, the result of the nonholonomic ICA shows that there are still some distortions in the separated signals. Finally, regarding the results of the proposed method, there is a considerable reduction of CD. This indicates that the proposed BSD algorithm can successfully achieve the separation and deconvolution for a convolutive mixture of temporally correlated signals. This can also be confirmed from the results of the power spectra in Figs. 5.12 and 5.13; the power spectra of the original source and the proposed BSD are almost the same.

5.6 Conclusion

In this chapter, we described a new framework of SIMO-model-based ICA and its applicability to acoustic signal processing. The SIMO-ICA can separate the mixed signals into SIMO-model-based signals which can maintain the spatial qualities. As the good examples of SIMO-ICA's application, we apply SIMO-ICA to the blind source separation problem of the binaural sounds. The experimental results reveal that the performance of the proposed SIMO-ICA is superior to that of the conventional ICA-based method, and the separated signals of SIMO-ICA maintain the spatial qualities of each binaural sound source. This can show the promising applicability of the SIMO-ICA to various high-fidelity audio systems.

In very recent years, it has been reported that SIMO-ICA can be combined with nonlinear postprocessing such as (time–frequency) binary masking [21]. It should be emphasized that the proposed two-stage method has important property, i.e., applicability to *real-time* processing. In general, SIMO-ICA-based BSS methods require enormous calculations, but binary masking needs very low computational complexities. Therefore, because of the introduction of binary masking into SIMO-ICA, the proposed combination can function as a real-time system. We have already built a pocket-size real-time BSS module, where the proposed two-stage BSS algorithm can work on a general-purpose DSP (TEXAS INSTRUMENTS TMS320C6713; 200 MHz clock, 100 kB program size, 1 MB working memory) as shown in Fig. 5.14. Figure 5.15 shows a configuration of a real-time implementation for the proposed two-stage BSS. Many applications of the SIMO-ICA-based BSS are very promising.

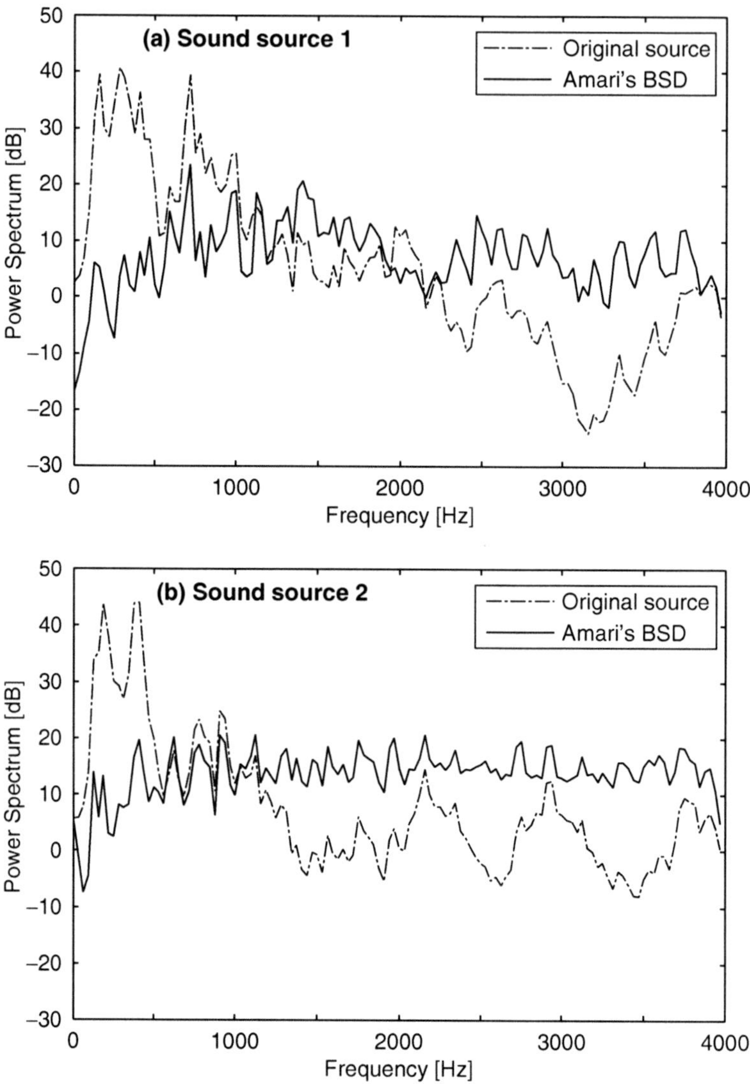

Fig. 5.12. Long-term averaged power spectra of (a) sound source 1 and (b) sound source 2 in original sources and Amari's BSD. The short-term analysis with the 256-th order FFT is applied to each speech segment, and all of the short-term spectra are averaged along with the whole data.

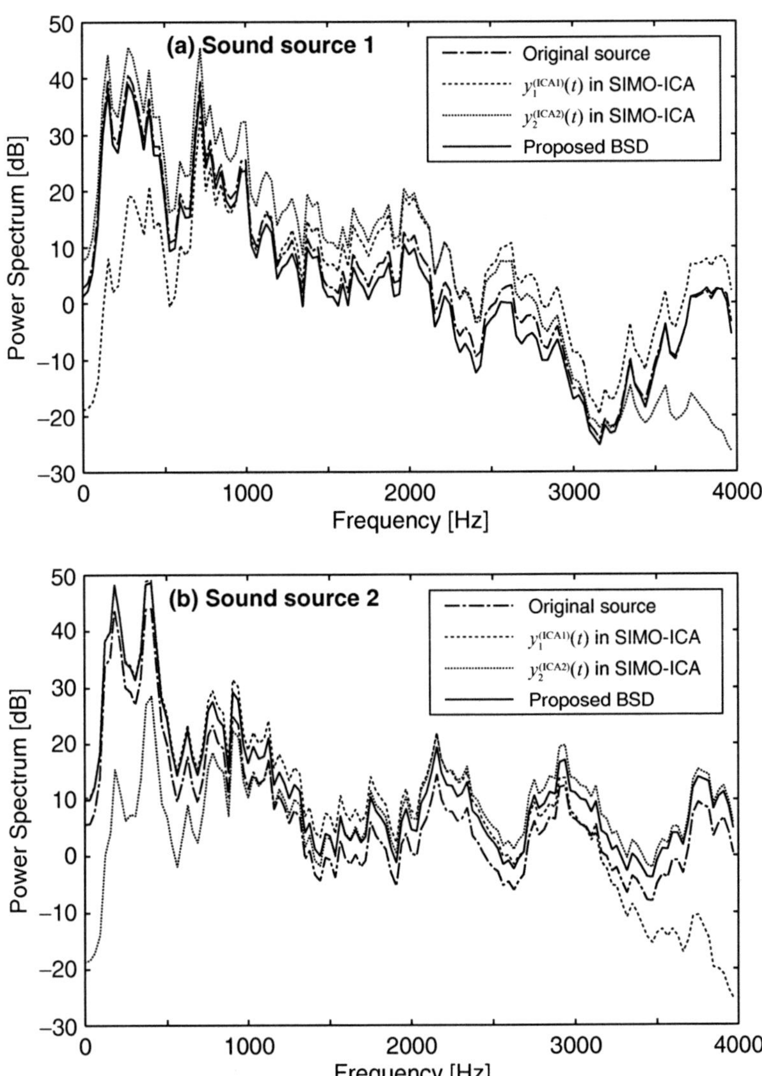

Fig. 5.13. Long-term averaged power spectra of (a) sound source 1 and (b) sound source 2 in original sources, SIMO-ICA, and proposed BSD. The short-term analysis with the 256-th order FFT is applied to each speech segment, and all of the short-term spectra are averaged along with the whole data.

Fig. 5.14. Overview of pocket-size real-time BSS module, where SIMO-ICA-based two-stage BSS algorithm [21] works on TEXAS INSTRUMENTS TMS320C6713 DSP.

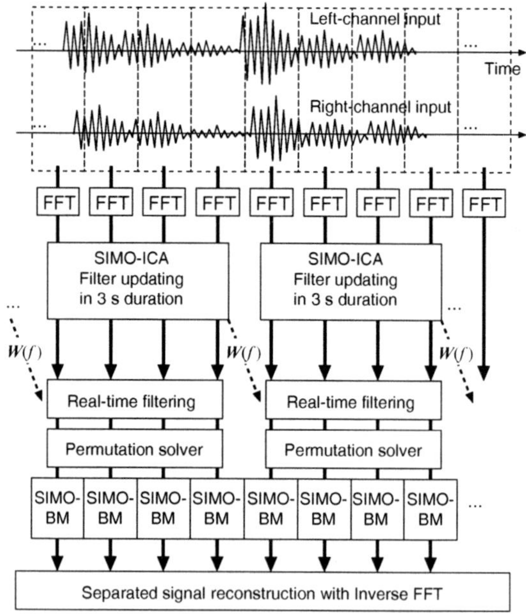

Fig. 5.15. Signal flow in real-time implementation of SIMO-ICA-based two-stage BSS [21].

References

1. G. W. Elko, "Microphone array systems for hands-free telecommunication," *Speech Communication*, vol. 20, pp. 229–240, 1996.
2. J. L. Flanagan, J. D. Johnston, R. Zahn, and G. W. Elko, "Computer-steered microphone arrays for sound transduction in large rooms," *J. Acoust. Soc. Am.*, vol. 78, pp. 1508–1518, 1985.
3. P. Comon, "Independent component analysis, a new concept?" *Signal Processing,* vol. 36, pp. 287–314, 1994.
4. S. Amari, S. Douglas, A. Cichocki, and H. Yang, "Multichannel blind deconvolution and equalization using the natural gradient, *Proc. Int. Workshop on Wireless Commun.*, pp. 101–104, 1997.
5. P. Smaragdis, "Blind separation of convolved mixtures in the frequency domain," *Neurocomputing*, vol. 22, pp. 21–34, 1998.
6. S. Choi, S. Amari, A. Cichocki, and R. Liu, "Natural gradient learning with a nonholonomic constraint for blind deconvolution of multiple channels," *Proc. Int. Workshop on ICA and BSS*, pp. 371–376, 1999.
7. S. Araki, R. Mukai, S. Makino, T. Nishikawa, and H. Saruwatari, "The fundamental limitation of frequency domain blind source separation for convolutive mixtures of speech," *IEEE Transactions on Speech and Audio Processing*, vol. 11, no. 2, pp. 109–116, 2003.
8. H. Saruwatari, S. Kurita, K. Takeda, F. Itakura, T. Nishikawa, and K. Shikano, "Blind source separation combining independent component analysis and beamforming," *EURASIP Journal on Applied Signal Processing* vol. 2003, no. 11, pp. 1135–1146, 2003.
9. H. Saruwatari, T. Kawamura, T. Nishikawa, A. Lee, and K. Shikano, "Blind source separation based on a fast-convergence algorithm combining ICA and beamforming," *IEEE Transactions on Speech and Audio Processing*, vol. 14, no. 2, pp. 666–678, 2006.
10. J. Blauert, *Spatial Hearing (revised edition).* Cambridge, MA: The MIT Press, 1997.
11. Y. Tatekura, H. Saruwatari, and K. Shikano, "Sound reproduction system including adaptive compensation of temperature fluctuation effect for broad-band sound control," *IEICE Trans. Fundamentals,* vol. E85–A, no. 8, pp. 1851–1860, Aug. 2002.
12. T. Takatani, T. Nishikawa, H. Saruwatari, and K. Shikano, "High-fidelity blind separation of acoustic signals using SIMO-model-based independent component analysis," *IEICE Trans. Fundamentals*, vol. E87-A, no. 8, pp. 2063–2072, 2004.
13. T. Takatani, T. Nishikawa, H. Saruwatari, and K. Shikano, "High-fidelity blind source separation of acoustic signals using SIMO-model-based ICA with information-geometric learning," *Proc. IWAENC2003,* pp. 251–254, 2003.
14. T. Takatani, T. Nishikawa, H. Saruwatari, and K. Shikano, "Blind separation of binaural sound mixtures using SIMO-model-based independent component analysis," *ICASSP2004*, vol. IV, pp. 113–116, May 2004.
15. T. Takatani, S. Ukai, T. Nishikawa, H. Saruwatari, and K. Shikano, "A self-generator method for initial filters of SIMO-ICA applied to blind separation of binaural sound mixtures," *IEICE Trans. Fundamentals*, vol. E88-A, no. 7, pp. 1673–1682, 2005.

16. H. Saruwatari, H. Yamajo, T. Takatani, T. Nishikawa, and K. Shikano, "Blind separation and deconvolution for convolutive mixture of speech combining SIMO-model-based ICA and multichannel inverse filtering," *IEICE Trans. Fundamentals*, vol. E88-A, no. 9, pp. 2387–2400, 2005.
17. K. Matsuoka and S. Nakashima, "Minimal distortion principle for blind source separation," *Proc. International Conference on Independent Component Analysis and Blind Signal Separation*, pp. 722–727, Dec. 2001.
18. H. Xu and L. Tong, "A deterministic approach to blind identification of multichannel FIR systems," *Proc. ICASSP94*, vol. IV, pp. 581–584, 1994.
19. K. Furuya and Y. Kaneda, "Two-channel blind deconvolution of nonminimum phase FIR system," *IEICE Trans. Fundamentals*, vol. E80-A, no. 5, pp. 804–808, 1997.
20. Z. Ding and Y. Li, *Blind Equalization and Identification*, Marcel Dekker, Inc., New York, 2001.
21. Y. Mori, H. Saruwatari, T. Takatani, S. Ukai, K. Shikano, T. Hiekata, Y. Ikeda, H. Hashimoto, and T. Morita, "Blind separation of acoustic signals combining SIMO-model-based independent component analysis and binary masking," *EURASIP Journal on Applied Signal Processing*, vol. 2006, Article ID 34970, 17 pages, 2006.

6 Independent Vector Analysis for Convolutive Blind Speech Separation

Intae Lee, Taesu Kim, and Te-Won Lee

Institute for Neural Computation, University of California, San Diego
9500 Gilman Drive Dept. 0523 La Jolla, California 92093-0523
E-mail: intelli@ucsd.edu

Abstract. As a method to tackle blind source separation (BSS) in the frequency domain, we introduce independent vector analysis (IVA), an extension of independent component analysis (ICA) from univariate components to multivariate components. Given a mixture of statistically independent multivariate sources where the mixing is constrained to be component-wise, ICA needs to be followed by an additional algorithmic scheme in order to correct the permutation disorder that occurs after the component-wise separation, whereas IVA utilizes the inner dependency of the multivariate components and separates the fellow source components together. The efficiency of this new formulation in solving the permutation problem has been proven in its application to convolutive mixture of independent speech signals. Maximum likelihood (ML) approaches or information theoretic approaches have been employed where the time–frequency model of speech has been modelled by several multivariate joint densities, and natural gradient or Newton method algorithms have been derived. Here, we present a gentle tutorial on IVA for the separation of speech signals in the frequency domain.

6.1 Introduction

Blind source separation (BSS), or blind signal separation, refers to a set of problems that aim to separate individual source signals from their mixtures. In those problems, source signals are mixed in the recording of the sensors, and sensor noise that is not considered as an individual source signal can be added to the mixtures in the model but is ignored in many cases. BSS can be categorized by the relative number of sensors to the number of sources: underdetermined BSS when the number of sensors is less than the number of sources and (over)determined BSS otherwise. For sure, (over)determined BSS is the easier type of problem and there are many algorithms that can solve the (over)determined problems especially when the mixture is instantaneous, i.e. when the mixture and the separation can be denoted respectively as

$$\mathbf{x}[t] = \mathbf{A}\mathbf{s}[t], \tag{6.1}$$

$$\mathbf{y}[t] = \mathbf{W}\mathbf{x}[t], \tag{6.2}$$

where $\mathbf{x}[t]$ $\left(= \left[x_1[t], x_2[t], \ldots, x_l[t]\right]^{\mathrm{T}}\right)$ is the array signal observed in the sensors, $\mathbf{s}[t]$ $\left(= \left[s_1[t], s_2[t], \ldots, s_k[t]\right]^{\mathrm{T}}\right)$ is the source array signal, $\mathbf{y}[t]$ $\left(= \left[y_1[t], y_2[t], \ldots, y_k[t]\right]^{\mathrm{T}}\right)$ is the output array signal which is also the estimated

S. Makino et al. (eds.), Blind Speech Separation, 169–192.

source array signal, and $\mathbf{A}$ and $\mathbf{W}$ are respectively the $l \times k$ ($l \geq k$ in (over)determined case) mixing matrix and the $k \times l$ unmixing matrix which we mostly assume to be full rank and invariant over time. Since overdetermined BSS can be reduced to problems when $k = l$ using dimensionality reduction techniques, for convenience we will assume l equals k and that the mixing matrices are invertible from here on.

When the source signals are acoustic, the mixture is convolutive rather than instantaneous, i.e. there are propagation time delay and reverberation in the signal recording, and thus the mixture can be denoted approximately as

$$\mathbf{x}[t] = \sum_{\tau} \mathbf{A}[\tau]\mathbf{s}[t-\tau], \tag{6.3}$$

where $\mathbf{A}[t]$ ($t = 0, 1, \ldots$) is the transfer function matrix. By applying short-time Fourier transforms to this time-domain convolutive mixture, we can convert it to an instantaneous mixture problem in the frequency domain, which we also call time–frequency (T–F) domain since time structure is also involved. In each frequency bin, the mixture and its separation can be denoted as in (6.1);

$$\mathbf{x}^f[n] = \mathbf{A}^f\mathbf{s}^f[n], \tag{6.4}$$

$$\mathbf{y}^f[n] = \mathbf{W}^f\mathbf{x}^f[n], \tag{6.5}$$

for $f = 1, 2, \ldots, F$ where $\mathbf{x}^f[n]$, $\mathbf{s}^f[n]$, $\mathbf{y}^f[n]$, $\mathbf{A}^f$, and $\mathbf{W}^f$ denote respectively the observation signals, the source signals, the output signals, the mixing matrix, and the unmixing matrix in the f-th frequency bin, and F is the number of frequency bins. Note that in the T–F domain, n, different from t, is the number of the frame where the n-th short time Fourier transform is applied and the signals and the mixing matrices are complex-valued.

One of the most popular methods that solves the overdetermined BSS problem of instantaneous mixture is independent component analysis (ICA). ICA takes advantage of the assumption that source signals are statistically independent and thus learns the unmixing matrix by maximizing the independence among the output signals. When applying standard ICA methods to frequency-domain BSS, however, there arises the problem of grouping the fellow frequency components together, the well-known permutation problem. This problem is due to the permutation indeterminacy of ICA and has resulted in extensive works that proposed techniques to correct the permutation disorder. Smoothing the frequency-domain filter was one approach [1, 2] and for colored signals the inter-frequency correlation between the signal envelopes was utilized [3, 4]. Also, direction of arrival estimation was exploited [5, 6].

A new ICA formulation termed independent vector analysis (IVA) has been very successful in solving the permutation problem of frequency-domain BSS, especially with the application of speech signals [7, 9, 11]. IVA is an extension of ICA from univariate components to multivariate components and it utilizes not only the statistical independence among multivariate signals but also the statistical inner dependency of each multivariate signal. Hence, new objective functions, also known as contrast functions, that measure the independence among multivariate signals were proposed and speech was modelled in the T–F domain with multivariate probability density functions (PDF). A similar work had been proposed in [10] where the author used the term frequency coupling. The fraternal frequency components were modelled with sparse PDFs and that they share the same scale parameter which is estimated from the data. This resembles, and can be generalized in, the IVA framework. In the following sections, we introduce ICA and the IVA formulation in more detail.

6.2 Contrast Functions of Independent Component Analysis

Here, we present the contrast functions of ICA for a better understanding. Most ICA algorithms ignore the time structure of the signals and deal with them as i.i.d. samples of random variables (RV) and thus for convenience we will mostly omit the time structure "$[t]$" and "$[n]$" and express the signals as RVs.

6.2.1 Mutual Information

Since ICA aims at maximizing the statistical independence among output signals, mutual information has been employed as one of the general objective functions of ICA. The mutual information of the output signal y_i's is denoted by

$$\mathrm{I}(\mathbf{y}) = \mathrm{D}\big(\mathrm{f}_{\mathbf{y}} || \prod_i \mathrm{f}_{y_i}\big) \tag{6.6}$$

$$= \int_{\Re^{\dim(\mathbf{z})}} \mathrm{f}_{\mathbf{y}}(\mathbf{z}) \log \frac{\mathrm{f}_{\mathbf{y}}(\mathbf{z})}{\prod_i \mathrm{f}_{y_i}(z_i)} \mathrm{d}\mathbf{z}, \tag{6.7}$$

where $\mathrm{D}(\cdot||\cdot)$ denotes the Kullback–Leibler (KL) divergence, $\mathrm{f}_{\mathbf{y}}(\cdot)$ denotes the PDF of an RV $\mathbf{y}$, and $\mathbf{z}$ $\big(= [z_1, z_2, \ldots]^{\mathrm{T}}\big)$ is a dummy variable. Mutual information can be regarded as a metric that measures the distance from a given joint PDF to the product of its marginal PDFs. (However, note that KL divergence has no symmetric property: $\mathrm{D}(\mathrm{p}||\mathrm{q}) \neq \mathrm{D}(\mathrm{q}||\mathrm{p})$.) Hence mutual information as a contrast function measures how independent the components of the output data are and is minimized to zero if and only if the output component y_i's are mutually independent.

6.2.2 Likelihood Function

Likelihood function is also a very popular contrast function that has been employed in ICA algorithms such as infomax [12, 13]. In case the outcome is given while the model parameter is unknown, maximum likelihood (ML) approach learns the model parameter such that the likelihood function, i.e. the probability that the corresponding model yields the specific outcome, is maximized. In ICA formulation, the likelihood function can be denoted as

$$\Pr(\mathbf{D_x}|\mathbf{A}) = \prod_n \Pr(\mathbf{x}[n]|\mathbf{A}) \tag{6.8}$$

$$= \prod_n \int \Pr(\mathbf{x}[n]|\mathbf{A}, \mathbf{s}[n] = \mathbf{z}) \mathrm{f}_\mathbf{s}(\mathbf{z}) d\mathbf{z} \tag{6.9}$$

$$= \prod_n \int \delta(\mathbf{x}[n] - \mathbf{A}\mathbf{z}) \mathrm{f}_\mathbf{s}(\mathbf{z}) d\mathbf{z} \tag{6.10}$$

$$= \prod_n |\det(\mathbf{W})| \mathrm{f}_\mathbf{s}(\mathbf{W}\mathbf{x}[n]) \tag{6.11}$$

$$= |\det(\mathbf{W})|^N \prod_n \mathrm{f}_\mathbf{s}(\mathbf{y}[n]), \tag{6.12}$$

where $\mathbf{D_x}$ and $\delta(\cdot)$ denote the given observation data and the Dirac delta function, respectively. Note that the matrices $\mathbf{A}$ and $\mathbf{W}$ are the model parameters and are related to each other by $\mathbf{W} = (\mathbf{A})^{-1}$. Also note that, in the likelihood function of ICA, a source prior $\mathrm{f}_\mathbf{s}(\cdot)$ is needed. The normalized log-likelihood can be modified as

$$\frac{1}{N} \log \Pr(\mathbf{D_x}|\mathbf{A}) = \log\left(|\det(\mathbf{W})|\right) + \tilde{\mathrm{E}}\left[\log \mathrm{f}_\mathbf{s}(\mathbf{y})\right] \tag{6.13}$$

$$= \log\left(|\det(\mathbf{W})|\right) + \mathrm{H}(\mathrm{f}_\mathbf{y}) - \mathrm{D}(\mathrm{f}_\mathbf{y}||\mathrm{f}_\mathbf{s}) \tag{6.14}$$

$$= \mathrm{H}(\mathrm{f}_\mathbf{x}) - \mathrm{D}(\mathrm{f}_\mathbf{y}||\mathrm{f}_\mathbf{s}), \tag{6.15}$$

where $\tilde{\mathrm{E}}[\ldots]$ denotes $\frac{1}{N}\sum_n \cdots$ and $\mathrm{H}(\cdot)$ denotes the entropy function;

$$\mathrm{H}(\mathrm{f}_\mathbf{y}) \ \left(= \mathrm{H}(\mathbf{y})\right) = -\int_{\Re^{\dim(\mathbf{z})}} \mathrm{f}_\mathbf{y}(\mathbf{z}) \log \mathrm{f}_\mathbf{y}(\mathbf{z}) d\mathbf{z}. \tag{6.16}$$

Note that in (6.15), $\mathrm{H}(\mathrm{f}_\mathbf{x})$ is a constant term and thus the likelihood function is equivalent to $\mathrm{D}(\mathrm{f}_\mathbf{y}||\mathrm{f}_\mathbf{s})$, i.e. ML is equivalent to minimizing the KL divergence from the output joint density $\mathrm{f}_\mathbf{y}(\cdot)$ to the source prior $\mathrm{f}_\mathbf{s}(\cdot)$. For this reason, $\mathrm{f}_\mathbf{s}(\cdot)$ is also called the source target (that $\mathrm{f}_\mathbf{y}(\cdot)$ is aimed at). Since the source prior, or source target, is usually assumed a priori, to distinguish it from the real source distribution we will denote it as $\hat{\mathrm{f}}_\mathbf{s}(\cdot)$ from here on. Since ICA assumes statistical independence among source signals, the source targets are products of marginal PDFs;

$$\hat{\mathrm{f}}_\mathbf{s}(\cdot) = \prod_i \hat{\mathrm{f}}_{s_i}(\cdot). \tag{6.17}$$

6.2.3 Pictorial View in Information Geometry

By expressing mutual information and likelihood contrasts of ICA using KL divergence, it can be seen that they are very similar: $\mathrm{D}(\mathrm{f}_{\mathbf{y}}||\prod_i \mathrm{f}_{y_i})$ and $\mathrm{D}(\mathrm{f}_{\mathbf{y}}||\prod_i \hat{\mathrm{f}}_{s_i})$, respectively. Yet, it should be noted that they are not the same, i.e. the PDF $\prod_i \mathrm{f}_{y_i}(\cdot)$ in mutual information is the product of the output signals' marginal PDFs while the PDF $\prod_i \hat{\mathrm{f}}_{s_i}(\cdot)$ in likelihood is a fixed source target.

This can be better understood when they are, although roughly, illustrated in information geometry as it was done in [15]. Information geometry is the space of PDFs where each point in the space corresponds to a PDF. Let's define product manifold as the set of PDFs that are products of their marginals and draw it as a hyperplane in the space. Then the original source PDF $\prod_i \mathrm{f}_{s_i}(\cdot)$ will lie somewhere on it and also the source target $\prod_i \hat{\mathrm{f}}_{s_i}(\cdot)$ can be chosen to be a point on it.

In Fig. 6.1, the points that correspond to the PDF of the source array $\mathbf{s}$, the observation array $\mathbf{x}$, and the output array $\mathbf{y}$, product manifold, and mutual information are depicted. $\prod_i \mathrm{f}_{s_i}(\cdot)$ and $\mathrm{f}_{\mathbf{x}}(\cdot)$ are fixed points in the space and $\mathrm{f}_{\mathbf{y}}(\cdot)$ moves around in the space with respect to the unmixing matrix $\mathbf{W}$. In the figure, two different $\mathrm{f}_{\mathbf{y}}(\cdot)$'s are depicted respectively to $\mathbf{W} = \mathbf{W}_1$ and $\mathbf{W} = \mathbf{W}_2$. Note that $\mathbf{x}$ and $\mathbf{y}$ are constrained to be linearly transformed array signals of $\mathbf{s}$ and thus the points that $\mathrm{f}_{\mathbf{y}}(\cdot)$ can occupy are restricted, which is depicted as a solid curve in the figure. Since $\prod_i \mathrm{f}_{y_i}(\cdot)$

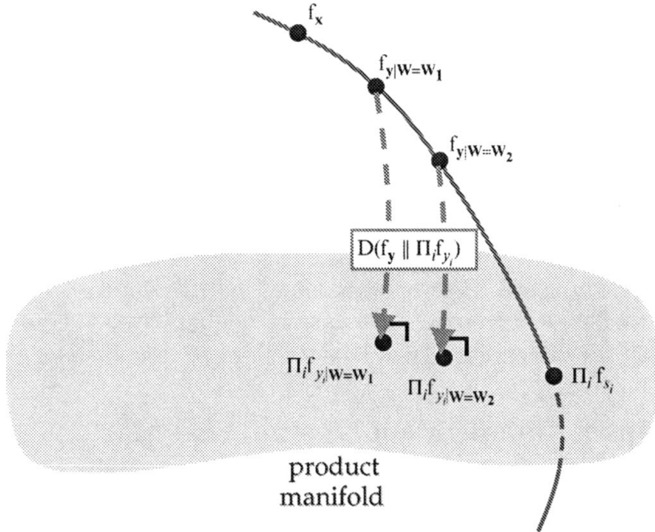

Fig. 6.1. Mutual information is depicted in information geometry. Mutual information is the KL distance from the joint PDF of the output signals to a hyperplane, i.e. the product manifold.

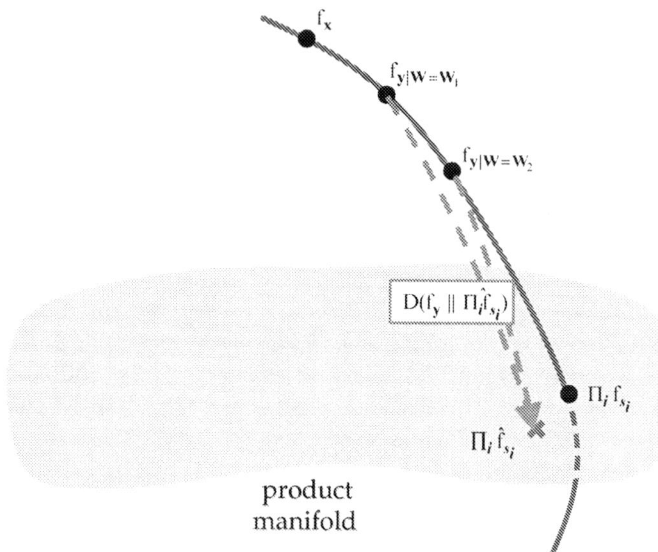

Fig. 6.2. Likelihood function is depicted in information geometry. Likelihood can be regarded as the KL distance from the joint PDF of the output signals to a fixed point, i.e. the source target.

is the closest point on the product manifold from $f_{\mathbf{y}}(\cdot)$ in terms of KL divergence, mutual information can be regarded as a distance from a point, i.e. the joint PDF of the output signals, to a hyperplane, i.e. the product manifold.

In Fig. 6.2, instead of mutual information, the likelihood function is depicted. Here, likelihood measures the distance from a point, i.e. the joint PDF of the output signals, to another fixed point, i.e. the source target.

The goal of employing mutual information or likelihood function as a contrast function of ICA is to put $f_{\mathbf{y}}(\cdot)$ as close as possible to the product manifold. Mutual information can directly maximize the independence among the output components regardless of the source types unless there are more than one Gaussian sources. However, there is a need to estimate the density of each output signal y_i from the data. This results in heavy computation and slow learning speed. On the other hand, in order to use likelihood it is important to choose a proper source target. The importance becomes more significant as the dimensionality of individual source signals rises.

6.3 What is Independent Vector Analysis?

The model of IVA consists of a set of the standard ICA models (Fig. 6.3) where the univariate sources across different layers are dependent such that they can be aligned and grouped together as a multivariate variable, or vector. In Fig. 6.4, a 2 × 2 IVA mixture model is depicted where $\mathbf{s}_1 \left(= [s_1^1, s_1^2, \ldots, s_1^F]^{\mathrm{T}} \right)$ and $\mathbf{s}_2 \left(= [s_2^1, s_2^2, \ldots, s_2^F]^{\mathrm{T}} \right)$ denote the multivariate sources

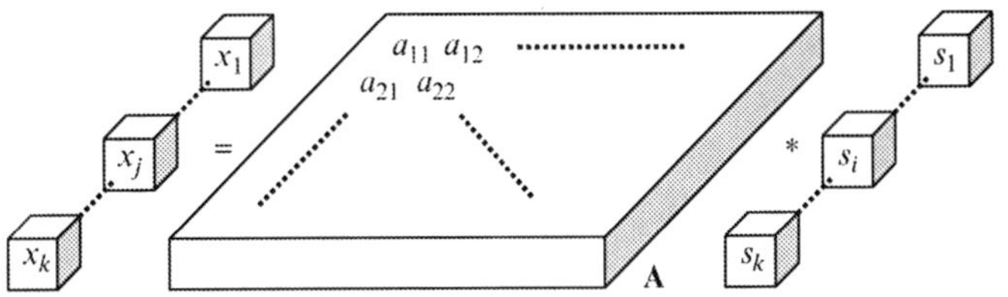

Fig. 6.3. Source mixture model of ICA.

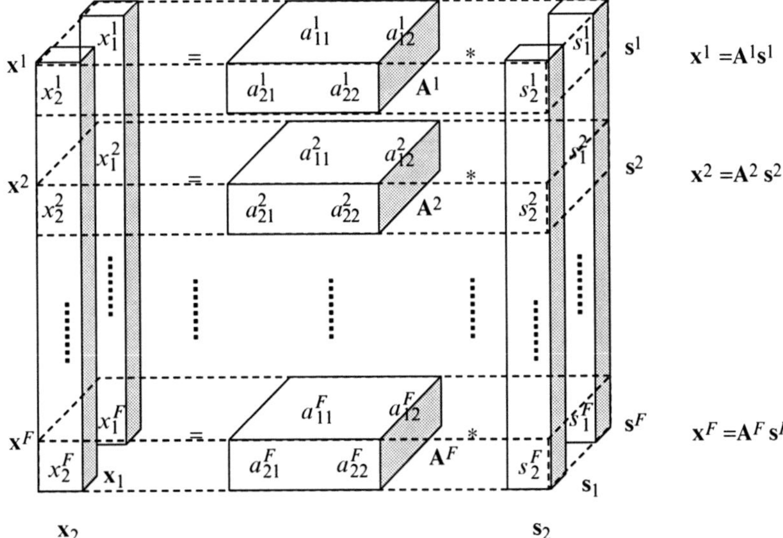

Fig. 6.4. The mixture model of IVA. ICA is extended to a formulation with multivariate variables, or vectors, where the components of each vector are dependent. The difference of IVA from multidimensional ICA or subspace ICA is that the mixing process is restricted to the source components on the same horizontal layer to form a standard ICA mixture.

and $\mathbf{x}_1$ $\left(= [x_1^1, x_1^2, \ldots, x_1^F]^{\mathrm{T}} \right)$ and $\mathbf{x}_2$ $\left(= [x_2^1, x_2^2, \ldots, x_2^F]^{\mathrm{T}} \right)$ denote the observed multivariate mixtures. As it can be seen, the mixing of the multivariate sources is constrained component-wise forming ICA mixture models in each layer. The difference of applying IVA from applying multiple ICA's to such a scenario is that IVA groups the dependent sources as a multivariate variable and learns each group as a whole.

The idea of grouping dependent sources together as a multidimensional source in ICA was first proposed by J.-F. Cardoso [16] with the name of multidimensional ICA (MICA). Also, A. Hyvärinen [17] proposed a maximum likelihood algorithm of MICA, which is subspace ICA, or independent subspace analysis (ISA). While IVA closely resembles those models and while MICA (including ISA) and IVA share the same contrasts, they differ in their

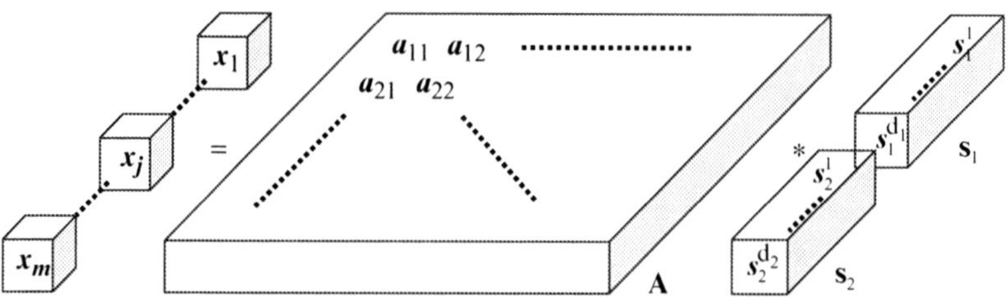

Fig. 6.5. The mixture model of MICA and ISA. The model consists of a single mixture layer where all source components are mixed together.

mixture models. For comparison, the mixture model of MICA is depicted in Fig. 6.5. While IVA consists of multiple standard ICA layers where the component-wise mixing keeps the dependent source components unmixed, MICA consists of a single mixture layer which cannot be decomposed into multiple ICA mixtures, since it is assumed that the dependent sources are also mixed.

The IVA model fits the model of the frequency-domain overdetermined BSS problem. The individual mixture layer in IVA corresponds to the instantaneous mixture in each frequency bin and the dependent sources grouped together as a multivariate variable correspond to the (short-time) frequency components of a time signal.

6.4 Contrast Functions of IVA (or MICA): Spatially White Data

Many ICA algorithms keep the output data zero-mean and spatially white since uncorrelatedness is a necessary condition for independence and since it increases the learning speed. While it can also be done in MICA algorithms including ISA, in IVA, however, it is only feasible for the output data in each layer, $\mathbf{y}^f$ $\left(= [y_1^f, y_2^f, \ldots]^{\mathrm{T}}\right)$, since mixing and unmixing in IVA are restricted to each layer. Here, we keep $\mathbf{y}^f$ zero-mean and white by preprocessing the observation data $\mathbf{x}^f$ $\left(= [x_1^f, x_2^f, \ldots]^{\mathrm{T}}\right)$ to be zero-mean and white, and by constraining the unmixing matrix $\mathbf{W}^f$'s to be orthogonal,

$$\mathrm{E}[\mathbf{x}^f(\mathbf{x}^f)^{\mathrm{H}}] = \mathbf{I}, \tag{6.18}$$

$$\mathbf{W}^f(\mathbf{W}^f)^{\mathrm{H}} = \mathbf{I}, \tag{6.19}$$

for $f = 1, 2, \ldots, F$. For convenience, we will assume that $\mathbf{x}^f$ is already preprocessed to be zero-mean and white from here on.

6.4.1 Entropic Contrasts

Since the basic idea of separating mutually independent multivariate sources is the same for MICA and IVA, they share the same contrast functions.

Similar to ICA, the contrast functions of IVA can be represented by the mutual information among the multidimensional variable $\mathbf{y}_i$'s,

$$\mathrm{D}\big(\mathrm{f}_{\mathbf{y}} \| \prod_i \mathrm{f}_{\mathbf{y}_i}\big) = \sum_i \mathrm{H}(\mathbf{y}_i) - \mathrm{H}(\mathbf{y}). \tag{6.20}$$

Please note that the output signal $\mathbf{y}_i \ \big(= [y_1^1, y_1^2, \ldots, y_1^F]^{\mathrm{T}}\big)$ is a vector now. This contrast reduces to the minimum, zero, if and only if $\mathbf{y}_i$'s are mutually independent. Note that the term $\mathrm{H}(\mathbf{y})$ in (6.20) is constant since $\log|\det(\mathbf{W}^f)| = 0$ for any value of f, which follows from (6.19), and hence, minimizing the KL divergence term in (6.20) is equivalent to minimizing the sum of the entropies of the multidimensional variables, i.e.

$$\arg\min_{\{\mathbf{W}^f\}} \mathrm{D}\big(\mathrm{f}_{\mathbf{y}} \| \prod_i \mathrm{f}_{\mathbf{y}_i}\big) = \arg\min_{\{\mathbf{W}^f\}} \sum_i \mathrm{H}(\mathbf{y}_i) \tag{6.21}$$

with the constraint of orthogonal $\mathbf{W}^f$'s.

In order to make the discussion that will follow simpler, we assume, in addition to the mutual independence of $\mathbf{s}_i \ \big(= [s_i^1, \ldots, s_i^F]\big)$'s,

$$\mathrm{E}\big[\mathbf{s}_i(\mathbf{s}_i)^{\mathrm{H}}\big] = \mathbf{I}, \tag{6.22}$$

i.e. the components of each multidimensional source are, although dependent, uncorrelated and the variance of each component is set to one which can be assumed owing to the scaling indeterminacy in ICA. Then, by the constraint of zero-mean and white $\mathbf{y}^f$, the whole output signal $\mathbf{y} \ \big(= [\mathbf{y}^1; \mathbf{y}^2; \ldots; \mathbf{y}^F]\big)$ is also kept zero-mean and white. (The notation $[\mathbf{a}; \mathbf{b}; \ldots]$ is meant to denote $[\mathbf{a}^{\mathrm{T}}, \mathbf{b}^{\mathrm{T}}, \ldots]^{\mathrm{T}}$ for column vectors $\mathbf{a}$ and $\mathbf{b}$.)

$$\mathrm{E}[\mathbf{y}\mathbf{y}^{\mathrm{H}}] = \mathbf{I}. \tag{6.23}$$

As in ICA, negentropy can be employed and another entropic contrast of IVA that is equivalent to the other entropic contrasts can be obtained. Negentropy is defined as follows:

$$\mathrm{N}(\mathbf{y}) = \mathrm{D}\big(\mathrm{f}_{\mathbf{y}} \| \mathrm{f}_{\mathbf{y}^{\mathcal{G}}}\big), \tag{6.24}$$

where $\mathbf{y}^{\mathcal{G}}$ denotes the (multidimensional) Gaussian RV that has the same mean vector and the same covariance matrix with RV $\mathbf{y}$. From the following Pythagorean relation in information geometry,

$$\mathrm{N}(\mathbf{y}_i) = \mathrm{H}(\mathbf{y}_i^{\mathcal{G}}) - \mathrm{H}(\mathbf{y}_i), \tag{6.25}$$

note that $\mathrm{H}(\mathbf{y}_i^{\mathcal{G}})$ in (6.25) is a constant term since, from (6.23), $\mathbf{y}_i$ is zero-mean and white such that $\mathbf{y}_i^{\mathcal{G}}$ is always fixed to be the unique zero-mean and white multidimensional Gaussian RV. Hence, by plugging in the relation of (6.25) into the right-hand-side of (6.21), we can easily see that,

$$\arg\min_{\{\mathbf{W}^f\}} \sum_i \mathrm{H}(\mathbf{y}_i) = \arg\max_{\{\mathbf{W}^f\}} \sum_i \mathrm{N}(\mathbf{y}_i), \tag{6.26}$$

with the constraint of orthogonal $\mathbf{W}^f$'s. It is clear that the entropic contrasts of IVA in (6.21) and (6.26) are extensions of the classic entropic contrasts of ICA from univariate variables to multivariate variables, since by replacing y_i for the $\mathbf{y}_i$ terms, we come up with the entropic contrasts of ICA.

As it was done for ICA (J.-F. Cardoso, [15]), the relation between the mutual information contrast (6.21) and the sum of negentropies (6.26) can be visually seen in information geometry. By defining the Gaussian manifold and independent vector manifold as the set of multivariate Gaussian PDFs and the set of PDFs where the multivariate $\mathbf{y}_i$'s are mutually independent, respectively, the mutual information (6.21) and the sum of negentropies (6.26) can be regarded as the distances from a point $\mathrm{f}_{\mathbf{y}}(\cdot)$ to the Gaussian manifold and to the independent vector manifold, respectively.

In Fig. 6.6, the points, manifolds, and the entropic contrasts are depicted and indicated where the PDFs are for zero-mean and white RVs. Note that the Gaussian manifold is drawn as a single point since a zero-mean and white multivariate Gaussian PDF is unique. When updating the unmixing matrix $\mathbf{W}^f$'s, the point $\mathrm{f}_{\mathbf{y}}(\cdot)$ moves around correspondingly while keeping the distance

$$\mathrm{N}(\mathbf{y}) = \mathrm{D}\big(\mathrm{f}_{\mathbf{y}} \| \prod_i \mathrm{f}_{\mathbf{y}_i}\big) + \sum_i \mathrm{N}(\mathbf{y}_i) \tag{6.27}$$

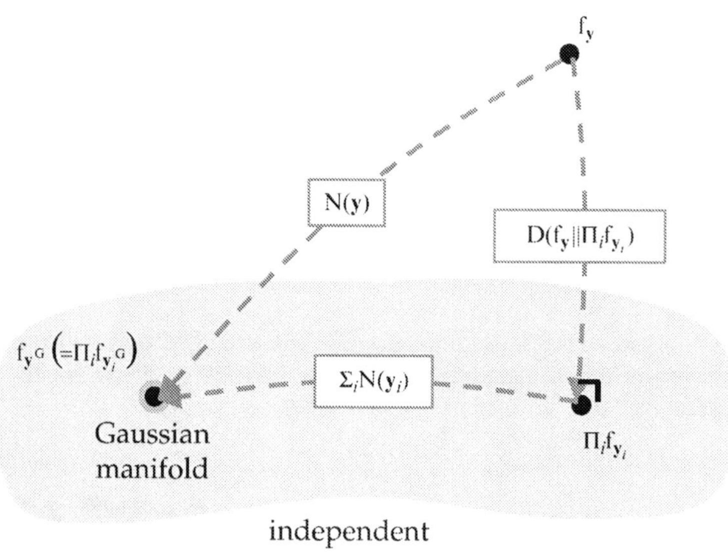

Fig. 6.6. Entropic contrasts of multidimensional ICA or IVA in the information geometry of zero-mean white PDFs.

constant. Hence our goal is either to minimize the distance from $\mathrm{f}_{\mathbf{y}}(\cdot)$ to the independent vector manifold, $\mathrm{D}\big(\mathrm{f}_{\mathbf{y}} \| \prod_i \mathrm{f}_{\mathbf{y}_i}\big)$, or to maximize the distance from $\prod_i \mathrm{f}_{\mathbf{y}_i}(\cdot)$ to the Gaussian manifold, $\sum_i \mathrm{N}(\mathbf{y}_i)$.

Please note that the assumption on the multivariate source components in (6.22) is not essential for the derivation of the negentropy contrast in (6.26) or the Pythagorean relation in (6.27). However, the motivation of the assumption is because the source distributions we later assume are spherically symmetric multivariate distributions which imply uncorrelatedness in the multivariate source components, and also because the explanation becomes much simpler with the assumption given.

6.4.2 Likelihood Function

As $\mathbf{y}$ is $[\mathbf{y}^1; \mathbf{y}^2; \ldots; \mathbf{y}^F]$, we will let $\mathbf{s}\ (= [\mathbf{s}^1; \mathbf{s}^2; \ldots; \mathbf{s}^F])$ and $\mathbf{x}\ (= [\mathbf{x}^1; \mathbf{x}^2; \ldots; \mathbf{x}^F])$ denote the whole source signal and the whole observation signal, respectively. The (normalized) log-likelihood contrast of IVA is in the form of

$$\tilde{\mathrm{E}}\big[\log\big(\hat{\mathrm{f}}_{\mathbf{s}}(\mathbf{y})\big)\big] + \sum_f \log|\det(\mathbf{W}^f)| = \sum_i \tilde{\mathrm{E}}\big[\log\big(\hat{\mathrm{f}}_{\mathbf{s}_i}(\mathbf{y}_i)\big)\big], \tag{6.28}$$

since we keep the unmixing matrices orthogonal, such that

$$\log|\det(\mathbf{W}^f)| = 0, \quad f = 1, 2, \ldots, F. \tag{6.29}$$

6.5 Modelling the Frequency Components of Speech

In applying IVA to frequency-domain BSS problems, we come to deal with complex-valued variables. Because real-valued contrast functions of complex-valued variables are not analytic, in a number of signal processing applications there arises the question of how to deal with complex variables. There are standard ways to handle the problem. The first is to write every complex variable in terms of two real variables, i.e. replace each complex variable z with two real variables which are the real an imaginary parts of z, and then use the tools for real-valued variables. An alternative is to follow the lines of [18, 19], which is equivalent to the first method described, but cleaner in notations. These methods basically carry out the optimization in the real field with respect to the real and imaginary parts of the complex variables.

In most scenarios it is known that complex RVs show circular symmetry around the origin. Hence, as the relation between the separated real variables, we assume circularity in the source variables such that

$$\tilde{\mathrm{E}}[\mathbf{s}\mathbf{s}^{\mathrm{T}}] = \mathbf{O}. \tag{6.30}$$

And hence,

$$\tilde{\mathrm{E}}[\mathbf{x}\mathbf{x}^{\mathrm{T}}] = \mathbf{O}. \tag{6.31}$$

With the circularity assumption on the frequency components of speech, they need to be modelled with a proper multivariate PDF. Some researchers observed that speech has the property of spherical invariance and they modelled bandlimited speech with spherically invariant random process (SIRP). For details and relevant topics, see [20, 21] and the references therein. The frequency models of speech that have been proposed in IVA formulation are closely related to SIRP.

6.5.1 L^p-Norm-Invariant Joint Densities

A group of PDFs which can be represented in the form of

$$\hat{\mathrm{f}}_{\mathbf{s}_i}(\mathbf{z}) \propto \mathrm{e}^{-\frac{1}{\sigma}(\|\mathbf{z}\|_p)^{\frac{1}{m}}} \tag{6.32}$$

$$= \mathrm{e}^{-\frac{1}{\sigma}\left(\sum_f |z^f|^p\right)^{\frac{1}{pm}}}, \tag{6.33}$$

where σ is the constant that controls the variance, has been proposed in [22]. The densities are l^p-norm-invariant for the given value of p and the sparseness can be controlled by the parameter m. Also the contour shape of the PDF varies by choosing different p values. Note that each complex variable in (6.33) is always circular (i.e. circularly symmetric in its PDF) even though the PDF of the whole multivariate variable is not spherically symmetric for $p \neq 2$. For convenience, we will denote each PDF with the given p and m values in (6.33) by N-I$_{(p,m)}$. For example, independent Laplace and corresponds to N-I$_{(1,1)}$. When N-I$_{(p,m)}$ is used as the source target in the likelihood contrast function (6.28), we will denote the contrast function by ML$_{\text{N-I}(p,m)}$.

In speech separation, the best performances were obtained for (p, m) being around $(1.9, 7)$ and for p being from 1.8 to 2.0 the performances were similar, which justifies the efficiency of using sparse and spherical joint densities for modelling speech in the frequency domain.

6.5.2 Sparse and Spherically Symmetric Joint Densities

In [7, 9, 11], several sparse and spherically symmetric joint PDFs were proposed for modelling the frequency components of speech. One, which we will call spherically symmetric Laplace distribution (SSL) is

$$\hat{\mathrm{f}}_{\mathbf{s}_i}(\mathbf{z}) \propto \mathrm{e}^{-\frac{1}{\sigma}\|\mathbf{z}\|_2}. \tag{6.34}$$

SSL was first used in ISA [17] in order to extract groups of features. Note N-I$_{(2,1)}$ equals SSL. Also a spherically symmetric exponential (l^2-)norm distribution (SEND) was proposed:

$$\hat{f}_{\mathbf{s}_i}(\mathbf{z}) \propto \frac{e^{-\frac{1}{\sigma}\|\mathbf{z}\|_2}}{\|\mathbf{z}\|_2^{d-1}}, \tag{6.35}$$

where d is the total dimension in real field. It should be noted that the PDFs of complex variables are in fact functions of real variables which are the real and imaginary parts of the complex variables and thus the PDFs are functions of $2F$ real variables ($d = 2F$). SEND was derived such that it has spherical contours and the l^2-norm of the variables $\|\mathbf{s}_i\|_2 \left(= \sqrt{\sum_f |s_i^f|^2}\right)$ follows exponential distribution. Each distribution can be regarded as an extension of Laplace distribution, a.k.a. double exponential distribution. These distributions shrink to the Laplace distribution if $d = 1$. As $\mathrm{ML}_{\mathrm{N-I}(p,m)}$, we will denote the likelihood contrast function with the source target of SSL and SEND by $\mathrm{ML}_{\mathrm{SSL}}$ and $\mathrm{ML}_{\mathrm{SEND}}$, respectively.

For better understanding of spherically symmetric sparse PDFs, the density plots and the contour plots of the two-dimensional independent Laplace distribution and SSL are compared in Fig. 6.7 where the variance of each

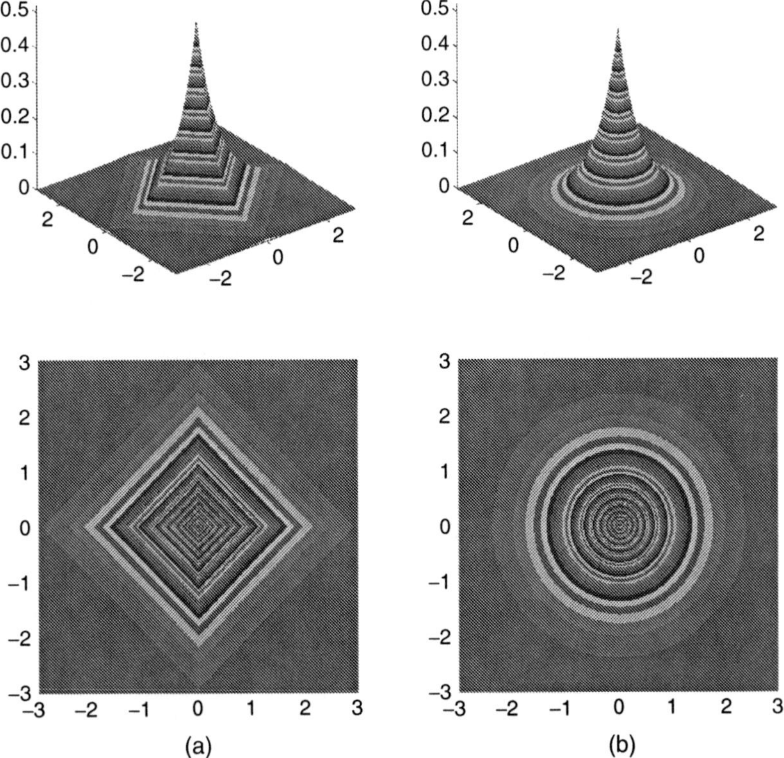

Fig. 6.7. The two-dimensional density plot and the contour plot of (a) independent Laplace distribution and (b) spherically symmetric Laplace (SSL) distribution. When being compared with independent Laplace distribution, it can be seen that SSL has some variance dependency.

marginal density is set to be one. By comparing the density of the points around the diagonals in the contour plots, it can be seen that the ones in SSL have higher density than the ones in independent Laplace distribution which implies that the variables of SSL have some variance dependency.

When we replace $\hat{\mathrm{f}}_{\mathbf{s}_i}(\cdot)$ in the likelihood contrast (6.28) with either SEND or SSL, the likelihood contrast can be written the form of

$$\sum_i \tilde{\mathrm{E}}\big[\mathrm{G}\big(\sum_f |y_i^f|^2\big)\big] \tag{6.36}$$

with the relation of

$$\mathrm{G}\big(\sum_f |y_i^f|^2\big) = -\log \hat{\mathrm{f}}_{\mathbf{s}_i}(\mathbf{y}_i), \tag{6.37}$$

where $y_i^f = (\mathbf{w}_i^f)^{\mathrm{H}}\mathbf{x}^f$ and $(\mathbf{w}_i^f)^{\mathrm{H}}$ denotes the i-th row of the unmixing matrix $\mathbf{W}^f$ with the normalization constraint of $(\mathbf{w}_i^f)^{\mathrm{H}}\mathbf{w}_i^f = 1$. Note that the contrast has changed its sign to be negative likelihood.

6.5.3 A Semi-Nonparametric Contrast Function for Spherical Multivariate Sources

A semi-nonparametric objective function for spherical multivariate sources was proposed in [10, 22]:

$$\arg\min_{\{\mathbf{W}^f\}} \sum_i \tilde{\mathrm{E}}\Big[\log\big(\sum_f |y_i^f|^2\big)\Big] \tag{6.38}$$

with the constraint of orthogonal $\mathbf{W}^f$'s. This contrast will be denoted as SNP. Although this contrast was derived from nonparametric perspective, objective functions that are almost equivalent to it can be derived from ML perspective using a set of spherically symmetric multivariate source targets in which the l^2-norm of the multidimensional variable ($\|\mathbf{s}_i\|_2$) follows

$$\mathrm{f}_{\|\mathbf{s}_i\|_2}(r) \propto r^m, \quad \epsilon < r, \tag{6.39}$$

for any value of m in $(-\infty, -1)$ where ϵ is an arbitrarily small positive number, or

$$\mathrm{f}_{\|\mathbf{s}_i\|_2}(r) \propto r^{-1}, \quad \epsilon < r < L, \tag{6.40}$$

where L is an arbitrarily large number, or

$$\mathrm{f}_{\|\mathbf{s}_i\|_2}(r) \propto r^m, \quad 0 < r < L, \tag{6.41}$$

for any value of m in $(-1, d-1)$.

Note that the contrast SNP (6.38) also falls into the form in (6.36) with a corresponding $\mathrm{G}(\cdot)$. The nonlinearity function $\mathrm{G}(\cdot)$'s that correspond to individual contrast functions are summarized in Table. 6.1.

Experiments have shown that the performance of IVA with SNP contrast is undistinguishable from the performance of IVA with $\mathrm{ML}_{\mathrm{SEND}}$ when applied to frequency-domain blind speech separation.

Table 6.1. Nonlinearity functions and their derivatives.

Contrast	G(z)	G′(z)	G″(z)
ML$_{\text{SSL}}$	$\sqrt{z}$	$\frac{1}{2\sqrt{z}}$	$-\frac{1}{4\sqrt{z^3}}$
ML$_{\text{SEND}}$	$\sqrt{\frac{2}{F}}\sqrt{z}+(F-\frac{1}{2})\log z$	$\frac{1}{\sqrt{2F}\sqrt{z}}+(F-\frac{1}{2})\frac{1}{z}$	$-\frac{1}{\sqrt{8F}\sqrt{z^3}}-(F-\frac{1}{2})\frac{1}{z^2}$
SNP	$\log z$	$\frac{1}{z}$	$-\frac{1}{z^2}$

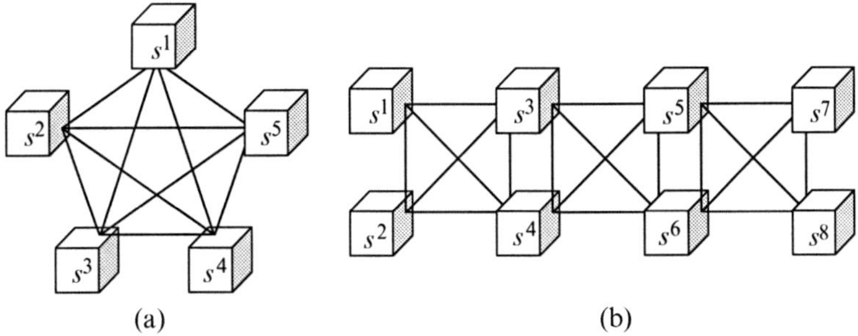

Fig. 6.8. Undirected graphs for IVA dependency models. Here, the line connections of each clique represent a fixed spherical dependency weight. (a) A global clique to represent spherical dependency. (b) A chain of cliques to represent the proposed dependency. Here the dependency propagates through the overlaps of the chains and hence, the dependency between two components weakens while the distance between them increases.

6.5.4 Sparse and Locally Spherical Joint Densities

The assumption of spherical symmetry and sparseness seemed valid for speech separation and also resulted in decent separation results. However, when compared to the result of frequency domain ICA followed by perfect permutation correction, the separation results of IVA using spherical joint densities are slightly inferior. This suggests that such source priors do not model speech exactly and that the performance of IVA for speech separation can be improved by finding better dependency models.

As an undirected graph, a spherical dependency model can be depicted as a total clique where, roughly speaking, all of the line connections represent the same kind, or same weight, of dependency. The undirected graph for a global clique is depicted in Fig. 6.8 (a). In real world speech, however, it seems unreasonable to assign the same kind of dependency to neighboring frequency components and to frequency components that reside far apart, because the dependency of neighboring frequency components is much stronger than that of frequency components being far apart.

Hence, in [23], a dependency model was proposed for acoustic signal model that is locally spherical and the dependency among the source components is propagated through overlaps of cliques so that the dependency between components weakens while the distance between them increases. Such an example is drawn as an undirected graph in Fig. 6.8 (b). The corresponding multivariate PDF is given in the form of

$$\hat{\mathrm{f}}_{\mathbf{s}_i}(\mathbf{z}) \propto \mathrm{e}^{-\frac{1}{\sigma_1}\sqrt{\sum_{f=f_1^{\mathrm{b}}}^{f_1^{\mathrm{e}}}|z^f|^2}-\frac{1}{\sigma_2}\sqrt{\sum_{f=f_2^{\mathrm{b}}}^{f_2^{\mathrm{e}}}|z^f|^2}-\ldots-\frac{1}{\sigma_K}\sqrt{\sum_{f=f_K^{\mathrm{b}}}^{f_K^{\mathrm{e}}}|z^f|^2}}, \quad (6.42)$$

where f_k^{b} and f_k^{e} are the beginning and the end indices of the k-th clique, respectively, and K is the number of cliques. Note that we have large flexibility in modelling the size of each clique and also the size of overlaps.

When we replace $\hat{\mathrm{f}}_{\mathbf{s}_i}(\cdot)$ in the likelihood contrast (6.28) with (6.42), the likelihood contrast can be written in the form of

$$\sum_i \tilde{\mathrm{E}}\left[\sum_k \frac{1}{\sigma_k}\sqrt{\sum_{f=f_k^{\mathrm{b}}}^{f_k^{\mathrm{e}}}|y_i^f|^2}\right]. \quad (6.43)$$

Some good choice of the cliques resulted in consistently better performance than the ones with globally spherical joint densities.

6.6 Contrast Optimization and Scaling

ICA algorithms mainly consist of two parts. One is the choice of contrast function and the other is the choice of the optimization method. For optimization, a number of ICA algorithms use gradient descent update rule including natural gradient [24], or relative gradient [25]. Also fixed-point iteration [26], Newton's method update rule [27], or exhaustive search by rotation [28] have been applied. Since the contrast functions were selected, proper optimization methods need to be chosen for the IVA algorithms to be derived. Since gradient descent methods are straightforward, here we employ the Newton's method update rule. Newton's method update, when compared to other gradient descent methods, converges fast and is free from selecting the learning rate. In addition, Newton's method would find rather stationary points than specified maxima or minima, and thus it has more flexibility in separating sources; for instance, a FastICA algorithm which is derived from ML perspective and is designed for super-Gaussian sources can also find sub-Gaussian sources.

6.6.1 Quadratic Taylor Expansion in Complex Notations

In order to use complex notations while applying Newton's method to the contrasts, we employ the definitions of complex gradient and complex Hessian in [18] and [19]. It was shown that a complex variable and its complex conjugate must be considered as separate variables and that the complex gradient and Hessian correspond to their real counterparts one-to-one by simple linear transformations [19]. That is, for

$$\mathbf{w}\ (=[w_1, w_2, \ldots]^{\mathrm{T}}) = \mathbf{u} + \mathrm{j}\mathbf{v}\ (=[u_1, u_2, \ldots]^{\mathrm{T}} + j[v_1, v_2, \ldots]^{\mathrm{T}}), \tag{6.44}$$

$$\mathbf{z} = [u_1, v_1, u_2, v_2, \ldots]^{\mathrm{T}}, \tag{6.45}$$

$$\mathbf{w}_{\mathrm{D}} = [w_1, w_1^*, w_2, w_2^*, \ldots]^{\mathrm{T}}, \tag{6.46}$$

$$\mathrm{g}(\mathbf{w}) = \mathrm{g}_{\mathrm{D}}(\mathbf{w}, \mathbf{w}^*) \tag{6.47}$$

$$= \mathrm{h}(\mathbf{u}, \mathbf{v}), \tag{6.48}$$

where $\mathbf{u}$ and $\mathbf{v}$ are real-valued vectors, it was shown that the quadratic Taylor polynomial of $\mathrm{h}(\mathbf{u}, \mathbf{v})$ with respect to $\mathbf{z}$ around the point $\mathbf{z} = \mathbf{O}$ is equivalent to a Taylor polynomial form in complex notations as the following:

$$\mathrm{h}(\mathbf{O}, \mathbf{O}) + \frac{\partial \mathrm{h}(\mathbf{O}, \mathbf{O})}{\partial \mathbf{z}^{\mathrm{T}}}\mathbf{z} + \frac{1}{2}\mathbf{z}^{\mathrm{T}}\frac{\partial^2 \mathrm{h}(\mathbf{O}, \mathbf{O})}{\partial \mathbf{z}\partial \mathbf{z}^{\mathrm{T}}}\mathbf{z} \tag{6.49}$$

$$= \mathrm{g}_{\mathrm{D}}(\mathbf{O}, \mathbf{O}) + \frac{\partial \mathrm{g}_{\mathrm{D}}(\mathbf{O}, \mathbf{O})}{\partial \mathbf{w}_{\mathrm{D}}^{\mathrm{T}}}\mathbf{w}_{\mathrm{D}} + \frac{1}{2}\mathbf{w}_{\mathrm{D}}^{\mathrm{T}}\frac{\partial^2 \mathrm{g}_{\mathrm{D}}(\mathbf{O}, \mathbf{O})}{\partial \mathbf{w}_{\mathrm{D}}\partial \mathbf{w}_{\mathrm{D}}^{\mathrm{T}}}\mathbf{w}_{\mathrm{D}} \tag{6.50}$$

if (and only if),

$$\frac{\partial}{\partial w_i} = \frac{1}{2}\left(\frac{\partial}{\partial u_i} - j\frac{\partial}{\partial v_i}\right), \quad \frac{\partial}{\partial w_i^*} = \frac{1}{2}\left(\frac{\partial}{\partial u_i} + j\frac{\partial}{\partial v_i}\right), \tag{6.51}$$

$$\frac{\partial^2}{\partial a \partial b} = \frac{\partial}{\partial a}\left(\frac{\partial}{\partial b}\right) = \frac{\partial}{\partial b}\left(\frac{\partial}{\partial a}\right) = \frac{\partial^2}{\partial b \partial a}, \quad a, b \in \{w_{i_1}, w_{i_2}^*\}. \tag{6.52}$$

However, although in complex notations, it is still cumbersome to apply Newton's method to a contrast using the quadratic Taylor polynomial form on the right-hand-side of (6.50). Here, we introduce a quadratic Taylor polynomial in notations of the original complex variables $\mathbf{w}$ and $\mathbf{w}^*$ instead of $\mathbf{w}_{\mathrm{D}}$. By changing the order of the components in $\mathbf{w}_{\mathrm{D}}$ such that $\mathbf{w}_{\mathrm{D}} = [\mathbf{w}; \mathbf{w}^*]$ and plugging it into the right-hand-side of (6.50), it can be seen that the expression decomposes into

$$\begin{aligned}
\mathrm{g}(\mathbf{w}) \approx\ & \mathrm{g}(\mathbf{w}_{\mathrm{o}}) + \frac{\partial \mathrm{g}(\mathbf{w}_{\mathrm{o}})}{\partial \mathbf{w}^{\mathrm{T}}}(\mathbf{w} - \mathbf{w}_{\mathrm{o}}) + \frac{\partial \mathrm{g}(\mathbf{w}_{\mathrm{o}})}{\partial \mathbf{w}^{\mathrm{H}}}(\mathbf{w} - \mathbf{w}_{\mathrm{o}})^* \\
& + \frac{1}{2}(\mathbf{w} - \mathbf{w}_{\mathrm{o}})^{\mathrm{T}}\frac{\partial^2 \mathrm{g}(\mathbf{w}_{\mathrm{o}})}{\partial \mathbf{w}\partial \mathbf{w}^{\mathrm{T}}}(\mathbf{w} - \mathbf{w}_{\mathrm{o}}) \\
& + \frac{1}{2}(\mathbf{w} - \mathbf{w}_{\mathrm{o}})^{\mathrm{H}}\frac{\partial^2 \mathrm{g}(\mathbf{w}_{\mathrm{o}})}{\partial \mathbf{w}^*\partial \mathbf{w}^{\mathrm{H}}}(\mathbf{w} - \mathbf{w}_{\mathrm{o}})^* \\
& + (\mathbf{w} - \mathbf{w}_{\mathrm{o}})^{\mathrm{H}}\frac{\partial^2 \mathrm{g}(\mathbf{w}_{\mathrm{o}})}{\partial \mathbf{w}^*\partial \mathbf{w}^{\mathrm{T}}}(\mathbf{w} - \mathbf{w}_{\mathrm{o}}),
\end{aligned} \tag{6.53}$$

where the point of Taylor expansion has changed from $\mathbf{w} = \mathbf{O}$ to $\mathbf{w} = \mathbf{w}_{\mathrm{o}}$. The above notations we used for gradient and Hessian with vectors are defined as

$$\frac{\partial}{\partial \mathbf{w}} = \left[\frac{\partial}{\partial w_1}, \frac{\partial}{\partial w_2}, \ldots\right]^{\mathrm{T}}, \quad \frac{\partial}{\partial \mathbf{w}^{\mathrm{T}}} = \left(\frac{\partial}{\partial \mathbf{w}}\right)^{\mathrm{T}}, \tag{6.54}$$

$$\frac{\partial}{\partial \mathbf{w}^*} = \left[\frac{\partial}{\partial w_1^*}, \frac{\partial}{\partial w_2^*}, \ldots\right]^{\mathrm{T}}, \quad \frac{\partial}{\partial \mathbf{w}^{\mathrm{H}}} = \left(\frac{\partial}{\partial \mathbf{w}^*}\right)^{\mathrm{T}}, \tag{6.55}$$

$$\frac{\partial^2}{\partial \mathbf{a} \partial \mathbf{b}^{\mathrm{T}}} = \frac{\partial}{\partial \mathbf{a}} \left(\frac{\partial}{\partial \mathbf{b}^{\mathrm{T}}} \right) = \frac{\partial}{\partial \mathbf{b}^{\mathrm{T}}} \left(\frac{\partial}{\partial \mathbf{a}} \right) = \frac{\partial^2}{\partial \mathbf{b}^{\mathrm{T}} \partial \mathbf{a}}, \quad \mathbf{a}, \mathbf{b} \in \{\mathbf{w}, \mathbf{w}^*\}. \tag{6.56}$$

Although this form is enough for the derivation of a Newton's method update as it was done in [8, 11], to be strict we can employ the quadratic Taylor expansion form with respect to the whole set of $\mathbf{w}_i^f$'s instead of a single $\mathbf{w}_i^f$ since the contrast of IVA is a function of the whole $\mathbf{w}_i^f$'s;

$$\begin{aligned}
&\mathrm{g}(\{\mathbf{w}_i^f\}) \approx \\
&\mathrm{g}(\{\mathbf{w}_{i,\mathrm{o}}^f\}) \\
&+ \sum_{k_1} \frac{\partial \mathrm{g}(\{\mathbf{w}_{i,\mathrm{o}}^f\})}{\partial (\mathbf{w}_i^{k_1})^{\mathrm{T}}} (\mathbf{w}_i^{k_1} - \mathbf{w}_{i,\mathrm{o}}^{k_1}) + \sum_{k_1} \frac{\partial \mathrm{g}(\{\mathbf{w}_{i,\mathrm{o}}^f\})}{\partial (\mathbf{w}_i^{k_1})^{\mathrm{H}}} (\mathbf{w}_i^{k_1} - \mathbf{w}_{i,\mathrm{o}}^{k_1})^* \\
&+ \sum_{k_1,k_2} \frac{1}{2} (\mathbf{w}_i^{k_1} - \mathbf{w}_{i,\mathrm{o}}^{k_1})^{\mathrm{T}} \frac{\partial^2 \mathrm{g}(\{\mathbf{w}_{i,\mathrm{o}}^f\})}{\partial \mathbf{w}_i^{k_1} \partial (\mathbf{w}_i^{k_2})^{\mathrm{T}}} (\mathbf{w}_i^{k_2} - \mathbf{w}_{i,\mathrm{o}}^{k_2}) \\
&+ \sum_{k_1,k_2} \frac{1}{2} (\mathbf{w}_i^{k_1} - \mathbf{w}_{i,\mathrm{o}}^{k_1})^{\mathrm{H}} \frac{\partial^2 \mathrm{g}(\{\mathbf{w}_{i,\mathrm{o}}^f\})}{\partial (\mathbf{w}_i^{k_1})^* \partial (\mathbf{w}_i^{k_2})^{\mathrm{H}}} (\mathbf{w}_i^{k_2} - \mathbf{w}_{i,\mathrm{o}}^{k_2})^* \\
&+ \sum_{k_1,k_2} (\mathbf{w}_i^{k_1} - \mathbf{w}_{i,\mathrm{o}}^{k_1})^{\mathrm{H}} \frac{\partial^2 \mathrm{g}(\{\mathbf{w}_{i,\mathrm{o}}^f\})}{\partial (\mathbf{w}_i^{k_1})^* \partial (\mathbf{w}_i^{k_2})^{\mathrm{T}}} (\mathbf{w}_i^{k_2} - \mathbf{w}_{i,\mathrm{o}}^{k_2}).
\end{aligned} \tag{6.57}$$

6.6.2 Newton's Method

Let's set $\mathrm{g}(\{\mathbf{w}_i^f\})$ to be the summation term of the contrast in (6.36) as

$$\mathrm{g}(\{\mathbf{w}_i^f\}) = \tilde{\mathrm{E}}\Big[\mathrm{G}\big(\sum_f |(\mathbf{w}_i^f)^{\mathrm{H}} \mathbf{x}^f|^2\big)\Big] - \sum_f \lambda_i^f \left((\mathbf{w}_i^f)^{\mathrm{H}} \mathbf{w}_i^f - 1\right), \tag{6.58}$$

where the normalization constraint is added with Lagrange multiplier λ_i^f's. Please note that the other contrast functions, i.e. $\mathrm{ML}_{\mathrm{N-I}}(p, m)$ and (6.38), can also be optimized by Newton's method in the same manner. The $\mathbf{w}_i^{k_1}$ that optimizes the function $\mathrm{g}(\{\mathbf{w}_i^f\})$ will set the gradient $\frac{\partial \mathrm{g}(\{\mathbf{w}_i^f\})}{\partial (\mathbf{w}_i^{k_1})^*}$ to be zero and hence from (6.57),

$$\begin{aligned}
&\frac{\partial \mathrm{g}(\{\mathbf{w}_{i,o}^f\})}{\partial (\mathbf{w}_i^{k_1})^*} + \sum_{k_2} \frac{\partial^2 \mathrm{g}(\{\mathbf{w}_{i,o}^f\})}{\partial (\mathbf{w}_i^{k_1})^* \partial (\mathbf{w}_i^{k_2})^{\mathrm{T}}} (\mathbf{w}_i^{k_2} - \mathbf{w}_{i,o}^{k_2}) \\
&+ \sum_{k_2} \frac{\partial^2 \mathrm{g}(\{\mathbf{w}_{i,o}^f\})}{\partial (\mathbf{w}_i^{k_1})^* \partial (\mathbf{w}_i^{k_2})^{\mathrm{H}}} (\mathbf{w}_i^{k_2} - \mathbf{w}_{i,o}^{k_2})^* \\
&\equiv \mathbf{O}.
\end{aligned} \tag{6.59}$$

Then the derivative terms in (6.59) become

$$\frac{\partial \mathrm{g}(\{\mathbf{w}_{i,\mathrm{o}}^{f}\})}{\partial (\mathbf{w}_{i}^{k_1})^{*}} = \tilde{\mathrm{E}}\Big[(y_{i,\mathrm{o}}^{k_1})^{*}\,\mathrm{G}'\big(\sum_{f}|y_{i,\mathrm{o}}^{f}|^2\big)\mathbf{x}^{k_1}\Big] - \lambda_i^{k_1}\mathbf{w}_{i,\mathrm{o}}^{k_1}, \tag{6.60}$$

$$\frac{\partial^2 \mathrm{g}(\{\mathbf{w}_{i,\mathrm{o}}^{f}\})}{\partial (\mathbf{w}_{i}^{k_1})^{*}\partial (\mathbf{w}_{i}^{k_1})^{\mathrm{T}}} \tag{6.61}$$

$$= \tilde{\mathrm{E}}\Big[\Big(\mathrm{G}'\big(\sum_{f}|y_{i,\mathrm{o}}^{f}|^2\big) + |y_{i,\mathrm{o}}^{k_1}|^2\,\mathrm{G}''\big(\sum_{f}|y_{i,o}^{f}|^2\big)\Big)\mathbf{x}^{k_1}(\mathbf{x}^{k_1})^{\mathrm{H}}\Big] - \lambda_i^{f}\mathbf{I} \tag{6.62}$$

$$\approx \tilde{\mathrm{E}}\Big[\mathrm{G}'\big(\sum_{f}|y_{i,\mathrm{o}}^{f}|^2\big) + |y_{i,\mathrm{o}}^{k_1}|^2\,\mathrm{G}''\big(\sum_{f}|y_{i,\mathrm{o}}^{f}|^2\big)\Big]\tilde{\mathrm{E}}\big[\mathbf{x}^{k_1}(\mathbf{x}^{k_1})^{\mathrm{H}}\big] - \lambda_i^{f}\mathbf{I} \tag{6.63}$$

$$= \Big(\tilde{\mathrm{E}}\Big[\mathrm{G}'\big(\sum_{f}|y_{i,\mathrm{o}}^{f}|^2\big) + |y_{i,o}^{k_1}|^2\,\mathrm{G}''\big(\sum_{f}|y_{i,\mathrm{o}}^{f}|^2\big)\Big] - \lambda_i^{f}\Big)\mathbf{I}, \tag{6.64}$$

$$\frac{\partial^2 \mathrm{g}(\{\mathbf{w}_{i,\mathrm{o}}^{f}\})}{\partial (\mathbf{w}_{i}^{k_1})^{*}\partial (\mathbf{w}_{i}^{k_2})^{\mathrm{T}}} \qquad (k_1 \neq k_2) \tag{6.65}$$

$$= \tilde{\mathrm{E}}\Big[(y_{i,\mathrm{o}}^{k_1})^{*}y_{i,\mathrm{o}}^{k_2}\,\mathrm{G}''\big(\sum_{f}|y_{i,\mathrm{o}}^{f}|^2\big)\mathbf{x}^{k_1}(\mathbf{x}^{k_2})^{\mathrm{H}}\Big] \tag{6.66}$$

$$\approx \tilde{\mathrm{E}}\Big[(y_{i,\mathrm{o}}^{k_1})^{*}y_{i,\mathrm{o}}^{k_2}\,\mathrm{G}''\big(\sum_{f}|y_{i,\mathrm{o}}^{f}|^2\big)\Big]\tilde{\mathrm{E}}\big[\mathbf{x}^{f}(\mathbf{x}^{k})^{\mathrm{H}}\big] \tag{6.67}$$

$$= \tilde{\mathrm{E}}\Big[(y_{i,\mathrm{o}}^{k_1})^{*}y_{i,\mathrm{o}}^{k_2}\,\mathrm{G}''\big(\sum_{f}|y_{i,\mathrm{o}}^{f}|^2\big)\Big]\mathbf{O} \tag{6.68}$$

$$= \mathbf{O}, \tag{6.69}$$

$$\frac{\partial^2 \mathrm{g}(\{\mathbf{w}_{i,\mathrm{o}}^{f}\})}{\partial (\mathbf{w}_{i}^{k_1})^{*}\partial (\mathbf{w}_{i}^{k_1})^{\mathrm{H}}} \tag{6.70}$$

$$= \tilde{\mathrm{E}}\Big[\big((y_{i,\mathrm{o}}^{k_1})^{*}\big)^2\,\mathrm{G}''\big(\sum_{f}|y_{i,\mathrm{o}}^{f}|^2\big)\mathbf{x}^{k_1}(\mathbf{x}^{k_1})^{\mathrm{T}}\Big] \tag{6.71}$$

$$\approx \tilde{\mathrm{E}}\Big[\big((y_{i,\mathrm{o}}^{k_1})^{*}\big)^2\,\mathrm{G}''\big(\sum_{f}|y_{i,\mathrm{o}}^{f}|^2\big)\Big]\,\tilde{\mathrm{E}}\big[\mathbf{x}^{k_1}(\mathbf{x}^{k_1})^{\mathrm{T}}\big] \tag{6.72}$$

$$= \mathbf{O}, \tag{6.73}$$

$$\frac{\partial^2 \mathrm{g}(\{\mathbf{w}_{i,\mathrm{o}}^f\})}{\partial(\mathbf{w}_i^{k_1})^* \partial(\mathbf{w}_i^{k_2})^{\mathrm{H}}} \qquad (k_1 \neq k_2) \tag{6.74}$$

$$= \tilde{\mathrm{E}}\Big[(y_{i,\mathrm{o}}^{k_1})^* (y_{i,\mathrm{o}}^{k_2})^* \, \mathrm{G}''\big(\sum_f |y_{i,\mathrm{o}}^f|^2\big) \mathbf{x}^{k_1} (\mathbf{x}^{k_2})^{\mathrm{T}}\Big] \tag{6.75}$$

$$\approx \tilde{\mathrm{E}}\Big[(y_{i,\mathrm{o}}^{k_1})^* (y_{i,\mathrm{o}}^{k_2})^* \, \mathrm{G}''\big(\sum_f |y_{i,\mathrm{o}}^f|^2\big)\Big] \, \tilde{\mathrm{E}}\big[\mathbf{x}^{k_1} (\mathbf{x}^{k_2})^{\mathrm{T}}\big] \tag{6.76}$$

$$= \mathbf{O}, \tag{6.77}$$

where $y_{i,\mathrm{o}}^f = (\mathbf{w}_{i,\mathrm{o}}^f)^{\mathrm{H}} \mathbf{x}^f$ and some approximations by separation of expectations were done in (6.63), (6.67), (6.72), and (6.76) as was done in [14]. (6.64), (6.68), (6.73), and (6.77) follow from (6.18) and (6.31), respectively.

By putting all results into the Newton's method in (6.59), our corresponding iterative algorithm becomes as follows,

$$\mathbf{w}_i^{k_1} \leftarrow \mathbf{w}_{i,\mathrm{o}}^{k_1} - \frac{\tilde{\mathrm{E}}\Big[(y_{i,\mathrm{o}}^{k_1})^* \, \mathrm{G}'\big(\sum_f |y_{i,\mathrm{o}}^f|^2\big)\mathbf{x}^{k_1}\Big] - \lambda_i^{k_1} \mathbf{w}_{i,\mathrm{o}}^{k_1}}{\tilde{\mathrm{E}}\Big[\mathrm{G}'\big(\sum_f |y_{i,\mathrm{o}}^{k_1}|^2\big) + |y_{i,\mathrm{o}}^{k_1}|^2 \, \mathrm{G}''\big(\sum_f |y_{i,\mathrm{o}}^f|^2\big)\Big] - \lambda_i^{k_1}}, \tag{6.78}$$

where it can be easily evaluated that the Lagrange multiplier λ_i^f should be

$$\lambda_i^{k_1} = \tilde{\mathrm{E}}\Big[|y_{i,\mathrm{o}}^{k_1}|^2 \, \mathrm{G}'\big(\sum_f |y_{i,\mathrm{o}}^f|^2\big)\Big]. \tag{6.79}$$

Although we derived the Newton step with respect to the whole set of $\mathbf{w}_i^f$'s, it is to be seen that the result is the same as the one in [8, 11] where the Newton step was derived with respect to a single $\mathbf{w}_i^f$. As it can be seen from the derivation, it resulted from spatially white source data.

Also, instead of evaluating λ_i^f, we can remove it by multiplying the numerator in (6.78) on both sides of the equation. Hence, with the need of normalization, the learning rule becomes

$$\mathbf{w}_i^{k_1} \leftarrow \tilde{\mathrm{E}}\Big[\mathrm{G}'\big(\sum_f |y_{i,\mathrm{o}}^f|^2\big) + |y_{i,\mathrm{o}}^{k_1}|^2 \, \mathrm{G}''\big(\sum_f |y_{i,\mathrm{o}}^f|^2\big)\Big] \mathbf{w}_{i,\mathrm{o}}^{k_1}$$
$$- \tilde{\mathrm{E}}\Big[(y_{i,\mathrm{o}}^{k_1})^* \, \mathrm{G}'\big(\sum_f |y_{i,\mathrm{o}}^f|^2\big) \, \mathbf{x}^{k_1}\Big]. \tag{6.80}$$

In addition to normalization, the rows of the unmixing matrix $\mathbf{W}^f$'s need to be decorrelated. Since the contrasts are not one-unit contrast functions, the symmetric decorrelation scheme

$$\mathbf{W}^f \leftarrow (\mathbf{W}^f (\mathbf{W}^f)^{\mathrm{H}})^{-\frac{1}{2}} \mathbf{W}^f \tag{6.81}$$

fits better than the deflationary decorrelation scheme.

It is known that many local optima exist in a cost function framework which results from wrong permutation [10, 29]. In our algorithms we employ identity matrix as the initial unmixing matrix in all frequency bins in order to be helped avoiding those local optima.

6.6.3 Scaling Problem

In frequency-domain BSS, other than the permutation problem, the problem of assigning proper variance to each frequency component with respect to its fellow frequency components has to be solved. This scaling problem arises because of the scaling indeterminacy of ICA, and also IVA. Please note that the scaling indeterminacy in the complex field includes phase ambiguity also. Here, we employ the well-known method of minimal distortion principle [30] to fix the scaling problem. After all unmixing matrices, i.e. $\mathbf{W}^f$'s, are learned, we further multiply the diagonal of $(\mathbf{W}^f)^{-1}$ to the left of $\mathbf{W}^f$, for $f = 1, 2, \ldots, F$. The reason this can solve the scaling problem is as follows. Let's say

$$\mathbf{W}^f \mathbf{A}^f = \mathbf{D}^f \mathbf{P}^f, \quad f = 1, \ldots, F, \tag{6.82}$$

where $\mathbf{D}^f$ is a diagonal scaling matrix and $\mathbf{P}^f$ is a permutation matrix. Note that these two matrices $\mathbf{D}^f$ and $\mathbf{P}^f$ represent respectively the scaling indeterminacy and the permutation indeterminacy of ICA. Then,

$$\operatorname{diag}\big((\mathbf{W}^f)^{-1}\big) = \operatorname{diag}\big(\mathbf{A}^f (\mathbf{P}^f)^{\mathrm{T}} (\mathbf{D}^f)^{-1}\big) \tag{6.83}$$

$$= \operatorname{diag}\big(\mathbf{A}^f (\mathbf{P}^f)^{\mathrm{T}}\big) (\mathbf{D}^f)^{-1}. \tag{6.84}$$

Hence,

$$\operatorname{diag}\big((\mathbf{W}^f)^{-1}\big) \mathbf{W}^f \mathbf{A}^f = \operatorname{diag}\big(\mathbf{A}^f (\mathbf{P}^f)^{\mathrm{T}}\big) (\mathbf{D}^f)^{-1} \mathbf{D}^f \mathbf{P}^f \tag{6.85}$$

$$= \operatorname{diag}\big(\mathbf{A}^f (\mathbf{P}^f)^{\mathrm{T}}\big) \mathbf{P}^f, \tag{6.86}$$

where there is no effect of the scaling matrix $\mathbf{D}^f$.

6.7 Experiments

Here we show the speech separation result of IVA algorithms when applied to real data that are recorded in a real office environment. Detailed experimental results with simulated data can be seen in [11, 22, 23]. Three human voices and a piece of hip-hop music played from a stationary speaker were captured by four equidistantly aligned microphones. The sources were located approximately 1 m $\sim$ 2 m away from the microphones. 8-second long data were used. Also, 2048-point FFT and a 2048-tab Hanning window with the shift size of 512 samples were chosen.

Figures 6.9(b) and (c) show the recorded observation signals in the microphones and the source signals separated by an ML IVA algorithm that has SEND as its source prior, respectively. Note that the first plot in Fig. 6.9 (c) is the separated piece of music, and that, IVA is successful and very robust even in a challenging scenario.

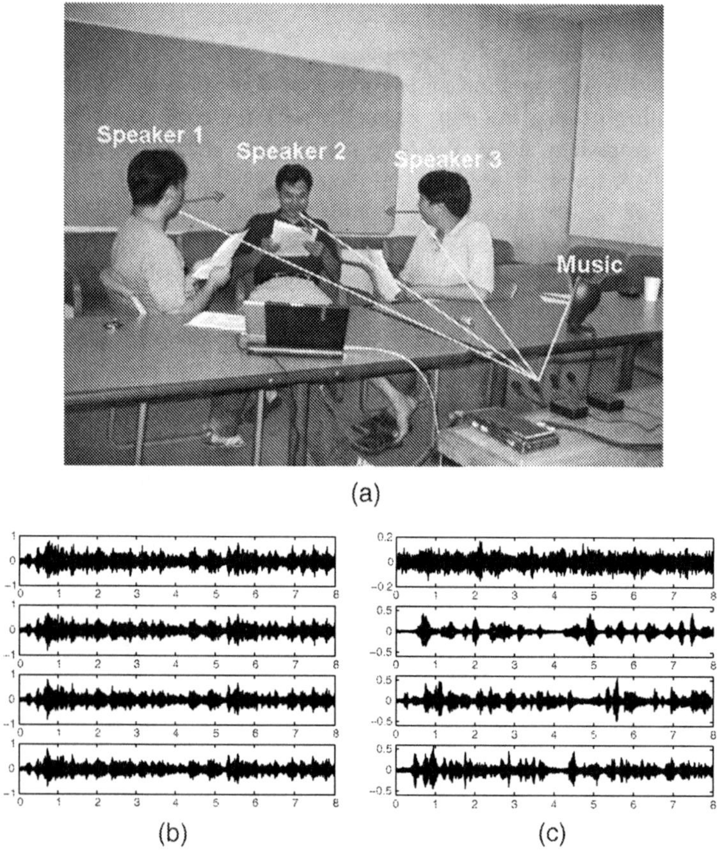

Fig. 6.9. Experiments with real acoustic recordings. (a) The real recording environment. (b) The recorded observation signals in the microphones. (c) The separation result of an IVA algorithm with $\mathrm{ML}_{\mathrm{SEND}}$ contrast.

6.8 Conclusions

In this chapter, we introduced independent vector analysis, a new formulation of ICA for multivariate components, which indeed presents a strong solution to the permutation problem of frequency-domain BSS that has been studied for a long time. In ML approach, IVA uses a multidimensional source prior that captures the inter-frequency dependencies. The multidimensional source prior is a vector representation of the ICA where the multivariate sources are assumed independent while the elements in the source are dependent. This adds a source constraint to the ICA learning so that each source is learned with the inter-frequency dependency relation and hence the permutation problem can be solved. In IVA, similar to ICA, we have certain flexibilities in defining or modelling the multidimensional source prior of speech. For speech, while modelling the source prior in a more accurate manner will lead to better performance, joint PDFs that are sparse and spherically symmetric or alike have proven to be proper and efficient models.

Acknowledgment

This material is based upon work supported by the national science foundation (NSF) under grant no. 0535251.

References

1. P. Smaragdis, "Blind separation of convolved mixtures in the frequency domain," *Neurocomputing*, vol. 22, pp. 21–34, 1998.
2. L. Parra and C. Spence, "Convolutive blind separation of non-stationary sources," *IEEE Trans. on Speech and Audio Processing*, vol. 8, no. 3, pp. 320–327, 2000.
3. J. Anemueller and B. Kollmeier, "Amplitude modulation decorrelation for convolutive blind source separation," *Proc. Int. Conf. on Independent Component Analysis and Blind Source Separation*, pp. 215–220, 2000.
4. N. Murata, S. Ikeda, and A. Ziehe, "An approach to blind source separation based on temporal structure of speech signals," *Neurocomputing*, vol. 41, pp. 1–24, 2001.
5. M. Z. Ikram and D. R. Morgan, "A beamforming approach to permutation alignment for multichannel frequency-domain blind speech separation," *Proc. IEEE Int. Conf. on Acoustics, Speech, and Signal Processing*, pp. 881–884, 2002.
6. H. Sawada, R. Mukai , S. Araki, and S. Makino, "A robust and precise method for solving the permutation problem of frequency-domain blind source separation," *Proc. Int. Conf. on Independent Component Analysis and Blind Source Separation*, pp. 505–510, 2003.
7. T. Kim, H. T. Attias, S.-Y. Lee, and T.-W. Lee, "Blind source separation exploiting higher-order frequency dependencies," *IEEE Trans. on Speech and Audio Processing*, vol. 15, no. 1, pp. 70–79, 2007.
8. I. Lee, T. Kim, and T.-W. Lee, "Complex FastIVA: a robust maximum likelihood approach of MICA for convolutive BSS," *Lecture Notes in Computer Science*, vol. 3889, pp. 625–632, 2006.
9. A. Hiroe, "Solution of permutation problem in frequency domain ICA, using multivariate probability density functions," *Lecture Notes in Computer Science*, vol. 3889, pp. 601–608, 2006.
10. M. Davies, "Audio Source Separation," *Mathematics in Signal Processing 5*, Oxford University Press, pp. 57–68, 2002.
11. I. Lee, T. Kim, and T.-W. Lee, "Fast Fixed-Point Independent Vector Analysis Algorithms for Convolutive Blind Source Separation", *Signal Processing*, vol. 87, no. 8, pp. 1859–1871, 2007.
12. A. J. Bell and T. J. Sejnowski, "An information maximization approach to blind separation and blind deconvolution," *Neural Computation*, vol. 7, no. 6, pp. 1129–1159, 1995.

13. J.-F. Cardoso, "Infomax and maximum likelihood for source separation," *IEEE Signal Processing Letters*, vol. 4, no. 4, 1997.
14. E. Bingham and A. Hyvärinen, "A fast fixed-point algorithm for independent componenet analysis of complex-valued signals," *Int. J. of Neural Systems*, vol. 10, no. 1, pp. 1–8, 2000.
15. J.-F. Cardoso, "Entropic contrasts for source separation: Geometry and stability," S. Haykin, Ed. *Unsupervised Adaptive Filtering*, vol. 1, pp. 139–189, John Wiley and Sons, 2000.
16. J.-F. Cardoso, "Multidimensional independent component analysis," *Proc. IEEE Int. Conf. on Acoustics, Speech, and Signal Processing*, pp. 1941–1944, 1998.
17. A. Hyvärinen and P. O. Hoyer, "Emergence of phase and shift invariant features by decomposition of natural images into independent feature subspaces," *Neural Computation*, vol. 12, no. 7, pp. 1705–1720, 2000.
18. D. H. Brandwood, "A complex gradient operator and its application in adaptive array theory," *IEE Proc. F and H*, vol. 130, no. 1, pp. 11–16, 1983.
19. A. van den Bos, "Complex gradient and Hessian," *IEE Proceedings on Vision, Image and Signal Processing*, vol. 141, pp. 380–382, 1994.
20. H. Brehm and W. Stammler, "Description and generation of spherically invariant speech-model signals," *Signal Processing*, no. 12, pp. 119–141, 1987.
21. H. Buchner, R. Aichner, and W. Kellermann, "Blind source separation for convolutive mixtures: A unified treatment," *Audio Signal Processing for Next-Generation Multimedia Communication Systems*, Y. Huang and J. Benesty Ed., Kluwer Academic Publishers, Boston, pp. 255–293, 2004.
22. I. Lee and T.-W. Lee, "On the Assumption of Spherical Symmetry and Sparseness for the Frequency-Domain Speech Model", *IEEE Trans. on Speech, Audio and Language Processing*, vol. 15, no. 5, pp. 1521–1528, 2007.
23. G.-J. Jang, I. Lee, and T.-W. Lee, "Independent Vector Analysis using Non-Spherical Joint Densities for the Separation of Speech Signals", *Proc. IEEE Int. Conf. on Acoustics, Speech, and Signal Processing*, vol. 2, pp. 629–632, 2007.
24. S.-I. Amari, A. Cichocki, and H. H. Yang, "A new learning algorithm for blind signal separation," *Adv. Neural Information Processing Systems*, vol. 8, pp. 757–763, 1996.
25. J.-F. Cardoso, "The invariant approach to source separation," *Proc. International Symposium on Nonlinear Theory and Applications (NOLTA)*, vol. 1, pp. 55–60, 1995.
26. A. Hyvärinen and E. Oja, "A Fast fixed-Point algorithm for independent component analysis," *Neural Computation*, vol. 9, no. 7, pp. 1483–1492, 1997.
27. A. Hyvärinen, "Fast and Robust fixed-point algorithms for independent componenet analysis, *IEEE Trans. on Neural Networks*, vol. 10, no. 3, pp. 626–634, 1999.
28. E. G. Learned-Miller and J. W. Fisher III, "ICA using spacings estimates of entropy," *J. of Machine Learning Research*, vol. 4, pp. 1271–1295, 2003.
29. M. Z. Ikram and D. R. Morgan, "Exploring permutation inconsistency in blind separation of signals in a reverberant environment," *Proc. IEEE Int. Conf. on Acoustics, Speech, and Signal Processing*, pp. 1041–1044, 2000.
30. K. Matsuoka and S. Nakashima, "Minimal distortion principle for blind source separation," *Proc. Int. Conf. on Independent Component Analysis and Blind Source Separation*, pp. 722–727, 2001.

7 Relative Newton and Smoothing Multiplier Optimization Methods for Blind Source Separation

Michael Zibulevsky

Department of Computer Science, Technion, Haifa 32000, Israel
E-mail: mzib@cs.technion.ac.il

Abstract. We study a relative optimization framework for quasi-maximum likelihood blind source separation and relative Newton method as its particular instance. The structure of the Hessian allows its fast approximate inversion. In the second part we present Smoothing Method of Multipliers (SMOM) for minimization of sum of pairwise maxima of smooth functions, in particular sum of absolute value terms. Incorporating Lagrange multiplier into a smooth approximation of max-type function, we obtain an extended notion of nonquadratic augmented Lagrangian. Our approach does not require artificial variables, and preserves the sparse structure of Hessian. Convergence of the method is further accelerated by the Frozen Hessian strategy. We demonstrate efficiency of this approach on an example of blind separation of sparse sources. The nonlinearity in this case is based on the absolute value function, which provides superefficient source separation.

7.1 Introduction

In this chapter we study quasi-maximum likelihood blind source separation (quasi-ML BSS) [1, 2] in batch mode, without orthogonality constraint. This criterion provides improved separation quality [3, 4], and is particularly useful in separation of sparse sources. We will present optimization methods, which produce quasi-ML BSS efficiently.

7.1.1 Quasi-ML Blind Source Separation (BSS)

Consider the BSS problem, where an N-channel sensor signal $x(t)$ arises from N unknown scalar source signals $s_i(t)$, $i = 1, \ldots, N$, linearly mixed together by an unknown $N \times N$ matrix A

$$x(t) = As(t). \tag{7.1}$$

We wish to estimate the mixing matrix A and the N-dimensional source signal $s(t)$. In the discrete time case $t = 1, 2, \ldots, T$ we use matrix notation $X = AS$, where X and S are $N \times T$ matrices with the signals $x_i(t)$ and $s_i(t)$ in the corresponding rows. We also denote the unmixing matrix $W = A^{-1}$.

S. Makino et al. (eds.), Blind Speech Separation, 193–214.

When the sources are *i.i.d*, stationary and white, the normalized minus-log-likelihood of the observed data X is (see for example [4])

$$L(W;X) = -\log|\det W| + \frac{1}{T}\sum_{i,t} h\big(W_i x(t)\big), \tag{7.2}$$

where W_i is i-th row of W, $h(\cdot) = -\log f(\cdot)$, and $f(\cdot)$ is the probability density function (pdf) of the sources. Consistent estimator can be obtained by minimization of (7.2), also when $h(\cdot)$ is not exactly equal to $-\log f(\cdot)$. Such *quasi-ML estimation* is practical when the source pdf is unknown, or is not well-suited for optimization. For example, when the sources are sparse or sparsely representable, the absolute value function or its smooth approximation is a good choice for $h(\cdot)$ [5–10]. Here we will use a family of convex smooth approximations to the absolute value

$$h_1(c) = |c| - \log(1+|c|) \tag{7.3}$$

$$h_\lambda(c) = \lambda h_1(c/\lambda) \tag{7.4}$$

with λ a proximity parameter: $h_\lambda(c) \to |c|$ as $\lambda \to 0^+$. Widely accepted natural gradient method does not work well when the approximation of the absolute value becomes too sharp. In this work we consider the relative Newton method, which overcomes this obstacle.

The Newton equations considered in this work are similar in part to those obtained by Pham and Garat [1], using different considerations. However, the algorithm given in [1], is not used in practice, because of a possibility of convergence to spurious solutions. We overcome this difficulty using line search and forcing positive definiteness of the Hessian.

Several other Newton-like BSS methods have been studied in the literature. They are based on negentropy approximation with orthogonality constraint [11], cumulant model [12, 13] and joint diagonalization of correlation matrices [14–17].

The relative Newton method presented here is dedicated to quasi-ML BSS in general (not only to the sparse source case).

7.1.2 Smoothing Method of Multipliers (SMOM) for Sum-Max Problems

In the second part we present a method for minimization of a sum of pairwise maxima of smooth functions, in particular sum of absolute value terms, arising in quasi-ML BSS.

Methods of multipliers, involving *nonquadratic augmented Lagrangians* [18–27] successfully compete with the interior-point methods in nonlinear and semidefinite programming. They are especially efficient when a very high accuracy of solution is required. This success is explained by the fact, that due to iterative update of multipliers, the penalty parameter does not need to become extremely small in the neighborhood of solution.

Direct application of the augmented Lagrangian approach to the sum-max problem requires introduction of artificial variables, one per element of the sum, that significantly increases the problem size and the computational burden. Alternatively, one can use a smooth approximation of the max-type function [28, 29], which will keep the size of the problem unchanged, but will require the smoothing parameter to become extremely small in order to get accurate solution.

In this work we incorporate multiplier into a smooth approximation of the max-type function, obtaining an extended notion of augmented Lagrangian. This allows us to keep the size of the problem unchanged, and achieve a very accurate solution under a moderate value of the smoothing parameter. Convergence of the method is further accelerated by the Frozen Hessian strategy.

We demonstrate the efficiency of this approach on an example of blind separation of sparse sources with the absolute value nonlinearity. It preserves sparse structure of the Hessian, and achieves 12–15 digits of source separation accuracy.

7.2 Relative Optimization Algorithm

We consider the following algorithm for minimization of the quasi-ML function (7.2):

- Start with an initial estimate W_1 of the separation matrix;
- **For** $k = 1, 2, ...,$ until convergence
  1. Compute the current source estimate $U_k = W_k X$;
  2. Starting with $V{=}I$, get V_{k+1} by one or few steps of a conventional optimization method, decreasing sufficiently $L(V; U_k)$;
  3. Update the estimated separation matrix $W_{k+1} = V_{k+1} W_k$;
- **End**

The relative (natural) gradient method [30–32] is a particular instance of this approach, when a standard gradient descent step is used in Item 2. The following remarkable property of the relative gradient is also preserved in general: *given current source estimate U, the progress of the method does not depend on the original mixing matrix.* This means that even nearly ill-conditioned mixing matrix influences the convergence of the method not more than a starting point. Convergence analysis of the Relative Optimization algorithm is presented in Appendix A. In the following we will use a Newton step in Item 2 of the method.

7.3 Hessian Evaluation

The likelihood $L(W; X)$ is a function of a matrix argument W. The corresponding gradient is also a matrix

$$G(W) = \nabla L(W; X) = -W^{-T} + \frac{1}{T} h'(WX) X^T, \tag{7.5}$$

where $h'(WX)$ is a matrix with the elements $h'\big((WX)_{ij}\big)$. The Hessian of $L(W;X)$ is a linear mapping $\mathcal{H}$ defined via the differential of the gradient

$$dG = \mathcal{H}dW. \tag{7.6}$$

We can also express the Hessian in standard matrix form converting W into a long vector $w = \text{vec}(W)$ using row stacking. We will denote the reverse conversion $W = \text{mat}(w)$. Let

$$\hat{L}(w, X) \equiv L(\text{mat}(w), X), \tag{7.7}$$

so that the gradient

$$g(w) = \nabla\hat{L}(w;X) = \text{vec}(G(W)). \tag{7.8}$$

Then

$$dg = Hdw, \tag{7.9}$$

where H is an $N^2 \times N^2$ Hessian matrix. We also have

$$dg = \text{vec}(dG). \tag{7.10}$$

7.3.1 Hessian of $-\log\det W$

Using the expression

$$d(W^{-1}) = -W^{-1}(dW)W^{-1},$$

which follows from the equality

$$0 = d(WW^{-1}) = (dW)W^{-1} + Wd(W^{-1}),$$

we obtain the differential of the first term in (7.5)

$$dG = d(-W^{-T}) = A^T(dW^T)A^T, \tag{7.11}$$

where $A = W^{-1}$. A particular element of the differential is determined using A^i and A_j, i-th column and j-th row of A respectively:

$$dG_{ij} = {A^i}^T dW^T A_j^T = \text{Trace}{A^i}^T dW^T A_j^T = \text{Trace}A_j^T {A^i}^T dW^T, \tag{7.12}$$

or

$$dG_{ij} = \text{Trace}A^i A_j dW = \langle (A^i A_j)^T, dW\rangle. \tag{7.13}$$

On the other hand, (7.9) and (7.10) gives us

$$dG_{ij} = dg_k = \langle H_k, dw\rangle, \tag{7.14}$$

where H_k – k-th row of H, $k = (i-1)N + j$. Comparing the last two equations, we conclude that the k-th row of H contains the matrix $(A^i A_j)^T$ stacked row-wise

$$H_k = (\text{vec}(A^i A_j)^T)^T. \tag{7.15}$$

7.3.2 Hessian of $\frac{1}{T}\sum_{m,t} h(W_m x(t))$

It is easy to see that the Hessian of the second term in $\hat{L}(w, X)$ is a block-diagonal matrix with the following $N \times N$ blocks:

$$B^m = \frac{1}{T}\sum_t h''(W_m x(t))x(t)x^T(t), \quad m = 1, \ldots, N. \tag{7.16}$$

7.4 Newton Method

Newton method is an efficient tool of unconstrained optimization. It often converges fast and provides quadratic rate of convergence. However, its iteration may be costly, because of the necessity to compute the Hessian matrix and solve the corresponding system of equations. In the next section we will see that this difficulty can be overcome using the relative Newton method.

First, let us consider the standard Newton approach, in which the direction is given by solution of the linear equation

$$Hy = -\nabla\hat{L}(w; X), \tag{7.17}$$

where $H = \nabla^2\hat{L}(w; X)$ is the Hessian of (7.7). In order to guarantee descent direction in the case of nonconvex objective function, we use modified Cholesky factorization[1] [33], which automatically finds a diagonal matrix R such that the matrix $H + R$ is positive definite, providing a solution to the modified system

$$(H + R)y = -\nabla\hat{L}(w; X). \tag{7.18}$$

After the direction y is found, the new iterate w^+ is given by

$$w^+ = w + \alpha y, \tag{7.19}$$

where the step size α is determined by exact line search

$$\alpha = \arg\min_{\alpha} \hat{L}(w + \alpha y; X) \tag{7.20}$$

or by a backtracking line search [33]:

$\alpha := 1$
While $\hat{L}(w + \alpha y; X) > \hat{L}(w; X) + \beta\alpha\nabla\hat{L}(w; X)^T d$
    $\alpha := \gamma\alpha$
end
where $0 < \alpha < 1$ and $0 < \gamma < 1$. The use of line search guarantees

[1] We use the MATLAB code of modified Cholesky factorization by Brian Borchers, available at http://www.nmt.edu/~borchers/ldlt.html

monotone decrease of the objective function at every Newton iteration. In our computations the line search constants were $\beta = \gamma = 0.3$. It may also be reasonable to give β a small value, like 0.01.

Computational complexity. The Hessian is an $N^2 \times N^2$ matrix; its computation requires N^4 operations in (7.15) and $N^3 T$ operations in (7.16). Solution of the Newton system (7.18) using modified Cholesky decomposition, requires $N^6/6$ operations for decomposition and N^4 operations for back/forward substitution. Altogether, we need

$$2N^4 + N^3 T + N^6/6$$

operations for one Newton step. Comparing this to the cost of the gradient evaluation (7.5), which is equal to $N^2 T$, we conclude that the Newton step costs about N gradient steps when the number of sources is small (say, up to 20). Otherwise, the third term become dominating, and the complexity grows as N^6.

7.5 Relative Newton Method

In order to make the Newton algorithm invariant to the value of mixing matrix, we use the relative Newton method, which is a particular instance of the Relative Optimization algorithm. This approach simplifies the Hessian computation and the solution of the Newton system.

7.5.1 Basic Relative Newton Step

The optimization in Item 2 of the Relative Optimization algorithm is produced by a single Newton-like iteration with exact or backtracking line search. The Hessian of $L(I; U)$ has a special structure, which permits fast solution of the Newton system. First, the Hessian of $-\log \det W$ given by (7.15), becomes very simple and sparse, when $W = A = I$: each row of H

$$H_k = \text{vec}^T(e_i e_j^T), \qquad (7.21)$$

contains only one nonzero element, which is equal to 1. Here e_j is an N-element standard basis vector, containing 1 at j-th position. The remaining part of the Hessian is block-diagonal. There are various techniques for solving sparse symmetric systems. For example, one can use sparse modified Cholesky factorization for direct solution, or alternatively, conjugate gradient-type methods, possibly preconditioned by incomplete Cholesky factor, for iterative solution. In both cases, the Cholesky factor is often not as sparse as the original matrix, but it becomes sparser, when appropriate matrix permutation is applied before factorization (see for example MATLAB functions CHOLINC and SYMAMD.)

7.5.2 Fast Relative Newton Step

Further simplification of the Hessian is obtained by considering its structure at the solution point $U_k = S$. The elements of the m-th block of the second term of $\nabla^2 L(I; S)$ given by (7.16), are equal to

$$B_{ij}^m = \frac{1}{T} \sum_t h''(s_m(t)) s_i(t) s_j(t), \quad i, j = 1, \ldots, N.$$

When the sources are independent and zero mean, we have the following zero expectation

$$E\{h''(s_m(t)) s_i(t) s_j(t)\} = 0, \quad m, i \neq j,$$

hence the off-diagonal elements B_{ij}^m converge to zero as sample size grows. Therefore we use a diagonal approximation of this part of the Hessian

$$B_{ii}^m = \frac{1}{T} \sum_t h''(u_m(t)) u_i^2(t), i = 1, \ldots, N; \quad m = 1, \ldots, N, \tag{7.22}$$

where $u_m(t)$ are current estimates of the sources. In order to solve the simplified Newton system, let us return to the matrix-space form (7.6) of the Hessian operator. Let us pack the diagonal of the Hessian given by (7.22) into $N \times N$ matrix D, row-by-row. Taking into account that $A = I$ in (7.11), we will obtain the following expression for the differential of the gradient:

$$dG = \mathcal{H} dW = dW^T + D \odot dW, \tag{7.23}$$

where "$\odot$" denotes element-wise multiplication of matrices. For an arbitrary matrix Y,

$$\mathcal{H} Y = Y^T + D \odot Y. \tag{7.24}$$

In order to solve the Newton system

$$Y^T + D \odot Y = G, \tag{7.25}$$

we need to solve $N(N-1)/2$ systems of size 2×2 with respect to Y_{ij} and Y_{ji}

$$\begin{aligned} D_{ij} Y_{ij} + Y_{ji} &= G_{ij}, \quad i = 1, \ldots, N; \ j = 1, \ldots, i-1 \\ D_{ji} Y_{ji} + Y_{ij} &= G_{ji}. \end{aligned} \tag{7.26}$$

The diagonal elements Y_{ii} can be found directly from the set of single equations

$$D_{ii} Y_{ii} + Y_{ii} = G_{ii}. \tag{7.27}$$

In order to guarantee a descent direction and avoid saddle points, we modify the Newton system (7.26), changing the sign of the negative eigenvalues

[33]. Namely, we compute analytically the eigenvectors and the eigenvalues of 2×2 matrices

$$\begin{pmatrix} D_{ij} & 1 \\ 1 & D_{ji} \end{pmatrix},$$

invert the sign of the negative eigenvalues, and force small eigenvalues to be above some threshold (say, 10^{-8} of the maximal one in the pair). Than we solve the modified system, using the eigenvectors already obtained and the modified eigenvalues.

Computational complexity. Computing the diagonal of the Hessian by (7.22) requires N^2T operations, which is equal to the cost of the gradient computation. Solution cost of the set of 2×2 linear equations (7.26) is about $15N^2$ operations, which is negligible compared to the gradient cost.

7.6 Sequential Optimization

When the sources are sparse, the quality of separation greatly improves with reduction of smoothing parameter λ in the absolute value approximation (7.4). On the other hand, the optimization of the likelihood function becomes more difficult for small λ. Therefore, we use sequential optimization with gradual reduction of λ. Denote

$$L(W; X, \lambda) = -\log|\det W| + \frac{1}{T}\sum_{i,t} h_\lambda\big(W_i x(t)\big), \tag{7.28}$$

where $h_\lambda(\cdot)$ is given by (7.3–7.4).

Sequential optimization algorithm

1. Start with λ_1 and W_1;
2. **For** $k = 1, 2, .., K$,
   (a) Compute current source estimate $U_k = W_k X$;
   (b) Find $V_{k+1} = \arg\min_V L(V, U_k, \lambda_k)$, using $V = I$ as a starting point;
   (c) Update the separation matrix $W_{k+1} = V_{k+1} W_k$;
   (d) Update the smoothing parameter $\lambda_{k+1} = \mu\lambda_k$.
3. **End**

In our computations we choose the parameters $\lambda_1 = 1$ and $\mu = 0.01$. Note that step (b) includes the whole loop of unconstrained optimization, which can be performed, for example, by the relative Newton method.

7.7 Smoothing Method of Multipliers (SMOM)

Even gradual reduction of the smoothing parameter may require significant number of Newton steps after each update of λ. More efficient way to achieve an accurate solution of a problem involving a sum of absolute value functions is to use SMOM method presented in this section. This method is an extension of augmented Lagrangian technique [19, 20, 25] used in constrained optimization. It allows to obtain accurate solution without forcing the smoothing parameter λ to go to zero. In this work we combine the SMOM with relative optimization.

Consider non-smooth optimization problem

$$\min_w \left\{ F(w) = f_0(w) + \sum_{i=1}^{m} \max\left[\alpha_i f_i(w),\ \beta_i f_i(w)\right] \right\} , \tag{7.29}$$

where $f_i(w),\ i = 0, \ldots, m$ are smooth functions, $\alpha_i < \beta_i$ are certain constants. In particular, when $\alpha_i = -1$ and $\beta_i = 1$, we get sum of absolute value terms, like in the quasi-ML BSS:

$$F(w) = f_0(w) + \sum_{i=1}^{m} |f_i(w)|. \tag{7.30}$$

This kind of problem arises in many other areas of signal and image processing, in the context of sparse representations, total variation regularization, etc. (see for example [5]).

7.7.1 Smoothing the Max-Function

Consider a maximum function shown in Fig. 7.1,

$$r(t) = \max(\alpha t, \beta t) .$$

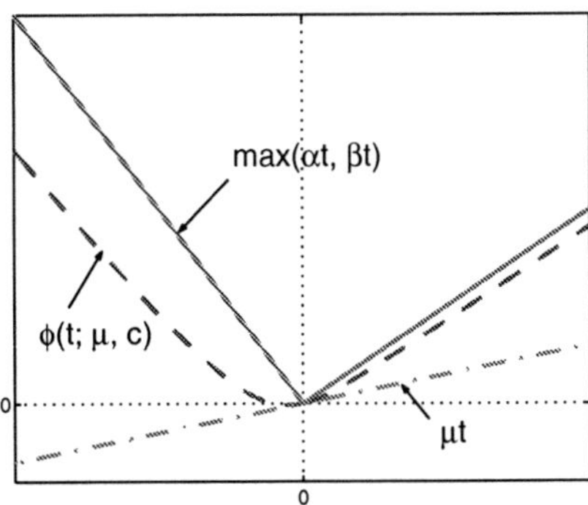

Fig. 7.1. Max$(\alpha t, \beta t)$ – solid line; smoothing function $\varphi(t; \mu, \lambda)$ – dashed line; linear support μt – dot-dashed line.

We introduce its smooth approximation $\varphi(t;\mu,\lambda)$, $\alpha < \mu < \beta$, $\lambda > 0$, with two parameters: μ and λ. The parameter λ defines the accuracy of the approximation of the max-function $r(\cdot)$, becoming perfect as $\lambda \to 0$. The parameter μ determines the derivative of φ at $t = 0$; it will serve as a Lagrange multiplier. The graph of the linear function μt is tangent to the plot of $\varphi(\cdot;\mu,\lambda)$ at the origin.

The function φ possesses the following properties:

- $\varphi(t;\mu,\lambda)$ is convex in t;
- $\varphi(0;\mu,\lambda) = 0$;
- $\varphi_t'(0;\mu,\lambda) = \mu$;
- $\lim_{t\to-\infty} \varphi_t'(t;\mu,\lambda) = \alpha$;
- $\lim_{t\to+\infty} \varphi_t'(t;\mu,\lambda) = \beta$;
- $\lim_{\lambda\to 0} \varphi(t;\mu,\lambda) = r(t)$.

The particular form of the function φ we introduce and prefer to use in our computations consists of three smoothly connected branches:

$$\varphi(t,\mu,\lambda) = \begin{cases} \alpha t - p_1 \log \frac{t}{\tau_1} + s_1, & t < \tau_1 \le 0 \\ \frac{t^2}{2\lambda} + \mu t, & \tau_1 \le t \le \tau_2 \\ \beta t - p_2 \log \frac{t}{\tau_2} + s_2, & t > \tau_2 \ge 0\ . \end{cases} \tag{7.31}$$

The coefficients $p_1, p_2, s_1, s_2, \tau_1, \tau_2$ are chosen to make the function φ continuous and twice differentiable at the joint points τ_1 and τ_2:

$$\begin{aligned} \tau_1 &= \frac{\lambda(\alpha-\mu)}{2}; \quad \tau_2 = \frac{\lambda(\beta-\mu)}{2}; \\ p_1 &= \tau_1^2/\lambda; \qquad p_2 = \tau_2^2/\lambda; \\ s_1 &= \frac{\tau_1^2}{2\lambda} + (\mu-\alpha)\tau_1; \\ s_2 &= \frac{\tau_2^2}{2\lambda} + (\mu-\beta)\tau_2. \end{aligned}$$

When $\tau_i = 0$, we put $p_i \log \frac{t}{\tau_i} = \lambda \tau_i^2 \log \frac{t}{\tau_i} = 0$. One can note, that when $\alpha \to 0$ and $\beta \to \infty$, $\varphi(\cdot)$ becomes a *quadratic-logarithmic* penalty for inequality constraint in nonquadratic augmented Lagrangian [25]. On the other hand, when $\alpha \to -\infty$ and $\beta \to \infty$, $\varphi(\cdot)$ becomes a quadratic penalty for equality constraint in a standard quadratic augmented Lagrangian [34]. In this way our approach generalizes known augmented Lagrangian techniques.

7.7.2 Generalized Lagrangian and Augmented Lagrangian

A standard way to introduce augmented Lagrangian for the sum-max problem would be to reformulate it as a smooth constrained optimization problem using artificial variables, which will increase the problem size significantly. Instead we introduce extended notions of Lagrangian and augmented

Lagrangian, which keep the problem size unchanged, preserving at the same time many important classical properties of these functions. We propose the following extended notion of Lagrangian:

$$L(w,u) = f_0(w) + \sum_{i=1}^{m} u_i f_i(w), \quad \alpha \le u_i \le \beta, \tag{7.32}$$

and corresponding extended notion of augmented Lagrangian

$$M(w,u,\lambda) = f_0(w) + \sum_{i=1}^{m} \varphi(f_i(w), u_i, \lambda), \quad \alpha \le u_i \le \beta. \tag{7.33}$$

As we show in [35], the Lagrangian saddle point theorem and duality theory can be extended to our case. As an important consequence of this, *there exists a vector of multipliers* u^**, such that an unconstrained minimizer* w^* *of the augmented Lagrangian provides an optimal solution to Problem (7.29), i.e.* we do not need to force the smoothing parameter λ toward zero in order to solve the problem. This property serves as a basis for the method of multipliers presented below.

7.7.3 SMOM Algorithm

We introduce the method of multipliers, which performs the following steps at each outer iteration:

1. Minimize augmented Lagrangian in w, starting from the point w^k given by the previous iteration
$$w^{k+1} = \arg\min_{w} M(w, u^k, \lambda_k); \tag{7.34}$$
2. Update the multipliers using the derivative of φ with respect to the first argument
$$u_i^{k+1} = \varphi'(f_i(w^{k+1}), u_i^k, c_k), \quad i = 1, \dots, m; \tag{7.35}$$
3. Update the smoothing parameter (optionally)
$$\lambda_{k+1} = \gamma \lambda_k, \qquad 0 < \gamma < 1. \tag{7.36}$$

The multiplier update rule (7.35) is motivated by the fact that in this way w^{k+1} becomes a minimizer of the Lagrangian: $\nabla_w L(w^{k+1}, u^{k+1}) = 0$. Therefore the optimal solution (w^*, u^*) is a fixed point of the algorithm. Similar considerations are used in standard augmented Lagrangian algorithm (see for example [25]).

In practical implementation we restrict the relative change of the multipliers to some bounds in order to stabilize the method:

$$\gamma_1 < \frac{u_i^{k+1} - \alpha}{u_i^k - \alpha} < \gamma_2, \tag{7.37}$$

$$\gamma_1 < \frac{\beta - u_i^{k+1}}{\beta_i - u_i^k} < \gamma_2, \tag{7.38}$$

$$\alpha + \delta < u_i^{k+1} < \beta - \delta. \tag{7.39}$$

We also restrict the smoothing parameter to remain above some minimal value λ_{min}. We usually put $\gamma = 0.5$, $\gamma_1 = \frac{1}{\gamma_2} = 2$, $\delta = 10^{-6}$, $\lambda_{min} = 10^{-3}$. In general, the algorithm is rather insensitive to changes in the parameters in order of magnitude or more. Convergence analysis of the method in convex case is presented in [35]. In practice, it works well also with nonconvex problems, as we will see on example of quasi-ML BSS. In the later case we use the relative Newton method at the inner optimization stage (7.34).

7.7.4 Frozen Hessian Strategy

A useful fact is that changes in the Hessian of the augmented Lagrangian become very small at late outer iterations. This happens because changes in primal variables and multipliers become small toward convergence to solution, while the smoothing parameter λ remains constant. Therefore we can reuse the inverse Hessian (or its Cholesky factor) from the previous iterations [26], unless the number of steps in the current unconstrained optimization exceeds a predefined limit (say, 3–10 steps). Often 5–7 last outer iterations require only one Newton step each, without recomputing Hessian at all (see the experimental section).

7.8 Computational Experiments

Two data sets were used. The first group of sources was artificial sparse data with Bernoulli–Gaussian distribution

$$f(s) = p\delta(s) + (1-p)\frac{1}{\sqrt{2\pi\sigma^2}}\exp(-s^2/2\sigma^2),$$

generated by the MATLAB function SPRANDN. We used the parameters $p = 0.5$ and $\sigma = 1$. The second group of sources were four natural images from [36]. The mixing matrix was generated randomly with uniform *i.i.d.* entries.

7.8.1 Relative Newton Method

In all experiments we used a backtracking line search with the constants $\beta = \gamma = 0.3$. Figure 7.2 shows the typical progress of different methods applied to the artificial data with 5 mixtures of 10k samples. The fast relative Newton method converges in about the same number of iterations as the relative Newton with exact Hessian, but significantly outperforms it in time. Natural gradient in batch mode requires much more iterations, and has a difficulty to converge when the smoothing parameter λ in (7.4) becomes too small.

In the second experiment, we demonstrate the advantage of the batch-mode quasi-ML separation, when dealing with sparse sources. We compared the the fast relative Newton method with stochastic natural gradient [30–32], Fast ICA [11] and JADE [37]. All three codes were obtained from public web sites [38–40]. Stochastic natural gradient and Fast ICA used tanh nonlinearity. Figure 7.3 shows separation of artificial stochastic sparse data: 5 sources

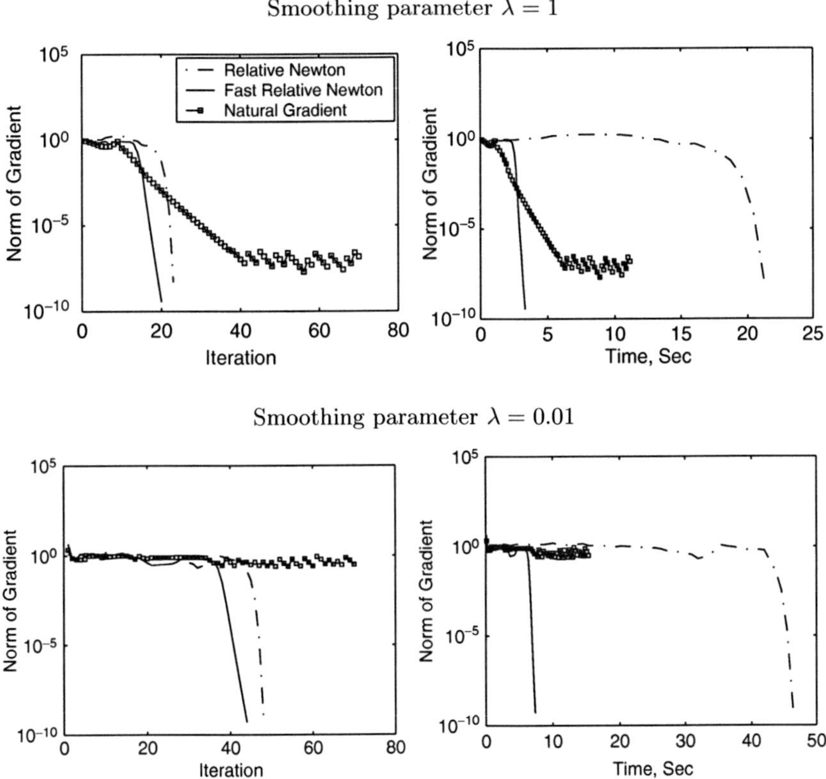

Fig. 7.2. Separation of artificial sparse data with 5 mixtures by 10k samples. Relative Newton with exact Hessian – dashed line, fast relative Newton – continuous line, natural gradient in batch mode – squares.

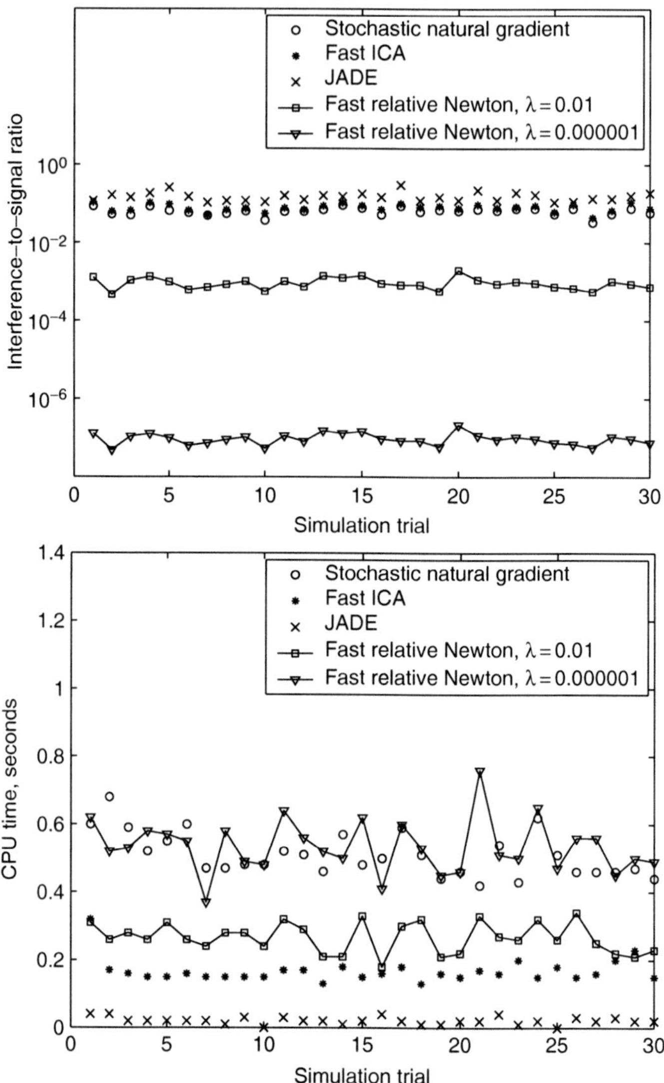

Fig. 7.3. Separation of stochastic sparse data: 5 sources of 500 samples, 30 simulation trials. Top – interference-to-signal ratio, bottom – CPU time.

of 500 samples, 30 simulation trials. The quality of separation is measured by interference-to-signal ratio (ISR) in amplitude units. As we see, fast relative Newton significantly outperforms other methods, providing practically ideal separation with the smoothing parameter $\lambda = 10^{-6}$ (sequential update of the smoothing parameter was used here). Timing is of about the same order for all the methods, except of JADE, which is known to be much faster with relatively small matrices.

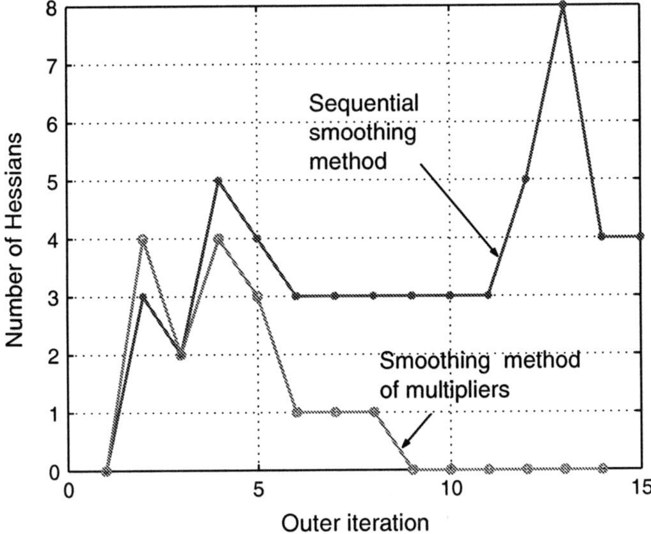

Fig. 7.4. Relative Newton method with frozen Hessian: number of Hessian evaluations per outer iteration (the first data set).

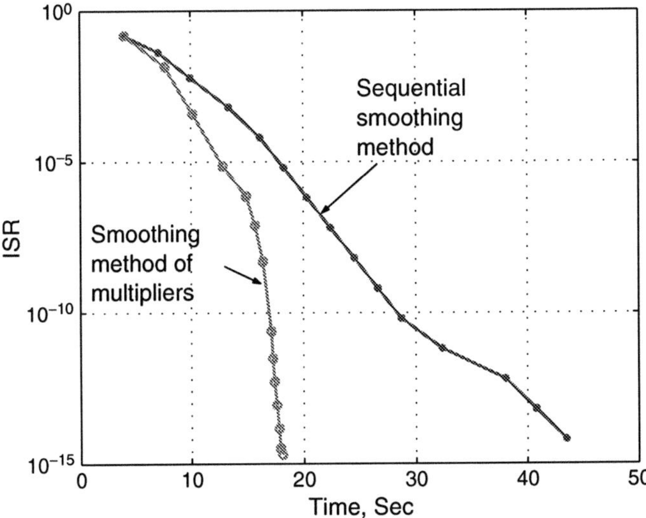

Fig. 7.5. Interference-to-signal ratio: progress with iterations/CPU time. Dots correspond to outer iterations.

7.8.2 SMOM Combined with the Relative Newton Method

In the third experiment we have used the first stochastic sparse data set: 5 mixtures, 10k samples. Figure 7.5 demonstrates advantage of the SMOM

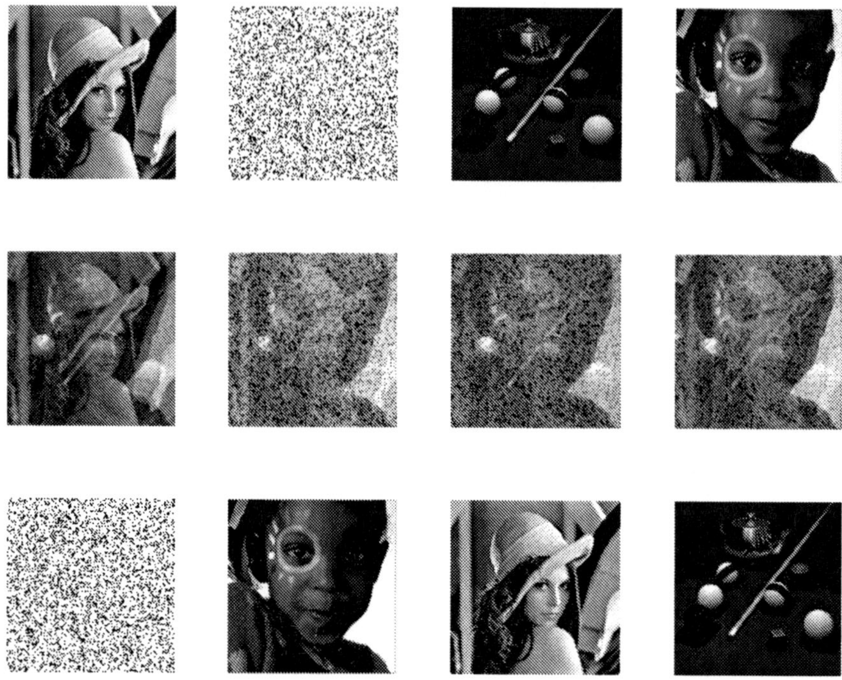

Fig. 7.6. Separation of images with preprocessing by differentiation. Top – sources, middle – mixtures, bottom – separated.

combined with the frozen Hessian strategy. As we see, the last six outer iterations does not require new Hessian evaluations. At the same time the sequential smoothing method without Lagrange multipliers, requires 3 to 8 Hessian evaluations per outer iterations toward end. As a consequence the method of multipliers converges much faster.

In the fourth experiment, we separated four natural images [36], presented in Fig. 7.6. Sparseness of images can be achieved via various wavelet-type transforms [8–10], but even simple differentiation can be used for this purpose, since natural images often have sparse edges. Here we used the stack of horizontal and vertical derivatives of the mixture images as an input to separation algorithms. Figure 7.7 shows the separation quality achieved by stochastic natural gradient, Fast ICA, JADE, the fast relative Newton method with $\lambda = 10^{-2}$ and the the SMOM. Like in the previous experiments, SMOM provides practically ideal separation with ISR of about 10^{-12}. It outperforms the other methods by several orders of magnitude.

In the fifth experiment we separated five musical sound recordings (one second, 10k samples), presented in Fig. 7.8. Preprocessing (sparsification) was produced by the short time Fourier transform (STFT) of the mixtures using time windows of 2048 samples. Real and imaginary parts of the coefficients were concatenated into long vectors. As we see from Fig. 7.9, the use of

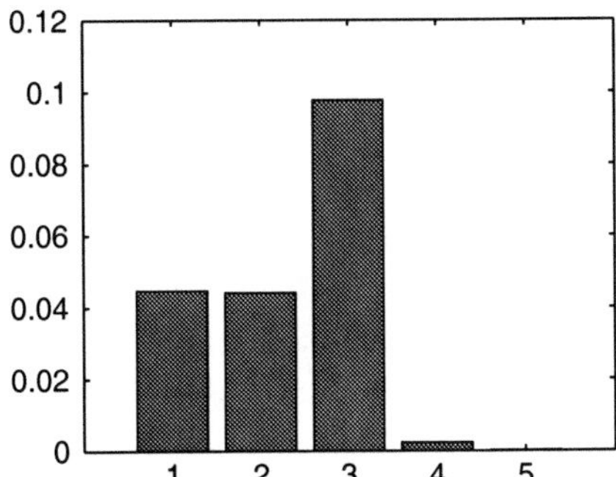

Fig. 7.7. Interference-to-signal ratio (ISR) of image separation: 1 – stochastic natural gradient; 2 – Fast ICA; 3 – JADE; 4 – Relative Newton with $\lambda = 10^{-2}$; 5 – SMOM (bar 5 is not visible because of very small ISR, of order 10^{-12}).

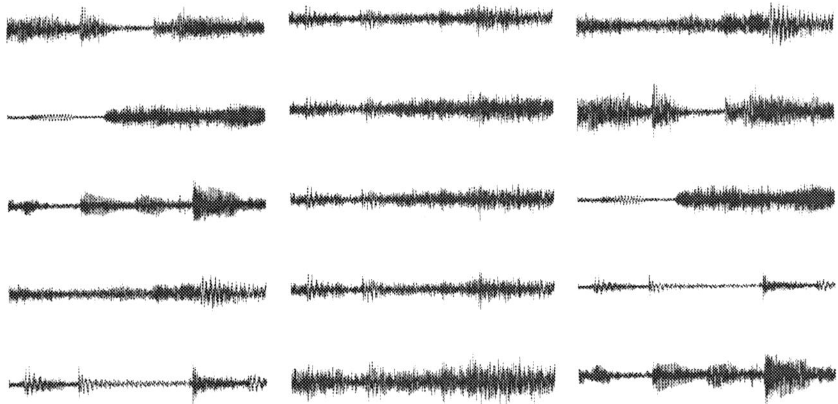

Fig. 7.8. Separation of musical sounds (one second, 10k samples) using Relative Newton method: Left – sources, middle – mixtures, right – separated.

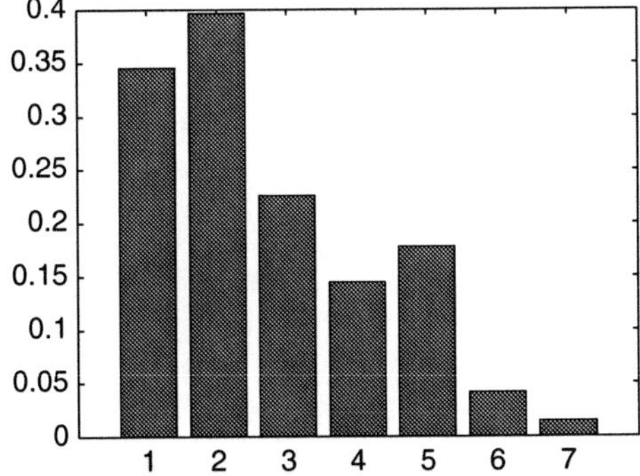

Fig. 7.9. Interference-to-signal ratio (ISR) of sound separation: 1 – JADE; 2 – Fast ICA; 3 – stochastic natural gradient; The remaning bars – separation using STFT transform of the mixed signals: 4 – JADE; 5 – Fast ICA; 6 – stochastic natural gradient; 7 – Relative Newton with $\lambda = 10^{-3}$ (SMOM gives similar results).

STFT significantly improves quality of separation, but even then the Relative Newton method with $\lambda = 10^{-3}$ and SMOM provide much better quality then the other methods.

7.9 Conclusions

We have presented the relative optimization framework for quasi-ML BSS, and studied the relative Newton method as its particular instance. Gradient-type computational cost of the Newton iteration makes it especially attractive.

We also presented SMOM method for minimization of sum of pairwise maxima of smooth functions (in particular sum of absolute value terms, like used in quasi-ML separation of sparse sources). Incorporating Lagrange multiplier into a smooth approximation of max-type function, we obtained an extended notion of nonquadratic augmented Lagrangian. This approach does not require artificial variables, and preserves sparse structure of Hessian.

We apply the Frozen Hessian strategy, using the fact that changes in the Hessian of the augmented Lagrangian become very small at late outer iterations of SMOM. In our experiments 5–7 last outer iterations require only one Newton step each, without recomputing Hessian at all.

Experiments with sparsely representable artificial data, natural images and sounds show that quasi-ML separation is almost perfect when the nonlinearity approaches the absolute value function.

Currently we are conducting more experiments with non-sparse source distributions and various kinds of nonlinearities. Preliminary results confirm fast convergence of the relative Newton method.

Appendix A
Convergence Analysis of Relative Optimization Algorithm

Definition 1 *We say that a function f sufficiently decreases at iteration k, if for any $\epsilon > 0$ there exists $\delta > 0$ such that from $\|x_k - x^*\| > \epsilon$ it follows that $f(x_k) - f(x_{k+1}) > \delta$. Here x^* is a local minimum closest to x_k.*

Suppose the following properties of the function $h(\cdot)$ in (7.2)

(h1) $h(\cdot)$ is bounded below;
(h2) $h(\cdot)$ grows faster than $\log(|\cdot|)$ toward $\pm\infty$

$$\lim_{\alpha \to \pm\infty} h(\alpha)/\log(|\alpha|) = \infty. \tag{7.40}$$

Proposition 1 *The sequence $L(W_k; X)$ generated by Relative Optimization algorithm is monotone decreasing at each step by the value*

$$L(W_k; X) - L(W_{k+1}; X) = L(I; U_k) - L(V_k; U_k) > 0. \tag{7.41}$$

Proof Optimization step 2 reduces the function value

$$L(V_k; U_k) < L(I; U_k).$$

Taking into account that by (7.2)

$$L(W_k; X) = -\log|\det W_k| + L(I; U_k), \tag{7.42}$$

$$L(W_{k+1}; X) = -\log|\det W_k| + L(V_k; U_k), \tag{7.43}$$

we get (7.41). □

Proposition 2 *The likelihood function (7.2) is bounded from below and has bounded level sets.*

Proof is based on the properties (h1 – h2). We need to show that the function $L(W; X)$ in (7.2) has an infinite growth along any radial direction

$$\lim_{\alpha \to \infty} L(\alpha W_0) = \infty$$

for any invertible W_0. This is an obvious consequence of (7.40). □

Lemma 1 *The sequence W_k generated by the Relative Optimization algorithm, has limit point[s]; any limit point belongs to a local minimum of the likelihood function (7.2).*

Proof The sequence of the function values $L(W_k; X)$ generated by the Relative Optimization algorithm is monotone decreasing (by Proposition 1), so all iterates W_k belong to the level set $\{W : L(W; X) < L(W_0; X)\}$, which is bounded according to Proposition 2. Therefore the sequence of the iterates W_k has limit point[s].

The second part of the proof is continued by contradiction. Let $\bar{W}$ be a limit point, which is not equal to the closest local minimum W^*, *i.e.* for any point W_k from a small neighborhood of $\bar{W}$, $W_k \in \mathcal{N}(\bar{W})$,

$$\|I - W^* W_k^{-1}\| > \epsilon > 0. \tag{7.44}$$

Let V_k^* be a local minimizer of $L(\cdot; U_k)$, so that $W^* = V_k^* W_k$. It follows from (7.44) that

$$\|I - V_k^*\| > \epsilon, \tag{7.45}$$

therefore step 2 of the Relative Optimization algorithm provides significant decrease of the objective function (see Definition 1)

$$L(I; U_k) - L(V_k; U_k) > \delta. \tag{7.46}$$

Since $\bar{W}$ is a concentration point, there are infinite number of iterates $W_k \in \mathcal{N}(\bar{W})$ satisfying (7.44–7.46). Taking into account (7.41), we conclude that the function $L(W_k; X)$ will decrease infinitely, which contradicts its below boundedness, stated by Proposition 2. □

References

1. D. Pham and P. Garat, "Blind separation of a mixture of independent sources through a quasi-maximum likelihood approach," *IEEE Transactions on Signal Processing*, vol. 45, no. 7, pp. 1712–1725, 1997.
2. A. J. Bell and T. J. Sejnowski, "An information-maximization approach to blind separation and blind deconvolution," *Neural Computation*, vol. 7, no. 6, pp. 1129–1159, 1995.
3. J.-F. Cardoso, "On the performance of orthogonal source separation algorithms," in *EUSIPCO*, Edinburgh, Sept. 1994, pp. 776–779.
4. ——, "Blind signal separation: statistical principles," *Proceedings of the IEEE*, vol. 9, no. 10, pp. 2009–2025, Oct. 1998. [Online]. Available: ftp://sig.enst.fr/pub/jfc/Papers/ProcIEEE.us.ps.gz
5. S. S. Chen, D. L. Donoho, and M. A. Saunders, "Atomic decomposition by basis pursuit," *SIAM J. Sci. Comput.*, vol. 20, no. 1, pp. 33–61, 1998.
6. B. A. Olshausen and D. J. Field, "Sparse coding with an overcomplete basis set: A strategy employed by v1?" *Vision Research*, vol. 37, pp. 3311–3325, 1997.
7. M. S. Lewicki and B. A. Olshausen, "A probabilistic framework for the adaptation and comparison of image codes," *Journal of the Optical Society of America*, vol. 16, no. 7, pp. 1587–1601, 1999.
8. M. Zibulevsky and B. A. Pearlmutter, "Blind source separation by sparse decomposition in a signal dictionary," *Neural Computations*, vol. 13, no. 4, pp. 863–882, 2001.
9. M. Zibulevsky, B. A. Pearlmutter, P. Bofill, and P. Kisilev, "Blind source separation by sparse decomposition," in *Independent Components Analysis: Princeiples and Practice*, S. J. Roberts and R. M. Everson, Eds. Cambridge University Press, 2001.
10. M. Zibulevsky, P. Kisilev, Y. Y. Zeevi, and B. A. Pearlmutter, "Blind source separation via multinode sparse representation," in *Advances in Neural Information Processing Systems 12*. MIT Press, 2002.
11. A. Hyvärinen, "Fast and robust fixed-point algorithms for independent component analysis," *IEEE Transactions on Neural Networks*, vol. 10, no. 3, pp. 626–634, 1999.
12. T. Akuzawa and N. Murata, "Multiplicative nonholonomic Newton-like algorithm," *Chaos, Solitons and Fractals*, vol. 12, p. 785, 2001.
13. T. Akuzawa, "Extended quasi-Newton method for the ICA," Laboratory for Mathematical Neuroscience, RIKEN Brain Science Institute, Tech. Rep., 2000, http://www.mns.brain.riken.go.jp/~akuzawa/publ.html.
14. D. Pham, "Joint approximate diagonalization of positive definite matrices," *SIAM J. on Matrix Anal. and Appl.*, vol. 22, no. 4, pp. 1136–1152, 2001.
15. D. Pham and J.-F. Cardoso, "Blind separation of instantaneous mixtures of non-stationary sources," *IEEE Transactions on Signal Processing*, vol. 49, no. 9, pp. 1837–1848, 2001.
16. M. Joho and K. Rahbar, "Joint diagonalization of correlation matrices by using Newton methods with application to blind signal separation," *SAM 2002*, 2002, http://www.phonak.uiuc.edu/~joho/research/publications/sam_2002_2.pdf.
17. A. Ziehe, P. Laskov, G. Nolte, and K.-R. Mueller, "A fast algorithm for joint diagonalization with non-orthogonal transformations and its application to blind source separation," *Journal of Machine Learning Research*, vol. 5, pp. 801–818, July 2004.

18. B. Kort and D. Bertsekas, "Multiplier methods for convex programming," *Proc 1073 IEEE Conf. Decision Control, San-Diego, Calif.*, pp. 428–432, 1973.
19. D. Bertsekas, *Constrained Optimization and Lagrange Multiplier Methods.* New York: Academic Press, 1982.
20. R. Polyak, "Modified barrier functions: Theory and methods," *Math. Programming*, vol. 54, pp. 177–222, 1992.
21. A. Ben-Tal, I. Yuzefovich, and M. Zibulevsky, "Penalty/barrier multiplier methods for min-max and constrained smooth convex programs," Opt. Lab., Dept. of Indust. Eng., Technion, Haifa, Israel, Tech. Rep. 9, 1992.
22. P. Tseng and D. Bertsekas, "Convergence of the exponential multiplier method for convex programming," *Math. Programming*, vol. 60, pp. 1–19, 1993.
23. M. G. Breitfeld and D. Shanno, "Computational experience with penalty/barrier methods for nonlinear programming," *Annals of Operations Research*, vol. 62, pp. 439–464, 1996.
24. M. Zibulevsky, "Penalty/barrier multiplier methods for large-scale nonlinear and semidefinite programming," Ph.D. dissertation, Technion – Israel Institute of Technology, 1996, http://ie.technion.ac.il/~mcib/.
25. A. Ben-Tal and M. Zibulevsky, "Penalty/barrier multiplier methods for convex programming problems," *SIAM Journal on Optimization*, vol. 7, no. 2, pp. 347–366, 1997.
26. L. Mosheyev and M. Zibulevsky, "Penalty/barrier multiplier algorithm for semidefinite programming," *Optimization Methods and Software*, vol. 13, no. 4, pp. 235–261, 2000.
27. M. Kocvara and M. Stingl, "PENNON – a code for convex nonlinear and semidefinite programming," *Optimization Methods and Softwarte*, vol. 18(3), pp. 317–333, 2003.
28. A. Ben-Tal and M. Teboulle, "A smoothing technique for nondifferentiable optimization problems," *Fifth French German Conference, Lecture Notes in Math. 1405, Springer-Verlag, New York*, pp. 1–11, 1989.
29. C. Chen and O. L. Mangasarian, "A class of smoothing functions for nonlinear and mixed complementarity problems," *Computational Optimization and Applications*, vol. 5, pp. 97–138, 1996.
30. A. Cichocki, R. Unbehauen, and E. Rummert, "Robust learning algorithm for blind separation of signals," *Electronics Letters*, vol. 30, no. 17, pp. 1386–1387, 1994.
31. S. Amari, A. Cichocki, and H. H. Yang, "A new learning algorithm for blind signal separation," in *Advances in Neural Information Processing Systems 8.* MIT Press, 1996. [Online]. Available: http://www.cs.cmu.edu/Groups/NIPS/NIPS95/Papers.html
32. J.-F. Cardoso and B. Laheld, "Equivariant adaptive source separation," *IEEE Transactions on Signal Processing*, vol. 44, no. 12, pp. 3017–3030, 1996.
33. P. E. Gill, W. Murray, and M. H. Wright, *Practical Optimization.* New York: Academic Press, 1981.
34. R. Rockafellar, *Convex Analysis.* Princeton, NJ: Princeton University Press, 1970.
35. M. Zibulevsky, "Smoothing method of multipliers for sum-max problems," Dept. of Elec. Eng., Technion, Tech. Rep., 2003, http://ie.technion.ac.il/~mcib/.

36. A. Cichocki, S. Amari, and K. Siwek, "ICALAB toolbox for image processing – benchmarks," 2002, http://www.bsp.brain.riken.go.jp/ICALAB/ICALABImageProc/benchmarks/.
37. J.-F. Cardoso, "High-order contrasts for independent component analysis," *Neural Computation*, vol. 11, no. 1, pp. 157–192, 1999.
38. S. Makeig, "ICA toolbox for psychophysiological research," Computational Neurobiology Laboratory, the Salk Institute for Biological Studies, 1998, http://www.cnl.salk.edu/˜ ica.html.
39. A. Hyvärinen, "The Fast-ICA MATLAB package," 1998, http://www.cis.hut.fi/˜aapo/.
40. J.-F. Cardoso, "JADE for real-valued data," 1999, http://sig.enst.fr:80/˜cardoso/guidesepsou.html.

Part II

Underdetermined Blind Speech Separation with Sparseness

8 The DUET Blind Source Separation Algorithm*

Scott Rickard

University College Dublin
Belfield, Dublin 4, Ireland
E-mail: scott.rickard@ucd.ie

Abstract. This chapter presents a tutorial on the DUET Blind Source Separation method which can separate any number of sources using only two mixtures. The method is valid when sources are W-disjoint orthogonal, that is, when the supports of the windowed Fourier transform of the signals in the mixture are disjoint. For anechoic mixtures of attenuated and delayed sources, the method allows one to estimate the mixing parameters by clustering relative attenuation-delay pairs extracted from the ratios of the time–frequency representations of the mixtures. The estimates of the mixing parameters are then used to partition the time–frequency representation of one mixture to recover the original sources. The technique is valid even in the case when the number of sources is larger than the number of mixtures. The method is particularly well suited to speech mixtures because the time–frequency representation of speech is sparse and this leads to W-disjoint orthogonality. The algorithm is easily coded and a simple MATLAB® implementation is presented[1]. Additionally in this chapter, two strategies which allow DUET to be applied to situations where the microphones are far apart are presented; this removes a major limitation of the original method.

8.1 Introduction

In the field of blind source separation (BSS), assumptions on the statistical properties of the sources usually provide a basis for the demixing algorithm [1]. Some common assumptions are that the sources are statistically independent [2, 3], are statistically orthogonal [4], are nonstationary [5], or can be generated by finite dimensional model spaces [6]. The independence and orthogonality assumptions can be verified experimentally for speech signals. Some of these methods work well for instantaneous demixing, but fail if propagation delays are present. Additionally, many algorithms are computationally intensive as they require the estimation of higher-order statistical moments or the optimization of a nonlinear cost function.

One area of research in blind source separation that is particularly challenging is when there are more sources than mixtures. We refer to such a case

*This material is based upon work supported by the Science Foundation Ireland under the PIYRA Programme.

[1]The author is happy to provide the MATLAB® code implementation of DUET presented here.

S. Makino et al. (eds.), Blind Speech Separation, 217–241.

as *degenerate*. Degenerate blind source separation poses a challenge because the mixing matrix is not invertible. Thus the traditional method of demixing by estimating the inverse mixing matrix does not work. As a result, most BSS research has focussed on the square or overdetermined (nondegenerate) case.

Despite the difficulties, there are several approaches for dealing with degenerate mixtures. For example, [7] estimates an arbitrary number of sources from a single mixture by modeling the signals as autoregressive processes. However, this is achieved at a price of approximating signals by autoregressive stochastic processes, which can be too restrictive. Another example of degenerate separation uses higher order statistics to demix three sources from two mixtures [8]. This approach is not feasible however for a large number of sources since the use of higher order statistics of mixtures leads to an explosion in computational complexity. Similar in spirit to DUET, van Hulle employed a clustering method for relative amplitude parameter estimation and degenerate demixing [9]. The assumptions used by van Hulle were that only one signal at a given time is nonzero and that mixing is instantaneous, that is, there is only a relative amplitude mixing parameter associated with each source. In real world acoustic environments, these assumptions are not valid.

DUET, the Degenerate Unmixing Estimation Technique, solves the degenerate demixing problem in an efficient and robust manner. The underlying principle behind DUET can be summarized in one sentence:

> It is possible to blindly separate an arbitrary number of sources given just two anechoic mixtures provided the time–frequency representations of the sources do not overlap too much, which is true for speech.

The way that DUET separates degenerate mixtures is by partitioning the time–frequency representation of one of the mixtures. In other words, DUET assumes the sources are already 'separate' in that, in the time–frequency plane, the sources are *disjoint*. The 'demixing' process is then simply a partitioning of the time–frequency plane. Although the assumption of disjointness may seem unreasonable for simultaneous speech, it is approximately true. By approximately, we mean that the time–frequency points which contain significant contributions to the average energy of the mixture are very likely to be dominated by a contribution from only one source. Stated another way, two people rarely excite the same frequency at the same time.

This chapter has the following structure. In Sect. 8.2 we discuss the assumptions of anechoic mixing, W-disjoint orthogonality, local stationarity, closely spaced microphones, and different source spatial signatures which lead to the main observation. In Sect. 8.3 we describe the construction of the 2D weighted histogram which is the key component of the mixing parameter estimation in DUET and we describe the DUET algorithm. In Sect. 8.4

we propose two possible extensions to DUET which eliminate the requirement that the microphones be close together. In Sect. 8.5 we discuss a proposed measure of disjointness. After the conclusions in Sect. 8.6, we provide the MATLAB® utility functions used in the earlier sections.

8.2 Assumptions

8.2.1 Anechoic Mixing

Consider the mixtures of N source signals, $s_j(t), j = 1, \dots, N$, being received at a pair of microphones where only the direct path is present. In this case, without loss of generality, we can absorb the attenuation and delay parameters of the first mixture, $x_1(t)$, into the definition of the sources. The two **anechoic mixtures** can thus be expressed as,

$$x_1(t) = \sum_{j=1}^{N} s_j(t), \tag{8.1}$$

$$x_2(t) = \sum_{j=1}^{N} a_j s_j(t - \delta_j), \tag{8.2}$$

where N is the number of sources, δ_j is the arrival delay between the sensors, and a_j is a relative attenuation factor corresponding to the ratio of the attenuations of the paths between sources and sensors. We use Δ to denote the maximal possible delay between sensors, and thus, $|\delta_j| \leq \Delta, \forall j$. The anechoic mixing model is not realistic in that it does not represent echoes, that is, multiple paths from each source to each mixture. However, in spite of this limitation, the DUET method, which is based on the anechoic model, has proven to be quite robust even when applied to echoic mixtures.

8.2.2 W-Disjoint Orthogonality

We call two functions $s_j(t)$ and $s_k(t)$ **W-disjoint orthogonal** if, for a given windowing function $W(t)$, the supports of the windowed Fourier transforms of $s_j(t)$ and $s_k(t)$ are disjoint. The windowed Fourier transform of $s_j(t)$ is defined,

$$\hat{s}_j(\tau, \omega) := F^W[s_j](\tau, \omega) := \frac{1}{\sqrt{2\pi}} \int_{-\infty}^{\infty} W(t - \tau) s_j(t) e^{-i\omega t} dt. \tag{8.3}$$

The W-disjoint orthogonality assumption can be stated concisely,

$$\hat{s}_j(\tau,\omega)\hat{s}_k(\tau,\omega) = 0, \quad \forall \tau,\omega, \quad \forall j \neq k. \tag{8.4}$$

This assumption is the mathematical idealization of the condition that it is likely that every time–frequency point in the mixture with significant energy is dominated by the contribution of one source. Note that, if $W(t) \equiv 1$, $\hat{s}_j(\tau,\omega)$ becomes the Fourier transform of $s_j(t)$, which we will denote $\hat{s}_j(\omega)$. In this case, W-disjoint orthogonality can be expressed,

$$\hat{s}_j(\omega)\hat{s}_k(\omega) = 0, \forall j \neq k, \forall \omega, \tag{8.5}$$

which we call **disjoint orthogonality**.

W-disjoint orthogonality is crucial to DUET because it allows for the separation of a mixture into its component sources using a binary mask. Consider the mask which is the indicator function for the support of $\hat{s}_j$,

$$M_j(\tau,\omega) := \begin{cases} 1 & \hat{s}_j(\tau,\omega) \neq 0 \\ 0 & \text{otherwise.} \end{cases} \tag{8.6}$$

M_j separates $\hat{s}_j$ from the mixture via

$$\hat{s}_j(\tau,\omega) = M_j(\tau,\omega)\hat{x}_1(\tau,\omega), \forall \tau,\omega. \tag{8.7}$$

So if we could determine the masks which are the indicator functions for each source, we can separate the sources by partitioning. The question is: how do we determine the masks? As we will see shortly, the answer is we label each time–frequency point with the delay and attenuation differences that explain the time–frequency magnitude and phase between the two mixtures, and these delay-attenuation pairs cluster into groups, one group for each source.

8.2.3 Local Stationarity

A well-known Fourier transform pair is:

$$s_j(t-\delta) \leftrightarrow e^{-i\omega\delta}\hat{s}_j(\omega). \tag{8.8}$$

We can state this using the notation of (8.3) as,

$$F^W[s_j(\cdot - \delta)](\tau,\omega) = e^{-i\omega\delta}F^W[s_j(\cdot)](\tau,\omega), \tag{8.9}$$

when $W(t) \equiv 1$. The above equation is not necessarily true, however, when $W(t)$ is a windowing function. For example, if the windowing function were a Hamming window of length 40 ms, there is no reason to believe that two 40 ms windows of speech separated by, say, several seconds are related by a phase shift. However, for shifts which are small relative to the window size, (8.9) will hold even if $W(t)$ has finite support. This can be thought of as a form of a narrowband assumption in array processing [10], but this label is perhaps misleading in that speech is not narrowband, and *local stationarity* seems a more appropriate moniker. What is necessary for DUET is that (8.9) holds for all δ, $|\delta| \leq \Delta$, even when $W(t)$ has finite support, where Δ is the maximum time difference possible in the mixing model (the microphone separation divided by the speed of signal propagation). We formally state the **local stationarity** assumption as,

$$F^W[s_j(\cdot - \delta)](\omega, \tau) = e^{-i\omega\delta} F^W[s_j(\cdot)](\omega, \tau), \quad \forall \delta, |\delta| \leq \Delta, \tag{8.10}$$

where the change from (8.9) is the inclusion of the limitation of the range of δ for which the equality holds.

8.2.4 Microphones Close Together

Additionally, one crucial issue is that DUET is based, as we shall soon see, on the extraction of attenuation and delay mixing parameters estimates from each time–frequency point. We will utilize the local stationarity assumption to turn the delay in time into a multiplicative factor in time–frequency. Of course, this multiplicative factor $e^{-i\omega\delta}$ only uniquely specifies δ if $|\omega\delta| < \pi$ as otherwise we have an ambiguity due to phase-wrap. So we require,

$$|\omega\delta_j| < \pi, \quad \forall \omega, \forall j, \tag{8.11}$$

to avoid phase ambiguity. This is guaranteed when the microphones are separated by less than $\pi c/\omega_m$ where ω_m is the maximum frequency present in the sources and c is the speed of sound. For example, for a maximum frequency of 16 kHz the microphones must be placed within approximately 1 cm of each other, and for a maximum frequency of 8 kHz the microphones must be placed within approximately 2 cm of each other.

8.2.5 Different Spatial Signatures

In the anechoic mixing model described by (8.1) and (8.2), if two sources have identical spatial signatures, that is identical relative attenuation and relative delay mixing parameters, then they can be combined into one source without changing the model. In this case, the DUET technique will fail to separate them from each other because DUET uses only the relative attenuation and relative delay mixing parameters to identify the components of each source. Note that physical separation of the sources does not necessarily result in different spatial signatures. For omnidirectional microphones, each attenuation-delay pair is associated with a circle (in 3-space) of possible physical locations. In the no-attenuation zero-delay case, the ambiguity circle becomes a plane. For directional microphones, the incidence of spatial ambiguities can be reduced [11]. We will thus assume that the sources have different spatial signatures,

$$(a_j \neq a_k) \text{ or } (\delta_j \neq \delta_k), \quad \forall j \neq k. \tag{8.12}$$

8.3 DUET

8.3.1 Main Observation and Outline

The assumptions of anechoic mixing and local stationarity allow us to rewrite the mixing equations (8.1) and (8.2) in the time–frequency domain as,

$$\begin{bmatrix} \hat{x}_1(\tau,\omega) \\ \hat{x}_2(\tau,\omega) \end{bmatrix} = \begin{bmatrix} 1 & \dots & 1 \\ a_1 e^{-i\omega\delta_1} & \dots & a_N e^{-i\omega\delta_N} \end{bmatrix} \begin{bmatrix} \hat{s}_1(\tau,\omega) \\ \vdots \\ \hat{s}_N(\tau,\omega) \end{bmatrix}. \tag{8.13}$$

With the further assumption of W-disjoint orthogonality, at most one source is active at every (τ,ω), and the mixing process can be described,

$$\text{for each } (\tau,\omega), \quad \begin{bmatrix} \hat{x}_1(\tau,\omega) \\ \hat{x}_2(\tau,\omega) \end{bmatrix} = \begin{bmatrix} 1 \\ a_j e^{-i\omega\delta_j} \end{bmatrix} \hat{s}_j(\tau,\omega) \quad \text{for some } j. \tag{8.14}$$

Of course, in the above equation, j depends on (τ,ω) in that j is the index of the source active at (τ,ω). The main observation that DUET leverages is that the ratio of the time–frequency representations of the mixtures does not depend on the source components but only on the mixing parameters associated with the active source component:

$$\forall(\tau,\omega) \in \Omega_j, \quad \frac{\hat{x}_2(\tau,\omega)}{\hat{x}_1(\tau,\omega)} = a_j e^{-i\omega\delta_j}, \tag{8.15}$$

where

$$\Omega_j := \{(\tau,\omega) : \hat{s}_j(\tau,\omega) \neq 0\}. \tag{8.16}$$

The mixing parameters associated with each time–frequency point can be calculated:

$$\tilde{a}(\tau,\omega) := \left|\hat{x}_2(\tau,\omega)/\hat{x}_1(\tau,\omega)\right|, \tag{8.17}$$

$$\tilde{\delta}(\tau,\omega) := (-1/\omega)\angle(\hat{x}_2(\tau,\omega)/\hat{x}_1(\tau,\omega)). \tag{8.18}$$

Under the assumption that the microphones are sufficiently close together so that the delay estimate is not incorrect due to wrap-around, the *local attenuation estimator* $\tilde{a}(\tau,\omega)$ and the *local delay estimator* $\tilde{\delta}(\tau,\omega)$ can only take on the values of the actual mixing parameters. Thus, the union of the $(\tilde{a}(\tau,\omega),\tilde{\delta}(\tau,\omega))$ pairs taken over all (τ,ω) is the set of mixing parameters (a_j,δ_j):

$$\bigcup_{(\tau,\omega)} \{(\tilde{a}(\tau,\omega),\tilde{\delta}(\tau,\omega))\} = \{(a_j,\delta_j) \,:\, j=1,\ldots,N\}. \tag{8.19}$$

So, we now know the set of mixing parameter pairs. As we saw in (8.7), we can demix via binary masking if we can determine the indicator function of each source. We can now determine the indicator functions via

$$M_j(\tau,\omega) = \begin{cases} 1 & (\tilde{a}(\tau,\omega),\tilde{\delta}(\tau,\omega)) = (a_j,\delta_j) \\ 0 & \text{otherwise} \end{cases} \tag{8.20}$$

and then demix using the masks.

In summary, the essentials to the DUET method are:

1. Construct the time–frequency representation of both mixtures.
2. Take the ratio of the two mixtures and extract local mixing parameter estimates.
3. Combine the set of local mixing parameter estimates into N pairings corresponding to the true mixing parameter pairings.
4. Generate one binary mask for each determined mixing parameter pair corresponding to the time–frequency points which yield that particular mixing parameter pair.
5. Demix the sources by multiplying each mask with one of the mixtures.
6. Return each demixed time–frequency representation to the time domain.

In practice, because not all of the assumptions are strictly satisfied, the local mixing parameter estimates will not be precisely the mixing parameters. In this case, we can replace the definition of Ω_j with

$$\Omega_j := \{(\tau,\omega) : |\hat{s}_j(\tau,\omega)| >> |\hat{s}_k(\tau,\omega)|, \quad \forall k \neq j\} \tag{8.21}$$

and then

$$\forall(\tau,\omega) \in \Omega_j, \quad \frac{\hat{x}_2(\tau,\omega)}{\hat{x}_1(\tau,\omega)} \approx a_j e^{-i\omega\delta_j} \tag{8.22}$$

and the estimates will cluster around the mixing parameters. These clusters will be sufficiently far apart to be identifiable as long as our spatially separate assumption is satisfied. The determination of these clusters is the topic of the next section.

8.3.2 Two-Dimensional Smoothed Weighted Histogram

In order to account for the fact that our assumptions made previously will not be satisfied in a strict sense, we need a mechanism for clustering the relative attenuation-delay estimates. In [12], we considered the maximum-likelihood (ML) estimators for a_j and δ_j in the following mixing model:

$$\begin{bmatrix} \hat{x}_1(\tau,\omega) \\ \hat{x}_2(\tau,\omega) \end{bmatrix} = \begin{bmatrix} 1 \\ a_j e^{-i\omega\delta_j} \end{bmatrix} \hat{s}_j(\tau,\omega) + \begin{bmatrix} \hat{n}_1(\tau,\omega) \\ \hat{n}_2(\tau,\omega) \end{bmatrix}, \quad \forall(\tau,\omega) \in \Omega_j, \tag{8.23}$$

where $\hat{n}_1$ and $\hat{n}_2$ are noise terms which represent the assumption inaccuracies. Rather than estimating a_j, we estimate

$$\alpha_j := a_j - \frac{1}{a_j}, \tag{8.24}$$

which we call the *symmetric attenuation* because it has the property that if the microphone signals are swapped, the attenuation is reflected symmetrically about a center point ($\alpha = 0$). That is, swapping the microphone signals changes α_j to $-\alpha_j$. Signals that are louder on microphone 1 will have $\alpha_j < 0$ and signals that are louder on microphone 2 will have $\alpha_j > 0$. In contrast, there is an inequity of treatment if we estimate a_j directly. Swapping the microphone signals changes a_j to $1/a_j$, and this results in signals louder on microphone 1 occupying $1 > a_j > 0$ and signals louder on microphone 2 occupying $\infty > a_j > 1$. We, therefore, define the local symmetric attenuation estimate,

$$\tilde{\alpha}(\tau,\omega) := \left|\frac{\hat{x}_2(\tau,\omega)}{\hat{x}_1(\tau,\omega)}\right| - \left|\frac{\hat{x}_1(\tau,\omega)}{\hat{x}_2(\tau,\omega)}\right|. \tag{8.25}$$

Motivated by the form of the ML estimators [12], the following pair of estimators emerge:

$$\tilde{\alpha}_j = \frac{\displaystyle\iint_{(\tau,\omega)\in\Omega_j} |\hat{x}_1(\tau,\omega)\hat{x}_2(\tau,\omega)|^p \, \omega^q \tilde{\alpha}(\tau,\omega) \mathrm{d}\tau \mathrm{d}\omega}{\displaystyle\iint_{(\tau,\omega)\in\Omega_j} |\hat{x}_1(\tau,\omega)\hat{x}_2(\tau,\omega)|^p \, \omega^q \mathrm{d}\tau \mathrm{d}\omega} \tag{8.26}$$

and

$$\tilde{\delta}_j = \frac{\displaystyle\iint_{(\tau,\omega)\in\Omega_j} |\hat{x}_1(\tau,\omega)\hat{x}_2(\tau,\omega)|^p \, \omega^q \tilde{\delta}(\tau,\omega) \mathrm{d}\tau \mathrm{d}\omega}{\displaystyle\iint_{(\tau,\omega)\in\Omega_j} |\hat{x}_1(\tau,\omega)\hat{x}_2(\tau,\omega)|^p \, \omega^q \mathrm{d}\tau \mathrm{d}\omega}, \tag{8.27}$$

which have been parameterized by p and q. Note that the form of each estimator is that of a weighted average of the local symmetric attenuation and local delay estimators with the weight for a given time–frequency point being $|\hat{x}_1(\tau,\omega)\hat{x}_2(\tau,\omega)|^p \, \omega^q$. Various choices for p and q are noteworthy:

- $p = 0$, $q = 0$: the counting histogram proposed in the original DUET algorithm [13]
- $p = 1$, $q = 0$: motivated by the ML symmetric attenuation estimator [12]
- $p = 1$, $q = 2$: motivated by the ML delay estimator [12]
- $p = 2$, $q = 0$: in order to reduce delay estimator bias [12]
- $p = 2$, $q = 2$: for low signal-to-noise ratio or speech mixtures [14]

The reason why the delay estimator gives more weight to higher frequencies is that the delay is calculated from a noisy phase estimate $\omega\delta + n$ by dividing the phase by ω. So in effect, the delay estimate is $\delta + n/\omega$ and the higher the frequency, the smaller the noise term. Our practical experience with DUET suggests that $p = 1$, $q = 0$ is a good default choice. When the sources are not equal power, we would suggest $p = 0.5$, $q = 0$ as it prevents the dominant source from hiding the smaller source peaks in the histogram.

Regardless of the choice of p and q, the difficulty with the estimators is that they require knowledge of the time–frequency supports of each source, but that is exactly what we are solving for. Based on the observation in the previous section that the local symmetric attenuation and delay estimates will cluster around the actual symmetric attenuation and delay mixing parameters of the original sources, we need a mechanism for determining these clusters. The estimators (8.26) and (8.27) suggest the construction of a two-dimensional weighted histogram to determine the clusters and the estimate mixing parameters (a_j, δ_j). The histogram is the key structure used for localization and separation. By using $(\tilde{\alpha}(\tau,\omega), \tilde{\delta}(\tau,\omega))$ pairs to indicate the indices into the histogram and using $|\hat{x}_1(\tau,\omega)\hat{x}_2(\tau,\omega)|^p \omega^q$ for the weight, clusters of weight will emerge centered on the actual mixing parameter pairs (a_j, δ_j). Assuming the (a_j, δ_j) pairs are reasonably separated from each other, smoothing the histogram with a window which is large enough to capture all the contributions of one source without capturing the contributions of multiple sources will result in N distinct peaks emerging with centers $(a_j, \delta_j), j = 1, \ldots, N$. Thus:

The weighted histogram separates and clusters the parameter estimates of each source. The number of peaks reveals the number of sources, and the peak locations reveal the associated source's anechoic mixing parameters.

We now formally define the two-dimensional weighted histogram by first defining the set of points which will contribute to a given location in the histogram,

$$I(\alpha, \delta) := \{(\tau,\omega) : |\tilde{\alpha}(\tau,\omega) - \alpha| < \Delta_\alpha, \, |\tilde{\delta}(\tau,\omega) - \delta| < \Delta_\delta\}, \tag{8.28}$$

where Δ_α and Δ_δ are the smoothing resolution widths. The two-dimensional smoothed weighted histogram is constructed,

$$H(\alpha, \delta) := \iint_{(\tau,\omega)\in I(\alpha,\delta)} |\hat{x}_1(\tau,\omega)\hat{x}_2(\tau,\omega)|^p \, \omega^q \mathrm{d}\tau \mathrm{d}\omega. \tag{8.29}$$

All the weight associated with time–frequency points yielding local estimates $(\tilde{\alpha}(\tau,\omega), \tilde{\delta}(\tau,\omega))$ within $(\Delta_\alpha, \Delta_\delta)$ of (α, δ) contributes to the histogram at (α, δ). If all the assumptions are satisfied, the histogram will have the form,

$$H(\alpha,\delta) = \begin{cases} \iint |\hat{s}_j(\tau,\omega)|^{2p}\, \omega^q \mathrm{d}\tau \mathrm{d}\omega & |\alpha_j - \alpha| < \Delta_\alpha, |\delta_j - \delta| < \Delta_\delta \\ 0 & \text{otherwise,} \end{cases} \tag{8.30}$$

and the N peaks corresponding to the N sources will be clearly visible.

A MATLAB® implementation of the histogram calculation is shown in Fig. 8.5 (see Appendix). While the presentation in this chapter is in continuous time, the end implementation is in discrete time. The conversion of the presented concepts into discrete time is straightforward. One slight difference in the implementation is that the delay calculation results in the unit of the delay being samples, which can easily be converted into seconds by multiplying by the sampling rate. It is convenient for the purposes of demonstration that the delay be in samples, so we will use this convention in the coded implementation. The code in Fig. 8.5 was run on two anechoic mixtures of five speech sources. The five 16 kHz speech signals were taken from the TIMIT database and looped so that each signal has length six seconds. The sum of the five signals was used as the first mixture. The second mixture contained the five sources with relative amplitudes $(1.1, 0.9, 1.0, 1.1, 0.9)$ and sample delays $(-2, -2, 0, 2, 2)$. Figure 8.1 shows the five peaks associated with five original speech sources. The peak locations are exactly the mixing parameter pairs for each of the sources.

8.3.3 Separating the Sources

The mixing parameters can be extracted by locating the peaks in the histogram. We have investigated several different automatic peak enumeration/identification methods including weighted k-means, model-based peak removal, and peak tracking [15], but no single technique appears to be appropriate in all settings and often we resort to manual identification of the peaks. Once the peaks have been identified, our goal is to determine the time–frequency masks which will separate each source from the mixtures. This is a trivial task once the mixing parameters have been determined. We simply assign each time–frequency point to the peak location which is closest to the local parameter estimates extracted from the time–frequency point. In [12] we proposed using the likelihood function to produce a measure of closeness. Given histogram peak centers $(\tilde{\alpha}_j, \tilde{\delta}_j), j = 1, \ldots, N$, we convert the symmetric attenuation back to attenuation via

$$\tilde{a}_j = \frac{\tilde{\alpha}_j + \sqrt{\tilde{\alpha}_j^2 + 4}}{2} \tag{8.31}$$

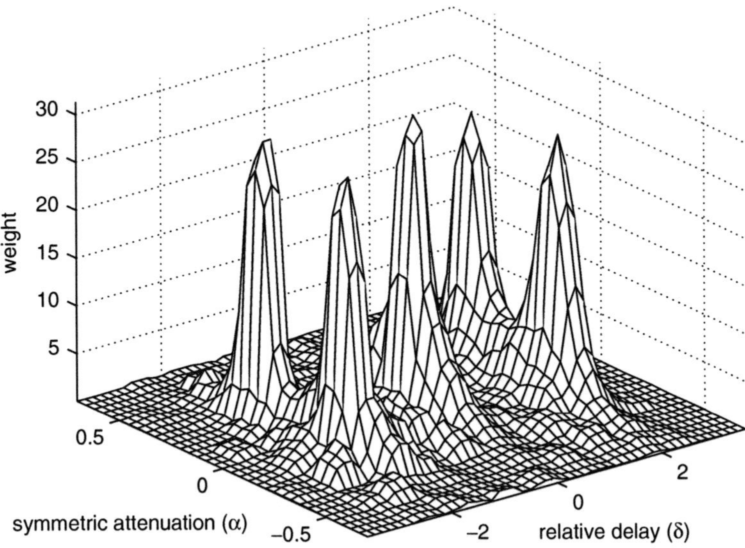

Fig. 8.1. DUET two-dimensional cross power weighted ($p = 1, q = 0$) histogram of symmetric attenuation ($a - 1/a$) and delay estimate pairs from two mixtures of five sources. Each peak corresponds to one source and the peak locations reveal the source mixing parameters.

assign a peak to each time–frequency point via

$$J(\tau,\omega) := \underset{k}{\operatorname{argmin}} \frac{\left|\tilde{a}_k e^{-i\tilde{\delta}_k\omega}\hat{x}_1(\tau,\omega) - \hat{x}_2(\tau,\omega)\right|^2}{1+\tilde{a}_k^2} \tag{8.32}$$

and then assign each time–frequency point to a mixing parameter estimate via

$$\tilde{M}_j(\tau,\omega) := \begin{cases} 1 & J(\tau,\omega) = j \\ 0 & \text{otherwise.} \end{cases} \tag{8.33}$$

Essentially, (8.32) and (8.33) assign each time–frequency point to the mixing parameter pair which best explains the mixtures at that particular time–frequency point. We demix via masking and ML combining [12],

$$\tilde{\hat{s}}_j(\tau,\omega) = \tilde{M}_j(\tau,\omega)\left(\frac{\hat{x}_1(\tau,\omega) + \tilde{a}_j e^{i\tilde{\delta}_j\omega}\hat{x}_2(\tau,\omega)}{1+\tilde{a}_j^2}\right). \tag{8.34}$$

We can then reconstruct the sources from their time–frequency representations by converting back into the time domain. We are now ready to summarize the DUET algorithm.

8.3.4 The DUET BSS Algorithm

1. Construct time–frequency representations $\hat{x}_1(\tau, \omega)$ and $\hat{x}_2(\tau, \omega)$ from mixtures $x_1(t)$ and $x_2(t)$
2. Calculate $\left(\left| \frac{\hat{x}_2(\tau,\omega)}{\hat{x}_1(\tau,\omega)} \right| - \left| \frac{\hat{x}_1(\tau,\omega)}{\hat{x}_2(\tau,\omega)} \right|, \frac{-1}{\omega} \angle \left(\frac{\hat{x}_2(\tau,\omega)}{\hat{x}_1(\tau,\omega)} \right) \right)$
3. Construct 2D smoothed weighted histogram $H(\alpha, \delta)$ as in (8.29)
4. Locate peaks and peak centers which determine the mixing parameter estimates
5. Construct time–frequency binary masks for each peak center $(\tilde{\alpha}_j, \tilde{\delta}_j)$ as in (8.33)
6. Apply each mask to the appropriately aligned mixtures as in (8.34)
7. Convert each estimated source time–frequency representation back into the time domain

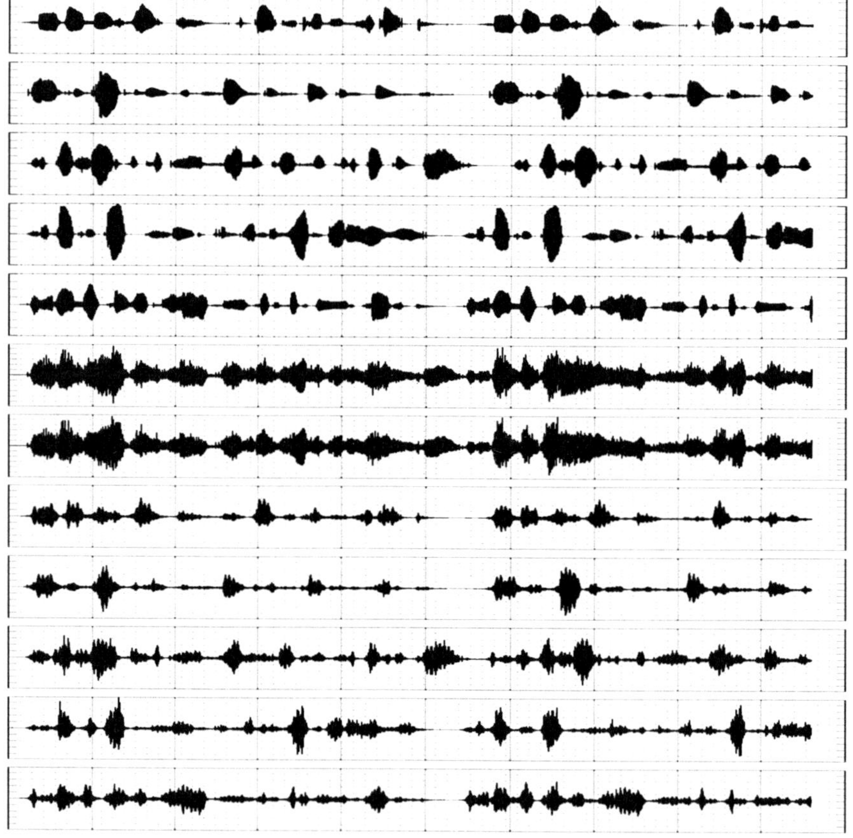

Fig. 8.2. Five original sources, two mixtures, and the five estimates of the original sources.

For the mixtures used to create the histogram in Fig. 8.1, the peaks were identified and the code presented in Fig. 8.8 was used to implement steps 5–7 in the algorithm. The original sources, mixtures, and demixed sources are presented in Fig. 8.2.

8.4 Big Delay DUET

The DUET technique presented in the previous section is limited to being able to estimate the mixing parameters and separate sources that arrive within an intra-mixture delay of less than $\frac{1}{2f_m}$ where f_m is the highest frequency of interest in the sources. This can be seen from the condition for the accurate determination of δ from $e^{i\omega_m\delta}$ requires $|\omega_m\delta| < \pi$. With $\omega_m = 2\pi f_m$, we arrive at $\delta < \frac{1}{2f_m}$. If we are sampling at the Nyquist rate $(2f_m)$, then δ must be less than one sampling interval. If we are sampling at twice Nyquist, then δ must be less than two sampling intervals. Thus, DUET is applicable when the sensors are separated by at most $\frac{c}{2f_m}$ meters where c is the speed of propagation of the signals. For example, for voice mixtures where the highest frequency of interest is 4000 Hz and the speed of sound is 340 m/s, the microphones must be separated by less than 4.25 cm in order for DUET to be able to localize and separate the sources correctly for all frequencies of interest for all possible source locations. In some applications, microphones cannot be placed so closely together. In this section, we present two possible extensions to DUET that allow for arbitrary microphone spacing [16].

The first extension involves analyzing the phase difference between frequency adjacent time–frequency ratios to estimate the delay parameter. This technique increases the maximum possible separation between sensors from $\frac{1}{2f_m}$ to $\frac{1}{2\Delta f}$ where Δf is the frequency spacing between adjacent frequency bins in the time–frequency representation. As one can choose Δf, this effectively removes the sensor spacing constraint.

The second extension involves progressively delaying one mixture against the second and constructing a histogram for each delay. When the delaying of one mixture moves the intersensor delay of a source to less than $\frac{1}{2f_m}$, the delay estimates will align and a peak will emerge. When the inter-sensor delay of a source is larger than $\frac{1}{2f_m}$, the delay estimates will spread and no dominant peak will be visible. The histograms are then tiled to produce a histogram which covers a large range of possible delays and the true mixing parameter peaks should be dominant in this larger histogram.

We now describe each method in more detail.

8.4.1 Method One: Differential

The first method makes the additional assumption that,

$$|\hat{s}_j(\tau,\omega)| \approx |\hat{s}_j(\tau,\omega+\Delta\omega)|, \forall j, \forall\omega, \tau. \tag{8.35}$$

That is, the power in the time–frequency domain of each source is a smooth function of frequency. From before we have $\forall(\tau,\omega) \in \Omega_j$,

$$\begin{bmatrix} \hat{x}_1(\tau,\omega) \\ \hat{x}_2(\tau,\omega) \end{bmatrix} = \begin{bmatrix} 1 \\ a_j e^{-i\omega\delta_j} \end{bmatrix} \hat{s}_j(\tau,\omega), \tag{8.36}$$

and now, in addition, we have,

$$\begin{bmatrix} \hat{x}_1(\tau,\omega+\Delta\omega) \\ \hat{x}_2(\tau,\omega+\Delta\omega) \end{bmatrix} = \begin{bmatrix} 1 \\ a_j e^{-i(\omega+\Delta\omega)\delta_j} \end{bmatrix} \hat{s}_j(\tau,\omega+\Delta\omega). \tag{8.37}$$

Thus,

$$\left(\frac{\hat{x}_2(\tau,\omega)}{\hat{x}_1(\tau,\omega)}\right) \overline{\frac{\hat{x}_2(\tau,\omega+\Delta\omega)}{\hat{x}_1(\tau,\omega+\Delta\omega)}} = (a_j e^{-i\omega\delta_j})(a_j e^{i(\omega+\Delta\omega)\delta_j}) = a_j^2 e^{i\Delta\omega\delta_j}, \tag{8.38}$$

and the $|\omega\delta_j| < \pi$ constraint has been relaxed to $|\Delta\omega\delta_j| < \pi$. Note that $\Delta\omega$ is a parameter that can be made arbitrarily small by oversampling along the frequency axis. As the estimation of the delay from (8.38) is essentially the estimation of the derivative of a noisy function, results can be improved by averaging delay estimates over a local time–frequency region.

8.4.2 Method Two: Tiling

The second method constructs a number of histograms by iteratively delaying one mixture against the other. The histograms are appropriately overlapped corresponding to the delays used and summed to form one large histogram with the delay range of the summation histogram much larger than the delay range of the individual histograms. Specifically, consider the histogram created from the two mixtures where the second mixture has been delayed by β (that is, the second mixture is replaced with $x_2(t-\beta)$ in effect, changing each δ_j to $\delta_j + \beta$),

$$H_\beta(\alpha,\delta) := \begin{cases} \iint_{(\tau,\omega)\in I(\alpha,\delta)} |\hat{x}_1(\tau,\omega)\hat{x}_2(\tau,\omega)|^p \, \omega^q \mathrm{d}\tau \mathrm{d}\omega & \delta \in (-m_\delta, m_\delta) \\ 0 & \text{otherwise,} \end{cases} \tag{8.39}$$

where $2m_\delta$ is the delay width of the histogram. If we happen to choose β such that $|w_m(\delta_j + \beta)| < \pi$, then there will be no phase-wrap issue, all the delay estimates extracted will correspond to a delay of $\delta_j + \beta$, and a clear peak should emerge in the histogram for that source. After peak identification, we can subtract back out the shift by β for the delay estimate to determine the true δ_j.

We construct the summary tiled histogram T by appropriately overlapping and adding a series of local histograms H_β constructed for linearly spaced values of β,

$$T(\alpha,\delta) := \sum_{k=-K}^{K} H_{k\delta_m}(\alpha,\delta-k\delta_m), \quad \delta \in (-K\delta_m-m_\delta, K\delta_m+m_\delta). \tag{8.40}$$

The peaks that emerge in the overall histogram will correspond to the true delays. Demixing is accomplished using the standard DUET demixing by using the histogram tile that contains the source peak to be separated. As the interference from other sources will tend to be separated at zero delay, it is preferred to use a histogram tile where the peak is not centered at zero for separation.

Figure 8.3 shows a standard DUET histogram and a tiled DUET histogram for a five source mixing example. The same speech files were used as in Fig. 8.5, the only difference being that the sample delays used this time were $(-170,-100,0,50,150)$. The standard DUET histogram fails to resolve any source mixing parameter pairs and simply has one peak at the origin. The tiled DUET histogram, on the other hand, has a clear peak located at each of the mixing parameter pairings. The differential method has similar performance to the tiled method. Thus, DUET can be extended to the case when large delays are present and the close microphone limitation has been eliminated.

8.5 Approximate W-Disjoint Orthogonality

In this section we discuss a quantitative measure of W-disjoint orthogonality as the W-disjoint orthogonality assumption is not strictly satisfied for our signals of interest. In order to measure to what degree the condition is approximately satisfied, we consider the following generalization, which has been discussed in [17–19]. We propose the normalized difference between the signal energy maintained in masking and the interference energy maintained in masking as a measure of the approximate disjoint orthogonality associated with a particular mask, $M(\tau,\omega)$, $0 \le M(\tau,\omega) \le 1$:

$$D_j(M) := \frac{\iint |M(\tau,\omega)\hat{s}_j(\tau,\omega)|^2 \, \mathrm{d}\tau \mathrm{d}\omega - \iint |M(\tau,\omega)\hat{y}_j(\tau,\omega)|^2 \, \mathrm{d}\tau \mathrm{d}\omega}{\iint |\hat{s}_j(\tau,\omega)|^2 \, \mathrm{d}\tau \mathrm{d}\omega}, \tag{8.41}$$

where

$$\hat{y}_j(\tau,\omega) := \sum_{\substack{k=1 \\ k\neq j}}^{N} \hat{s}_k(\tau,\omega) \tag{8.42}$$

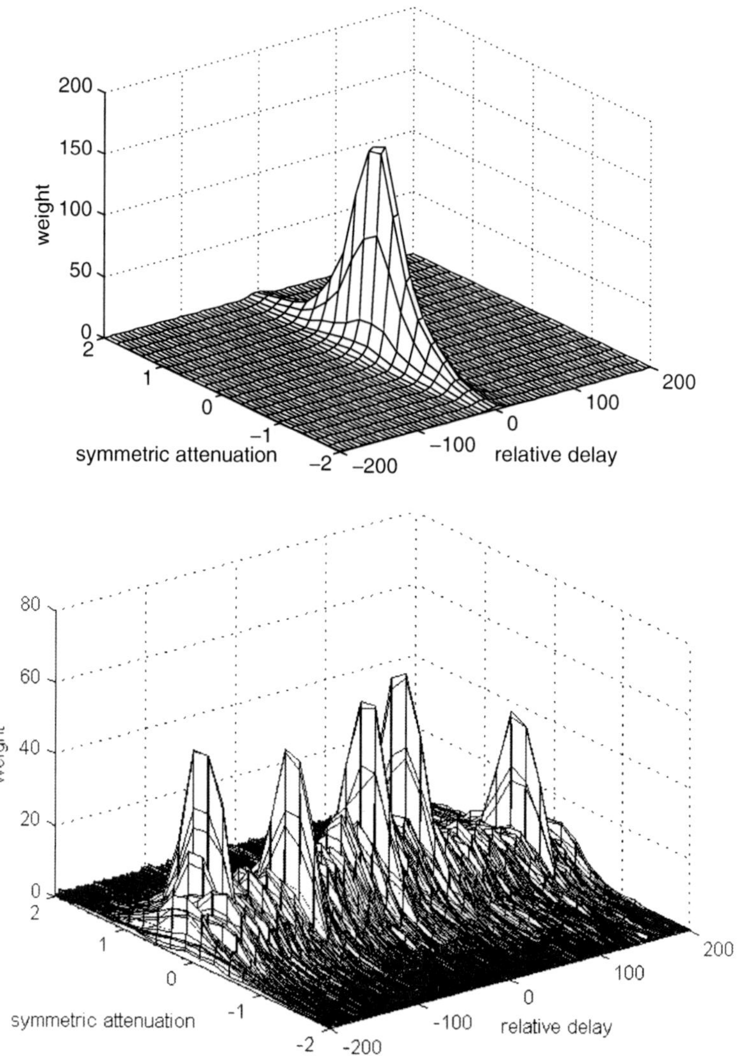

Fig. 8.3. Comparison of standard DUET (top) and tiled DUET (bottom) in a five source mixing example with large relative delays: $(-170, -100, 0, 50, 150)$ samples. Standard DUET fails when the delays are large, but the extension, tiled DUET, succeeds in identifying the number of sources and their associated mixing parameters.

so that $\hat{y}_j(\tau, \omega)$ is the summation of the sources interfering with the jth source. $D_j(M)$ combines two important performance criteria: (1) how well a demixing mask preserves the source of interest, and (2) how well a demixing mask suppresses the interfering sources. It is easily shown that $D_j(M)$ is bounded by $-\infty$ and 1. If it is desirable that the W-disjoint orthogonality measure be bounded between 0 and 1, then we suggest the following mapping:

$$d_j(M) := 2^{D_j(M)-1}, \tag{8.43}$$

which has the properties:

1. $d_j(M) = 1$ implies that s_j is W-disjoint orthogonal with all interfering signals,
2. $d_j(M) = 1/2$ implies that application of mask M results in a demixture with equal source of interest and interference energies, and
3. $d_j(M) \approx 0$ implies that the mask M results in a demixture dominated by interference energy.

Now that we have a quantitative measure which describes the demixing performance of masking, we ask the question which mask should we use in order to maximize (8.43). It follows from (8.41) that

$$M_j^*(\tau,\omega) := \begin{cases} 1 & |\hat{s}_j(\tau,\omega)| > |\hat{y}_j(\tau,\omega)| \\ 0 & |\hat{s}_j(\tau,\omega)| \leq |\hat{y}_j(\tau,\omega)| \end{cases} \tag{8.44}$$

maximizes $d_j(M)$ as it turns 'on' signal coefficients where the source of interest dominates the interference and turns 'off' the remaining coefficients. The terms of equal magnitude in (8.44) we have arbitrarily turned 'off', but including them or excluding them makes no difference to the W-disjoint orthogonal measure as the terms cancel. The mask M_j^* is the optimal mask for demixing from a W-disjoint orthogonal performance standpoint. In order to determine the level of approximate W-disjoint orthogonality in a given mixture, we construct the optimal masks as in (8.44) and examine $d_j(M)$ for each source. We perform this analysis for mixtures of simultaneous speech in the next section. We use the results to determine the appropriate window size for use in the DUET algorithm.

8.5.1 Approximate W-Disjoint Orthogonality of Speech

In Fig. 8.4 we measure the W-disjoint orthogonality for pairwise (and 3-way, 4-way, up to 8-way) mixing as a function of window size. For the tests, N speech files, sampled at 16 kHz, were selected at random from the TIMIT database. One signal was randomly selected to be the target and the remaining $N-1$ signals were summed to form an interference signal. Both the target signal and interference signal were then transformed into the time–frequency domain using a Hamming window of size $\{2^0, 2^1, \ldots, 2^{15}\}$. The magnitude of the coefficients of a target source was compared to the sum of the remaining sources to generate the mask M_j^*. Using the mask, $d_{M_j^*}$ was calculated. Over 300 mixtures were generated and the results averaged to form each data point shown in the figure. In all three cases the Hamming window of size 1024 produced the representation that was the most W-disjoint orthogonal. A similar

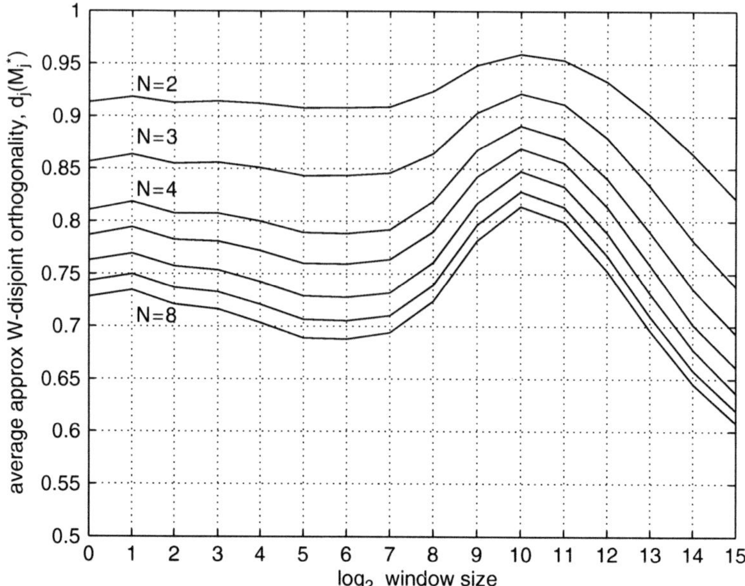

Fig. 8.4. W-disjoint orthogonality for time–frequency representations of mixtures of $N = 2, 3, \ldots, 8$ speech sources as a function of window size used in the time–frequency transformation. Speech is most W-disjoint orthogonal when a window of 1024 samples is used, corresponding to 64 ms length.

conclusion regarding the optimal time–frequency resolution of a window for speech separation was arrived at in [20]. Note that even when the window size is 1 (i.e., time domain), the mixtures still exhibit a high level of W-disjoint orthogonality. This fact was exploited by those methods that used the time-disjoint nature of speech [9, 21–23]. Figure 8.4 clearly shows the advantage of moving from the time domain to the time–frequency domain: the speech signals are more disjoint in the time–frequency domain provided the window size is sufficiently large. Choosing the window size too large, however, results in reduced W-disjoint orthogonality.

8.6 Conclusions

In this chapter we presented a tutorial overview of the DUET blind source separation method. DUET assumes that the source signals are disjoint in the time–frequency domain (W-disjoint orthogonal) and exploits the fact that the ratio of the time–frequency representations of the mixtures can be used to partition the mixtures into the original sources. The key construct in DUET is the two-dimensional smoothed weighted histogram which is used to cluster the mixing parameter estimates. By assuming an anechoic mixing model, all time–frequency points provide input to the histogram as we can eliminate the frequency variable and extract the delay. The fact that both estimation and separation can be done when the number of sources is larger than the number of mixtures without significant computational complexity, as is demonstrated by the MATLAB® code in Sect. 8.6, is testimony to the usefulness of the technique.

The extraction of the delay relies on the assumption that the microphone must be close together, which limits the environments in which DUET can be applied. In Sect. 8.4 we presented two possible extensions to DUET which remove this limitation and demonstrated that the extended DUET can estimate and demix sources regardless of the microphone separation.

Additionally, in Sect. 8.5 we verified that speech signals satisfy W-disjoint orthogonality enough to allow for mixing parameter estimation and source separation. Our own experience with DUET over the past eight years shows that multiple speakers talking simultaneously can be demixed with two microphones with high fidelity of recovered signals using DUET.

References

1. A. Cichocki and S. Amari, *Adaptive Blind Signal and Image Processing.* Wiley, 2002.
2. A. Bell and T. Sejnowski, "An information-maximization approach to blind separation and blind deconvolution," *Neural Computation*, vol. 7, pp. 1129–1159, 1995.
3. J. Cardoso, "Blind signal separation: Statistical principles," *Proceedings of IEEE, Special Issue on Blind System Identification and Estimation*, pp. 2009–2025, Oct. 1998.
4. E. Weinstein, M. Feder, and A. Oppenheim, "Multi-channel signal separation by decorrelation," *IEEE Trans. on Speech and Audio Processing*, vol. 1, no. 4, pp. 405–413, Oct. 1993.
5. L. Parra and C. Spence, "Convolutive blind source separation of non-stationary sources," *IEEE Transactions on Speech and Audio Processing*, pp. 320–327, May 2000.
6. H. Broman, U. Lindgren, H. Sahlin, and P. Stoica, "Source separation: A TITO system identification approach," *Signal Processing*, vol. 73, pp. 169–183, 1999.
7. R. Balan, A. Jourjine, and J. Rosca, "A particular case of the singular multivariate AR identification and BSS problems," in *1st International Conference on Independent Component Analysis*, Assuis, France, 1999.
8. P. Comon, "Blind channel identification and extraction of more sources than sensors," in *SPIE Conference*, San Diego, July 19–24 1998, pp. 2–13.
9. M. V. Hulle, "Clustering approach to square and non-square blind source separation," in *IEEE Workshop on Neural Networks for Signal Processing (NNSP)*, Madison, Wisconsin, Aug. 23–25 1999, pp. 315–323.
10. H. Krim and M. Viberg, "Two Decades of Array Signal Processing Research, The Parametric Approach," *IEEE Signal Processing Magazine*, pp. 67–94, July 1996.

11. B. Coleman and S. Rickard, "Cardioid microphones and DUET," in *IEE Irish Signals and Systems Conference (ISSC2004)*, July 2004, pp. 264–269.
12. O. Yilmaz and S. Rickard, "Blind separation of speech mixtures via time–frequency masking," *IEEE Transactions on Signal Processing*, vol. 52, no. 7, pp. 1830–1847, July 2004.
13. A. Jourjine, S. Rickard, and O. Yılmaz, "Blind Separation of Disjoint Orthogonal Signals: Demixing N Sources from 2 Mixtures," in *Proc. ICASSP2000, June 5–9, 2000, Istanbul, Turkey*, June 2000.
14. T. Melia, "Underdetermined blind source separation in echoic environments using linear arrays and sparse representtions," Ph.D. dissertation, University College Dublin, Dublin, Ireland, Mar. 2007.
15. S. Rickard, R. Balan, and J. Rosca, "Real-time time–frequency based blind source separation," in *3rd International Conference on Independent Component Analysis and Blind Source Separation (ICA2001)*, Dec. 2001.
16. S. Rickard and R. Balan, "Method for estimating mixing parameters and separating multiple sources from signal mixtures," Dec. 2003, US Patent Application no. 20030233227.
17. S. Rickard and O. Yilmaz, "On the approximate W-disjoint orthogonality of speech," in *IEEE International Conference on Acoustics, Speech, and Signal Processing (ICASSP)*, Orlando, Florida, USA, May 2002, pp. 529–532.
18. O. Yilmaz and S. Rickard, "Blind separation of speech mixtures via time–frequency masking," *IEEE Transactions on Signal Processing*, vol. 52, no. 7, pp. 1830–1847, July 2004.
19. S. Rickard, "Sparse sources are separated sources," in *14th European Signal Processing Conference (EUSIPCO)*, Sept. 2006.
20. M. Aoki, M. Okamoto, S. Aoki, H. Matsui, T. Sakurai, and Y. Kaneda, "Sound source segregation based on estimating incident angle of each frequency component of input signals acquired by multiple microphones," *Acoustical Science and Technology*, vol. 22, no. 2, pp. 149–157, 2001.
21. J.-K. Lin, D. G. Grier, and J. D. Cowan, "Feature extraction approach to blind source separation," in *IEEE Workshop on Neural Networks for Signal Processing (NNSP)*, Amelia Island Plantation, Florida, Sept. 24–26 1997, pp. 398–405.
22. T.-W. Lee, M. Lewicki, M. Girolami, and T. Sejnowski, "Blind source separation of more sources than mixtures using overcomplete representations," *IEEE Signal Processing Letters*, vol. 6, no. 4, pp. 87–90, Apr. 1999.
23. L. Vielva, D. Erdogmus, C. Pantaleon, I. Santamaria, J. Pereda, and J. C. Principe, "Underdetermined blind source separation in a time-varying environment," in *IEEE International Conference on Acoustics, Speech, and Signal Processing (ICASSP)*, vol. 3, Orlando, Florida, USA, May 13–17 2002, pp. 3049–3052.

Appendix: MATLAB functions

```
%%%%%%%%%%%%%%%%%%%%%%%%%%%%%%%%%%%%%%%%%%%%%%%%%%%%%%%%%
% 1. analyze the signals - STFT
wlen = 1024; timestep = 512; numfreq = 1024;
awin = hamming(wlen); % analysis window is a Hamming window

tf1 = tfanalysis(x1,awin,timestep,numfreq); % time-freq domain
tf2 = tfanalysis(x2,awin,timestep,numfreq); % time-freq domain
tf1(1,:)=[]; tf2(1,:)=[]; % remove dc component from mixtures
% to avoid dividing by zero frequency in the delay estimation

% calculate pos/neg frequencies for later use in delay calc
freq = [(1:numfreq/2) ((-numfreq/2)+1:-1)]*(2*pi/(numfreq));
fmat = freq(ones(size(tf1,2),1),:)';
%%%%%%%%%%%%%%%%%%%%%%%%%%%%%%%%%%%%%%%%%%%%%%%%%%%%%%%%%
% 2. calculate alpha and delta for each t-f point
R21 = (tf2+eps)./(tf1+eps); % time-freq ratio of the mixtures

%%% 2.1 HERE WE ESTIMATE THE RELATIVE ATTENUATION (alpha) %%%
a = abs(R21); % relative attenuation between the two mixtures
alpha = a - 1./a; % 'alpha' (symmetric attenuation)
%%% 2.2 HERE WE ESTIMATE THE RELATIVE DELAY (delta) %%%%
delta = -imag(log(R21))./fmat; % 'delta' relative delay
%%%%%%%%%%%%%%%%%%%%%%%%%%%%%%%%%%%%%%%%%%%%%%%%%%%%%%%%%
% 3. calculate weighted histogram
p = 1; q = 0; % powers used to weight histogram
tfweight = (abs(tf1).*abs(tf2)).^p .* abs(fmat).^q; % weights
maxa = 0.7; maxd = 3.6; % hist boundaries
abins = 35; dbins = 50; % number of hist bins
% only consider time-freq points yielding estimates in bounds
amask=(abs(alpha)<maxa)&(abs(delta)<maxd);
alpha_vec = alpha(amask);
delta_vec = delta(amask);
tfweight = tfweight(amask);
% determine histogram indices
alpha_ind = round(1+(abins-1)*(alpha_vec+maxa)/(2*maxa));
delta_ind = round(1+(dbins-1)*(delta_vec+maxd)/(2*maxd));

% FULL-SPARSE TRICK TO CREATE 2D WEIGHTED HISTOGRAM
A = full(sparse(alpha_ind,delta_ind,tfweight,abins,dbins));
% smooth the histogram - local average 3-by-3 neighboring bins
A = twoDsmooth(A,3);
% plot 2-D histogram
mesh(linspace(-maxd,maxd,dbins),linspace(-maxa,maxa,abins),A);
```

Fig. 8.5. MATLAB® code which constructs the two-dimensional weighted histogram.

```
function tfmat = tfanalysis(x,awin,timestep,numfreq)
% time-frequency analysis
% X is the time domain signal
% AWIN is an analysis window
% TIMESTEP is the # of samples between adjacent time windows.
% NUMFREQ is the # of frequency components per time point.
%
% TFMAT complex matrix time-freq representation

x = x(:); awin = awin(:); % make inputs go columnwise

nsamp = length(x); wlen = length(awin);

% calc size and init output t-f matrix
numtime = ceil((nsamp-wlen+1)/timestep);
tfmat = zeros(numfreq,numtime+1);

for i = 1:numtime
    sind = ((i-1)*timestep)+1;
    tfmat(:,i) = fft(x(sind:(sind+wlen-1)).*awin,numfreq);
end
i = i+1;
sind = ((i-1)*timestep)+1;
lasts = min(sind,length(x));
laste = min((sind+wlen-1),length(x));
tfmat(:,end) = fft([x(lasts:laste) ;
                   zeros(wlen-(laste-lasts+1),1)].*awin,numfreq);
```

Fig. 8.6. The *tfanalysis.m* function.

```
function smat = twoDsmooth(mat,ker)
%TWO2SMOOTH - Smooth 2D matrix.
%
% smat = twoDsmooth(mat,ker)
%
% MAT is the 2D matrix to be smoothed.
% KER is either
%  (1) a scalar, in which case a ker-by-ker matrix of
%       1/ker^2 is used as the matrix averaging kernel
%  (2) a matrix which is used as the averaging kernel.
%
% SMAT is the smoothed matrix (same size as mat).

if prod(size(ker))==1, % if ker is a scalar
    kmat = ones(ker,ker)/ker^2;
else
    kmat = ker;
end

% make kmat have odd dimensions
[kr kc] = size(kmat); if rem(kr,2) == 0,
    kmat = conv2(kmat,ones(2,1))/2;
    kr = kr + 1;
end if rem(kc,2) == 0,
    kmat = conv2(kmat,ones(1,2))/2;
    kc = kc + 1;
end

[mr mc] = size(mat);
fkr = floor(kr/2); % number of rows to copy on top and bottom
fkc = floor(kc/2); % number of columns to copy on either side
smat = conv2(...
    [mat(1,1)*ones(fkr,fkc) ones(fkr,1)*mat(1,:) ...
    mat(1,mc)*ones(fkr,fkc);
    mat(:,1)*ones(1,fkc) mat mat(:,mc)*ones(1,fkc)
    mat(mr,1)*ones(fkr,fkc) ones(fkr,1)*mat(mr,:) ...
    mat(mr,mc)*ones(fkr,fkc)], ...
    flipud(fliplr(kmat)),'valid');
```

Fig. 8.7. The *twoDsmooth.m* function.

```
%%%%%%%%%%%%%%%%%%%%%%%%%%%%%%%%%%%%%%%%%%%%%%%%%%%
% 4. peak centers (determined from histogram)
numsources = 5;
peakdelta = [-2 -2 0 2 2];
peakalpha = [.19 -.21 0 .19 -.21];

% convert alpha to a
peaka = (peakalpha+sqrt(peakalpha.^2+4))/2;

%%%%%%%%%%%%%%%%%%%%%%%%%%%%%%%%%%%%%%%%%%%%%%%%%%%
% 5. determine masks for separation
bestsofar=Inf*ones(size(tf1));
bestind=zeros(size(tf1));
for i = 1:length(peakalpha)
    score = abs(peaka(i)*exp(-sqrt(-1)*fmat*peakdelta(i)) ...
        .*tf1-tf2).^2/(1+peaka(i)^2);
    mask = (score < bestsofar);
    bestind(mask) = i;
    bestsofar(mask) = score(mask);
end

%%%%%%%%%%%%%%%%%%%%%%%%%%%%%%%%%%%%%%%%%%%%%%%%%%%
% 6. & 7. demix with ML alignment and convert to time domain
est = zeros(numsources,length(x1)); % demixtures
for i=1:numsources
    mask = (bestind==i);
    esti = tfsynthesis([zeros(1,size(tf1,2)) ;
      ((tf1+peaka(i)*exp(sqrt(-1)*fmat*peakdelta(i)).*tf2) ...
        ./(1+peaka(i)^2)).*mask], ...
        sqrt(2)*awin/1024, timestep, numfreq);
    est(i,:) = esti(1:length(x1))';
    % add back into the demix a little bit of the mixture
    % as that eliminates most of the masking artifacts
    soundsc(est(i,:)+0.05*x1,fs); pause; % play demixture
end
```

Fig. 8.8. MATLAB® code for separating the sources from the mixtures, given estimates of the mixing parameters.

```
function x = tfsynthesis(timefreqmat,swin,timestep,numfreq)
% time-frequency synthesis
% TIMEFREQMAT is the complex matrix time-freq representation
% SWIN is the synthesis window
% TIMESTEP is the # of samples between adjacent time windows.
% NUMFREQ is the # of frequency components per time point.
%
% X contains the reconstructed signal.

swin = swin(:); % make synthesis window go columnwise

winlen = length(swin);
[numfreq numtime] = size(timefreqmat);
ind = rem((1:winlen)-1, numfreq) +1;
x = zeros((numtime-1)*timestep + winlen,1);

for i = 1:numtime % overlap, window, and add
    temp = numfreq*real(ifft(timefreqmat(:,i)));
    sind = ((i-1)*timestep);
    rind = (sind+1):(sind+winlen);
    x(rind) = x(rind) + temp(ind).*swin;
end
```

Fig. 8.9. The *tfsynthesis.m* function.

9 K-means Based Underdetermined Blind Speech Separation

Shoko Araki, Hiroshi Sawada, and Shoji Makino

NTT Communication Science Laboratories, NTT Corporation, 2-4 Hikaridai Seika-cho, Soraku-gun, Kyoto 619-0237, Japan
E-mail: {shoko,sawada,maki}@cslab.kecl.ntt.co.jp

Abstract. This chapter addresses a blind sparse source separation method that can employ arbitrarily arranged multiple microphones. Some sparse source separation methods, which rely on source sparseness and an anechoic mixing model, have already been proposed. The validity of the sparseness and anechoic assumptions will be investigated in this chapter. As most of the existing methods utilize a stereo (two sensors) system, they limit the separation ability to a 2-dimensional half-plane. This chapter describes a method for multiple microphones. This method employs the k-means algorithm, which is an efficient clustering algorithm. The method can be easily applied to three or more sensors arranged nonlinearly. Promising results were obtained for 2- and 3-dimensionally distributed speech signals with nonlinear/non-uniform sensor arrays in a real room even in underdetermined situations.

9.1 Introduction

Blind source separation (BSS) [1] is an approach for estimating source signals that uses only the mixed signal information observed at each sensor. The BSS technique for speech dealt with in this chapter has many applications including hands-free teleconference systems and automatic conference minute generators.

Suppose that sources $\mathrm{s}_1, \ldots, \mathrm{s}_N$ are convolutively mixed and observed at M sensors

$$\mathrm{x}_j(t) = \sum_{k=1}^{N} \sum_l \mathrm{h}_{jk}(l)\, \mathrm{s}_k(t-l),\ j=1, \ldots, M, \tag{9.1}$$

where $\mathrm{h}_{jk}(l)$ represents the impulse response from source k to sensor j. This chapter assumes that N and M are known, and that the sensor spacing is sufficiently small to avoid the spatial aliasing problem. In this chapter, we focus particularly on an underdetermined situation where the number of sources N can exceed the number of sensors M ($N > M$). The goal for BSS is to obtain separated signals $\mathrm{y}_k(t)$ that are estimations of s_k solely from M observations.

Two approaches have been studied and employed to solve the BSS problem; one is based on independent component analysis (ICA) (e.g., [2]) and the

S. Makino et al. (eds.), Blind Speech Separation, 243–270.

other relies on the sparseness of source signals (e.g., [3]). Recently, many ICA methods have been proposed even for the convolutive BSS problem [2, 4–10]. ICA works well even in a reverberant condition when the number of sources N is less than or equal to the number of sensors M. On the other hand, the sparseness based approaches are attractive because they can handle the underdetermined problem, i.e., $N > M$.

The sparseness based approaches can be divided into two main categories. One method is based on MAP estimation, where the sources are estimated after mixing matrix estimation [11–17], and the other extracts each signal with a time–frequency binary mask [3, 18–20]. The former method includes mixing matrix estimation and l_1-norm minimization in the frequency domain (i.e., for complex numbers), both of which still present difficulties [16]. The latter, namely the binary mask approach, has the advantage of being implemented in real-time [21]. This chapter focuses on the binary mask approach.

With the binary mask approach, we assume that signals are sufficiently sparse, and therefore, we can also assume that at most one source is dominant at each time–frequency point. If the sparseness assumption holds, and if we can assume an anechoic situation, we can estimate the geometrical information about the dominant source at each time–frequency point. The geometrical information is estimated by using the level and phase differences between observations. When we consider this information for all time–frequency points, the points can be grouped into N clusters [3, 18, 20]. Because an individual cluster corresponds to an individual source, we can separate each signal by selecting the observation signal at time–frequency points in each cluster with a binary mask. The best-known approach may be the Degenerate Unmixing Estimation Technique (DUET) [3, 18, 21]. In this chapter we will assess the sparseness and anechoic assumptions, which the binary mask approach relies on, for speech signals under different reverberant conditions.

Previously, clustering for estimating N clusters was performed manually [3, 18], by using kernel density estimation [20], or with an ML based gradient method [21]. On the other hand, if clustering could be performed with the k-means clustering algorithm [22], it would be automated and accelerated. Employing the k-means algorithm is the prime motivation of this chapter.

The second motive behind this chapter is to employ more than three sensors arranged 2- or 3-dimensionally, which could have a nonlinear/nonuniform alignment. Only a few authors have generalized [16, 17, 23] a method for more than two sensors. The authors of [23] used up to eight sensors, however, their sensors were still linearly arranged. The paper [24] proposed a multichannel DUET by combining the sparseness assumption and the Estimation of Signal Parameters via Rotational Invariance Technique (ESPRIT). However, as the method is based on ESPRIT, it still limits the array shape to a linear array or two sets of congruent arrays. A stereo system and a linear sensor array limits the separation ability on a 2-dimensional half-plane, e.g., they cannot separate sources placed in a mirror image arrangement. To allow the free location of sources, we need more than three sensors arranged 2- or 3-dimensionally.

Based on these two motivations, this chapter describes a new binary mask approach called MENUET (Multiple sENsor dUET), which employs the k-means clustering algorithm. The level ratios and phase differences between multiple observations are employed as features for the clustering. To employ the k-means algorithm successfully, the variances of level ratios and phase differences are set at a comparable level. Moreover, MENUET does not require sensor location information. This allows us to employ freely arranged multiple sensors easily. Therefore, MENUET can separate signals that are distributed 2- or 3-dimensionally.

The organization of this chapter is as follows. Section 9.2 presents the basic framework of the binary mask based BSS method. In Sect. 9.3, we investigate the relevance of the sparseness and anechoic assumptions which the binary mask approach relies on. In Sect. 9.4, we describe some features for clustering, and test how each feature will be clustered by the k-means clustering algorithm. The proposed method MENUET with k-means clustering and multiple sensors is described in Sect. 9.4.2. Sect. 9.5 reports some experimental results obtained with nonlinearly arranged sensors in underdetermined scenarios. Even when the sources and sensors were distributed 2- or 3-dimensionally, we obtained good separation results with the k-means algorithm for each scenario under reverberant conditions ($\mathrm{RT}_{60} = 128$ and 300 ms). The final section concludes this chapter.

9.2 Separation Procedures

This section reviews the procedures used with the binary mask based method. Figure 9.1 shows the flow of the binary mask approach.

[Step 1] Signal transformation to the time–frequency domain: The binary mask approach usually employs a time–frequency domain representation. First, time-domain signals $\mathrm{x}_j(t)$ (9.1) sampled at frequency f_s are converted into frequency-domain time-series signals $x_j(f,t)$ with a T-point short-time Fourier transform (STFT):

$$x_j(f,t) \leftarrow \sum_{r=0}^{T-1} \mathrm{x}_j(r+tS)\,\mathrm{win}(r)\,e^{-\jmath 2\pi f r}, \tag{9.2}$$

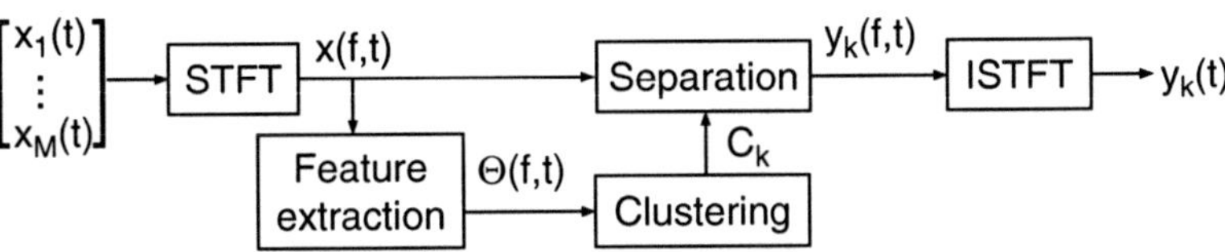

Fig. 9.1. Basic scheme of binary mask approach.

where $f \in \{0, \frac{1}{T}f_s, \ldots, \frac{T-1}{T}f_s\}$ is a frequency, win(r) is a window that tapers smoothly to zero at each end, t is a new index representing time, and S is the window shift size. We utilized a Hanning window $\frac{1}{2}(1-\cos\frac{2\pi r}{T})$ $(r = 0, \ldots, T-1)$ as the window win(r).

There are two advantages to working in the time–frequency domain. First, convolutive mixtures (9.1) can be approximated as instantaneous mixtures at each frequency:

$$x_j(f,t) \approx \sum_{k=1}^{N} h_{jk}(f) s_k(f,t), \tag{9.3}$$

or in a vector notation,

$$\mathbf{x}(f,t) \approx \sum_{k=1}^{N} \mathbf{h}_k(f) s_k(f,t), \tag{9.4}$$

where $h_{jk}(f)$ is the frequency response from source k to sensor j, and $s_k(f,t)$ is a frequency-domain time-series signal of $\mathrm{s}_k(t)$ obtained by the same operation as (9.2), $\mathbf{x} = [x_1, \ldots, x_M]^T$, and $\mathbf{h}_k = [h_{1k}, \ldots, h_{Mk}]^T$ is a mixing vector that consists of the frequency responses from source s_k to all sensors.

The second advantage is that the sparseness of a source signal becomes prominent in the time–frequency domain [12, 19], if the source is colored and nonstationary such as speech. The possibility of $s_k(f,t)$ being close to zero is much higher than that of $\mathrm{s}_k(t)$. When the signals are sufficiently sparse in the time–frequency domain, we can assume that the sources rarely overlap and, (9.3) and (9.4), respectively, can be approximated as

$$x_j(f,t) \approx h_{jk}(f) s_k(f,t), \quad {}^{\exists}k \in \{1, \ldots, N\}, \tag{9.5}$$

$$\mathbf{x}(f,t) \approx \mathbf{h}_k(f) s_k(f,t), \quad {}^{\exists}k \in \{1, \ldots, N\}, \tag{9.6}$$

where $s_k(f,t)$ is a dominant source at the time–frequency point (f,t). For instance this is approximately true for speech signals [3, 15]. Figure 9.2 (a) shows example spectra of three speech sources, in which we can see their temporal/frequency sparseness. We will assess the sparseness of speech signals in the next section.

[Step 2] Feature extraction: If the sources $s_k(f,t)$ are sufficiently sparse, separation can be realized by gathering the time–frequency points (f,t) where only one signal s_k is estimated to be dominant. To estimate such time–frequency points, some features $\boldsymbol{\Theta}(f,t)$ are calculated by using the frequency-domain observation signals $\mathbf{x}(f,t)$.

Most existing methods utilize the level ratio and/or phase difference between *two* observations as their features $\boldsymbol{\Theta}(f,t)$. The features can be summarized as

$$\boldsymbol{\Theta}(f,t) = \left[\frac{|x_2(f,t)|}{|x_1(f,t)|}, \frac{1}{2\pi f}\arg\left[\frac{x_2(f,t)}{x_1(f,t)}\right]\right]^T. \tag{9.7}$$

In the phase difference term, the frequency dependence is normalized by being divided with f. Thanks to the frequency normalization, we can handle

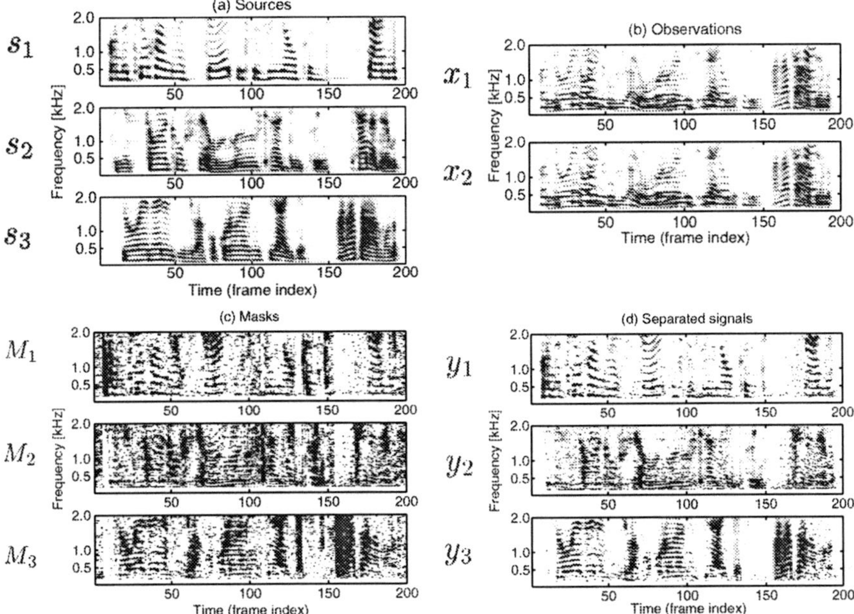

Fig. 9.2. Example spectra of (a) speech sources, (b) observations, (c) masks and (d) separated signals ($N = 3$, $M = 2$).

all time–frequency points simultaneously in the next step. If we do not use such frequency normalization, we have to solve the permutation problem among frequencies after the clustering step [16, 17]. Some features are listed in Table 9.2 and discussed in Sect. 9.4.

Such features (9.7) represent geometric information on sources and sensors, if the sources are sufficiently sparse. Let us assume that the mixing process is expressed solely by the attenuation $\lambda_{jk} \geq 0$ and time delay τ_{jk} from source k to sensor j:

$$h_{jk}(f) \approx \lambda_{jk} \exp\left[-\jmath 2\pi f \tau_{jk}\right]. \tag{9.8}$$

If there is no reverberation (i.e., an anechoic situation), λ_{jk} and τ_{jk} are determined solely by the geometric condition of the sources and sensors (Fig. 9.3). That is, the binary mask approach also assumes an *anechoic* situation. We check the relevance of the anechoic model (9.8) in Sect. 9.3.

If the sources are sparse (9.5), the feature vector (9.7) becomes

$$\boldsymbol{\Theta}(f,t) = \left[\frac{\lambda_{2k}}{\lambda_{1k}}, -(\tau_{2k} - \tau_{1k})\right]^T, \quad {}^{\exists}k. \tag{9.9}$$

That is, the feature $\boldsymbol{\Theta}(f,t)$ contains geometric information on the dominant source s_k at each time–frequency point (f,t).

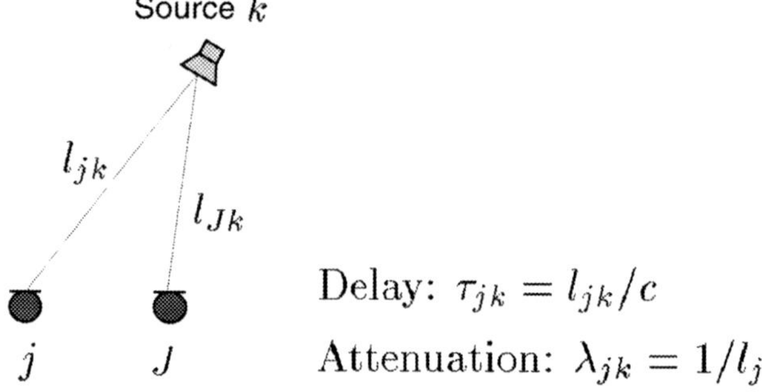

Fig. 9.3. Mixing process model.

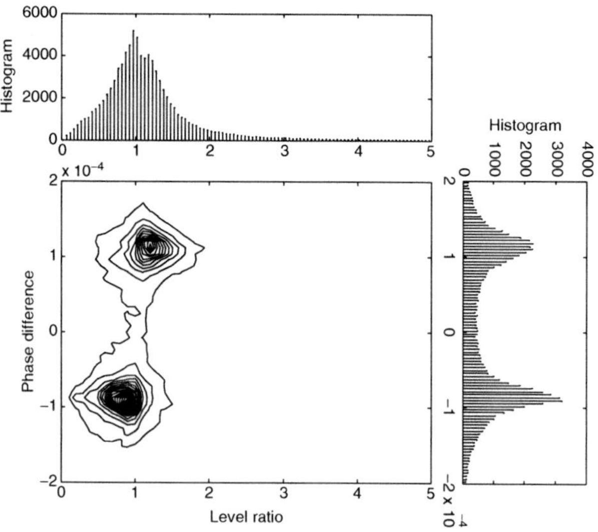

Fig. 9.4. Example histogram of (9.7) when $N = 2$ and $M = 2$. Top: histogram of the level ratio, bottom left: the contour plot of the 2-D histogram, bottom right: histogram of the phase difference. Sources were set at 45° and 120°, the reverberation time $\mathrm{RT}_{60} = 128$ ms and the distance between sources and sensors was $R = 50$ cm (see Fig. 9.9). Outliers are delated for better visualization.

Figure 9.4 shows an example histogram of the features (9.7) for all time–frequency points of two 5-second speech signals in a weak reverberant condition. In Fig. 9.4, the features are well localized, and we can see two clusters, which correspond to each source. Therefore, we can separate signals by picking out the time–frequency points in each cluster. This is the basic idea of the binary mask approach.

[Step 3] Clustering: As shown in Fig. 9.4, if we can find clusters, each cluster corresponds to an individual source. Therefore, the next step is the clustering of the features $\boldsymbol{\Theta}(f,t)$. With an appropriate clustering algorithm, the features $\boldsymbol{\Theta}(f,t)$ are grouped into N clusters $C_1, \ldots, C_N$, where N is the number of possible sources.

In this chapter, we employ the k-means clustering algorithm [22] due to its usability. The k-means algorithm is a well-known and very efficient clustering method. The implementation is easy: e.g., MATLAB has the function `kmeans`.

The clustering criterion of k-means clustering is to minimize the total sum $\mathcal{J}$ of the squared distances between cluster members and their centroids $\bar{\mathbf{c}}_k$:

$$\mathcal{J} = \sum_{k=1}^{M} \mathcal{J}_k, \; \mathcal{J}_k = \sum_{\boldsymbol{\Theta}(f,t) \in C_k} ||\boldsymbol{\Theta}(f,t) - \bar{\mathbf{c}}_k||^2. \tag{9.10}$$

After setting appropriate initial centroids $\bar{\mathbf{c}}_k$ $(k = 1, \ldots, N)$, $\mathcal{J}$ can be minimized by the following iterative updates:

$$C_k = \{\boldsymbol{\Theta}(f,t) \mid k = \operatorname{argmin}_k ||\boldsymbol{\Theta}(f,t) - \bar{\mathbf{c}}_k||^2\}, \tag{9.11}$$

$$\bar{\mathbf{c}}_k \leftarrow E[\boldsymbol{\Theta}(f,t)]_{\boldsymbol{\Theta} \in C_k}, \tag{9.12}$$

where $E[\cdot]_{\boldsymbol{\Theta} \in C_k}$ is a mean operator for the members of a cluster C_k. That is, the k-means clustering algorithm calculates (9.11) and (9.12) until the algorithm converges. The cluster members for each cluster are determined by (9.11). If the feature $\boldsymbol{\Theta}(f,t)$ is properly chosen, then the k-means works well and each cluster corresponds to an individual source.

Here, it should be noted that k-means clustering utilizes the squared distance $||\boldsymbol{\Theta}(f,t) - \bar{\mathbf{c}}_k||^2$, not the Mahalanobis distance $(\boldsymbol{\Theta}(f,t) - \bar{\mathbf{c}}_k)^T$ $\boldsymbol{\Sigma}_k^{-1}(\boldsymbol{\Theta}(f,t) - \bar{\mathbf{c}}_k)$, where $\boldsymbol{\Sigma}_k$ is the covariance matrix of the cluster k. That is, the k-means algorithm assumes clusters of a multivariate isotropic variance $\boldsymbol{\Sigma}_k = \boldsymbol{I}$ for all k, where $\boldsymbol{I}$ denotes an identity matrix.

[Step 4] Separation: Next, the separated signals $y_k(f,t)$ are estimated based on the clustering result. We design a time–frequency domain binary mask that extracts the time–frequency points of each cluster

$$M_k(f,t) = \begin{cases} 1 & \boldsymbol{\Theta}(f,t) \in C_k \\ 0 & \text{otherwise.} \end{cases} \tag{9.13}$$

Example binary mask spectra are shown in Fig. 9.2 (c). Then, applying the binary masks (Fig. 9.2 (c)) to one of the observations (Fig. 9.2 (b)) $x_J(f,t)$, we obtain separated signals (Fig. 9.2 (d)):

$$y_k(f,t) = M_k(f,t) x_J(f,t),$$

where J is a selected sensor index.

[Step 5] Separated signal reconstruction: At the end of the flow (Fig. 9.1), we obtain outputs $\mathrm{y}_k(t)$ by employing an inverse STFT (ISTFT) and the overlap-and-add method [25]:

$$\mathrm{y}_k(t) = \frac{1}{A} \sum_{l=0}^{S-1} \mathrm{y}_k^{m+l}(t), \tag{9.14}$$

where $A = \frac{1}{2}\frac{T}{S}$ is a constant for the Hanning window case,

$$\mathrm{y}_k^m(t) = \begin{cases} \sum_{f \in \{0, \frac{1}{T}f_s, \ldots, \frac{T-1}{T}f_s\}} y(f,m) e^{\jmath 2\pi f r} \\ \qquad\qquad (mS \leq t \leq mS + T - 1) \\ 0 \quad \text{(otherwise)}, \end{cases}$$

and $r = t - mS$.

9.3 Validity of Sparseness Assumption and Anechoic Assumption

The binary mask approach utilizes two assumptions:

1. Sparseness assumption (9.5): $x_j(f,t) \approx h_{jk}(f) s_k(f,t), \quad {}^{\exists}k \in \{1, \ldots, N\}$
2. Anechoic assumption (9.8): $h_{jk}(f) \approx \lambda_{jk} \exp\left[-\jmath 2\pi f \tau_{jk}\right]$.

In this section, we assess the relevance of these two assumptions using real speech signals and measured room impulse responses.

9.3.1 Sparseness Evaluation of Speech Signals

First, let us investigate the sparseness of speech signals.

It has already been shown that speech sources are sparser in the time–frequency domain than in the time-domain [12, 19]. Figure 9.5 shows an example histogram of signal amplitudes for a time-domain speech source signal (a male speech of 5 seconds) and its time–frequency domain representation at a frequency f of 1 kHz. We can see that the distribution of the time-domain signal has a heavier tail than that of the time–frequency domain signal. This means that the possibility of a time–frequency domain signal being close to zero is much higher than that of a time-domain signal.

Figure 9.6 provides another illustration showing that the time–frequency domain representation is sparser than the time-domain representation. Figure 9.6 is a scatter plot for two speech signals. The horizontal axis is the observation at sensor 1: $x_1 = h_{11}s_1 + h_{12}s_2$, the vertical axis shows the sensor 2 observation: $x_2 = h_{21}s_1 + h_{22}s_2$. Here, s_1 and s_2 were two female speech signals of 5 seconds each, and

$$\begin{bmatrix} h_{11} \; h_{12} \\ h_{21} \; h_{22} \end{bmatrix} = \begin{bmatrix} 1.0 \; 0.2 \\ 0.3 \; 0.9 \end{bmatrix} \tag{9.15}$$

was utilized for the simplicity. When sources are sparse enough, they rarely overlap, and therefore, we can see a clear slope $x_2/x_1 = h_{2k}/h_{1k}$ where k is the dominant source at each time or frame. No clear structure can be seen in the time-domain (Fig. 9.6 left). By contrast, we can see two slopes in the time–frequency representation (Fig. 9.6 right).

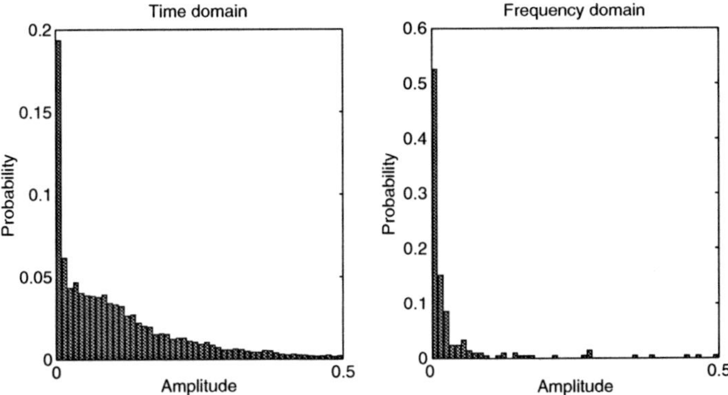

Fig. 9.5. Amplitude histogram for (left) a time-domain signal and (right) a time–frequency domain signal at $f = 1$ kHz. The frame length for STFT was $T = 512$ at 8 kHz sampling. Outliers are delated for better visualization.

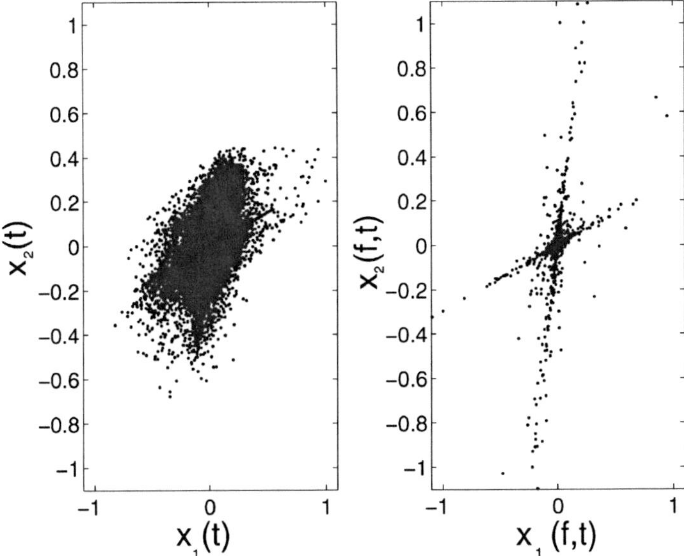

Fig. 9.6. Scatter plot for (left) a time-domain signal and (right) a time–frequency domain signal at $f = 1$ kHz. The frame length for STFT was $T = 512$ at 8 kHz sampling.

Now let us look more closely at the overlap of speech signals at each time–frequency point. Here we employ l_ϵ^0-norm to investigate the signal overlap.

l_ϵ^0-norm is defined as [26, 27]:

$$||s(f,t)||_{0,\epsilon(f)} = \#\{i, |s_i(f,t)| \geq \epsilon(f)\}. \tag{9.16}$$

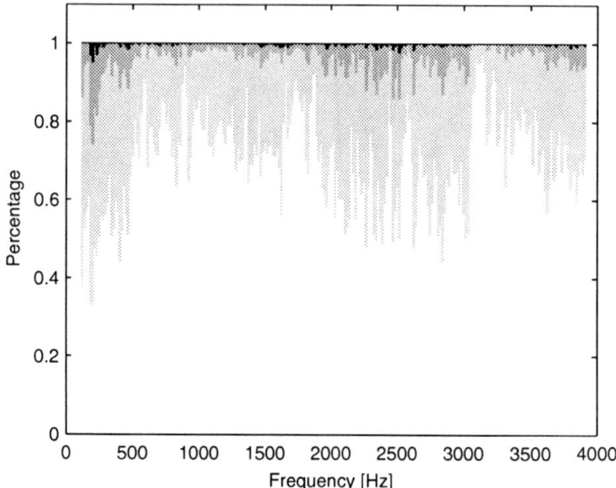

Fig. 9.7. Percentage of frames where l_ϵ^0 is $||s(f,t)||_{0,\epsilon(f)} = 0,\dots,N$ for $N=3$ source signals. Light gray: $||s(f,t)||_{0,\epsilon(f)} = 0$, Gray: $||s(f,t)||_{0,\epsilon(f)} = 1$, Dark gray: $||s(f,t)||_{0,\epsilon(f)} = 2$, Black: $||s(f,t)||_{0,\epsilon(f)} = 3$.

That is l_ϵ^0-norm means the number of sources that have a larger amplitude than a given threshold ϵ. Although ϵ is set at the noise level in the original l_ϵ^0-norm definition, the ϵ utilized here was

$$\epsilon(f) = \frac{1}{10}\mathrm{max}_i \mathrm{max}_t |s_i(f,t)|,$$

which gives us -20 dB of the maximum value at each frequency bin.

Figure 9.7 shows the percentage of frames where $||s(f,t)||_{0,\epsilon(f)} = 0,\dots,N$ for $N=3$ sources. At half of the frames, the speech sources are less than ϵ. Even in the time frames where the source(s) is (are) active, the frames where there is just one source are dominant. Only around 10% of the time frames have more than two active sources for $N=3$ sources. We can conclude that the speech sources are sufficiently sparse for us to employ the binary mask approach.

From the spectrograms of speech signals (e.g., Fig. 9.2 (a)), we can also see intuitively that the speech components seldom overlap.

9.3.2 Sparsest Representation with STFT

For sparseness based BSS, we should use as sparse a representation as possible. Let us look for a sparser representation with the STFT. When the frame size T for the STFT is too large, the possibility of several signals existing in a frame increases. On the other hand, if the frame size T is too short, then the frequency resolution becomes small. This means that a frequency bin contains a wide-band component, therefore, the possibility of multiple sources in each frequency bin again increases. Therefore, there should be an optimal frame size T.

For the sparseness evaluation, this chapter employs a measure called the approximate W-disjoint orthogonality (WDO) [28]:

$$r_k(z) = \frac{\sum_{(f,t)} |\Phi_{(k,z)}(f,t) x_{Jk}(f,t)|^2}{\sum_{(f,t)} |x_{Jk}(f,t)|^2} \times 100[\%]. \tag{9.17}$$

In (9.17), $x_{Jk}(f,t)$ means the short-time Fourier transformed observed signal k at a certain selected sensor J: $x_{Jk}(f,t)$ =STFT$[\sum_l \mathrm{h}_{Jk}(l)\, \mathrm{s}_k(t-l)]$. Moreover, in (9.17), $\Phi_{(k,z)}$ is a time–frequency binary mask that has a parameter z

$$\Phi_{(k,z)}(f,t) = \begin{cases} 1 & 20\log\left(|x_{Jk}(f,t)|/|\hat{x}_{Jk}(f,t)|\right) > z \\ 0 & \text{otherwise,} \end{cases} \tag{9.18}$$

where $\hat{x}_{Jk}(f,t)$ is the sum of the interference components at sensor J: $\hat{x}_{Jk}(f,t) = \mathrm{STFT}\left[\sum_{i=1, i\neq k}^{N} x_{Ji}(t)\right]$. The approximate WDO $r_k(z)$ indicates the percentage of the energy of source k for time–frequency points where it dominates the other sources by z dB. $r_k(z)$ is named as the preserved-signal ratio in [29]. That is, the approximate WDO $r_k(z)$ also measures the signal overlap at each time–frequency point. A larger (smaller) approximate WDO $r_k(z)$ means more (less) sparseness.

Figure 9.8 is the W-disjoint orthogonality for different frame sizes T for $N = 3, 4$ and 5 for an anechoic case. The sampling frequency was 8 kHz in the investigation. We found that a frame size T of 512 or 1024 (64 or 128 ms) provides us with the sparsest representation. This result is the same as that reported in previous research [19].

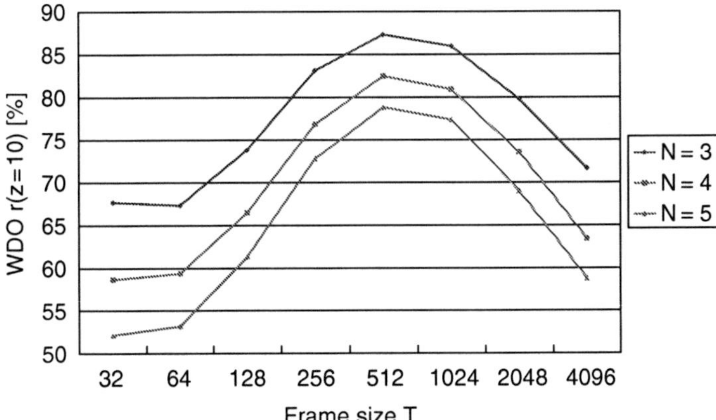

Fig. 9.8. Approximate WDO for each frame size T. $z = 10$. Anechoic.

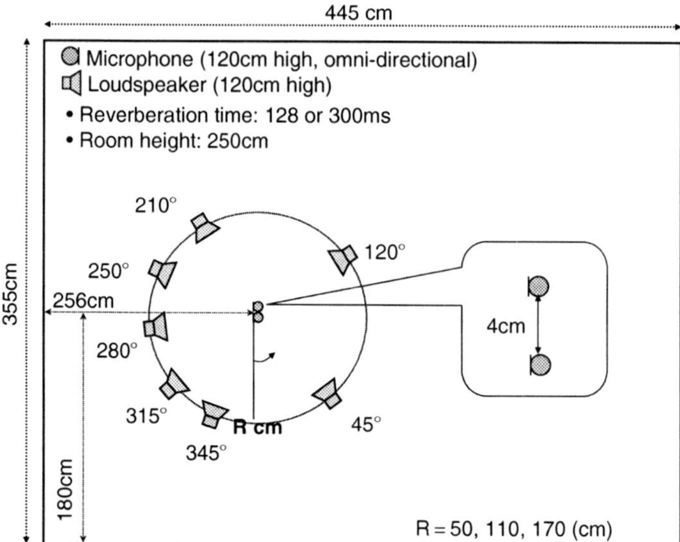

Fig. 9.9. Room setup ($M = 2$).

9.3.3 Sparseness of Reverberant Speech

It is difficult for the sparseness assumption (9.5) to hold when there is reverberation. In this subsection, the sparseness is investigated for three reverberation conditions: $\mathrm{RT}_{60} = 0, 128$ and 300 ms and three different distances R between the sensors and sources (see Fig 9.9). The room impulse responses h_{Jk} were measured in the room depicted in Fig. 9.9 and convolved with real speech signals.

Figure 9.10 is the approximate WDO under certain reverberant conditions. The sources were set at 45°, 120° and 315° when $N = 3$, 45°, 120°, 210° and 315° when $N = 4$, and 45°, 120°, 210°, 280° and 345° when $N = 5$. We used eight combinations of four male and four female speeches as the sources for each N condition. As seen in Fig. 9.10, the sparseness decreases with increases in both reverberation and distance R. That is, for reverberant and distant signals, it becomes difficult for the sparseness assumption (9.5) to hold.

9.3.4 Validity of Anechoic Assumption

In this section, let us assess the way in which reverberation and distance affect the anechoic assumption (9.8).

Figure 9.11 shows an example histogram of features (9.7) for two 5-second speech signals under a stronger reverberant condition than in Fig. 9.4. In Fig. 9.11 the reverberation time was $\mathrm{RT}_{60} = 300$ ms and the distance R between sensors and sources was 110 cm, whereas in Fig. 9.4 $\mathrm{RT}_{60} = 128$ ms

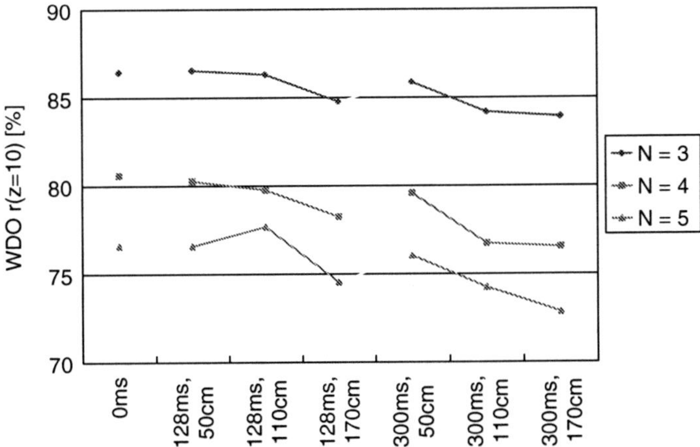

Fig. 9.10. Approximate WDO for some reverberant conditions. The x-axis e.g., "128 ms, 50 cm" indicates that the reverberation time was 128 ms, and the distance between the sources and sensors was 50 cm. $z = 10$.

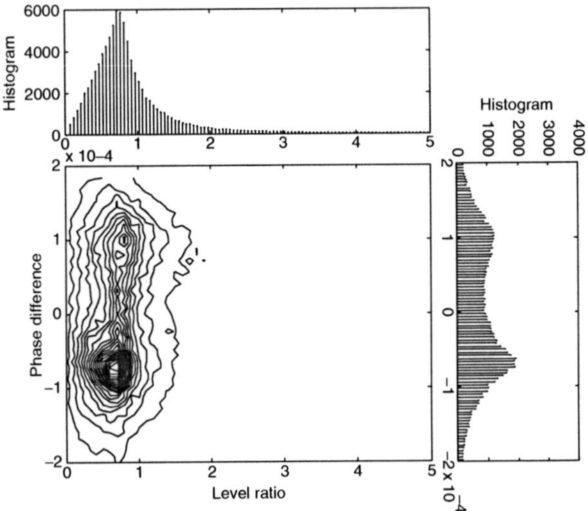

Fig. 9.11. Example histogram of (9.7) when $N = 2$ and $M = 2$. Top: histogram of the level ratio, bottom left: the contour plot of the 2-D histogram, bottom right: histogram of the phase difference. Sources were set at 45° and 120°, $\mathrm{RT}_{60} =$ 300 ms, $R = 110$ cm (see Fig. 9.9). Outliers are deleted for better visualization.

and $R = 50$ cm. The other conditions are the same as those of in Fig. 9.4. Compared with Fig. 9.4, each histogram in Fig. 9.11 is broadened due to the reverberation.

Table 9.1 shows the standard deviations σ_L and σ_P of the level ratio and the phase difference in (9.7), respectively. The table shows average values

Table 9.1. Averaged standard deviations of level ratio and phase difference in (9.9) for each cluster.

RT_{60}	128 ms			300 ms		
R [cm]	50	110	170	50	110	170
σ_L	0.60	1.05	2.06	0.87	1.68	2.79
σ_P [ms]	0.07	0.09	0.10	0.08	0.11	0.12

for two clusters and eight speaker combinations. Table 9.1 shows that the variance of the clusters becomes large as reverberation and distance increase. This means that it becomes difficult for the anechoic assumption (9.8) to hold, and therefore, clustering and separation become difficult in reverberant situations.

In real observations, it becomes difficult for the sparseness assumption (9.5) and the anechoic assumption (9.8) to hold. The binary mask approach has to handle such cases in a real separation.

9.4 Features for K-means Clustering Algorithm

Because the binary mask approach depends strongly on the clustering results of the feature vectors $\boldsymbol{\Theta}(f,t)$, the selection of an appropriate feature vector $\boldsymbol{\Theta}(f,t)$ is essential to this approach. In this section, we provide some variations of the features $\boldsymbol{\Theta}(f,t)$, and test how each feature will be clustered by the k-means algorithm.

Features $\boldsymbol{\Theta}(f,t)$ can be clustered by many methods. For example, the authors [3, 18] first clustered them manually, and [21] proposed an ML based gradient method. Gaussian mixture model (GMM) fitting can also be employed [30] if the number of Gaussians and the initial values for the mean and variance are set appropriately.

In contrast, this chapter attempts to employ the k-means clustering algorithm [22], which can both automate and simplify the clustering. In this section, we show that we need an appropriate normalization for the feature components (9.7) in order to utilize the k-means algorithm. This is because the k-means algorithm assumes that each cluster has a multivariate isotropic variance (see Sect. 9.2 [Step 3]). However, some features do not form clusters of isotropic variance, and therefore, they cannot be clustered well by the k-means algorithm. For instance, clusters with the feature (9.7) illustrated by Figs. 9.4 and 9.11 do not have isotropic variance: the phase difference is far smaller than the level ratio. Such a feature is difficult to cluster with the k-means algorithm. We will look for appropriate features for the k-means clustering in this section.

We could employ other clustering algorithms, e.g., with the Mahalanobis distance (Sect. 9.2 [Step 3]), or GMM fitting to handle clusters of anisotropic variance. Both require the adaptation of the covariance matrix $\boldsymbol{\Sigma}_k$ and centroid (mean) $\bar{\mathbf{c}}_k$ of each cluster k. Speaking from the authors' personal experience, such adaptations are more sensitive to their initial values than the k-means algorithm. Moreover, it could help us to perform a coordinate transformation that converts anisotropic distribution into a spherical (isotropic) distribution. This can be executed by using the covariance matrix of the feature vector $\boldsymbol{\Theta}(f,t)$. However, since in reality the feature $\boldsymbol{\Theta}(f,t)$ has many outliers, a well-considered covariance matrix calculation is required. To avoid such difficulities, this chapter attempts to employ the simple k-means algorithm with appropriately normalized features.

9.4.1 K-means Clustering for Features

Table 9.2 lists feature variations. Table 9.2 also shows the separation performance (the signal-to-interference ratio (SIR) improvement: see Sect. 2.2) when we cluster each feature with the k-means algorithm. Feature (9.7) corresponds to feature (A) in Table 9.2. Here we utilized two omnidirectional microphones for three speech sources set at 210°, 250° and 315° in Fig. 9.9 (RT_{60} = 128 ms and R = 50 cm). In Table 9.2, we also show the optimal results with known centroid values, which are calculated with known sources

Table 9.2. Separation performance (SIR improvement in [dB]) of typical features with the k-means algorithm. "optimum" shows the performance with known centroids. $N = 3$, $M = 2$.

	Feature $\boldsymbol{\Theta}(f,t)$		k-means	optimum
(A)	$\left[\frac{\lvert x_2(f,t)\rvert}{\lvert x_1(f,t)\rvert}, \frac{1}{2\pi f}\arg\left[\frac{x_2(f,t)}{x_1(f,t)}\right]\right]^T$	[18]	1.9	8.3
(B)	$\left[\frac{\lvert x_2(f,t)\rvert}{\lvert x_1(f,t)\rvert}, \frac{1}{2\pi f c^{-1} d}\arg\left[\frac{x_2(f,t)}{x_1(f,t)}\right]\right]^T$		5.7	14.1
(A)'	$\left[\frac{\lvert x_2(f,t)\rvert}{\lvert x_1(f,t)\rvert} - \frac{1}{\frac{\lvert x_2(f,t)\rvert}{\lvert x_1(f,t)\rvert}}, \frac{1}{2\pi f}\arg\left[\frac{x_2(f,t)}{x_1(f,t)}\right]\right]^T$	[3]	1.8	7.9
(C)	$\frac{1}{2\pi f}\arg\left[\frac{x_2(f,t)}{x_1(f,t)}\right]$		10.5	14.0
(D)	$\frac{1}{2\pi f c^{-1} d}\arg\left[\frac{x_2(f,t)}{x_1(f,t)}\right]$	[31]	11.6	14.0
(E)	$\left[\frac{\lvert x_1(f,t)\rvert}{A(f,t)}, \frac{\lvert x_2(f,t)\rvert}{A(f,t)}, \frac{1}{2\pi f}\arg\left[\frac{x_2(f,t)}{x_1(f,t)}\right]\right]^T$		5.2	7.9
(F)	$\left[\frac{\lvert x_1(f,t)\rvert}{A(f,t)}, \frac{\lvert x_2(f,t)\rvert}{A(f,t)}, \frac{1}{2\pi f c^{-1} d}\arg\left[\frac{x_2(f,t)}{x_1(f,t)}\right]\right]^T$		12.4	14.1
(G)	$\bar{\Theta}_j(f,t) = \lvert x_j(f,t)\rvert \exp\left[\jmath \frac{\arg[x_j(f,t)/x_J(f,t)]}{\alpha_j f}\right]$ $\boldsymbol{\Theta}(f,t) \leftarrow \bar{\boldsymbol{\Theta}}(f,t)\,/\,\lVert\bar{\boldsymbol{\Theta}}(f,t)\rVert$		12.2	14.1

$A(f,t) = \lVert\mathbf{x}(f,t)\rVert = \sqrt{\sum_{j=1}^{M} \lvert x_j(f,t)\rvert^2}$

α_j: weight parameter introduced in Sect. 9.4.2

and impulse responses (unblind). We can see that some features (A), (B) and (E) cannot achieve good separation performance with the k-means algorithm.

There are two main reasons for this. The first is that the phase term of feature (A) is too small as previously mentioned. Figure 9.12 (A) shows the histogram of feature (A) for $N = 3$ and $M = 2$. It can be seen that the phase term of feature (A) is far smaller than the level ratio. In Fig. 9.12 (A), $\times$ shows the centroids estimated by the k-means algorithm. They are not estimated correctly. The poor performance with (A) and (E) result from this imbalance between the level ratio and phase difference terms. With features (B) and (F), where the phase is divided by $2\pi f c^{-1} d$, the phase difference becomes larger, and the centroids are estimated precisely (see Fig. 9.12 (F)).

The second reason is related to the outliers of the level ratio $|x_2|/|x_1|$. In Fig. 9.12 (A), we can see several large values in the level ratio, although we used omnidirectional microphones where $|x_2|/|x_1| \approx 1$. The presence of the outliers in the level ratio means the feature (A) cannot be clustered correctly (Fig. 9.12 (A)), although the phase terms themselves can be clustered (Table 9.2 (C)). This is the reason for the poor performance with (A) and (B). Such outliers occur at too many time–frequency points for them to be removed without degrading the separated sound quality.

We found that when we normalize the level ratios as seen in features (E) and (F), they become ≤ 1 and prevent such outliers from occurring (Fig. 9.12 (F)). Therefore features (E) and (F) provide better performance than (A) and (B). Moreover, we found that feature (F), where both the level ratios and phase difference are normalized appropriately, achieves good performance (see Table 9.2) with the k-means algorithm.

In conclusion, when employing the k-means clustering algorithm, both the level and phase components of the feature vector $\boldsymbol{\Theta}(f,t)$ should be appropriately normalized.

9.4.2 Feature Vectors for Multiple Sensors

In the previous subsection, we found that clustering is successfully and effectively executed when we use normalized level ratios and phase differences such as feature (F) shown in Table 9.2:

$$\left[\frac{|x_1(f,t)|}{A(f,t)}, \frac{|x_2(f,t)|}{A(f,t)}, \frac{1}{2\pi f c^{-1} d} \arg\left[\frac{x_2(f,t)}{x_1(f,t)}\right]\right]^T,$$

where $A(f,t) = ||\mathbf{x}(f,t)|| = \sqrt{\sum_{j=1}^{M} |x_j(f,t)|^2}$. However, thus far, we have discussed the features of a stereo $M = 2$ system. A stereo system limits the separation ability on a 2-dimensional half-plane.

Therefore, this subsection expands the feature to a multiple sensor version based on the result described in the previous subsection. If we can utilize more than three sensors arranged nonlinearly, we can separate signals located

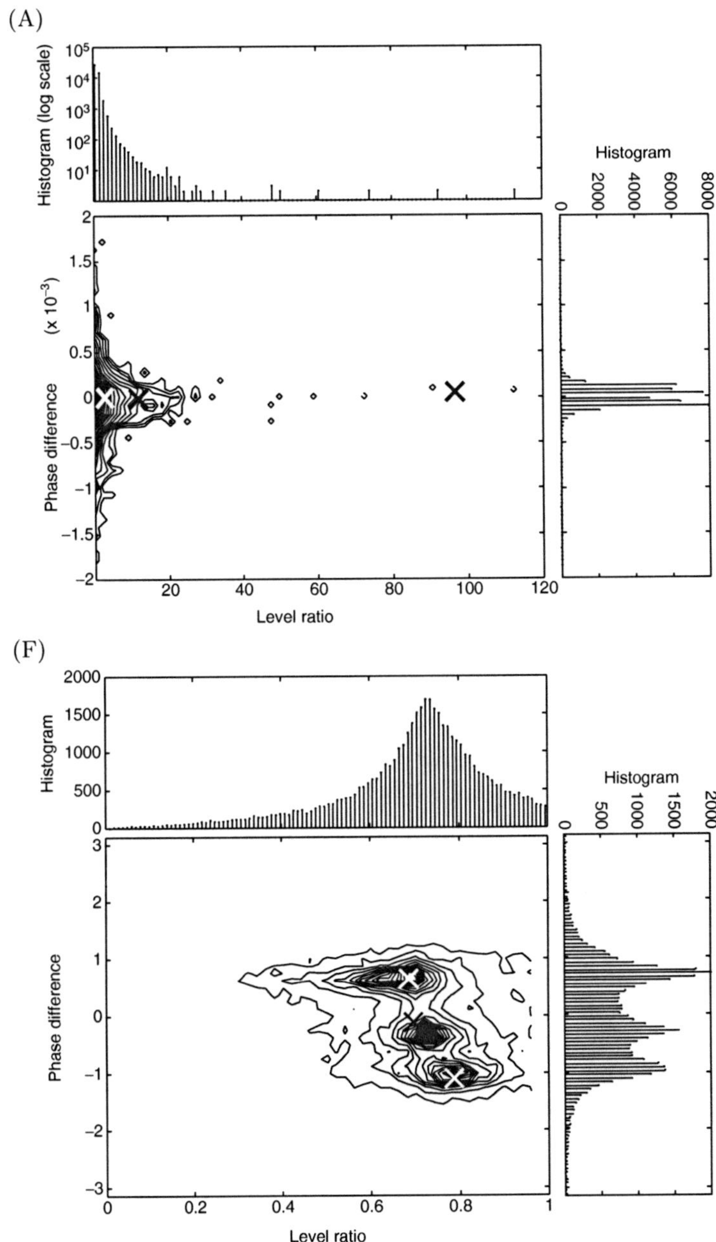

Fig. 9.12. Example histograms with features (A) and (F). For each feature, top: histogram of the level ratio term, bottom left: the contour plot of the 2-D histogram, bottom right: histogram of the phase difference term. X denotes the cluster centroids $\bar{\mathbf{c}}_k$ obtained by the k-means algorithm. In (F), we plot only two components $\left[\frac{|x_1(f,t)|}{A(f,t)}, \frac{1}{2\pi f c^{-1} d} \arg\left[\frac{x_2(f,t)}{x_1(f,t)}\right]\right]^T$. All data including outliers are plotted.

2- or 3-dimensionally. As the method can be considered an extension of DUET [3, 18, 21], we call the method Multiple sENsor dUET: MENUET.

Feature in MENUET: The feature in MENUET also employs the normalized level information and phase differences between multiple observations:

$$\boldsymbol{\Theta}(f,t) = [\boldsymbol{\Theta}^L(f,t), \boldsymbol{\Theta}^P(f,t)]^T, \tag{9.19}$$

where

$$\boldsymbol{\Theta}^L(f,t) = \left[\frac{|x_1(f,t)|}{A(f,t)}, \dots, \frac{|x_M(f,t)|}{A(f,t)}\right] \tag{9.20}$$

is the observation level information and

$$\boldsymbol{\Theta}^P(f,t) = \left[\frac{1}{\alpha_1 f} \arg\left[\frac{x_1(f,t)}{x_J(f,t)}\right], \dots, \frac{1}{\alpha_M f} \arg\left[\frac{x_M(f,t)}{x_J(f,t)}\right]\right] \tag{9.21}$$

is the phase difference information with respect to the phase of the J-th observation. In the above equations, $A(f,t) = \sqrt{\sum_{j=1}^{M} |x_j(f,t)|^2}$, J is the index of one of the sensors, and α_j ($j = 1, \dots, M$) is a positive weighting constant. By changing α_j, we can control the weights for the level and phase difference information of the observed signals; a larger α_j value puts weight on the level and a smaller value emphasizes the phase difference. This is the direct extention of feature (9.7).

The normalized level information has the property $0 \leq \Theta_j^L(f,t) \leq 1$, where Θ_j^L is the j-th component of $\boldsymbol{\Theta}_L$. This can prevent the occurrence of the outliers discussed in the previous section.

An appropriate value for the phase weight is $\alpha_j = \alpha = 4\pi c^{-1} d_{\max}$, where c is the propagation velocity and $d_{\max}$ is the maximum distance[1] between sensor J and sensor $^{\forall} j \in \{1, \dots, M\}$. Let us provide the reason for this. Here, we use the mixing model (9.8) and, without loss of generality, we assume that the delay parameter τ_{jk} is determined by

$$\tau_{jk} = l_{jk}/c, \tag{9.22}$$

where l_{jk} is the distance from source k to sensor j (see Fig. 9.3). We also use the fact that the maximum distance $d_{\max}$ between the sensors is greater than the maximum path difference:

$$\max_{j,k} |l_{jk} - l_{Jk}| \leq d_{\max}.$$

Using these assumptions and the mixing model (9.8), $\Theta_j^P(f,t)$, which is the j-th component of $\boldsymbol{\Theta}_P(f,t)$, becomes

$$\frac{1}{\alpha_j f} \arg\left[\frac{x_j(f,t)}{x_J(f,t)}\right] = \frac{2\pi c^{-1}(l_{jk} - l_{Jk})}{\alpha_j} \leq \frac{2\pi c^{-1} d_{\max}}{\alpha_j}. \tag{9.23}$$

[1] If we do not have an accurate value for $d_{\max}$, we may use a rough positive constant, as shown in Sect. 9.5.3.

Because the level information is normalized to have the range $0 \leq \boldsymbol{\Theta}^L(f,t) \leq 1$, the phase difference $\Theta_j^P(f,t)$ should also be normalized so that it has a similar range.

Here, as it is difficult to know the variance of each cluster in advance, we try to balance the ranges of $\boldsymbol{\Theta}^L(f,t)$ and $\boldsymbol{\Theta}^P(f,t)$. If we allow $\Theta_j^P(f,t)$ to have the range

$$-\frac{1}{2} \leq \boldsymbol{\Theta}^P(f,t) \leq \frac{1}{2}$$

(note $\Theta_j^P(f,t)$ can take a negative value), we have equality in (9.23) when $\alpha_j = \alpha = 4\pi c^{-1} d_{\max}$. That is, $\alpha = 4\pi c^{-1} d_{\max}$ realizes the same range as the level information.

Modified Feature: We can modify the extended feature (9.19) by using the complex representation,

$$\Theta_j(f,t) = \Theta_j^L(f,t) \exp[\jmath \Theta_j^P(f,t)], \tag{9.24}$$

where Θ_j^L and Θ_j^P are the j-th components of (9.20) and (9.21), respectively. This modification can also be realized by [30, 32]:

$$\bar{\Theta}_j(f,t) = |x_j(f,t)| \exp\left[\jmath \frac{\arg[x_j(f,t)/x_J(f,t)]}{\alpha_j f}\right], \tag{9.25}$$

$$\boldsymbol{\Theta}(f,t) \leftarrow \bar{\boldsymbol{\Theta}}(f,t) \,/\, ||\bar{\boldsymbol{\Theta}}(f,t)||, \tag{9.26}$$

where $\bar{\boldsymbol{\Theta}}(f,t) = [\bar{\Theta}_1(f,t), \ldots, \bar{\Theta}_M(f,t)]^T$. Feature (9.26) is the modified feature, where the phase difference information is held in the argument term (9.25), and the level information is normalized by the vector norm normalization (9.26). The weight parameter α_j has the same property as (9.19), however, $\alpha = 4c^{-1} d_{\max}$ should be the lower limit for successful clustering (see Appendix A).

Now the normalized vectors $\boldsymbol{\Theta}(f,t)$ (9.26) are M-dimensional complex vectors, and therefore the features will be clustered in an M-dimensional complex space. The unit-norm normalization (9.26) facilitates the distance calculation in the clustering (9.10), because it projects the vector on a hyper unit sphere. If the features $\boldsymbol{\Theta}(f,t)$ and the cluster centroid $\bar{\mathbf{c}}_k$ are on the unit sphere, i.e., $||\boldsymbol{\Theta}(f,t)|| = ||\bar{\mathbf{c}}_k|| = 1$, the square distance $||\boldsymbol{\Theta}(f,t) - \bar{\mathbf{c}}_k||^2 = 2(1 - Re(\bar{\mathbf{c}}_k^H \boldsymbol{\Theta}(f,t))$. That is the minimization of the distance $||\boldsymbol{\Theta}(f,t) - \bar{\mathbf{c}}_k||^2$ is equivalent to the maximization of the real part of the inner product $\bar{\mathbf{c}}_k^H \boldsymbol{\Theta}(f,t)$, whose calculation is less demanding in terms of computational complexity.

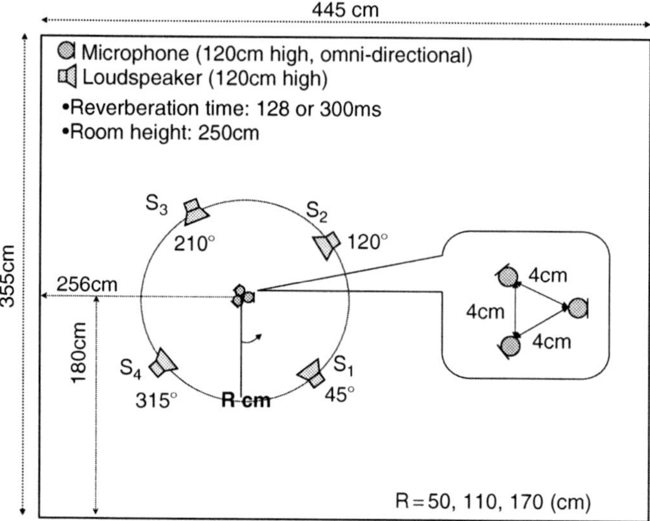

Fig. 9.13. Room setup ($N = 4, M = 3$).

9.5 Speech Separation Experiments

9.5.1 Experimental Conditions

We performed experiments with measured impulse responses $\mathrm{h}_{jk}(l)$ in a room as shown in Figs. 9.13 and 9.14. The room reverberation times RT_{60} were 128 and 300 ms. We used the same room for both reverberation times but changed the wall condition. We also changed the distance R between the sensors and sources. The distance variations were R = 50, 110, and 170 cm (see Figs. 9.13 and 9.14). Mixtures were made by convolving the measured impulse responses in the room and 5-second English speeches. For the anechoic test, we simulated the mixture by using the anechoic model (9.8) and the mixture model (9.1). The sampling rate was 8 kHz. The STFT frame size T was 512 and the window shift $S = T/4$.

Unless otherwise noted, we utilized modified feature (9.26) with $\alpha_j = \alpha = 4c^{-1}d_{\max}$ for the features, because the computational cost of distance calculation is small (see Sect. 9.4.2). We utilized the k-means algorithm for the clustering, where the number of sources N was given. We set the initial centroids of the k-means using a far-field model. The initialization method is described in Appendix B.

The separation performance was evaluated in terms of the SIR improvement and the signal to distortion ratio (SDR). Their definitions are given in Sect. 2.2. We investigated eight combinations of speakers and averaged the separation results.

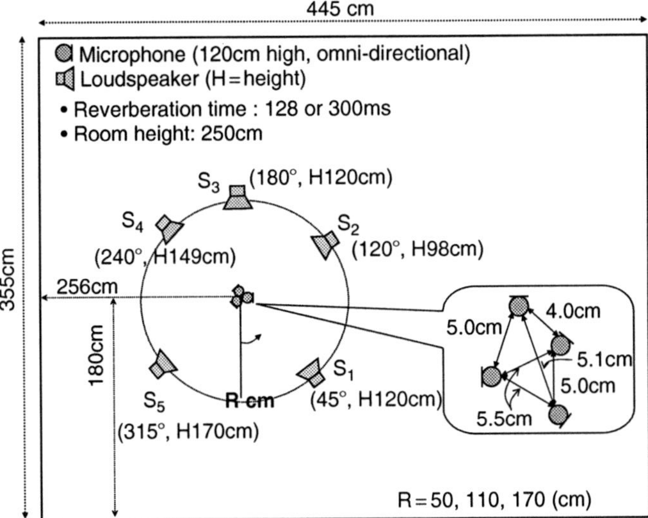

Fig. 9.14. Room setup ($N = 5, M = 4$).

9.5.2 Separation Results

With Two Sensors First, we tested the feature with two sensors under the condition described in Sect. 9.4.1, and compared the result with those for other features. Table 9.2 in Sect. 9.4.1 shows the result. The feature (9.19) corresponds to (F) and the modified feature (9.26) is shown as (G). We obtained better separation performance with these two features than with other features (A)–(E). A comparison with the MAP approach can be found in [16]. A comparison with the GMM fitting is provided in [30].

Note that two sensors/linear arrays do not work when the sources are placed at axisymmetrical locations with respect to the microphone array, because they have the equivalent features in (9.7).

With 2-D Three Sensors Here we show the separation results obtained with three sensors arranged 2-dimensionally (Fig. 9.13). Note that sources were also distributed 2-dimensionally.

Figure 9.15 (a) shows the separation result when $N = 4$ and $M = 3$. We can see that the proposed method achieved good separation performance with the nonlinear sensor arrangement. We also evaluated the performance for $N = 5$ and $M = 3$, where the source positions were 45°, 120°, 210°, 280° and 345°, and obtained good performance (Fig. 9.15 (b)).

With Four Sensors The proposed method MENUET can also be applied to a 3-dimensional sensor array arranged nonuniformly. The setup is shown in Fig. 9.14. Here, the system knew only the maximum distance $d_{\max}$ (5.5 cm)

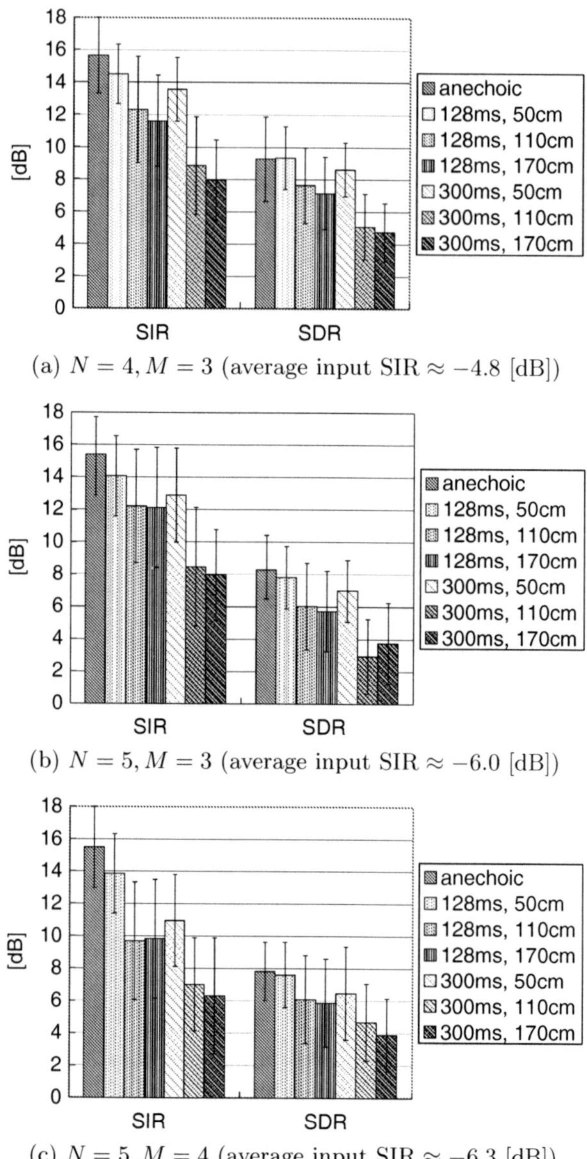

(a) $N = 4, M = 3$ (average input SIR ≈ -4.8 [dB])

(b) $N = 5, M = 3$ (average input SIR ≈ -6.0 [dB])

(c) $N = 5, M = 4$ (average input SIR ≈ -6.3 [dB])

Fig. 9.15. Average SIR improvement and SDR for each condition. The error bar shows the standard deviations for all outputs and combinations.

between the reference microphone and the others. To avoid the spatial aliasing problem, we utilized frequency bins up to 3100 Hz in this setup. Figure 9.15 (c) shows the separation result when $N = 5$ and $M = 4$. Figure 9.15 (c) shows that MENUET can be applied to such 3-dimensional microphone array systems.

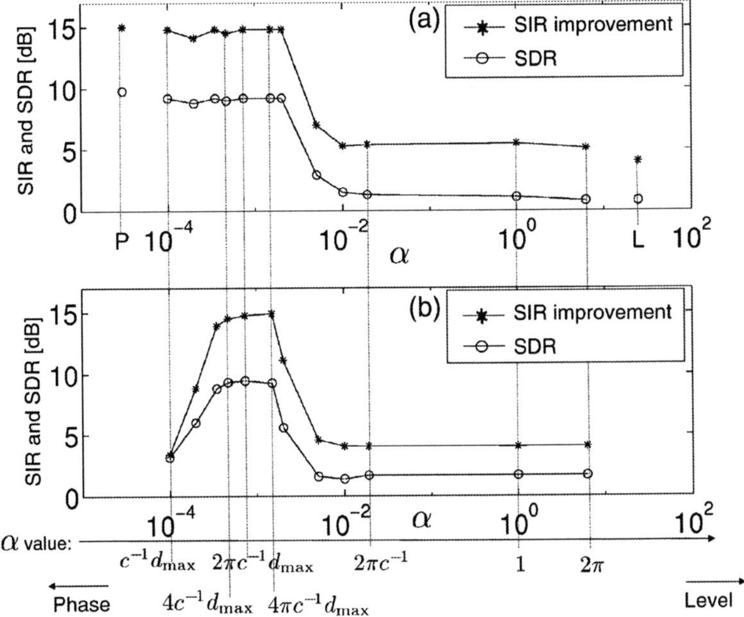

Fig. 9.16. Relationship between α and separation performance when $N = 4$, $M = 3$, $R = 50$ cm, and $\mathrm{RT}_{60} = 128$ ms. (a) With feature (9.19), (b) with modified feature (9.26). In (a), "P" and "L" denote the performance with $\boldsymbol{\Theta} = \boldsymbol{\Theta}^P$ and $\boldsymbol{\Theta} = \boldsymbol{\Theta}^L$, respectively.

9.5.3 Weight Parameter

Here we examine the relationship between the phase weight parameter α and the separation performance. As mentioned in Sect. 9.4.2, when α is large the level ratio is emphasized, and when α is small the phase difference is emphasized. Figure 9.16 shows the relationship when $N = 4$ and $M = 3$ (Fig. 9.13) with the feature (9.19) and the modified feature (9.26). Note that $\alpha = 2\pi$ corresponds to feature (9.7).

Figure 9.16 (a) shows the result with the feature (9.19). It achieved high performance when α was sufficiently small. This is because, the phase difference between the sensors was more reliable than the level ratio, owing to our microphone setup. As α became small, the performance saturated. On the other hand, the performance degraded as α became large. This is caused by the imbalance between the level ratio and phase difference terms, because the phase term becomes too small when α becomes large.

With modified feature (9.26), we obtained the best performance around $\alpha = 4c^{-1}d_{\max}$ (Fig. 9.16 (b)). This is because $\alpha = 4c^{-1}d_{\max}$ realizes the full use of the phase difference information (Appendix A), which is preferable for our sensor setting. As α became large, the performance degraded. Also when $\alpha < 4c^{-1}d_{\max}$ the performance worsened. It should be remembered that, with the modified feature, $\alpha = 4c^{-1}d_{\max}$ should be the lower limit (see Sect. 9.4.2). When $\alpha < 4c^{-1}d_{\max}$, the distance measure (9.10) for the clustering is not evaluated correctly (see Appendix A), and therefore, the clustering stage failed and the performance worsened.

We can also see from Fig. 9.16 that both features (9.19) and (9.26) achieved good performance over a wide α range. This means that we do not need the exact maximum sensor spacing $d_{\max}$. This allows us to utilize an arbitrarily arranged sensor array, although similar distances between pairs of sensors are preferable so that the k-means can use all sensor information optimally.

9.5.4 Discussion

In the experimental section, Fig. 9.15 also showed the performance tendency under reverberant conditions. The performance degrades as the reverberation time RT_{60} becomes long. Moreover, performance degradation was observed as the distance R became large. This is because, under long reverberation and/or large R conditions, it becomes difficult for the source sparseness (9.5) and anechoic assumptions (9.8) to hold as discussed in Sects. 9.3.3 and 9.3.4.

It is also important to mention nonlinear distortion in separated signals. There is nonlinear distortion (musical noise) in the outputs with our method, due to the winner-take-all property of the binary mask. The results of subjective tests with 10 listeners can be found in [32]. Some sound examples can be found at [33].

In this chapter, we described the sparse source separation procedure. In addition, using centroids estimated by clustering, we can also execute the source localization of sparse sources. This is because each centroid is the expectation value of the geometric information (9.9) of each source. As sparse source localization requires the sensor locations, it is not a blind process. The detailed procedures for sparse source localization are provided in [34]. As the method is based on MENUET, it allows us to employ an arbitrary sensor arrangement, and estimate the directions of sparse sources distributed 3-dimensionally.

9.6 Conclusion

This chapter described a novel sparse source separation method (MENUET). The employed features can be clustered easily by the k-means algorithm. Moreover, MENUET makes it easy to employ multiple sensors arranged in a nonlinear/nonuniform way.

We obtained promising experimental results in a room with weak reverberation even under underdetermined conditions. Although this chapter provided results solely for underdetermined cases, MENUET can also be applied to (over-)determined cases [32].

We also reported the performance under some reverberant conditions, where the sparseness and anechoic assumptions were deteriorating. The results showed that the direct and reverberant ratio is important for sparse signal processing. Sparse source separation in reverberant conditions is still an open problem.

A Weight Parameter for Modified Feature

In this section, we show the required condition for the phase weight parameter α for modified feature (9.25). Because the modified feature (9.25) is a complex vector, we have to consider the phase term when we perform clustering. When α in (9.25) is too large, the variance of the phase term becomes smaller than that of the level term. On the other hand, when α in (9.25) is too small, the phase changes too fast and causes a kind of aliasing problem. Moreover, it is important that the following condition holds for the distance measure (9.10) of the clustering: $|\Theta-\Theta'|$ increases monotonically as $|\arg[\Theta]-\arg[\Theta']|$ increases. However, if the phase term is larger than $\pi/2$, such a monotonic increase cannot hold. That is, the phase term should have the following relationship:

$$-\frac{\pi}{2} \leq \arg[\Theta] \leq \frac{\pi}{2}. \tag{9.27}$$

Recall (9.22) to model the mixing process (9.8), that is, the delay parameter τ_{jk} is determined by the distance l_{jk} from source k to sensor j:

$$\tau_{jk} = l_{jk}/c. \tag{9.28}$$

Substituting the mixing model (9.28) and (9.8), and the sparseness assumption (9.5) into (9.25) and (9.26) yields

$$\boldsymbol{\Theta}(f,t) \approx \frac{\lambda_{jk}}{D_k} \exp\left[-\jmath \frac{2\pi c^{-1}(l_{jk}-l_{Jk})}{\alpha_j}\right], \tag{9.29}$$

where $D_k = \sqrt{\sum_{j=1}^{M} \lambda_{jk}^2}$.

From the condition (9.27) and the model (9.29), the lower limit of α is given as

$$\begin{aligned} |\arg[\Theta]| &= \left|\frac{2\pi c^{-1}(l_{jk}-l_{Jk})}{\alpha_j}\right| \\ &\leq \left|\frac{2\pi c^{-1} d_{\max}}{\alpha_j}\right| \leq \frac{\pi}{2}, \qquad (9.30) \\ \alpha_j &\geq 4c^{-1} d_{\max}. \qquad (9.31) \end{aligned}$$

In (9.30), we used the fact that $\max_{j,k} |l_{jk} - l_{Jk}| \leq d_{\max}$.

From (9.31), we can conclude that the phase parameter $\alpha = 4c^{-1}d_{\max}$ should be the minimum value in order to maintain the relationship (9.27). In addition (9.27) has equality when $\alpha = 4c^{-1}d_{\max}$, which means that the phase difference information is most scattered. That is, the weight with $\alpha = 4c^{-1}d_{\max}$ allows us to make full use of the phase difference information. This is a preferable property for small sensor array systems (e.g., see Sect. 9.5), where phase differences between sensors are more reliable than level ratios for clustering.

B Initial Values for the *K*-means Clustering

The k-means algorithm is sensitive to the initial values of the centroids especially when the number of sources N is large and the reverberation time is long. Therefore we designed the initial centroids by using the far-field model, where the frequency response $h_{jk}(f)$ is given as

$$h_{jk}(f) \approx \exp\left[-\jmath 2\pi f c^{-1}\mathbf{d}_j^T\mathbf{q}_k\right],$$

and using the same normalization as each feature. Here, c is the propagation velocity of the signals, and the 3-dimensional vectors $\mathbf{d}_j$ and $\mathbf{q}_k$ represent the location of sensor j and the direction of source k, respectively (see Fig. 9.17 and [34]).

When designing the initial centroids, the sensor locations $\boldsymbol{d}_j$ ($j = 1, \ldots, M$) were on almost the same scale in each setup, and the initial directions $\mathbf{q}_k$ were set so that they were as scattered as possible. Concretely, in the experiments, we utilized the sensor vector $\mathbf{q}_k = [\cos\theta_k \cos\phi_k, \sin\theta_k \cos\phi_k, \sin\phi_k]^T$ (see Fig. 9.17). The azimuth of the k-th source was set at $\theta_k = \frac{2\pi}{N} \times k$ ($k = 1, \ldots, N$) for $M \geq 3$, and $\theta_k = \frac{\pi}{N} \times k$ ($k = 1, \ldots, N$) for $M = 2$. The elevation $\phi_k = 0$ for all sources k. Note that these initial values of $\boldsymbol{d}_j$ and $\mathbf{q}_k$ were not exactly the same in each setup.

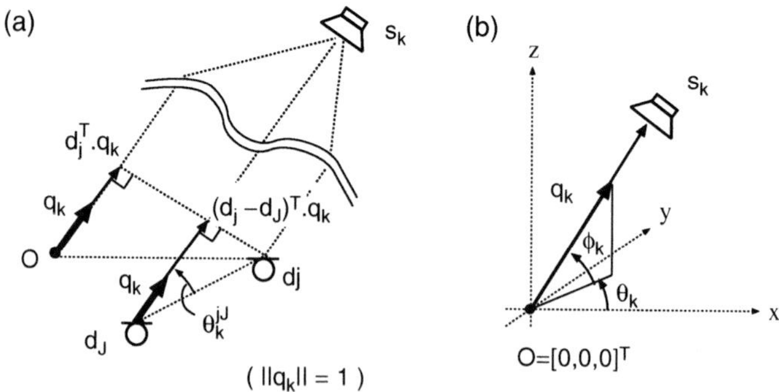

Fig. 9.17. (a) Far-field model, (b) definition of source direction.

References

1. S. Haykin, Ed., *Unsupervised Adaptive Filtering (Volume I: Blind Source Separation).* John Wiley & Sons, 2000.
2. A. Hyvärinen, J. Karhunen, and E. Oja, *Independent Component Analysis.* John Wiley & Sons, 2001.
3. Ö. Yılmaz and S. Rickard, "Blind separation of speech mixtures via time–frequency masking," *IEEE Trans. on SP*, vol. 52, no. 7, pp. 1830–1847, 2004.
4. H. Buchner, R. Aichner, and W. Kellermann, "Blind source separation for convolutive mixtures: A unified treatment," in *Audio Signal Processing for Next-Generation Multimedia Communication Systems*, Y. Huang and J. Benesty, Eds. Kluwer Academic Publishers, Feb. 2004, pp. 255–293.
5. H. Sawada, R. Mukai, S. Araki, and S. Makino, "Frequency-domain blind source separation," in *Speech Enhancement*, J. Benesty, S. Makino, and J. Chen, Eds. Springer, Mar. 2005, pp. 299–327.
6. S. Amari, S. Douglas, A. Cichocki, and H. Yang, "Multichannel blind deconvolution and equalization using the natural gradient," in *Proc. IEEE Workshop on Signal Processing Advances in Wireless Communications*, Apr. 1997, pp. 101–104.
7. P. Smaragdis, "Blind separation of convolved mixtures in the frequency domain," *Neurocomputing*, vol. 22, pp. 21–34, 1998.
8. L. Parra and C. Spence, "Convolutive blind separation of nonstationary sources," *IEEE Trans. Speech Audio Processing*, vol. 8, no. 3, pp. 320–327, May 2000.
9. J. Anemüller and B. Kollmeier, "Amplitude modulation decorrelation for convolutive blind source separation," in *Proc. ICA 2000*, June 2000, pp. 215–220.
10. S. Araki, R. Mukai, S. Makino, T. Nishikawa, and H. Saruwatari, "The fundamental limitation of frequency domain blind source separation for convolutive mixtures of speech," *IEEE Trans. Speech Audio Processing*, vol. 11, no. 2, pp. 109–116, 2003.
11. F. Theis, E. Lang, and C. Puntonet, "A geometric algorithm for overcomplete linear ICA," *Neurocomputing*, vol. 56, pp. 381–398, 2004.
12. P. Bofill and M. Zibulevsky, "Blind separation of more sources than mixtures using sparsity of their short-time Fourier transform," in *Proc. ICA2000*, 2000, pp. 87–92.
13. L. Vielva, D. Erdogmus, C. Pantaleon, I. Santamaria, J. Pereda, and J. C. Principe, "Underdetermined blind source separation in a time-varying environment," in *Proc. ICASSP2002*, 2002, pp. 3049–3052.
14. P. Bofill, "Underdetermined blind separation of delayed sound sources in the frequency domain," *Neurocomputing*, vol. 55, pp. 627–641, 2003.
15. A. Blin, S. Araki, and S. Makino, "Underdetermined blind separation of convolutive mixtures of speech using time–frequency mask and mixing matrix estimation," *IEICE Trans. Fundamentals*, vol. E88-A, no. 7, pp. 1693–1700, 2005.
16. S. Winter, W. Kellermann, H. Sawada, and S. Makino, "MAP-based underdetermined blind source separation of convolutive mixtures by hierarchical clustering and $l1$-norm minimization," *EURASIP Journal on Advances in Signal Processing*, Article ID 24717, 2007.
17. J. M. Peterson and S. Kadambe, "A probabilistic approach for blind source separation of underdetermined convolutive mixtures," in *Proc. ICASSP 2003*, vol. VI, 2003, pp. 581–584.

18. A. Jourjine, S. Rickard, and Ö. Yılmaz, "Blind separation of disjoint orthogonal signals: Demixing N sources from 2 mixtures," in *Proc. ICASSP2000*, vol. 12, 2000, pp. 2985–2988.
19. M. Aoki, M. Okamoto, S. Aoki, H. Matsui, T. Sakurai, and Y. Kaneda, "Sound source segregation based on estimating incident angle of each frequency component of input signals acquired by multiple microphones," *Acoustical Science and Technology*, vol. 22, no. 2, pp. 149–157, 2001.
20. N. Roman, D. Wang, and G. J. Brown, "Speech segregation based on sound localization," *Journal of Acoustical Society of America*, vol. 114, no. 4, pp. 2236–2252, Oct. 2003.
21. S. Rickard, R. Balan, and J. Rosca, "Real-time time–frequency based blind source separation," in *Proc. ICA2001*, Dec. 2001, pp. 651–656.
22. R. O. Duda, P. E. Hart, and D. G. Stork, *Pattern Classification*, 2nd ed. Wiley Interscience, 2000.
23. R. Balan, J. Rosca, and S. Rickard, "Non-square blind source separation under coherent noise by beamforming and time–frequency masking," in *Proc. ICA2003*, Apr. 2003, pp. 313–318.
24. T. Melia, S. Rickard, and C. Fearon, "Histogram-based blind source separation of more sources than sensors using a DUET-ESPRIT technique," in *Proc. EUSIPCO2005*, Sept. 2005.
25. S. Araki, S. Makino, H. Sawada, and R. Mukai, "Reducing musical noise by a fine-shift overlap-add method applied to source separation using a time–frequency mask," in *Proc. ICASSP2005*, vol. III, Mar. 2005, pp. 81–84.
26. J. Karvanen and A. Cichocki, "Measuring sparseness of noisy signals," in *Proc. ICA2003*, Apr. 2003, pp. 125–130.
27. S. Rickard, "Sparse sources are separated sources," in *Proc. EUSIPCO2006*, Sept. 2006.
28. S. Rickard and Ö. Yılmaz, "On the approximate W-disjoint orthogonality of speech," in *Proc. ICASSP2002*, vol. I, May 2002, pp. 529–532.
29. Ö. Yılmaz and S. Rickard, "Blind separation of speech mixtures via time–frequency masking," *IEEE Trans. Signal Processing*, vol. 52, no. 7, pp. 1830–1847, July 2004.
30. S. Araki, H. Sawada, R. Mukai, and S. Makino, "Underdetermined blind sparse source separation for arbitrarily arranged multiple sensors," *Signal Processing*, doi:10.1016/j.sigpro.2007.02.003, 2007.
31. S. Araki, S. Makino, A. Blin, R. Mukai, and H. Sawada, "Underdetermined blind separation for speech in real environments with sparseness and ICA," in *Proc. ICASSP 2004*, vol. III, May 2004, pp. 881–884.
32. ——, "A novel blind source separation method with observation vector clustering," in *Proc. 2005 International Workshop on Acoustic Echo and Noise Control (IWAENC 2005)*, Sept. 2005, pp. 117–120.
33. "http://www.kecl.ntt.co.jp/icl/signal/araki/xcluster_fine.html."
34. S. Araki, H. Sawada, R. Mukai, and S. Makino, "DOA estimation for multiple sparse sources with normalized observation vector clustering," in *Proc. ICASSP2006*, vol. 5, May 2006, pp. 33–36.

10 Underdetermined Blind Source Separation of Convolutive Mixtures by Hierarchical Clustering and L_1-Norm Minimization

Stefan Winter[1]*, Walter Kellermann[2], Hiroshi Sawada[1], and Shoji Makino[1]

[1] NTT Communication Science Laboratories, NTT Corporation
2-4 Hikaridai, Seika-cho, Soraku-gun, Kyoto 619-0237, Japan
E-mail: {wifan,sawada,maki}winter@lnt.de

[2] Chair of Multimedia Communication and Signal Processing,
University Erlangen-Nuremberg, Cauerstr. 7, 91058 Erlangen, Germany
E-mail: wk@lnt.de

Abstract. In this chapter we present a complete solution for underdetermined blind source separation (BSS) of convolutive speech mixtures based on two stages. In the first stage, the mixing system is estimated, for which we employ hierarchical clustering. Based on the estimated mixing system, the source signals are estimated in the second stage. The solution for the second stage utilizes the common assumption of independent and identically distributed sources. Modeling the sources by a Laplacian distribution leads to ℓ_1-norm minimization.

The ℓ_1-norm minimization has to deal with complex numbers due to the use of the time–frequency-domain approach. We compare a combinatorial approach initially only designed for real numbers with a second order cone programming (SOCP) approach suitable to deal with complex numbers. Although the former approach is not theoretically justified for complex numbers, its results are comparable to or even better than the SOCP solution. The advantage is a lower computational cost for problems with low numbers of sources and sensors.

The proposed approach is supported by experimental results and compared to a time–frequency masking approach.

10.1 Introduction

While the area of underdetermined blind source separation (BSS), i.e. BSS with more sources P than sensors Q ($P > Q$), has obtained more and more attention [1–23], it still remains a very challenging task. In contrast to well-determined BSS, i.e. BSS with the same or smaller number P of sources than sensors Q ($P \leq Q$), the solution to underdetermined BSS requires more than (inverse) system identification. In other words, even if the mixing system in terms of the mixing matrix is fully known, additional measures are necessary to actually separate the mixtures.

*Stefan Winter is on leave from the Department of Multimedia Communication and Signal Processing, University Erlangen-Nuremberg.

S. Makino et al. (eds.), Blind Speech Separation, 271–304.

Classical approaches for well-determined BSS based on independent component analysis (ICA) solely rely on the assumption of mutual statistical independence of the source signals and on rather general source models. However, the additional effort to separate the mixtures in underdetermined BSS requires more refined source models and further assumptions such as sparseness, which means that only a few instances of the original signals have a value significantly different from zero, or a finite alphabet [23].

Most existing approaches to underdetermined BSS were proposed for instantaneous mixtures as is shown in the survey of source separation methods in [24] and further examples such as [8, 9, 11, 25, 26]. Only a few recent approaches consider convolutive mixtures [6, 7, 27–29]. Some consider underdetermined BSS of convolutive mixtures but have certain limitations such as recovering only a limited number of sources [30] (though exactly), accounting only for anechoic recordings [5, 31, 32], performing only experiments with equal numbers of sources and sensors [33], or considering only separation but not estimation of the parameters for the mixing/unmixing system [13].

The reason that convolutive mixtures have not yet received as much attention in the context of underdetermined BSS as they did in the context of well-determined BSS is due to the fact that underdetermined BSS is not yet satisfactorily solved for the instantaneous mixing of independent and identically distributed (iid) sources and far from solving practical problems such as convolutive mixtures of nonwhite signals (e.g. speech).

The goal of this chapter is to present a complete underdetermined BSS system for convolutive mixtures. It is based on algorithms originally designed for instantaneous mixtures, which are adapted to convolutive mixtures by bin-wise BSS in the DFT domain. We adhere to a two-stage approach of blind system identification (BSI) followed by blind source recovery (BSR) in connection with an iid Laplacian source model [34, 35]. Figure 10.1 illustrates the two-stage structure of the unmixing system for $P = 3$ source signals and $Q = 2$ sensors.

Both stages are performed in the time–frequency-domain to increase sparseness and reduce the convolutive mixtures to instantaneous mixtures.

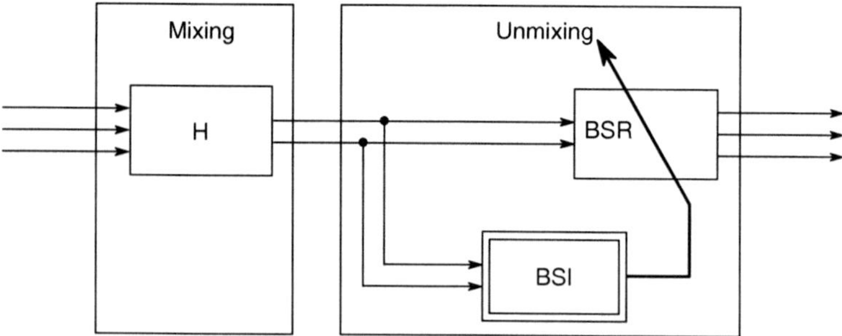

Fig. 10.1. Underdetermined BSS with two-stage unmixing approach.

For simplification we assume narrowband signals when deriving the solution. As a consequence, the DFT bins can be treated independently. However, we verify by experiments that it can be applied to wide-band signals such as speech signals in practical situations.

In BSI as the first stage we have to estimate the mixing matrix. We propose the use of hierarchical clustering for this task. It can be easily employed for complex-valued mixture samples that occur if BSS is performed in the time–frequency-domain. Furthermore, it does not limit the usable number of sensors as histogram-based methods do [1, 2, 6, 7], which utilize features such as direction-of-arrival (DOA) or exploit the amplitude relation between two sensor outputs. In addition, hierarchical clustering prevents convergence problems occuring with self-organizing maps (SOM), which are similar to algorithms such as GeoICA [9] or AICA [8].

In BSR as the second stage we separate the mixtures using the estimated mixing matrix from the first stage. We assume statistical independence for the sources [36]. Utilizing an iid Laplacian source model leads to a constrained ℓ_1-norm minimization. Since we have to deal with complex numbers for each DFT bin, we investigate the difference between real and complex-valued ℓ_1-norm minimization and its implication for underdetermined BSS of convolutive mixtures.

10.2 Underdetermined BSS for Acoustic Sources in Reverberant Environments

After providing a more detailed explanation of the problem of underdetermined BSS, this section offers an overview on existing strategies for solving it. Emphasis is placed on the two-stage approach of BSI followed by BSR.

10.2.1 Definitions

In our treatment of underdetermined BSS we assume a linear mixture model with negligible noise. In the following we introduce the underlying model and objective of underdetermined BSS.

Discrete-Time-Domain. With t denoting discrete time, $s_p(t) \in \mathbb{R}$ denoting the p-th real-valued unobservable source signal $(1 \leq p \leq P)$, and $h_{qp}(t) \in \mathbb{R}$ $(0 \leq t \leq L-1)$ denoting the L-tap impulse response from source p to sensor q $(1 \leq q \leq Q,\ Q < P)$, we observe mixtures $x_q(t) \in \mathbb{R}$ given by

$$x_q(t) = \sum_{p=1}^{P} \sum_{\kappa=0}^{L-1} s_p(t-\kappa) h_{qp}(\kappa). \tag{10.1}$$

DFT-Domain. Instead of solving the problem in the time-domain, we switch to the DFT-domain by applying a short-time Fourier transform (STFT) to the mixtures $x_q(t)$ assuming narrowband signals. While a wide-band approach would be desirable, extending the proposed solution to wide-band signals is not as straightforward as described for example in [37]. This is because the problem of underdetermined BSS has a different structure from traditional adaptive filtering problems as we show in the sequel.

Thus time-domain signals

$$\boldsymbol{s}(t) = [s_1(t), \ldots, s_P(t)]^T \tag{10.2}$$

and

$$\boldsymbol{x}(t) = [x_1(t), \ldots, x_Q(t)]^T \tag{10.3}$$

are converted into frequency-domain time-series

$$\boldsymbol{S}(f, \tau) = [S_1(f, \tau), \ldots, S_P(f, \tau)]^T \in \mathbb{C}^P \tag{10.4}$$

and

$$\boldsymbol{X}(f, \tau) = [X_1(f, \tau), \ldots, X_Q(f, \tau)]^T \in \mathbb{C}^Q \tag{10.5}$$

by an D-point STFT, respectively. Thereby $f = 0, f_s/D, \ldots, f_s(D–1)/D$ (f_s: sampling frequency; τ: time dependence). $\mathbb{C}$ denotes the complex numbers. Let us define $\boldsymbol{H}(f) \in \mathbb{C}^{Q \times P}$ as a matrix whose elements are the transformed impulse responses. We call the column vectors $\boldsymbol{h}_i(f)$ ($i = 1, \ldots, P$) mixing vectors and approximate the mixing process by

$$\boldsymbol{X}(f, \tau) = \boldsymbol{H}(f)\boldsymbol{S}(f, \tau). \tag{10.6}$$

This reduces the BSS problem from convolutive to instantaneous mixtures in each frequency bin f. For simplicity we will omit the arguments for the dependency on frequency f and time-frame τ in the following treatment.

Switching to the time–frequency-domain offers the additional advantage of facilitating the exploitation of the time–frequency sparseness of speech signals [4]. In general, sparseness of a signal means that only a few instances have a value significantly different from zero.

The sparseness of speech signals in the time–frequency-domain results from the fact that natural speech signals always exhibit pauses along the time and even during voice activity, some spectral regions exhibit only low intensity following the harmonic structure of speech spectra with energy concentrated in formants and at multiples of the pitch frequency. This means that in the time–frequency-domain, only a few frequency bins have significant energy at each time instance τ, while most frequency bins have values close to zero.

Note that sparseness itself is not yet sufficient to be helpful in underdetermined BSS. For example, if two speech signals are sparse but have their energy concentrated at identical time–frequency instances, the two signals would still overlap. Therefore the more precise concept of disjoint sparseness has been introduced, which in addition to sparseness requires that the involved signals do not have energy concentrated at identical time–frequency instances [38]. Speech signals fit the concept of disjoint sparseness since the pitch frequency depends on the speaker and also changes over time.

Using a sparse signal representation is very important as regards ensuring good separation performance since the separation is built on the assumption of sparse source signals. Intuitively, the more sparse the representation for each source, the less overlap between sources must be expected. Therefore, a signal representation maximizing sparseness with minimum loss of signal content is crucial.

The disadvantage of narrowband BSS in the time–frequency-domain is the internal permutation indeterminacy. It describes the problem that the order of the output signals is arbitrary and cannot be determined by the separation process, which results in incorrect frequency bin alignment.

In our framework we use a clustering-based method to mitigate the permutation indeterminacy [39, 40]. We also apply a modified version of the minimal distortion principle (MDP) [41] to alleviate the scaling problem. A more detailed structure of the unmixing system is illustrated in Fig. 10.2.

Objective in Underdetermined BSS. The ultimate objective in underdetermined BSS is the estimation of signals $y_u(t) \in \mathbb{R}$ or $Y_u(f,\tau) \in \mathbb{C}$, $(1 \leq u \leq P)$ that resemble the original signals $s_p(t)$ or $S_p(f,\tau)$, respectively, as closely as possible up to arbitrary permutation and scaling, if only the mixtures $x_q(t)$ are available. In contrast to well-determined BSS, here the knowledge or estimation of the impulse responses $h_{qp}(t)$ or their frequency-domain representation $\boldsymbol{H}$ is not sufficient. Even if the impulse responses are available, estimating the source signals from the observed signals poses a problem on its own as described in the following.

With well-determined BSS the mixing matrix $\boldsymbol{H}$ is quadratic ($Q = P$). Therefore, it can be inverted assuming that it is nonsingular, and (10.6) can be solved for $\boldsymbol{S}$. On the other hand, the mixing matrix $\boldsymbol{H}$ is not quadratic with underdetermined BSS ($Q < P$) but has more columns than rows, i.e. the left-side inverse does not exist. As a consequence, (10.6) cannot be solved

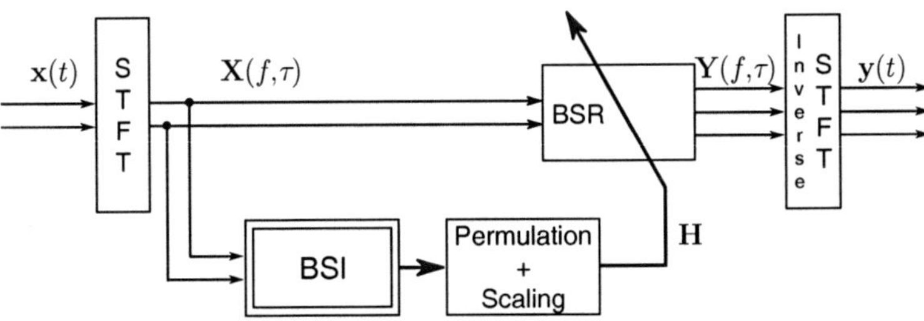

Fig. 10.2. Block diagram of unmixing system.

Table 10.1. Overview on existing approaches (continued in Table 10.2). Basic principles: [1]Sparseness; [2]Higher-order cumulants; [3]Bayesian inference; [4]Characteristic function.

#	Approach	Mixing model: Inst.	Mixing model: Anechoic	Mixing model: Echoic	Representation: Time	Representation: Frequency	Source model
				Multi-channel			
1	[29][1]			x		STFT	Sparseness
2	FOBIUM [10][2]	x				Complex-valued narrow-band	Sparseness
3	BIRTH [12][2]	x				Complex-valued narrow-band	Non-zero marginal 6th order cumulants + (cyclo-)stationary
4	[13][1]			x		x	Laplacian PDF
5	[15][3]	x	(post-nonlinear)		x		Generalized Gaussian
6	[16][1]	x				Overcomplete signal dictionary	Sparseness
7	[17][3]	x				MDCT	Sparseness
8	[18][4]	x			x		Non-Gaussian
9	[1][1]		x			STFT	Sparseness
10	[2][1]		x			STFT	Sparseness
11	[3][1]	x				Sparse domain	Sparseness
12	[4][1]	x				STFT	Sparseness
13	AICA [8][1]	x			x		Sparseness
14	[9][1]	x			x		Sparseness
15	[23][2]	x				Complex-valued narrow-band	Statistical independence
16	[33][3]			x	x		Autoregressive
17	[28][1]			x		STFT	Sparseness
18	[32][1]		x			Hilbert spectrum	Sparseness
19	TIFROM [25][1]	x				STFT	Partial sparseness
20	DUET [42][1]		x			STFT	Sparseness
21	[43][1]	x				STFT	Sparseness
22	[44][1]	x				Wigner-Ville	Sparseness
23	[38][1]	x				Wigner-Ville	Sparseness
24	[6][1]			x		STFT	Sparseness
25	[45][3]	x				Wavelets	Sparseness
26	Basis pursuit [46][1]			x	x		Sparseness
27	[31][1]		x			STFT	Sparseness
28	[26][1]	x				MDCT	Sparseness
29	[47][2]	x				Complex-valued	Finite set discrete-time signals
30	[48][4]	x			x		Non-Gaussian
31	[49][2]	x				Narrowband	Non-Gaussian
32	[5][1]		x			STFT	Sparseness
33	[50][3]	x			x		Gaussian Mixture Model
34	[27][1]			x			Sparseness
35	[51][3]	x				STFT	Gaussian scaled mixture model
36	[52][3]	x				DCT	Mixture of Gaussians

Table 10.2. Overview on existing approaches (continued from Table 10.1).

#	Unmixing: System identification	Unmixing: Separation	Comments
	Multi-channel		
1	ICA (separating into subspaces) + binary masking		
2	Forth order statistics	N/A	Experiments only with cyclostationary signals
3	Sixth-order statistics	N/A	
4	N/A	LP + SOCP	
5	Iterative MAP		
6	N/A	Modified Matching Pursuit	
7	MAP by EM		
8	Second characteristic function, PARAFAC based	N/A	Experiments with BPSK sources
9	Histogram	Time–frequency masking	
10	Histogram	Time–frequency masking	
11	Spectral estimation applied to histogram (ESPRIT)	N/A	
12	Clustering by triangular potential function	LP/shortest path	
13	Winner-takes-all clustering	LP	
14	Winner-takes-all clustering	Shortest-path Shortest-path	
15	Decomposition of higher-order cumulant tensors	N/A	Limited to 3x2, no experiments
16	Bayesian model, MCMC		only well-determined experiments
17	k-means clustering	MAP-based linear operation	
18	Time- and intensity-difference histogram	Binary masking	Only anechoic experiments
19	Ratios of mixtures	Sequential source elimination	Only partial separation
20	Clustering	Binary masking	
21	ICA + Time–frequency masking		Only music
22	N/A	Time–frequency masking	
23	Spatial clustering	Time–frequency masking	
24	Spatial clustering Spatial clustering	Time–frequency masking + ICA	
25	Joint optimization of MAP-based cost function		
26	N/A	Convex optimization of ℓ_1-norm	
27	Clustering	Constrained ℓ_p-norm minimization	
28	Fitting Laplacian mixture model to DOA histogram by EM	Time–frequency masking	
29	Higher-order cumulants	Virtual sensor	
30	Second characteristic function	N/A	Experiments with 4 binary sources
31	Decomposition of forth-order cumulant	N/A	No experiments
32	tensor Potential function based clustering	SOCP	
33	MAP-based iterative optimization		
34	Clustering	Time–frequency masking	
	Single-channel		
35	Bayesian inference		Prior training necessary
36	Bayesian inference		Prior training necessary

for $\boldsymbol{S}$. Instead, $\boldsymbol{S}$ can only be approximated based on the knowledge of the mixtures $\boldsymbol{X}$ and the mixing matrix $\boldsymbol{H}$. The exact result depends on the assumptions made for the approximation.

10.2.2 Two-Stage Approach to Underdetermined BSS

In this section we provide an overview on existing approaches to underdetermined BSS. They are usually based on few basic principles such as sparseness, higher-order cumulants, the characteristic function, and Bayesian inference. For the sake of completeness computational auditory scene analysis (CASA) [53, 54] should also be mentioned, although it is usually not considered as a BSS technique. For this reason it will not be further considered here. However, it has similarities with sparseness-based approaches and is also able to separate sound source mixtures without any knowledge of the mixing system. Besides the above mentioned principles, ICA [55] is sometimes also utilized in the context of underdetermined BSS [6, 29, 43]. However, so far ICA has only been used after the problem had been reduced to well-determined BSS by, for example, a sparseness based approach. Therefore, ICA will also not be further considered here.

Table 10.1 together with 10.2 provide a detailed list of various approaches and their classification according to different criteria.

Although combined approaches are possible, it is often convenient to distinguish between the two stages of system identification and source separation. They are denoted by blind system identification (BSI) [12, 47–49] and blind source recovery (BSR) [8, 56, 57], respectively. BSI is also referred to as blind mixing model recovery (BMMR) [8, 56]. Often, only some important parameters of the mixing system need to be identified, rather than impulse responses. For example, some (time–frequency-) masking-based approaches require only the knowledge of the DOA. In such cases BSI refers only to the required parameters of the mixing system.

Distinguishing between the two stages of BSI and BSR leads to the common two-stage structure of the unmixing system in Fig. 10.1 where the mixing matrix $\boldsymbol{H}$ estimated by BSI is used for BSR.

Blind System Identification. There are four principles that BSI in underdetermined BSS usually is based on. These are sparseness, higher-order cumulants, the characteristic function, or Bayesian inference.

Most approaches to BSI are based on the assumption of disjoint sparse sources [6, 31, 38, 45, 46, 58–62]. It is then assumed that the sparse signal representations exhibit little or no overlap in the time–frequency plane, so that the mixtures considered in small areas of the time–frequency plane are actually not underdetermined. Different properties such as phase difference or amplitude ratio are used to assign each instance of the mixtures to a limited number of sources. The result can be used in different ways: Either the sources can be estimated directly, or the result is utilized to estimate the mixing matrix or related information such as the DOA, which in turn is then used for estimating the source signals.

For obtaining sparse signal representations the STFT is the most common transformation. However, other transforms such as the discrete cosine transform or its modified version [17, 26, 63], the Wigner–Ville transform [44] or the Hilbert transform [32] are used, too. In addition, special overcomplete signal dictionaries are also used [46, 64]. While the MDCT offers the advantage of real-valued coefficients, the Wigner-Ville transform and the Hilbert transform can result in better resolution than the STFT. The advantage of the STFT is that convolutive mixtures become instantaneous ones.

Approaches that rely on higher-order cumulants exploit the non-Gaussianity of the source signals [65]. They often utilize tensors for concise derivations of the unmixing system. Their disadvantage is that they cannot be generalized to cumulants of arbitrary orders but have to be designed specifically for each order that is accounted for. In connection with discrete sources higher-order cumulants have been used to introduce the idea of virtual sensors [47, 66, 67]. By nonlinear functions additional virtual sensor signals are generated that eventually lead to a well-determined system. Closely related to approaches based on higher-order cumulants are approaches that are based on the characteristic function. Their advantage is a greater flexibility [18, 48].

Bayesian inference allows the most general approaches [68, 69]. Since probability density functions (PDFs) could theoretically be described by the characteristic function or all possible cumulants, both approaches are automatically accounted for if appropriate prior PDFs are chosen within the Bayesian framework. Sparseness can also be modeled by appropriate PDFs such as the Laplacian probability density function (PDF) where values close to zero have a high probability density and values away from zero have a low probability density. Therefore, sparseness can be treated within the framework of Bayesian inference. However, the sparseness assumption is not always necessary with approaches based on Bayesian inference as the example in [50] shows. The disadvantage of Bayesian inference-based approaches is that they can easily lead to analytically nontractable equations requiring computationally expensive numerical methods.

Blind Source Recovery. BSR builds on the results of the BSI stage. As with BSI, approaches for BSR can be based on different principles, most commonly, sparseness and Bayesian inference.

By assuming sparse sources each instant in the time–frequency plane can be assigned to few sources or ideally to only one source. Therefore, the problem is in some sense reduced to well-determined BSS. However, the separating filters are highly time-variant and therefore difficult to estimate. The assignment of each instant to the respective sources can be supported by the previously identified mixing system or its important parameters, respectively

[70]. Typically, masking approaches are built on this principle. However, in contrast to BSI, sparseness is not necessarily the most prevelant principle in BSR.

The most general approaches for estimating the unknown source signals are based on Bayesian inference. As a special case of Bayesian interference maximum a-posteriori (MAP) approaches are very common. They yield the posterior distribution of the desired parameters (here: source signals) by accounting for their prior distribution and the likelihood of the observed data (here: mixed signals). The generality of Bayesian inference stems from the fact that the prior distribution can contain all available knowledge about the desired parameters. As such, even sparseness is often encoded by appropriate PDFs. As a result, Bayesian inference often also leads to sparseness-based approaches [61]. In fact, several approaches are based on ideas derived from the general notion of Bayesian inference.

10.3 Blind System Identification

Several of previously proposed algorithms for the first stage BSI are based on PDFs represented by histograms and developed for only $Q = 2$ sensors [1–3]. Some could, in principle, be extended for larger numbers of sensors $Q > 2$. But the so-called curse of dimensionality [71] sets practical limits on the number of usable sensors with histograms. This problem becomes even worse with complex numbers, which double the histogram dimensions due to their real and imaginary part or amplitude and phase, respectively. Complex numbers are necessary if BSS is performed in the DFT-domain. Some approaches extract features such as the DOA or exploit the amplitude relation between two sensor outputs [1, 2, 6, 7]. In both cases only two sensors can contribute, no matter how many sensors are available.

Other algorithms such as GeoICA [9] or AICA [8] resemble SOM and could more easily be applied to convolutive mixtures. However, their convergence depends heavily on initial values [71]. Usually, countermeasures such as starting the algorithm several times with different initial values are computationally expensive.

Here we propose the use of hierarchical clustering for the estimation of the mixing matrix. This method operates directly on the complex-valued samples. While it does not limit the number of usable sensors, it also prevents the convergence problems which can occur with SOM-based algorithms.

10.3.1 Hierarchical Clustering

Hierarchical clustering is an unsupervised clustering technique that does not depend on initial conditions. It solely relies on a distance measure that corresponds to the dissimilarities between disjoint observation groups [71].

Among the most important advantages of the described hierarchical clustering algorithm is the fact that it operates directly on the sample data in any vector space of arbitrary dimensions. The only requirement is the definition of a distance measure for the considered vector space. Therefore, it can easily be applied to complex-valued data that occurs in time–frequency-domain convolutive BSS.

With hierarchical clustering, two strategies can be distinguished, namely the agglomerative (bottom-up) and divisive (top-down) strategy. The difference lies in the starting point of the clustering procedure. With the agglomerative strategy, the starting point are the single observation samples, considering them as clusters which contain only one object. From there clusters are combined, so that the number of clusters decreases while the average number of observations per cluster increases. With the divisive strategy, the starting point is a single cluster that contains all observations. From there, clusters are split to form new clusters. We utilize the agglomerative strategy that has been studied much more extensively than the divisive strategy.

The combination of clusters into new clusters with the agglomerative strategy is an iterative process and based on the distance between the current clusters. Starting from the 'single observation' clusters, the distance between each pair of clusters is calculated, resulting in a distance matrix. At each level of the iteration the two clusters with the least distance are combined and form a new cluster (Fig. 10.3). This process is called linking and repeated until the amount of clusters has decreased to a predetermined number c, $P \leq c \leq F$, where F denotes the total number of samples.

In the following we assume that the mixtures $\boldsymbol{X}$ are normalized with respect to phase and amplitude as follows:

$$\boldsymbol{X} \leftarrow \frac{\boldsymbol{X}}{|\boldsymbol{X}|_2} e^{-j\varphi_{X_1}} \tag{10.7}$$

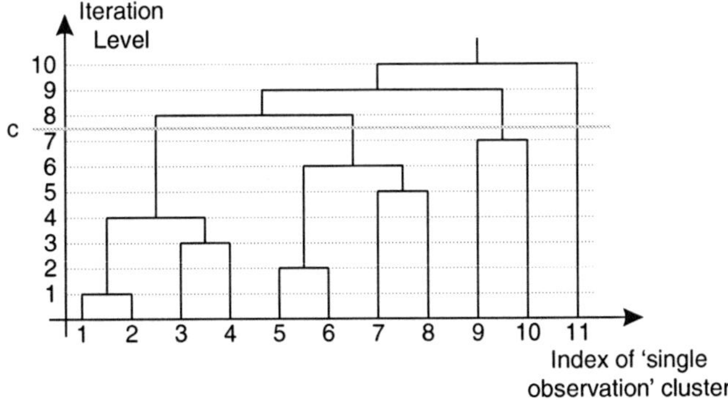

Fig. 10.3. Linking closest clusters.

where φ_{X_1} denotes the phase of the first vector component of $\boldsymbol{X}$ and $|\cdot|_p$ denotes the ℓ_p-norm defined by

$$|\boldsymbol{X}|_p = \left(\sum_i |X_i|^p\right)^{\frac{1}{p}} . \tag{10.8}$$

$|X_i|$ denotes the absolute value of X_i. For measuring the distance between clusters, we have to distinguish between two different problems. First, we need a distance measure $d(\boldsymbol{X}_{\tau_1}, \boldsymbol{X}_{\tau_2})$ that is applicable to Q-dimensional complex vector spaces. Among several possibilities, we use here the Euclidean distance based on the normalized samples, which is defined by

$$d(\boldsymbol{X}_{\tau_1}, \boldsymbol{X}_{\tau_2}) = |\boldsymbol{X}_{\tau_1} - \boldsymbol{X}_{\tau_2}|_2 . \tag{10.9}$$

Second, when a new cluster is formed, we need to enhance this distance measure to relate the new cluster to the other clusters. The method we employ here is called the nearest-neighbor technique. Let C_1 and C_2 denote two clusters as illustrated in Fig. 10.4. Then the distance $d(C_1, C_2)$ between these clusters is defined as the minimum distance between any pair of its samples by

$$d(C_1, C_2) = \min_{\boldsymbol{X}_{\tau_1} \in C_1,\ \boldsymbol{X}_{\tau_2} \in C_2} d(\boldsymbol{X}_{\tau_1}, \boldsymbol{X}_{\tau_2}) . \tag{10.10}$$

This approach requires distance calculations, but with a well-designed implementation as used here, the computational complexity can get as low as $O(n^2)$ [72], where n denotes the number of observations.

As mentioned earlier, most of the samples will cluster around the true mixing vectors $\boldsymbol{h}_i$, depending on the degree of disjoint sparseness of the original signals. Special attention must be paid to the remaining samples (outliers), which are randomly scattered in the space between the mixing vectors due to nonideal disjoint sparseness of the sources, reverberation, and noise if applicable. Usually they are far away from other samples and will be combined with other clusters only at higher levels of the clustering process (i.e. when only few clusters are left). This suggests to set the final number of clusters at a high number

$$c \gg P . \tag{10.11}$$

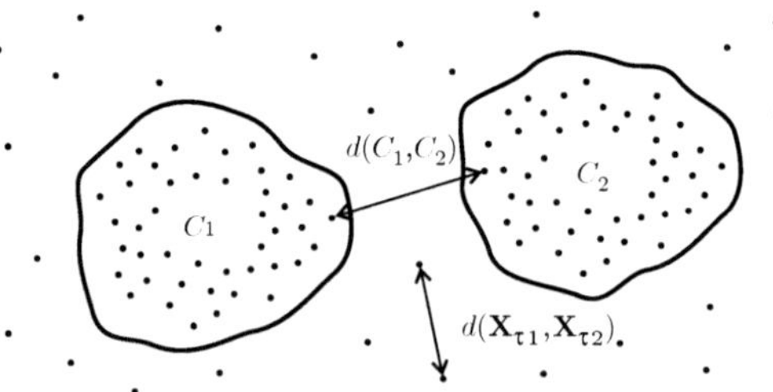

Fig. 10.4. Illustration of nearest neighbor technique.

By doing so, we prevent these outliers from being linked with the clusters around the mixing vectors $\boldsymbol{h}_i$, which would usually lead to distortions of the estimated mixing vector. This results in more robustness and has a similar effect as garbage models. More important, however, is the fact that we avoid combining desired clusters. Since the outliers are often far away from other clusters it might happen that desired clusters are closer to each other than to outliers. Experiments showed that the exact value of c does not matter as long as it was above 60 for $P \in \{3, 4, 5\}$ for the conditions given in Sect. 10.5.1. However, this value depends on the acoustic environment, ambient noise level, and the distance of the sources to the sensors.

Assuming that the clusters around the mixing vectors $\boldsymbol{h}_i$ include the largest numbers of samples we finally choose the P clusters with the largest numbers of samples as those representing the P. Thereby the number of sources P must be known. To obtain the mixing vectors, we average overall samples of each cluster

$$\mathbf{h}_i = \frac{1}{|C_i|} \sum_{\boldsymbol{X} \in C_i} \boldsymbol{X}, \qquad 1 \leq i \leq P, \tag{10.12}$$

where $|C_i|$ denotes the cardinality of cluster C_i. Thereby we assume that the contribution of other sources have zero mean.

An example for the resulting clusters is shown in Fig. 10.5. Here we chose $c = 100$ as we also did with the experiments in Sect. 10.5. An example where desired clusters were unintentionally combined because c was chosen too small is shown in Fig. 10.6. Further experimental details are given in Sect. 10.5.1.

No initial values for the mixing vectors $\boldsymbol{h}_i$ are required. This means, in particular, that if the assumption of clusters with many samples around the mixing vectors is true, then the algorithm converges to those clusters.

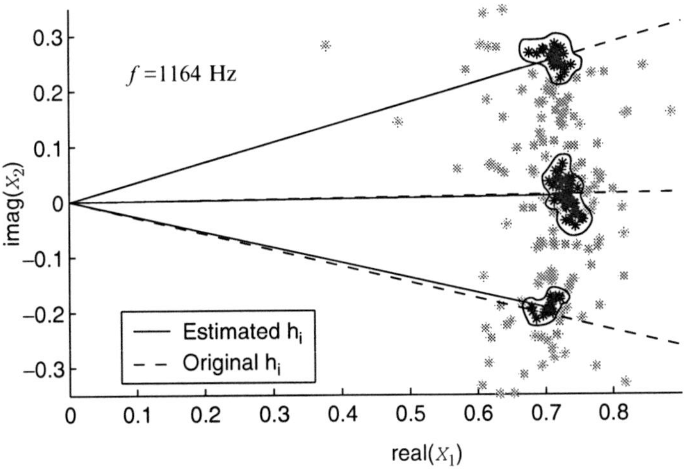

Fig. 10.5. Estimation of mixing vectors.

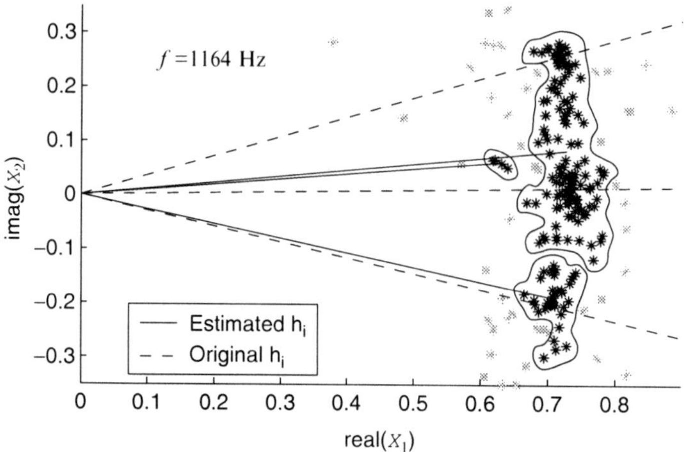

Fig. 10.6. Example for unintentionally combining desired clusters.

Besides choosing a distance measure, there is only the single parameter c that determines the number of clusters necessary to configure the clustering algorithm. Experiments have shown that the choice for this parameter is quite insensitive as long as it is above a certain limit that would combine desired clusters. Its choice is, in general, related to the sparseness of the sources and to the acoustic environment. The sparser the signals are or the less reverberant the acoustic environment is, respectively, the smaller the value of c can be chosen, because the number of outliers that must be avoided will be smaller.

While the considered signals must have some degree of sparseness in the time–frequency plane, to assure the assumption that the influence of the respective other signals can be disregarded for each cluster, they do not have to be statistically independent at this point to obtain useful clusters.

10.3.2 Permutation Indeterminacy

The permutation indeterminacy describes the problem that the order of the output signals is arbitrary and cannot be determined by the separation process. While this problem is often a minor problem when it comes to instantaneous mixtures, it becomes a very serious problem if convolutive mixtures are involved and separation is performed in the time–frequency-domain with the narrowband assumption. In this case, the output signals are not automatically aligned across the frequency bins. In order to align the output signals across the frequency bins, several principles such as clustering, DOA and correlation can be utilized. Here, we apply a clustering-based method to mitigate the permutation indeterminacy [40]. To this end, the mixing vectors are normalized such that they become independent of frequency. Similarly to the discussion in the section about hierarchical clustering for estimating the mixing vectors, the mixing vectors that correspond to the same source are assumed to form clusters. These clusters are determined and its members reordered accordingly so that their order in the mixing matrix reflects the cluster they belong to.

10.3.3 Scaling Indeterminacy

As a consequence of the scaling indeterminacy each output signal in each frequency bin can be multiplied independently by an arbitrary factor. In order to mitigate the effect of the scaling indeterminacy, we apply a normalization technique that can be derived by the so-called MDP [41]. By minimizing the distortion induced by the unmixing system this principle also prevents the unmixing system from whitening the signals.

The idea is to modify the unmixing system such that its output is as close as possible to the input that would be observed if only the source signal corresponding to the considered output was active. Note that the MDP does not dereverberate the mixed signals but only avoids linear distortion by the unmixing system. Originally, the MDP was introduced for well-determined BSS where each output was linked to one specific sensor. As a result, the normalization resulted in an unmixing system that minimizes the expected value

$$E\left\{\sqrt{\sum_{p=1}^{P}(Y_p - X_p)^2}\right\} \tag{10.13}$$

after the permutation indeterminacy has been solved. The expectation refers to time frames τ.

We extend it to underdetermined BSS by allowing an arbitrary sensor to serve as reference resulting in the minimization of

$$E\left\{\sqrt{\sum_{p=1}^{P}\left(Y_p - X_{r_p}\right)^2}\right\}, \qquad r_p \in \{1,\ldots,Q\} \tag{10.14}$$

after the permutation indeterminacy has been solved. As a special case only one sensor is necessary as reference for all output signals, i.e. $r_p = r \in \{1,\ldots,Q\}\,\forall p$. As a consequence, each output signal is multiplied by the component of the estimated mixing matrix that corresponds to the considered output signal and, for example, to the first sensor. As an alternative, an average of components that correspond to several sensors can be used instead of a single sensor. In both cases, the scaling factors can be applied directly to the estimated mixing matrix instead of the output signals.

10.4 Blind Source Recovery

For the second stage BSR we follow the common MAP principle. First, we outline how this principle can lead to constrained ℓ_1-norm minimization. Assuming the STFT for representing the mixture signals, we have to recover the sources from Q-dimensional complex-valued vectors for each DFT bin. Therefore, we continue with investigating the difference between real and complex-valued constrained ℓ_1-norm minimization and its implication for underdetermined BSS of convolutive mixtures.

10.4.1 Sparseness-Based Source Model

According to the MAP principle and Bayes' rule [73] we obtain an estimation $\boldsymbol{Y}$ of the source signals $\boldsymbol{S}$ by

$$\boldsymbol{Y} = \arg\max_{\boldsymbol{S}} P(\boldsymbol{S}|\boldsymbol{X}, \boldsymbol{H}) = \arg\max_{\boldsymbol{S}} P(\boldsymbol{X}|\boldsymbol{S}, \boldsymbol{H})P(\boldsymbol{S}) \tag{10.15}$$

once the mixing matrix $\boldsymbol{H}$ is known. $P(\boldsymbol{S}|\boldsymbol{X}, \boldsymbol{H})$ denotes the conditional a-posteriori PDF of the source signals given the mixed signals and the mixing matrix. $P(\boldsymbol{X}|\boldsymbol{S}, \boldsymbol{H})$ denotes the likelihood, which for the noiseless mixture model (10.6) is given by a Dirac impulse $\delta(\boldsymbol{X} - \boldsymbol{H}\boldsymbol{S})$. It requires the maximum of the conditional a-posteriori PDF to fulfill $\boldsymbol{X} = \boldsymbol{H}\boldsymbol{S}$, which essentially turns (10.15) into the contrained problem

$$\max_{\boldsymbol{S}} P(\boldsymbol{S}) \quad \text{s.t.} \quad \boldsymbol{H}\boldsymbol{S} = \boldsymbol{X}. \tag{10.16}$$

We further assume mutually independent source signals whose spectral components have Laplacian distributions

$$P(\boldsymbol{S}) = \prod_{p=1}^{P} P(S_p) \propto \prod_{p=1}^{P} \exp\left(-|S_p|\right), \tag{10.17}$$

whereby $|S_p|$ denotes the amplitude of S_p.

Since $\arg\max_{\boldsymbol{S}} P(\boldsymbol{S}) = \arg\max_{\boldsymbol{S}} \log\left(P(\boldsymbol{S})\right)$, (10.15) leads eventually to the minimization problem

$$\min_{\boldsymbol{S}} \sum_{p=1}^{P} |S_p| \quad \text{s.t.} \quad \boldsymbol{H}\boldsymbol{S} = \boldsymbol{X} \tag{10.18}$$

for each time instance τ and frequency bin f. In other words, (10.18) describes a constrained ℓ_1-norm minimization problem.

10.4.2 Constrained ℓ_1-Norm Minimization

For constrained optimization problems such as (10.18), techniques developed for linear programming (LP) , quadratic programming (QP), quadratically constrained quadratic programming (QCQP), second-order cone programming (SOCP) or semidefinite programming (SDP) can be used. Their relation with respect to complexity and generality is illustrated in Fig. 10.7. The more general problems include the simpler problems but result in increased computational complexity. Once optimiziation problems are transformed into the so-called standard form, efficient software packages such as SeDuMi [75] can be used to determine the optimal parameters.

Before investigating how to solve (10.18) we first introduce the minimization problems that can be solved by either linear programming (LP) or second-order cone programming (SOCP), respectively. Both LP and SOCP are in general concerned with minimization problems where both the objective function $\widehat{\boldsymbol{c}}^T\widehat{\boldsymbol{S}}$ and the constraints $\widehat{\boldsymbol{H}}\widehat{\boldsymbol{S}} = \widehat{\boldsymbol{X}}$ are linear functions of the parameters $\widehat{\boldsymbol{S}}$. Thereby $\widehat{\boldsymbol{c}} \in \mathbb{R}^{\widehat{P}}$, $\widehat{\boldsymbol{S}} \in \mathbb{R}^{\widehat{P}}$, $\widehat{\boldsymbol{X}} \in \mathbb{R}^{\widehat{Q}}$ and $\widehat{\boldsymbol{H}} \in \mathbb{R}^{\widehat{Q}\times\widehat{P}}$. LP and SOCP differ in the additional constraints they impose on the paramters $\widehat{\boldsymbol{S}}$.

With LP the parameters $\widehat{\boldsymbol{S}}$ are contrained to nonnegative values, i.e. LP solves minimization problems given by

$$\min \widehat{\boldsymbol{c}}^T\widehat{\boldsymbol{S}} \quad \text{s.t.} \quad \widehat{\boldsymbol{H}}\widehat{\boldsymbol{S}} = \widehat{\boldsymbol{X}},\ \widehat{S}_i \geq 0, \quad i = 1, \ldots, \widehat{P}. \tag{10.19}$$

On the other hand, with SOCP the nonnegative constraint of the parameters $\widehat{\boldsymbol{S}}$ of LP is replaced by second-order cone constraints, i.e. SOCP solves minimization problems given by [76–78]

$$\min_{\widehat{\boldsymbol{S}}} \widehat{\boldsymbol{c}}^T\widehat{\boldsymbol{S}} \quad \text{s.t.} \quad \widehat{\boldsymbol{H}}\widehat{\boldsymbol{S}} = \widehat{\boldsymbol{X}},\ \widehat{\boldsymbol{S}}_i \succeq 0, \quad i = 1, \ldots, I, \tag{10.20}$$

where the $\widehat{\boldsymbol{S}}_i$ represent a partition of $\widehat{\boldsymbol{S}}$ with I blocks and $\succeq$ denotes the second order constraint

$$\widehat{\boldsymbol{S}}_i^{(1)} \geq \left|\widehat{\boldsymbol{S}}_i^{(-1)}\right|_2 \tag{10.21}$$

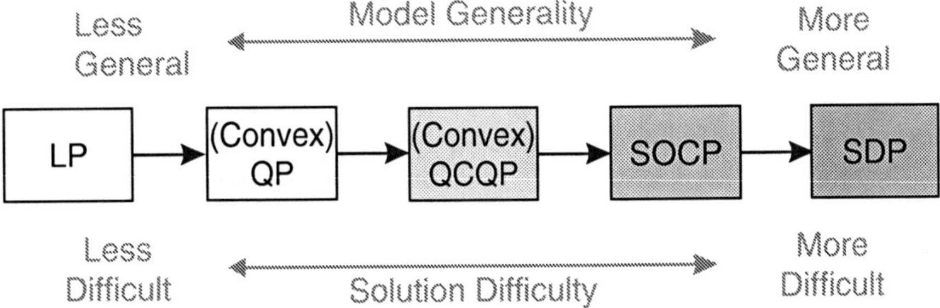

Fig. 10.7. Classification of SOCP (according to [74]).

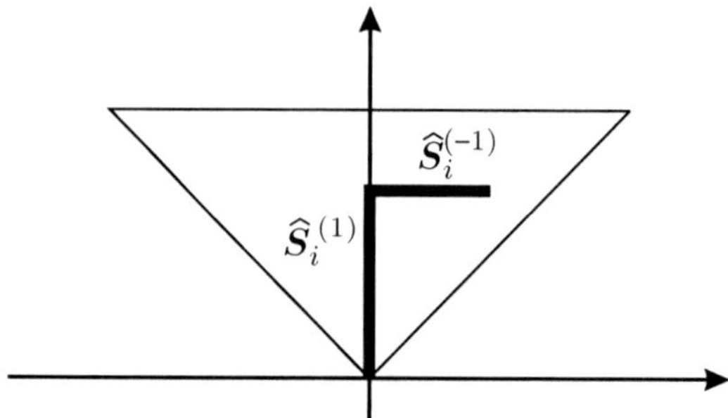

Fig. 10.8. Second-order cone.

meaning that each $\widehat{\boldsymbol{S}}_i$ is within a second order cone as illustrated in Fig. 10.8 for $\widehat{\boldsymbol{S}}_i \in \mathbb{R}^2$. Thereby $\widehat{\boldsymbol{S}}_i^{(1)}$ denotes the first component and $\widehat{\boldsymbol{S}}_i^{(-1)}$ all but the first component of $\widehat{\boldsymbol{S}}_i$.

Real Numbers If we had to consider only real-valued problems, (10.18) could be transformed into (10.19) by separating positive and negative values by

$$\widehat{\boldsymbol{S}} \leftarrow \begin{bmatrix} \boldsymbol{S}^+ \\ \boldsymbol{S}^- \end{bmatrix}, \ \hat{\boldsymbol{c}} \leftarrow \begin{bmatrix} \boldsymbol{1} \\ \boldsymbol{1} \end{bmatrix}, \ \widehat{\boldsymbol{H}} \leftarrow \begin{bmatrix} \boldsymbol{H} \\ -\boldsymbol{H} \end{bmatrix}, \ \widehat{\boldsymbol{X}} \leftarrow \boldsymbol{X}. \tag{10.22}$$

Here $\boldsymbol{1}$ stands for a unity matrix with appropriate dimensions. $\boldsymbol{S}^+$ and $\boldsymbol{S}^-$ are derived from $\boldsymbol{S}$ by setting all negative values or all positive values, respectively, to zero. Consequently, $\widehat{P} = 2P$ and $\widehat{Q} = 2Q$. Thus we could apply LP to solve (10.18) [79].

Although powerful algorithms for LP exist, they are still time consuming. Depending on the dimensions of the problem we can obtain a faster combinatorial algorithm if we use a certain property of the solution. It can be shown [9, 80] that the P-dimensional vector $\boldsymbol{Y}$, that solves (10.18), contains at least $P - Q$ zeros if the columns of $\boldsymbol{H}$ are normalized. The normalization can be assumed for BSS due to the scaling ambiguity.

The lower limit for the number of zeros can be considered a constraint on the solution of (10.18) and can easily be fulfilled by setting $P - Q$ elements of the solution to zero. Then we only have to determine the remaining Q elements. Assuming that we know where to place the zeros the remaining elements are found by multiplying the inverse of the quadratic matrix built by the remaining mixing vectors $\boldsymbol{h}_i$ with the constraining vector $\boldsymbol{X}$

$$\left[\boldsymbol{h}_{i_1} \ldots \boldsymbol{h}_{i_Q}\right]^{-1} \boldsymbol{X}, \quad i_1, \ldots, i_Q \in \{1, \ldots, P\}. \tag{10.23}$$

The correct placement of the zeros can be determined by combinatorially testing all possibilities and accepting the one with the smallest ℓ_1-norm. As a simple example let us consider

$$\boldsymbol{H} = \begin{bmatrix} 1 & 0.6 & -0.6 \\ 0 & 0.8 & 0.8 \end{bmatrix}, \qquad \boldsymbol{X} = \begin{bmatrix} 1 \\ 0.5 \end{bmatrix}. \tag{10.24}$$

According to the dimensions of the problem at least one element of the solution $\boldsymbol{Y}$ must be zero. The ℓ_1-norm of the possible solutions are

$$\left\| \begin{bmatrix} \begin{bmatrix} 1 & 0.6 \\ 0 & 0.8 \end{bmatrix}^{-1} \begin{bmatrix} 1 \\ 0.5 \end{bmatrix} \\ 0 \end{bmatrix} \right\|_1 = 1.25, \tag{10.25}$$

$$\left\| \begin{bmatrix} 0 \\ \begin{bmatrix} 1 & -0.6 \\ 0 & 0.8 \end{bmatrix}^{-1} \begin{bmatrix} 1 \\ 0.5 \end{bmatrix} \end{bmatrix} \right\|_1 = 2, \tag{10.26}$$

$$\left\| \begin{bmatrix} 0 \\ \begin{bmatrix} 0.6 & -0.6 \\ 0.8 & 0.8 \end{bmatrix}^{-1} \begin{bmatrix} 1 \\ 0.5 \end{bmatrix} \end{bmatrix} \right\|_1 = 1.\overline{6}. \tag{10.27}$$

The notation of (10.26) reflects the above description of setting one element to zero and inverting the remaining quadratic matrix. The chosen solution would be the one corresponding to (10.25).

This combinatorial method is based on the shortest-path algorithm [9] and the ℓ_0-norm that basically counts the number of nonzero elements. The combinatorial method stands in contrast to the approach in [81] where conditions are given for which the ℓ_0-norm can be calculated by an ℓ_p-norm with $0 < p \leq 1$.

Complex Numbers If complex numbers are involved then (10.22) can no longer be applied because such numbers possess a continuous phase in contrast to a discrete phase of real numbers. Thus we cannot use algorithms that solve LP problems for complex-valued problems. However, ℓ_1-norm minimization problems (10.18) with complex numbers can be transformed to SOCP problems in the following way.

Equation (10.18) is equivalent to

$$\min \beta \in \mathbb{R} \quad \text{s.t.} \quad \boldsymbol{X} = \boldsymbol{H}\boldsymbol{S}, \text{ and } |\boldsymbol{S}|_1 \leq \beta. \tag{10.28}$$

By decomposing $\beta = \sum_{i=1}^{P} \beta_i$, $\beta_i \in \mathbb{R}$, the second constraint $|\boldsymbol{S}|_1 \leq \beta$ can be expressed by

$$|\boldsymbol{S}|_1 = \sum_{i=1}^{P} \left\| \begin{bmatrix} \Re(S_i) \\ \Im(S_i) \end{bmatrix} \right\|_2 \leq \mathbf{1}^T \boldsymbol{\beta} = \mathbf{1}^T [\beta_1 \ \cdots \ \beta_P]^T = \beta, \tag{10.29}$$

where $\Re(\cdot)$ and $\Im(\cdot)$ denote the real and imaginary parts, respectively. Thus we can rewrite (10.18) as

$$\min_{\boldsymbol{\beta}} \mathbf{1}^T \boldsymbol{\beta} \quad \text{s.t.} \quad \boldsymbol{X} = \boldsymbol{H}\boldsymbol{S}, \quad \left\| \begin{bmatrix} \Re(S_i) \\ \Im(S_i) \end{bmatrix} \right\|_2 \leq \beta_i, \quad \forall i. \tag{10.30}$$

By defining

$$\widehat{\boldsymbol{S}} = \begin{bmatrix} \beta_1 \\ \Re(S_1) \\ \Im(S_1) \\ \vdots \\ \beta_P \\ \Re(S_P) \\ \Im(S_P) \end{bmatrix} \in \mathbb{R}^{3P}, \qquad \widehat{\boldsymbol{c}} = \begin{bmatrix} 1 \\ 0 \\ 0 \\ \vdots \\ 1 \\ 0 \\ 0 \end{bmatrix} \in \mathbb{R}^{3P}, \tag{10.31}$$

$$\widehat{\boldsymbol{X}} = \begin{bmatrix} \Re(\boldsymbol{X}) \\ \Im(\boldsymbol{X}) \end{bmatrix} \in \mathbb{R}^{2Q}, \tag{10.32}$$

$$\widehat{\boldsymbol{H}} = \begin{bmatrix} \mathbf{0} \; \Re(\boldsymbol{h}_1) \; -\Im(\boldsymbol{h}_1) \; \cdots \; \mathbf{0} \; \Re(\boldsymbol{h}_P) \; -\Im(\boldsymbol{h}_P) \\ \mathbf{0} \; \Im(\boldsymbol{h}_1) \;\; \Re(\boldsymbol{h}_1) \;\; \cdots \; \mathbf{0} \; \Im(\boldsymbol{h}_P) \;\; \Re(\boldsymbol{h}_P) \end{bmatrix} \in \mathbb{R}^{2Q \times 3P}, \tag{10.33}$$

we can write

$$\min_{\widehat{\boldsymbol{S}}} \widehat{\boldsymbol{c}}^T \widehat{\boldsymbol{S}} \quad \text{s.t.} \quad \widehat{\boldsymbol{X}} = \widehat{\boldsymbol{H}}\widehat{\boldsymbol{S}}, \quad \left\| \begin{bmatrix} \Re(S_i) \\ \Im(S_i) \end{bmatrix} \right\|_2 \leq \beta_i \quad \forall i. \tag{10.34}$$

The second constraint in (10.34) can be interpreted as a second order cone for each i.

Equation (10.34) describes an SOCP problem [74], which can be solved numerically for example with `SeDuMi` [75].

Analysis of Real and Complex-Valued ℓ_1-Norm Minimization In contrast to the real-valued ℓ_1-norm minimization problem where a minimum number of zeros can be guaranteed theoretically in the optimal solution, the number of zeros cannot be predicted with complex-valued problems as the following simple example shows. Let

$$\boldsymbol{H} = \begin{bmatrix} 1 \; 0.6 \; \frac{4}{\sqrt{17}} \\ 0 \; 0.8 \; \frac{0.8+j0.6}{\sqrt{17}} \end{bmatrix}, \qquad \boldsymbol{X} = \begin{bmatrix} 1 \\ 0.5 \end{bmatrix}. \tag{10.35}$$

Then the ℓ_1-norm of the solution obtained by SOCP is given by

$$|\boldsymbol{Y}_{socp}|_1 = \left\| \begin{bmatrix} 0.227 + 0.040i \\ 0.511 - 0.091i \\ 0.481 + 0.015i \end{bmatrix} \right\|_1 = 1.23. \tag{10.36}$$

It does not contain any zeros as we could expect with real numbers yet solves (10.18). In comparison, the ℓ_1-norm of the optimal combinatorial solution is given by

$$|\boldsymbol{Y}_{comb}|_1 = \left\| \begin{bmatrix} \begin{bmatrix} 0.6 & \frac{4}{\sqrt{17}} \\ 0.8 & \frac{0.8+j0.6}{\sqrt{17}} \end{bmatrix}^{-1} \begin{bmatrix} 1 \\ 0.5 \end{bmatrix} \\ 0 \end{bmatrix} \right\|_1 = 1.24. \tag{10.37}$$

This observation reveals a very important difference from real-valued problems and prevents the theoretical justification of a procedure similar to the combinatorial approach in Sect. 10.4.2. To better explain this difference between real and complex numbers we take a look at a general solution based on a combinatorial solution and the null-space $\mathcal{N}(\boldsymbol{H})$ of $\boldsymbol{H}$.

Even though the combinatorial solution $\boldsymbol{Y}_{comb}$ does not necessarily minimize the ℓ_1-norm, it fulfills just as well the SOCP solution $\boldsymbol{Y}_{socp}$

$$\boldsymbol{X} = \boldsymbol{H}\boldsymbol{Y}_{comb} = \boldsymbol{H}\boldsymbol{Y}_{socp}. \tag{10.38}$$

By defining the difference $\widetilde{\boldsymbol{Y}} = \boldsymbol{Y}_{socp} - \boldsymbol{Y}_{comb}$, (10.38) becomes

$$\boldsymbol{H}\boldsymbol{Y}_{comb} = \boldsymbol{H}\boldsymbol{Y}_{comb} + \underbrace{\boldsymbol{H}\widetilde{\boldsymbol{Y}}}_{=0}. \tag{10.39}$$

This means that if we have a combinatorial solution we can limit our search for the minimum ℓ_1-norm solution to the null-space $\mathcal{N}(\boldsymbol{H})$, i.e.

$$\min \left| \boldsymbol{Y}_{comb} + \widetilde{\boldsymbol{Y}} \right|_1 \tag{10.40}$$

with

$$\widetilde{\boldsymbol{Y}} \in \mathcal{N}(\boldsymbol{H}) \quad \Leftrightarrow \quad \widetilde{\boldsymbol{Y}} = (\boldsymbol{1} - \boldsymbol{H}^{-}\boldsymbol{H})\boldsymbol{z}, \quad \boldsymbol{z} \in \mathbb{C}^P, \tag{10.41}$$

where $\boldsymbol{H}^{-}$ is an arbitrary generalized inverse of $\boldsymbol{H}$. For $P = 3$ and $Q = 2$ we can express the combinatorial solution and the null-space without loss of generality by

$$\boldsymbol{Y}_{comb} = \begin{bmatrix} [\boldsymbol{h}_1\, \boldsymbol{h}_2]^{-1}\boldsymbol{X} \\ 0 \end{bmatrix}, \tag{10.42}$$

$$\mathcal{N}(\boldsymbol{H}) = \alpha \begin{bmatrix} [\boldsymbol{h}_1\, \boldsymbol{h}_2]^{-1}\boldsymbol{h}_3 \\ 1 \end{bmatrix}, \qquad \alpha \in \mathbb{C}. \tag{10.43}$$

With (10.42) and (10.43), the function to be minimized (10.40) can be written as

$$\begin{aligned} \left| \boldsymbol{Y}_{comb} + \alpha\widetilde{\boldsymbol{Y}} \right|_1 &= |f_{11}(\boldsymbol{H}, \boldsymbol{X}) + \alpha f_{12}(\boldsymbol{H}, \boldsymbol{X})| \\ &+ |f_{21}(\boldsymbol{H}, \boldsymbol{X}) + \alpha f_{22}(\boldsymbol{H}, \boldsymbol{X})| \\ &+ |f_{31}(\boldsymbol{H}, \boldsymbol{X}) + \alpha f_{32}(\boldsymbol{H}, \boldsymbol{X})|. \end{aligned} \tag{10.44}$$

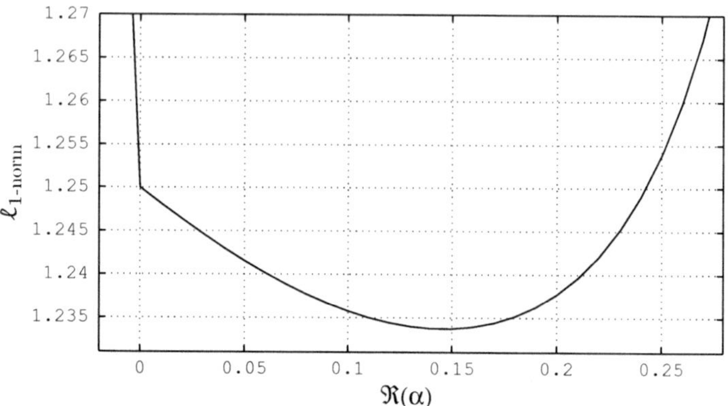

Fig. 10.9. Smooth slope.

Here $f_{ij}(\boldsymbol{H}, \boldsymbol{X})$ is a term that only depends on $\boldsymbol{H}$ and $\boldsymbol{X}$, which are constant for any given problem. If only real values are involved then (10.44) describes a piecewise linear function depending on α whose slope can only change a limited number of times in a discrete manner.

However, once complex numbers are involved, their imaginary part results in an inherent ℓ_2-norm, which leads to smooth slopes as they appear with second order or higher polynomials. This behavior becomes obvious in (10.20). There the ℓ_1-norm is changed from the sum of absolute values of real numbers to the sum of the ℓ_2-norms of the real and imaginary part. The introduction of the ℓ_2-norm explains the different behavior of complex-valued ℓ_1-norm minimization compared with its real counterpart. An example is shown in Fig. 10.9, where the dependence of the ℓ_1-norm on α is shown (here only the dependence on the real part of α is shown). The combinatorial solution that minimizes the ℓ_1-norm is given there for $\alpha = 0$. However, this is not the solution of (10.18), which is rather obtained for $\alpha \neq 0$.

10.5 Experiments

10.5.1 Experimental Conditions

Even though the combinatorial solution (CS) with a minimum number of zeros in Sect. 10.4.2 cannot be justified theoretically for complex numbers, in practice its performance is comparable to or even better than that of the SOCP solution. In our experiments we separated mixtures that we obtained from clean speech signals and recorded room impulse responses. We tested both approaches with both the estimated and the original mixing matrix with different numbers of sources ($P \in \{3, 4, 5\}$) and sensors ($Q \in \{2, 3\}$). We performed four experiments for each scenario. Each of the four experiments had a different combination of speakers selected from six male and female English speakers. Further experimental conditions are summarized in Table 10.3

Table 10.3. Experimental conditions.

Sensor distance	40 mm
Source signal length	7 seconds
Reverberation time T_R	120 ms
Sampling frequency f_s	8 kHz
Window type	von Hann
Filter length	1024 points
Shifting interval	256 points
Number of clusters c	100

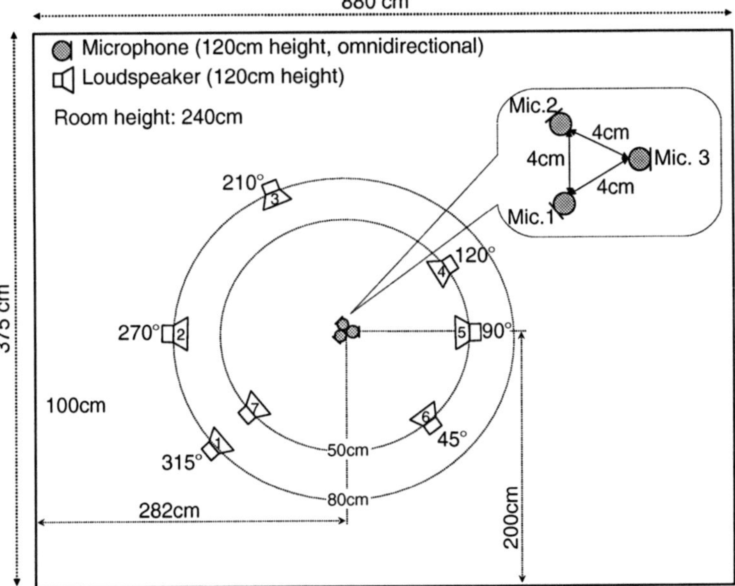

Fig. 10.10. Room setup.

and Fig. 10.10. For comparison, we also applied a time–frequency-masking approach to the same mixtures [27].

10.5.2 Performance Measures

To evaluate the separation performance , we decomposed an estimated signal y_p in the time-domain into an filtered version $y_{p,target}$ of the original signal s_p, an filtered mixture $e_{-p,interf}$ of the interfering signals and e_{artif}, which accounts for artifacts introduced by the separation algorithm

$$y_p = y_{p,target} + e_{-p,interf} + e_{artif}. \tag{10.45}$$

The filters are optimal in the sense that they project the estimated signal y_p into the subspace spanned by delayed versions of the original signal and the interfering signals, respectively [82, 83].

As performance measures we use the averaged source-to-distortion ratio (SDR)

$$\mathrm{SDR} = \frac{1}{P}\sum_p 10\log_{10}\frac{\sum y_{p,target}^2}{\sum (e_{-p,interf} + e_{artif})^2}, \tag{10.46}$$

the averaged source-to-interference ratio (SIR)

$$\mathrm{SIR} = \frac{1}{P}\sum_p 10\log_{10}\frac{\sum y_{p,target}^2}{\sum e_{-p,interf}^2}, \tag{10.47}$$

and the source-to-artifact ratio (SAR)

$$\mathrm{SAR} = \frac{1}{P}\sum_p 10\log_{10}\frac{\sum (y_{p,target} + e_{-p,interf})^2}{\sum e_{artif}^2}. \tag{10.48}$$

10.5.3 Results

The results are shown in Tables 10.4 through 10.7. The performance values of each combination give the average for the involved signals. The specific sources and sensors used in each scenario are indicated in the caption of each table following the numbering in Fig. 10.10.

To evaluate the performance improvement, we provide the input SDR, SIR and SAR measured at a single sensor in Table 10.8.

A subjective evaluation of the separated sources supports the result.

The SOCP solution and combinatorial solution yield similar results with the estimated mixing matrix. However, the combinatorial solution performs better with the optimal mixing matrix.

Although the difference in performance quality is negligible in practical applications with estimated mixing matrices, the computational complexity reveals great differences. The combinatorial solution has a low initial computational complexity but grows exponentially with the input dimension P. On the other hand, the SOCP solution has a high computational complexity even for low input dimensions P, but even in the worst case it grows only according to

$$O\left(\sqrt{P}\log(1/\epsilon)\right). \tag{10.49}$$

ϵ denotes the precision of the numerical algorithm [75]. Figure 10.11 illustrates this fact and shows on a logarithmic scale the time required by the two approaches to separate the sources in one frequency bin with 230 time frames

Table 10.4. Separation results for 3 sources (3-5-7), 2 mixtures (1-2).

Combi-	Original mixing matrix						Estimated mixing matrix						Time–frequency		
nation	CS			SOCP			CS			SOCP			masking		
	SDR	SIR	SAR	SDR	SIR	SAR	SDR	SIR	SAR	SDR	SIR	SAR	SDR	SIR	SAR
1	10.17	14.61	12.31	10.67	14.12	13.57	6.03	9.67	9.19	6.29	9.45	9.88	5.24	11.36	7.28
2	10.21	14.72	12.31	9.05	11.81	12.79	2.73	6.57	6.28	3.44	6.88	7.15	5.34	11.76	7.23
3	11.62	16.60	13.49	11.48	14.91	14.53	6.41	10.57	9.53	6.74	10.50	10.16	4.87	10.61	7.10
4	10.71	15.67	12.61	9.57	12.91	12.63	4.54	8.82	7.85	4.76	8.74	8.36	6.17	12.29	8.13
Average	**10.68**	**15.40**	**12.68**	**10.19**	**13.44**	**13.38**	**4.93**	**8.91**	**8.21**	**5.30**	**8.89**	**8.89**	**5.40**	**11.51**	**7.43**

Table 10.5. Separation results for 4 sources (1-3-4-6), 2 mixtures (1-2).

Combi-	Original mixing matrix						Estimated mixing matrix						Time–frequency		
nation	CS			SOCP			CS			SOCP			masking		
	SDR	SIR	SAR	SDR	SIR	SAR	SDR	SIR	SAR	SDR	SIR	SAR	SDR	SIR	SAR
1	4.91	8.73	7.84	4.32	7.33	8.98	−0.55	2.24	5.36	−0.26	2.16	6.18	1.33	5.80	4.70
2	5.73	9.97	8.25	4.96	8.18	9.10	−1.40	1.02	5.19	−0.36	1.96	5.87	2.01	7.40	5.05
3	5.58	9.57	8.32	4.13	7.00	8.66	−1.31	1.14	5.34	0.30	2.71	6.02	1.53	6.18	5.23
4	5.94	10.07	8.63	5.05	8.55	9.36	0.22	3.07	5.57	0.61	3.09	6.40	1.49	6.25	4.88
Average	**5.54**	**9.59**	**8.26**	**4.62**	**7.76**	**9.02**	**−0.76**	**1.87**	**5.36**	**0.07**	**2.48**	**6.12**	**1.59**	**6.41**	**4.96**

Table 10.6. Separation results for 4 sources (1-3-4-6), 3 mixtures (1-2-3).

Combi-nation	Original mixing matrix						Estimated mixing matrix						Time–frequency masking		
	CS			SOCP			CS			SOCP					
	SDR	SIR	SAR	SDR	SIR	SAR	SDR	SIR	SAR	SDR	SIR	SAR	SDR	SIR	SAR
1	13.93	18.45	15.9	13.38	16.71	16.24	9.64	13.15	12.46	9.76	12.93	12.88	6.30	13.56	7.50
2	14.15	18.77	16.07	14.36	17.92	17.00	5.66	8.41	9.91	7.36	10.19	11.11	7.15	14.25	8.34
3	14.66	20.01	16.21	14.64	18.73	16.86	11.35	15.38	13.71	11.58	15.16	14.26	6.69	13.66	7.96
4	14.58	19.25	16.46	14.48	18.26	16.96	10.23	13.12	13.67	10.75	13.36	14.46	7.01	14.12	8.23
Average	**14.33**	**19.12**	**16.16**	**14.22**	**17.91**	**16.76**	**9.22**	**12.51**	**12.44**	**9.86**	**12.91**	**13.18**	**6.79**	**13.89**	**8.01**

Table 10.7. Separation results for 5 sources (1-2-3-4-6), 3 mixtures (1-2-3).

Combi-nation	Original mixing matrix						Estimated mixing matrix						Time–frequency masking		
	CS			SOCP			CS			SOCP					
	SDR	SIR	SAR	SDR	SIR	SAR	SDR	SIR	SAR	SDR	SIR	SAR	SDR	SIR	SAR
1	9.80	13.86	12.17	10.12	13.46	13.03	6.31	9.81	9.35	6.63	9.73	10.03	4.62	10.65	6.39
2	10.00	14.03	12.39	10.38	13.71	13.30	6.02	9.57	9.08	6.37	9.58	9.71	4.97	11.35	6.52
3	10.23	14.27	12.61	10.43	13.48	13.66	6.08	9.28	9.52	6.33	9.19	10.12	4.74	10.87	6.47
4	9.68	13.67	12.12	10.30	13.67	13.20	6.39	9.89	9.43	6.71	9.85	10.08	4.03	10.39	5.73
Average	**9.93**	**13.95**	**12.32**	**10.31**	**13.58**	**13.30**	**6.20**	**9.64**	**9.35**	**6.51**	**9.59**	**9.99**	**4.59**	**10.81**	**6.28**

Table 10.8. Input SDR, SIR and SAR for different numbers N of sources.

Combination	3 sources			4 sources			5 sources		
	SDR	SIR	SAR	SDR	SIR	SAR	SDR	SIR	SAR
1	−3.11	−3.09	26.14	−4.52	−4.51	26.84	−5.57	−5.56	27.13
2	−2.79	−2.78	27.22	−4.35	−4.34	27.56	−5.69	−5.67	26.37
3	−2.79	−2.77	26.08	−4.46	−4.45	26.91	−5.59	−5.58	26.05
4	−2.80	−2.79	26.06	−4.53	−4.51	25.31	−5.83	−5.81	25.93
Average	**−2.87**	**−2.86**	**26.37**	**−4.47**	**−4.45**	**26.65**	**−5.67**	**−5.65**	**26.37**

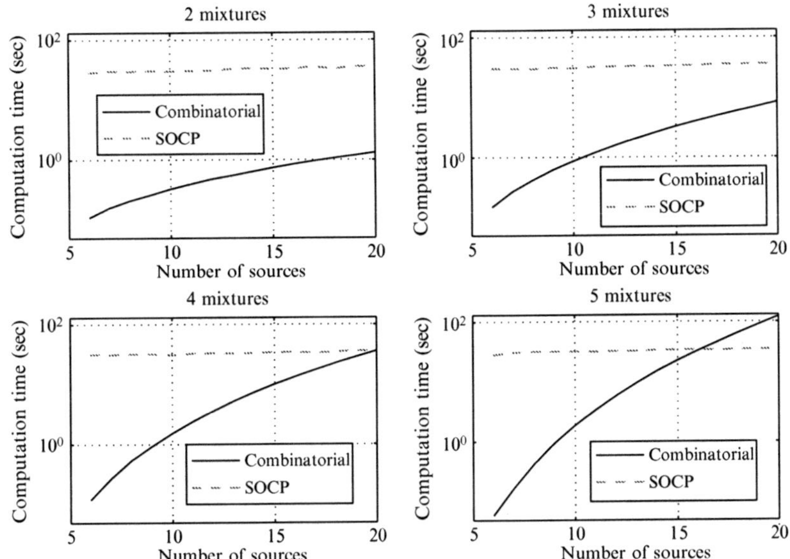

Fig. 10.11. Comparison of computational complexity.

for different numbers of sources and sensors. The simulations for Fig. 10.11 were performed on a 2.4 GHz PC based on random data and mixing matrices.

One reason for the big difference in the initial computational complexity can be found in the reusability of previous results. For underdetermined BSS in the time–frequency-domain the minimum ℓ_1-norm solution must be calculated several times with the same mixing matrix. The combinatorial solution is built on the inverses of selected mixing vectors. Once they are calculated they can be reused as long as the mixing matrix does not change. In contrast, SOCP cannot profit from the reuse of earlier results due to its algorithmic nature.

The time–frequency masking approach yields better separation in terms of the SIR than the proposed methods. This is because the time–frequency masking approach uses only time–frequency instances that originate from a single source with high confidence. In contrast, the proposed methods do not evaluate the confidence about the origin of a time–frequency instance but use all instances for separation in a uniform way. On the other hand, by using all time–frequency instances, the proposed methods result in fewer artifacts, as expressed by a higher SAR.

10.6 Conclusions

After providing an overview about existing solutions, we presented a complete underdetermined BSS system for convolutive mixtures. It is based on algorithms originally designed for instantaneous mixtures. We transfered and adapted them to convolutive mixtures by switching to the time–frequency-domain. We adhered to a two-stage approach of blind system identification (BSI) followed by blind source recovery (BSR) in connection with an iid Laplacian source model.

Both stages were performed in the time–frequency-domain to increase sparseness and reduce the convolutive mixtures to instantaneous mixtures. For simplification we assumed narrowband signals when deriving the solution but showed by experiments that it can be applied to speech signals in practical situations. Experimental results further confirmed that the assumption of sparseness in time–frequency and space and, therefore, clusters around the mixing vectors, is sufficiently fulfilled for convolutively mixed speech signals in the time–frequency-domain.

In BSI as the first stage we had to estimate the mixing matrix. We proposed the use of hierarchical clustering for this task. It can be easily employed for complex-valued mixture samples that occur if BSS is performed in the time–frequency-domain. Furthermore, it does not limit the usable number of sensors and prevents convergence problems.

In BSR as the second stage we separate the mixtures using the estimated mixing matrix from the first stage. We assumed statistical independence for the sources. This led to a constrained ℓ_1-norm minimization. We investigated the consequence of dealing with complex numbers as a result of the time–frequency-domain approach. Although the combinatorial solution with at least $P - Q$ zeros is not necessarily optimum for complex numbers, its performance is comparable to or even better than that of the second-order cone programming (SOCP) solution in terms of the common quadratic error criteria. In addition, the combinatorial solution has the advantage that it is faster for underdetermined BSS problems with low input/output dimensions.

References

1. O. Yilmaz and S. Rickard, “Blind separation of speech mixtures via time–frequency masking,” *IEEE Transactions on Signal Processing*, vol. 52, no. 7, pp. 1830–1847, July 2004. [Online]. Available: http://eleceng.ucd.ie/~srickard/bss.html
2. S. Rickard and O. Yilmaz, “On the approximate W-disjoint orthogonality of speech,” in *Proc. ICASSP 2002*, vol. 1, 2002, pp. 529–532.
3. L. Vielva, I. Santamaria, C. Pantaleon, J. Ibanez, and D. Erdogmus, “Estimation of the mixing matrix for underdetermined blind source separation using spectral estimation techniques,” in *Proc. EUSIPCO 2002*, vol. 1, Sept. 2002, pp. 557–560.
4. P. Bofill and M. Zibulevsky, “Blind separation of more sources than mixtures using sparsity of their short-time Fourier transform,” in *Proc. ICA 2000*, June 2000, pp. 87–92.
5. P. Bofill, “Underdetermined blind separation of delayed sound sources in the frequency domain,” *Neurocomputing*, vol. 55, no. 3–4, pp. 627–641, Oct. 2003.
6. S. Araki, S. Makino, A. Blin, R. Mukai, and H. Sawada, “Underdetermined blind separation for speech in real environments with sparseness and ICA,” in *Proc. ICASSP 2004*, vol. III, May 2004, pp. 881–884.
7. A. Blin, S. Araki, and S. Makino, “Underdetermined blind separation of convolutive mixtures of speech using time–frequency mask and mixing matrix estimation,” *IEICE Trans. Fundamentals*, vol. E88-A, no. 7, pp. 1693–1700, 2005.
8. K. Waheed and F. Salem, “Algebraic overcomplete independent component analysis,” in *Proc. ICA 2003*, 2003, pp. 1077–1082.
9. F. Theis, “Mathematics in independent component analysis,” Ph.D. dissertation, University of Regensburg, 2002. [Online]. Available: http://homepages.uni-regensburg.de/ thf11669/phdthesis.html
10. A. Ferréol, L. Albera, and P. Chevalier, “Fourth-order blind identification of underdetermined mixtures of sources (FOBIUM),” *IEEE Trans. on Signal Processing*, vol. 53, no. 5, pp. 1640–1653, May 2005.
11. L. D. Lathauwer and J. Castaing, “Second-order blind identification of underdetermined mixtures,” in *6th Int. Conference on Independent Component Analysis and Blind Signal Separation (ICA 2006)*, R. et al., Ed. Justinian P. Rosca, Deniz Erdogmus, José Carlos Príncipe, and Simon Haykin Charleston, SC, USA: Springer, Mar. 2006, pp. 40–47. [Online]. Available: http://publi-etis.ensea.fr/2006/LC06
12. L. Albera, P. Comon, P. Chevalier, and A. Ferrol, “Blind identification of underdetermined mixtures based on the hexacovariance,” in *Proc. ICASSP 2004*, vol. II, May 2004, pp. 29–32.
13. P. Bofill and E. Monte, “Underdetermined convoluted source reconstruction using lp and socp, and a neural approximator of the optimizer,” in *Independent Component Analysis and Blind Signal Separation*, ser. LNCS, vol. 3889. Springer, 2006, pp. 569–576.
14. Y. Deville, J. Chappuis, S. Hosseini, and J. Thomas, “Differential fast fixed-point bss for underdetermined linear instantaneous mixtures,” in *Independent Component Analysis and Blind Signal Separation*, ser. LNCS, vol. 3889. Springer, 2006, pp. 48–56.

15. C. Wei, L. Khor, W. Woo, and S. Dlay, "Post-nonlinear underdetermined ICA by Bayesian statistics," in *Independent Component Analysis and Blind Signal Separation*, ser. LNCS, vol. 3889. Springer, 2006, pp. 773–780.
16. S. Lesage, S. Krstulović, and R. Gribonval, "Under-determined source separation: Comparison of two approaches based on sparse decompositions," in *Independent Component Analysis and Blind Signal Separation*, ser. LNCS, vol. 3889. Springer, 2006, pp. 633–640.
17. C. Févotte and S. Godsill, "Blind separation of sparse sources using jeffrey's inverse prior and the em algorithm," in *Independent Component Analysis and Blind Signal Separation*, ser. LNCS, vol. 3889. Springer, 2006, pp. 593–600.
18. P. Comon and M. Rajih, "Blind identification of under-determined mixtures based on the characteristic function," in *ICASSP'05*, vol. IV, Mar. 2005, pp. 1005–1008.
19. L. Albera, A. Ferreol, P. Comon, and P. Chevalier, "Blind Identification of Overcomplete MixturEs of sources (BIOME)," *Linear Algebra Applications, Special Issue on Linear Algebra in Signal and Image Processing*, vol. 391C, pp. 3–30, Nov. 2004.
20. L. D. Lathauwer, "Simultaneous matrix diagonalization: the overcomplete case," in *Proc. of the Fourth International Symposium on Independent Component Analysis and Blind Signal Separation (ICA 2003)*, Apr. 2003, pp. 821–825.
21. L. D. Lathauwer, B. D. Moor, J. Vandewalle, and J.-F. Cardoso, "Independent component analysis of largely underdetermined mixtures," in *Proc. of the Fourth International Symposium on Independent Component Analysis and Blind Signal Separation (ICA 2003)*, Apr. 2003, pp. 29–34.
22. L. Vielva, D. Erdogmus, C. Pantaleon, I. Santamaria, J. Pereda, and J. Principe, "Underdetermined blind source separation in a time-varying environment," in *Proc. ICASSP 2002*, vol. 3, May 2002, pp. 3049–3052.
23. L. D. Lathauwer, P. Comon, B. D. Moor, and J. Vandewalle, "ICA algorithms for 3 sources and 2 sensors," in *Proc. IEEE Signal Processing Workshop on Higher-Order Statistics*, Caesarea, Israel, 1999, pp. 116–120.
24. P. OGrady, B. Pearlmutter, and S. Rickard, "Survey of sparse and non-sparse methods in source separation," *International Journal of Imaging Systems and Technology*, vol. 15, no. 1, pp. 18–33, July 2005.
25. F. Abrard and Y. Deville, "A time–frequency blind signal separation method applicable to underdetermined mixtures of dependent sources," *Signal Processing*, vol. 85, no. 7, pp. 1389–1403, July 2005. [Online]. Available: http://www.ast.obs-mip.fr/users/ydeville/papers/fa_yd_sigpro_2005_final%.pdf
26. N. Mitianoudis and T. Stathaki, "Overcomplete source separation using laplacian mixure models," *IEEE Signal Processing Letters*, vol. 12, no. 4, pp. 277–280, Apr. 2005.
27. S. Araki, H. Sawada, R. Mukai, and S. Makino, "A novel blind source separation method with observation vector clustering," in *Proc. IWAENC 2005*, Sept. 2005, pp. 117–120.
28. R. Olsson and L. Hansen, "Blind separation of more sources than sensors in convolutive mixtures," in *Proc. ICASSP 2006*, 2006.
29. M. Pedersen, D. Wang, J. Larsen, and U. Kjems, "Separating underdetermined convolutive speech mixtures," in *Independent Component Analysis and Blind Signal Separation*, ser. LNCS, vol. 3889. Springer, 2006, pp. 674–681.

30. Y. Li, J. Wang, and A. Cichocki, "Blind source extraction from convolutive mixtures in ill-conditioned multi-input multi-output channels," *IEEE Trans. on Circuits and Systems - I: Regular Papers*, vol. 51, no. 9, pp. 1814–1822, Sept. 2004.
31. R. Saab, O. Yilmaz, M. McKeown, and R. Abugharbieh, "Underdetermined sparse blind source separation with delays," in *Signal Processing with Adaptive Sparse Structured Representations Workshop (SPARS)*, 2005.
32. M. Molla, K. Hirose, and N. Minematsu, "Separation of mixed audio signals by source localization and binary masking with hilbert spectrum," in *Independent Component Analysis and Blind Signal Separation*, ser. LNCS, vol. 3889. Springer, 2006, pp. 641–648.
33. S. J. Godsill and C. Andrieu, "Bayesian separation and recovery of convolutively mixed autoregressive sources," in *Proc. ICASSP 1999*, vol. III, 1999, pp. 1733–1736. [Online]. Available: citeseer.csail.mit.edu/349030.html
34. S. Winter, H. Sawada, S. Araki, and S. Makino, "Overcomplete BSS for convolutive mixtures based on hierarchical clustering," in *Proc. ICA 2004*, Sept. 2004, pp. 652–660.
35. S. Winter, H. Sawada, and S. Makino, "On real and complex valued L1-norm minimization for overcomplete blind source separation," in *2005 IEEE Workshop on Applications of Signal Processing to Audio and Acoustics (WASPAA)*, New Paltz, NY, USA, 2005, pp. 86–89.
36. L. Vielva, D. Erdogmus, and J. C. Principe, "Underdetermined blind source separation using a probabilistic source sparsity model," in *Proc. ICA 2001*, 2001, pp. 675–679.
37. W. Kellermann and H. Buchner, "Wideband algorithms versus narrowband algorithms for adaptive filtering in the DFT domain," in *Proc. Asilomar Conf. on Signals, Systems, and Computers*, vol. 2, Nov. 2003, pp. 1278–1282.
38. N. Linh-Trung, A. Belouchrani, K. Abed-Meraim, and B. Boashash, "Separating more sources than sensors using time–frequency distributions," *EURASIP Journal on Applied Signal Processing*, vol. 2005, no. 17, pp. 2828–2847, 2005.
39. H. Sawada, S. Araki, R. Mukai, and S. Makino, "Blind extraction of a dominant source signal from mixtures of many sources," in *Proc. ICASSP 2005*, vol. III, 2005, pp. 61–64.
40. H. Sawada, R. Mukai, S. Araki, and S. Makino, "A robust and precise method for solving the permutation problem," *IEEE Trans. Speech and Audio Processing*, vol. 12, pp. 530–538, Sept. 2004.
41. K. Matsuoka, "Independent component analysis and its applications to sound signal separation," in *Proc. IWAENC 2003*, Kyoto, Sept. 2003, pp. 15–18.
42. A. Jourjine, S. Rickard, and O. Yilmaz, "Blind separation of disjoint orthogonal signals: Demixing n sources from 2 mixtures," *Proc. ICASSP 2000*, vol. 5, pp. 2985–2988, 2000.
43. M. Pedersen, T. Lehn-Schiøler, and J. Larsen, "BLUES from music: BLind Underdetermined Extraction of Sources from music," in *Independent Component Analysis and Blind Signal Separation*, ser. LNCS, vol. 3889. Springer, 2006, pp. 392–399.
44. A. Mansour, M. Kawamoto, and C. Puntonet, "A time–frequency approach to blind separation of underdetermine mixture of sources," in *Proc. IASTED International Conference Applied Simuation and Modelling*, Sept. 2003, pp. 413–418.

45. M. Zibulevsky and B. Pearlmutter, "Blind source separation by sparse decomposition," *Neural Computations*, vol. 13, no. 4, pp. 863–882, 2001. [Online]. Available: http://iew3.technion.ac.il/~mcib/
46. S. S. Chen, D. L. Donoho, and M. A. Saunders, "Atomic decomposition by basis pursuit," Dept. Stat., Stanford Univ, Stanford, CA, Tech. Rep., 1995. [Online]. Available: http://www-stat.stanford.edu/~donoho/Reports/1995/30401.pdf
47. P. Comon, "Blind channel identification and extraction of more sources than sensors," in *Proc. SPIE*, 1998, pp. 2–13, keynote address.
48. A. Taleb, "An algorithm for the blind identication of N independent signal with 2 sensors," in *Proc. ISSPA 01*, Aug. 2001, pp. 5–8.
49. J.-F. Cardoso, "Super-symmetric decomposition of the fourth-order cumulant tensor blind identification of more sources than sensors," in *Proc. ICASSP 91*, vol. V, 1991, pp. 3109–3112.
50. L. Khor, W. Woo, and S. Dlay, "Non-sparse approach to underdetermined blind signal estimation," in *Proc. ICASSP 2005*, 2005.
51. L. Benaroya, F. Bimbot, and R. Gribonval, "Audio source separation with a single sensor," *IEEE Trans. Audio, Speech and Language Processing*, vol. 14, no. 1, pp. 191–199, Jan. 2006.
52. T. Beierholm, B. Pedersen, and O. Winther, "Low complexity Bayesian single channel source separation," in *Proc. ICASSP 2003*, 2003.
53. D. Ellis, "Prediction-driven computational auditory scene analysis," Ph.D. dissertation, MIT, 1996.
54. J. Burred and T. Sikora, "On the use of auditory representations for sparsity-based sound source separation," in *Proc. IEEE Fifth Int. Conf. on Information, Communications and Signal Processing (ICICS)*, Bangkok, Thailand, Dec. 2005.
55. A. Hyvärinen, J. Karhunen, and E. Oja, *Independent Component Analysis*. New York: John Wiley & Sons, 2000.
56. F. Theis and E. Lang, "Formalization of the two-step approach to overcomplete BSS," in *Proc. of SIP 2002*, Kauai, Hawaii, USA, 2002, pp. 207–212. [Online]. Available: http://homepages.uni-regensburg.de/~thf11669/publications/theis02twostep_SIP02.pdf
57. K. Waheed, "Blind source recovery: state space formulations," Department of Electrical and Computer Engineering, Michigan State University, Tech. Rep., Sept. 2001.
58. P. Georgiev, P. G., D. Nuzillard, and A. Ralescu, "Sparse deflations in blind signal separation," in *Independent Component Analysis and Blind Signal Separation*, ser. LNCS, vol. 3889. Springer, 2006, pp. 807–814.
59. Y. Luo, W. Wang, J. Chambers, S. Lambotharan, and I. Proudler, "Exploitation of source nonstationarity in underdetermined blind source separation with advanced clustering techniques," *IEEE Trans. Signal Processing*, vol. 54, no. 6, pp. 2198–2212, June 2006.
60. C. Chang, P. C. Fung, and Y. S. Hung, "On a sparse component analysis approach to blind source separation," in *Independent Component Analysis and Blind Signal Separation*, ser. LNCS, vol. 3889. Springer, 2006, pp. 765–772.
61. B. A. Pearlmutter and V. K. Potluru, "Sparse separation: Principles and tricks," in *Proc SPIE*, vol. 5102, Apr. 2003, pp. 1–4.
62. I. Gorodnitsky and B. Rao, "Sparse signal reconstruction from limited data using FOCUSS: a re-weighted minimum norm algorithm," *IEEE Trans. Signal Processing*, vol. 45, no. 3, pp. 600–616, Mar. 1997.

63. T. Kristjansson, J. Hershey, and H. Attias, "Single microphone source separation using high resolution signal reconstruction," in *Proc. ICASSP 2004*, 2004.
64. A. Nesbit, M. Davies, M. Plumbley, and M. Sandler, "Source extraction from two-channel mixtures by joint cosine packet analysis," in *Proc. EUSICPO 2006*, 2006.
65. L. D. Lathauwer, B. D. Moor, and J. Vandewalle, "Ica techniques for more sources than sensors," in *Proc. HOS 99*, Caesarea, Israel, June 1999, pp. 121–124.
66. P. Comon and O. Grellier, "Non-linear inversion of underdetermined mixtures," in *Proc. ICA 99*, 1999, pp. 461–465.
67. P. Comon, "Blind identification and source separation in 2x3 under-determined mixtures," *IEEE Trans. Signal Processing*, vol. 52, no. 1, pp. 11–22, Jan. 2004.
68. C. M. Bishop, *Neural Networks for Pattern Recognition.* Oxford University Press, 1995.
69. A. Gelman, J. Carlin, H. Stern, and D. Rubin, *Bayesian Data Analysis.* Chapman & Hall, 1995.
70. D. Donoho and M. Elad, "Optimally-sparse representation in general (non-orthogonal) dictionaries via l1 minimization," *Proc. Nat. Aca. Sci*, vol. 100, no. 5, pp. 2197–2202, Mar. 2003. [Online]. Available: http://www.pnas.org/cgi/reprint/100/5/2197.pdf
71. T. Hastie, R. Tibshirani, and J. Friedman, *The Elements of Statistical Learning: Data Mining, Inference, and Prediction*, ser. Springer Series in Statistics. Springer-Verlag, 2002.
72. F. Murtagh, "Comments on 'Parallel algorithms for hierarchical clustering and cluster validity'," *IEEE Trans. on Pattern Analysis and Machine Intelligence*, vol. 14, no. 10, pp. 1056–1057, Oct. 1992.
73. A. Papoulis and S. Pillai, *Probability, Random Variables, and Stochastic Processes*, 4th ed. McGraw-Hill, 2002.
74. A. Pruessner, M. Bussieck, S. Dirkse, and A. Meeraus, "Conic programming in GAMS," in *INFORMS Annual Meeting*, Atlanta, Oct. 2003, pp. 19–22. [Online]. Available: http://www.gams.com/presentations/present_conic.pdf
75. J. Sturm, "Using SeDuMi 1.02, a MATLAB toolbox for optimization over symmetric cones," *Optimization Methods and Software*, vol. 11–12, pp. 625–653, 1999, special issue on Interior Point Methods. [Online]. Available: http://fewcal.kub.nl/sturm/software/sedumi.html
76. L. S. Lobo, L. Vandenberghe, S. Boyd, and H. Lebert, "Second order cone programming," *Linear Algebra and Its Applications*, vol. 284, pp. 193–228, 1998.
77. F. Alizadeh and D. Goldfarb, "Second-order cone programming," Rugers University, Tech. Rep., 2001.
78. S. Boyd and L. Vandenberghe, *Convex Optimization.* Cambridge University Press, 2004.
79. M. Lewicki and T. Sejnowski, "Learning overcomplete representations," *Neural Computation*, vol. 12, no. 2, pp. 337–365, 2000. [Online]. Available: citeseer.nj.nec.com/lewicki98learning.html
80. I. Takigawa, M. Kudo, and J. Toyama, "Performance analysis of minimum ℓ_1-norm solutions for underdetermined source separation," *IEEE Trans. Signal Processing*, vol. 52, no. 3, pp. 582–591, Mar. 2004.
81. D. Malioutov, M. Cetin, and A. Willsky, "Optimal sparse representations in general overcomplete bases," in *Proc. ICASSP 2004*, 2004, pp. 793–796.

82. E. Vincent, R. Gribonval, and C. Févotte, "Performance measurement in blind audio source separation," *IEEE Trans. Speech, Audio and Language Processing*, vol. 14, no. 4, pp. 1462-1469, Jul. 2006.
83. C. Févotte, R. Gribonval, and E. Vincent, "BSS_EVAL toolbox user guide – Revision 2.0," IRISA, Tech. Rep. 1706, Apr. 2005. [Online]. Available: http://bass-db.gforge.inria.fr/bss_eval/

11 Bayesian Audio Source Separation

Cédric Févotte

GET/Télécom Paris (ENST)
37–39, rue Dareau, 75014 Paris, France
E-mail: fevotte@tsi.enst.fr

Abstract. In this chapter we describe a Bayesian approach to audio source separation. The approach relies on probabilistic modeling of sound sources as (sparse) linear combinations of atoms from a dictionary and Markov chain Monte Carlo (MCMC) inference. Several prior distributions are considered for the source expansion coefficients. We first consider independent and identically distributed (iid) general priors with two choices of distributions. The first one is the Student t, which is a good model for sparsity when the shape parameter has a low value. The second one is a hierarchical mixture distribution; conditionally upon an indicator variable, one coefficient is either set to zero or given a normal distribution, whose variance is in turn given an inverted-Gamma distribution. Then, we consider more audio-specific models where both the identically distributed and independently distributed assumptions are lifted. Using a Modified Discrete Cosine Transform (MDCT) dictionary, a time–frequency orthonormal basis, we describe frequency-dependent structured priors which explicitly model the harmonic structure of sound, using a Markov hierarchical modeling of the expansion coefficients. Separation results are given for a stereophonic recording of three sources.

11.1 Introduction

In this chapter we take a Bayesian approach to blind source separation (BSS). We limit our study to the linear instantaneous problem, possibly underdetermined. Our notations are such that, for $t = 1, \dots, N$

$$\mathbf{x}_t = \mathbf{A}\,\mathbf{s}_t + \mathbf{e}_t \tag{11.1}$$

where $\mathbf{x}_t = [x_{1,t}, \dots, x_{m,t}]^T$ is a vector of size m containing the observations, $\mathbf{s}_t = [s_{1,t}, \dots, s_{n,t}]^T$ is a vector of size n containing the sources and $\mathbf{e}_t = [e_{1,t}, \dots, e_{m,t}]^T$ is a vector of size m containing additive noise/residual error. Variables without time index t will denote whole sequences of samples, e.g., $\mathbf{x} = [\mathbf{x}_1, \dots, \mathbf{x}_N]$ and $x_1 = [x_{1,1}, \dots, x_{1,N}]$. $\mathbf{A}$ is the mixing matrix of size $m \times n$ and all the latter variables are assumed to be real-valued.

11.1.1 The Bayesian Setting

One fundamental principle of the Bayesian approach is to consider all the unknown variables of the problem $\mathbf{A}$, $\mathbf{s}$, $\mathbf{e}$ to be the realizations of random variables. This approach hence underlies a *probabilistic modeling* of these parameters, as opposed to have a fixed, ground-truth value. Each of these variable is assumed to follow a model described by a set of parameters respectively

S. Makino et al. (eds.), Blind Speech Separation, 305–335.

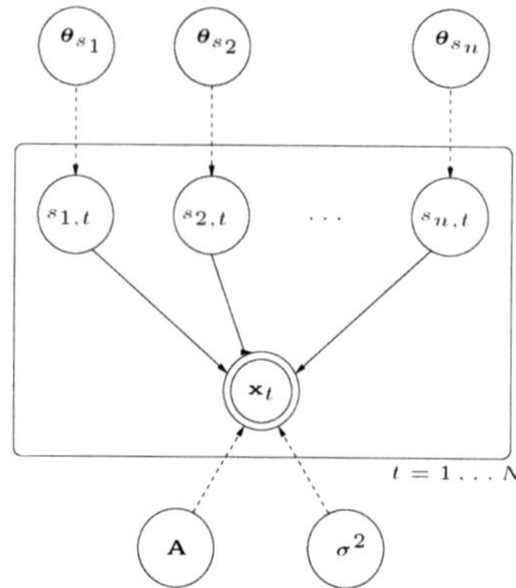

Fig. 11.1. Directed acyclic graph (DAG) representation of the linear instantaneous BSS problem.

denoted by $\boldsymbol{\theta}_{\mathbf{A}}$, $\boldsymbol{\theta}_{\mathbf{s}}$ and $\boldsymbol{\theta}_{\mathbf{e}}$. The parameters $\mathbf{A}$, $\mathbf{s}$, $\mathbf{e}$ are thus characterized by their *prior distributions* $p(\mathbf{A}|\boldsymbol{\theta}_{\mathbf{A}})$, $p(\mathbf{s}|\boldsymbol{\theta}_{\mathbf{s}})$ and $p(\mathbf{e}|\boldsymbol{\theta}_{\mathbf{e}})$. The parameters $\boldsymbol{\theta}_{\mathbf{A}}$, $\boldsymbol{\theta}_{\mathbf{s}}$ and $\boldsymbol{\theta}_{\mathbf{e}}$ are referred to as *hyperparameters*. Depending on the degree of prior knowledge about $\mathbf{A}$, $\mathbf{s}$ and $\mathbf{e}$ the hyperparameters can be either fixed or treated as unknown parameters themselves, to be estimated (or, in a machine learning parlance, to be *learnt*) from the data $\mathbf{x}$. In the following, the set of all the parameters is noted $\boldsymbol{\theta} = \{\mathbf{A}, \mathbf{s}, \boldsymbol{\theta}_{\mathbf{A}}, \boldsymbol{\theta}_{\mathbf{s}}, \boldsymbol{\theta}_{\mathbf{e}}\}$.[1] Figure 11.1 gives a graph representation of the model and shows the connections between the various parameters.

Given this statistical setting of the problem, Bayesian estimation revolves around the *posterior distribution* of the set of all unknown parameters $p(\boldsymbol{\theta}|\mathbf{x})$. Information about $\boldsymbol{\theta}$ or subsets of $\boldsymbol{\theta}$ is *inferred* from the data through manipulation of the posterior. As such, typical *point estimates* are the maximum a posteriori (MAP) estimate $\hat{\boldsymbol{\theta}}_{MAP} = \operatorname{argmax}_{\boldsymbol{\theta}} p(\boldsymbol{\theta}|\mathbf{x})$ and the minimum mean square error (MMSE) estimate $\hat{\boldsymbol{\theta}}_{MMSE} = \mathrm{E}\{\boldsymbol{\theta}|\mathbf{x}\} = \int \boldsymbol{\theta}\, p(\boldsymbol{\theta}|\mathbf{x})\, d\boldsymbol{\theta}$.

A wide literature now exists about Bayesian source separation, covering many source and mixture models, as well as many computational techniques. Early works include the ones of Knuth [1, 2] and Djafari [3]. These authors show how some standard independent component analysis (ICA) techniques (such as InfoMax [4]) can easily be rephrased in a Bayesian parlance, but also how the Bayesian formalism easily allows to go beyond the standard hypotheses of ICA – square mixing matrix, independent and identically distributed (iid) sources, no additive noise/residual error. As such, Knuth shows in [1, 2] how the Bayesian formalism allows to incorporate prior knowledge about the location of the sensors (the mixing matrix) in the separation process. In [3], Djafari presents MAP gradient approaches for

[1] $\mathbf{e}$ is omitted of the set as we simply have $\mathbf{e} = \mathbf{x} - \mathbf{A}\,\mathbf{s}$.

separation of possibly underdetermined and noisy mixtures of possibly spatially correlated and non-iid sources.

Bayesian approaches are precisely of utter interest in the underdetermined case. In the overdetermined non-noisy case, it is sufficient to estimate the mixing matrix and to apply its inverse to the observations to reconstruct the sources. In that case Cardoso shows in [5] that the sources need not to be accurately modeled to obtain a good estimation of the mixing matrix, and thus a good separation. Oppositely, the underdetermined case is an ill-posed problem because the mixing matrix is not invertible, and prior information about the sources is important to their reconstruction (the more information is available, the better the reconstruction). Prior information is also required in noisy scenarios. The Bayesian formalism is adapted to these problems because it allows to gather the available information about the sources in the prior $p(\mathbf{s}|\boldsymbol{\theta}_\mathbf{s})$. Note that there is a trade-off between the complexity of the source models and the complexity of the inference step to follow (consisting of minimizing a criterion or sampling a distribution). The source models should be built as to gather as much information as possible while keeping the inference step tractable.

11.1.2 Sparse Models: Analysis vs Synthesis

In this chapter, we address Bayesian separation of underdetermined noisy mixtures of audio sources. The source models that we use are based on *sparsity* properties. A sequence is said to be sparse when most of its samples take a value close to zero. The use of sparsity to handle the general linear instantaneous model, has arisen in several papers in the areas of learning [6–8] and source separation [9–12] (to name a few). For source separation, a linear transform is typically applied to each observation, transposing the problem in a *transform domain* where the sources become sparse. Most of the latter papers then take a Bayesian approach: the sources are assumed to be the realizations of a iid random process with distribution gathering most of its probability around zero and presenting heavy tails, thus modeling sparsity. This distribution is the Laplace distribution in [7–10], a mixture of Gaussians in [6], a mixture of Gaussian and Dirac in [11], a generalized Gaussian in [13]. Various inference strategies follow, all gradient-based or using the Expectation Maximization (EM) algorithm, and aiming at MAP or MMSE estimates of $\mathbf{A}$ and the sources.

The common ingredient of these methods is to work in a transform domain. There are two dual approaches for doing so. In many of the latter papers, a short-time Fourier transform is typically *applied* to the observations when dealing with audio, while a wavelet transform is used for images. This is an *analysis* approach. Given a dictionary $\boldsymbol{\Phi} \in \mathbb{R}^{K \times N}$ containing K waveforms, also referred to as atoms (with $K \geq N$), the analysis approach consists of computing the dot products of every atom of the dictionary with each observation, such that, $\forall j = 1, \ldots, m$

$$\tilde{x}_j^{an} = x_j \, \boldsymbol{\Phi}^T \tag{11.2}$$

and yielding

$$\tilde{\mathbf{x}}^{an} = \mathbf{A} \, \tilde{\mathbf{s}}^{an} + \tilde{\mathbf{e}}^{an} \tag{11.3}$$

where $\tilde{\mathbf{x}}^{an} = \mathbf{x} \, \boldsymbol{\Phi}^T$, $\tilde{\mathbf{s}}^{an} = \mathbf{s} \, \boldsymbol{\Phi}^T$ and $\tilde{\mathbf{e}}^{an} = \mathbf{e} \, \boldsymbol{\Phi}^T$. The analysis approach thus simply transforms the time domain linear instantaneous mixture in another linear instantaneous mixture. Any BSS method can be applied to the new mixture, and when an estimate $\widehat{\tilde{\mathbf{s}}^{an}}$ is obtained, a time domain estimate can be reconstructed through the pseudo-inverse of $\boldsymbol{\Phi}^T$ or, in a *frame* parlance, the dual operator [14], such that

$$\hat{\mathbf{s}} = \widehat{\tilde{\mathbf{s}}^{an}} \, \boldsymbol{\Phi} \, (\boldsymbol{\Phi}^T \, \boldsymbol{\Phi})^{-1} \tag{11.4}$$

Conversely, the *synthesis* approach models the sources as a linear combination of atoms from $\boldsymbol{\Phi}$, such that $\forall i = 1, \ldots, n$

$$s_i = \tilde{s}_i^{sy} \, \boldsymbol{\Phi} \tag{11.5}$$

and yielding

$$\mathbf{x} = \mathbf{A} \, \tilde{\mathbf{s}}^{sy} \, \boldsymbol{\Phi} + \mathbf{e} \tag{11.6}$$

The two approaches, analysis and synthesis, differ very much in nature. The analysis approach aims at *sparsifying* the data, and its underlying sources, while the synthesis approach intrinsically models the sources as a sparse linear combination of atoms. The motivation of the analysis approach is to come up with sources with lower entropy, which tends to cluster the data along the mixing matrix columns (see Chapter 7), and potentially yields estimates with lower error variance [5]. The synthesis approach is *generative* in essence, it allows to build source models taking into account the specificities of the physical phenomena which generated the signals. Note that, when $s = \tilde{s}^{sy} \, \boldsymbol{\Phi}$ and $\boldsymbol{\Phi}$ is overcomplete ($K > N$), the synthesis coefficients $\tilde{s}^{sy}$ are not retrieved through direct analysis, as we have $\tilde{s}^{an} = \tilde{s}^{sy} \, \boldsymbol{\Phi} \, \boldsymbol{\Phi}^T \neq \tilde{s}^{sy}$. Furthermore, the operator $\boldsymbol{\Phi} \, \boldsymbol{\Phi}^T$ creates, in general, a "blurring" effect and $\tilde{s}^{an}$ is potentially less sparse than $\tilde{s}^{sy}$ [15].

The *analysis* and *synthesis* coefficients coincide when $\boldsymbol{\Phi}$ is an orthonormal basis, i.e, $\boldsymbol{\Phi} \, \boldsymbol{\Phi}^T = \mathbf{I}_N$, which is the case we address in this chapter. We thus note $\tilde{s}_i = \tilde{s}_i^{an} = \tilde{s}_i^{sy}$. However, we will abide to a synthesis interpretation of the models described, and will mention the generalization to the overcomplete case when possible.

This chapter is organized as follows. In Sect. 11.2 we describe a general Bayesian framework for blind separation of sparse sources. Two source priors are considered in Sect. 11.2.1: the Student t prior and a hierarchical mixture prior. The first prior is a distribution with two parameters, one scale parameter and one shape parameter controlling the peakiness of the density (and thus the degree of sparsity). We will take advantage of its hierarchical formulation as a scale mixture of Gaussians (SMoG), which yields computational facilities. The second prior has a hierarchical structure too, it consists of a mixture of a Dirac distribution centered at 0 and a normal distribution, whose variance is in turn given a conjugate inverted-Gamma prior. We then derive in Sect. 11.2.3 a Gibbs sampler, a standard Markov chain Monte Carlo method, to generate samples from the posterior distribution $p(\boldsymbol{\theta}|\mathbf{x})$.

While the methodology presented in Sect. 11.2 is very general, in Sect. 11.3 we more specifically address audio sources. We describe improvements to the latter priors in order to take into account the specificities of audio signals. As such, using a MDCT time–frequency orthonormal basis, we show how to model simply the nonuniform energy distribution of audio signals and how to model structural constraints (corresponding to physical phenomena such as tonal parts). We present in Sect. 11.4 results of the separation of a stereophonic musical recording with three sources (singing voice, acoustic guitar, bass guitar) using the techniques described. Finally, conclusions are given in Sect. 11.5.

11.2 Bayesian Sparse Source Separation

11.2.1 Sparse Source Priors

In this section, we model each source coefficients sequence $\tilde{s}_i$ as an iid sequence, with two distributions considered, the Student t distribution and a hierarchical mixture prior. The sequences $\tilde{s}_1, \ldots, \tilde{s}_n$ are furthermore modeled as mutually independent, such that the prior distribution $p(\tilde{\mathbf{s}}|\boldsymbol{\theta}_{\tilde{\mathbf{s}}})$ factorizes as

$$p(\tilde{\mathbf{s}}|\boldsymbol{\theta}_{\mathbf{s}}) = \prod_{i=1}^{n} \prod_{k=1}^{N} p(\tilde{s}_{i,k}|\boldsymbol{\theta}_{s_i}) \tag{11.7}$$

Student t Prior The Student t prior is given by

$$p(\tilde{s}_{i,k}|\nu_i, \xi_i) = \frac{\Gamma(\frac{\nu_i+1}{2})}{\xi_i\,\sqrt{\nu_i\,\pi}\,\Gamma(\frac{\nu_i}{2})} \left(1 + \frac{1}{\nu_i}\left(\frac{\tilde{s}_{i,k}}{\xi_i}\right)^2\right)^{-\frac{\nu_i+1}{2}} \tag{11.8}$$

where ν_i is the degrees of freedom and ξ_i is a scale parameter. With $\xi_i = 1$ and $\nu_i = 1$, the Student t distribution is equal to the standard Cauchy distribution, and it tends to the standard Gaussian distribution as ν_i goes to infinity. Figure 11.2 plots Student t densities for several values of ν_i, with equal mode, i.e setting $\xi_i = \Gamma(\frac{\nu_i+1}{2})/\sqrt{\nu_i\,\pi}\,\Gamma(\frac{\nu_i}{2})$ for each density. Figure 11.2

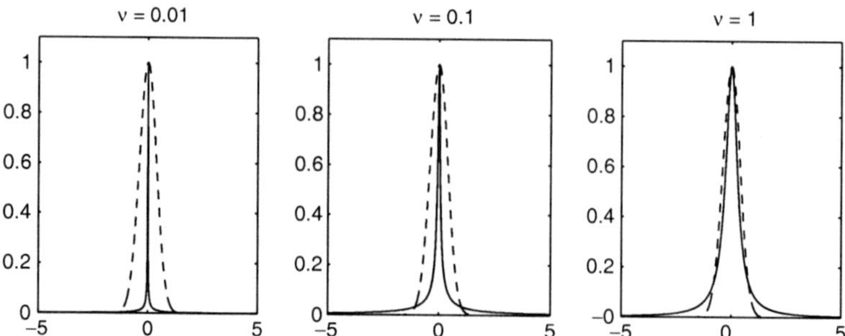

Fig. 11.2. Student t densities for $\nu \in \{0.01, 0.1, 1\}$ with equal value at the mode. The dash-lined plot is the Gaussian density with variance $1/2\,\pi$.

shows that for small ν_i, the Student t density gathers most of its probability mass around zero and exhibits "fatter tails" than the normal distribution. The Student t is thus a relevant model for sparsity.

The Student t distribution can be expressed as a SMoG [16], such that

$$p(\tilde{s}_{i,k}|\nu_i, \xi_i) = \int_0^{+\infty} \mathcal{N}(\tilde{s}_{i,k}|0, v_{i,k})\, \mathcal{IG}\,(v_{i,k}|\alpha_i, \lambda_i)\, dv_{i,k} \tag{11.9}$$

where

$$\alpha_i = \frac{\nu_i}{2} \quad \lambda_i = \frac{\nu_i\, \xi_i^2}{2} \tag{11.10}$$

and where $\mathcal{N}(x|\mu, v)$ and $\mathcal{IG}(x|\alpha, \beta)$ are the Gaussian and inverted-Gamma distributions, defined in Appendix A.1. The variance $v_{i,k}$ can be treated as an auxiliary variable and $p(\tilde{s}_{i,k}|\nu_i, \xi_i)$ can be thus interpreted as a marginal density of the joint distribution $p(\tilde{s}_{i,k}, v_{i,k}|\nu_i, \xi_i)$, defined by

$$p(\tilde{s}_{i,k}, v_{i,k}|\nu_i, \xi_i) = p(\tilde{s}_{i,k}|v_{i,k})\, p(v_{i,k}|\alpha_i, \lambda_i) \tag{11.11}$$

with

$$p(\tilde{s}_{i,k}|v_{i,k}) = \mathcal{N}(\tilde{s}_{i,k}|0, v_{i,k}) \quad \text{and} \quad p(v_{i,k}|\alpha_i, \lambda_i) = \mathcal{IG}\,(v_{i,k}|\alpha_i, \lambda_i) \tag{11.12}$$

This hierarchical formulation of the Student t, and the fact that the prior of $v_{i,k}$ is conjugate,[2] lead to easy Gibbs updates for both $\tilde{s}_{i,k}$ and $v_{i,k}$ as shown in Sect. 11.2.3. A graphical representation of the Student t source coefficient model is given at Fig. 11.3.

[2]If a parameter θ is observed through the observation x via the likelihood $p(x|\theta)$, the prior $p(\theta)$ is said to be conjugate when $p(\theta)$ and $p(\theta|x) \propto p(x|\theta)\, p(\theta)$ belong to the same family of distributions. Here, $v_{i,k}$ is observed through $\tilde{s}_{i,k}$ via $p(\tilde{s}_{i,k}|v_{i,k})$, its prior is $\mathcal{IG}(v_{i,k}|\alpha_i, \lambda_i)$ and its posterior $p(v_{i,k}|\tilde{s}_{i,k}, \alpha_i, \lambda_i)$, given at (11.38) is also inverted-Gamma.

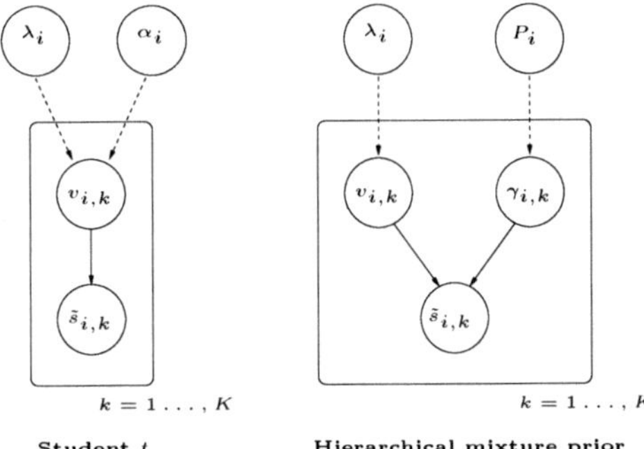

Fig. 11.3. Graphs of the source coefficients models. Left: Student t; Right: hierarchical mixture prior.

The source hyperparameters $\boldsymbol{\theta}_{s_i} = \{\alpha_i, \lambda_i\}$ can be given priors as well. Low values of ν_i, supporting sparsity, can be favored using and exponential prior

$$p(\alpha_i) = \beta_{\alpha_i} \exp(-\beta_{\alpha_i} \alpha_i). \tag{11.13}$$

A Gamma conjugate prior is chosen for λ_i, such that

$$p(\lambda_i) = \mathcal{G}(\lambda_i | \alpha_{\lambda_i}, \beta_{\lambda_i}) \tag{11.14}$$

The Student t can be interpreted as an infinite sum of Gaussians, which contrasts with the finite sums of Gaussians used in [6, 11]. The Laplace prior used in [7, 9] can also be expressed as a SMoG, with an exponential density on the variance $v_{i,k}$ [16]. However, the Student t prior has the advantage to offer a supplementary hyperparameter α_i which controls the sharpness of the distribution. The Laplace and the Student t belong to the more general family of general hyperbolic processes, with many shape parameters, which is used for source separation in [17]. However, the authors of [17] point out the limits of this class of priors, which is too general: too much flexibility does not in the end bring any prior information.

Hierarchical Mixture Prior The second prior we consider is a hierarchical mixture prior, given by

$$p(\tilde{s}_{i,k} | \gamma_{i,k}, v_{i,k}) = (1 - \gamma_{i,k})\, \delta_0(\tilde{s}_{i,k}) + \gamma_{i,k}\, \mathcal{N}(\tilde{s}_{i,k} | 0, v_{i,k}) \tag{11.15}$$

$$p(v_{i,k} | \alpha_i, \lambda_i) = \mathcal{IG}(v_{i,k} | \alpha_i, \lambda_i) \tag{11.16}$$

where $\delta_0(u)$ is the Dirac delta function and $\gamma_{i,k} \in \{0, 1\}$ is an indicator variable. When $\gamma_{i,k} = 0$, $\tilde{s}_{i,k}$ is set to zero; when $\gamma_{i,k} = 1$, $\tilde{s}_{i,k}$ has a normal distribution with zero mean and variance $v_{i,k}$, which is in turn assigned a conjugate inverted-Gamma prior. The set of indicator variables γ_i is so far modeled as iid, with Bernoulli prior

$$P(\gamma_{i,k} = 1|P_i) = P_i \quad P(\gamma_{i,k} = 0|P_i) = 1 - P_i \tag{11.17}$$

A graphical representation of the model is given at Fig. 11.3.

Contrary to the Student t, this prior *explicitly* models the possibility for a coefficient to be zero. Note that, conditionally upon $\gamma_{i,k} = 1$, the marginal $p(\tilde{s}_{i,k}|\gamma_k = 1, \alpha_i, \lambda_i)$ is Student t, with density $t(\tilde{s}_{i,k}|2\,\alpha_i, \sqrt{\lambda_i/\alpha_i})$. However, the degrees of freedom do not here play a key role like in previous paragraph, and in particular, it does not need to be given a small value. Here, the degree of sparsity is controlled by the parameter P_i, i.e the probability of one coefficient to be nonzero. The value of α_i is fixed to a value chosen by hand (see Sect. 11.4). Note that with the iid Bernoulli prior, $\gamma_{i,k}$ could be integrated out from (11.15), yielding simply

$$p(\tilde{s}_{i,k}|\boldsymbol{\theta}_{s_i}) = (1 - P_i)\,\delta_0(\tilde{s}_{i,k}) + P_i\, t(\tilde{s}_{i,k}|2\,\alpha_i, \sqrt{\lambda_i/\alpha_i}) \tag{11.18}$$

with $\theta_{s_i} = \{P_i, \lambda_i\}$. Like $v_{i,k}$, $\gamma_{i,k}$ is an auxiliary variable which, in the iid case, could be easily removed from the notations. It will play a more important role in Sect. 11.3.3 when used to model dependencies (structures) within the set of coefficients $\tilde{s}_i$.

The scale parameter λ_i is given a Gamma conjugate prior as before (see (11.14)). The probability P_i is given a Beta prior (defined in Appendix A.1), such that

$$p(P_i) = \mathcal{B}(P_i|\alpha_{P_i}, \beta_{P_i}) \tag{11.19}$$

The values of α_{P_i} and β_{P_i} can be adjusted so as to yield a prior for P_i favoring low values and thus sparsity.

11.2.2 Noise and Mixing Matrix Priors

We assume for simplicity the sequence $\mathbf{e}$ to be iid Gaussian, with covariance $\sigma^2\,\mathbf{I}_m$. However, different noise variances on every observation x_j could also easily be considered. When an orthonormal basis is used ($i.e$, $\mathbf{\Phi}^{-1} = \mathbf{\Phi}^T$), $\tilde{\mathbf{e}}$ is equivalently iid Gaussian with covariance $\sigma^2\,\mathbf{I}_m$, so that we have equivalence between time domain ($\mathbf{x}_t = \mathbf{A}\,\mathbf{s}_t + \mathbf{e}_t$) and transform domain ($\tilde{\mathbf{x}}_k = \mathbf{A}\,\tilde{\mathbf{s}}_k + \tilde{\mathbf{e}}_k$).

The variance σ^2 is treated as an unknown parameter ($\boldsymbol{\theta}_\mathrm{e} = \{\sigma^2\}$), and can be given a conjugate inverted-Gamma prior, such that

$$p(\sigma^2) = \mathcal{IG}(\sigma^2|\alpha_{\sigma^2}, \beta_{\sigma^2}) \tag{11.20}$$

The matrix $\mathbf{A}$ is treated in the following as a column vector $\mathbf{a}$ defined by

$$\mathbf{a} = \begin{bmatrix} \mathbf{r}_1 \\ \vdots \\ \mathbf{r}_n \end{bmatrix} \tag{11.21}$$

where $\mathbf{r}_1, \ldots, \mathbf{r}_n$ denote the transposed rows of $\mathbf{A}$. A Gaussian conjugate prior could be used for $\mathbf{a}$, however, in practice we will simply use a noninformative flat prior $p(\mathbf{a}) \propto 1$, so that $\boldsymbol{\theta}_{\mathbf{A}}$ is empty.

11.2.3 Markov Chain Monte Carlo Inference

We derive in the following a Gibbs sampler to generate samples (realizations) from the posterior distribution $p(\boldsymbol{\theta}|\mathbf{x})$. The obtained samples $\{\boldsymbol{\theta}^{(1)}, \ldots, \boldsymbol{\theta}^{(L)}\}$ then allow for computation of any required point estimate (as well as interval estimates). For example, the Minimum Mean Square Error (MMSE) estimate is approximated by

$$\hat{\boldsymbol{\theta}}_{MMSE} \approx \frac{1}{L} \sum_{l=1}^{L} \boldsymbol{\theta}^{(l)} \tag{11.22}$$

The Gibbs sampler only requires to be able to sample from the posterior distribution of certain subsets of parameters conditional upon data $\mathbf{x}$ and the remaining parameters [18, 19]. Let $\{\boldsymbol{\theta}_1, \ldots, \boldsymbol{\theta}_M\}$ denote a partition of $\boldsymbol{\theta}$. The Gibbs sampler is described by Algorithm 1. L_{bi} represents the number of iterations required before the Markov chain $\{\boldsymbol{\theta}^{(1)}, \boldsymbol{\theta}^{(2)}, \ldots\}$ reaches its stationary distribution $p(\boldsymbol{\theta}|\mathbf{x})$ (the *burn in* period). Thereafter, all samples are drawn from the desired stationary distribution. MCMC methods have the advantage to generate samples from the whole support of $p(\boldsymbol{\theta}|\mathbf{x})$, and thus to give an overall panorama of the posterior distribution of the parameters. When looking for a point estimate of $\boldsymbol{\theta}$, the MCMC approach theoretically prevents from falling into local modes of the posterior distribution, which is a common drawback of standard optimization methods, like Expectation Maximization or gradient type methods, which target point estimates (such as MAP estimates) directly.

The implementation of a Gibbs sampler thus requires to define a partition of the set of parameters $\boldsymbol{\theta}$, and then to sample each subset conditionally upon the others and the data. We will discuss different ways to partition $\boldsymbol{\theta}$ in the following, but as a general rule, fastest convergence is obtained when as many parameters as possible are sampled jointly [20, 21]. In that sense, Gibbs sampling can be thought of as a stochastic version of iterated relaxed gradient optimization where deterministic moves are replaced by random moves. The largest moves in the right direction, the fastest convergence.

Algorithm 1 Gibbs sampler

Initialize $\boldsymbol{\theta}^{(0)} = \{\boldsymbol{\theta}_1^{(0)}, \dots, \boldsymbol{\theta}_M^{(0)}\}$
for $l = 1 : L + L_{bi}$ **do**
$\boldsymbol{\theta}_1^{(l)} \sim p(\boldsymbol{\theta}_1 | \boldsymbol{\theta}_2^{(l-1)}, \dots, \boldsymbol{\theta}_M^{(l-1)}, \mathbf{x})$
$\boldsymbol{\theta}_2^{(l)} \sim p(\boldsymbol{\theta}_2 | \boldsymbol{\theta}_1^{(l)}, \boldsymbol{\theta}_3^{(l-1)}, \dots, \boldsymbol{\theta}_M^{(l-1)}, \mathbf{x})$
$\boldsymbol{\theta}_3^{(l)} \sim p(\boldsymbol{\theta}_3 | \boldsymbol{\theta}_1^{(l)}, \boldsymbol{\theta}_2^{(l)}, \boldsymbol{\theta}_4^{(l-1)} \dots, \boldsymbol{\theta}_M^{(l-1)}, \mathbf{x})$
$\vdots$
$\boldsymbol{\theta}_M^{(l)} \sim p(\boldsymbol{\theta}_M | \boldsymbol{\theta}_1^{(l)}, \boldsymbol{\theta}_2^{(l)}, \dots, \boldsymbol{\theta}_{M-1}^{(l)}, \mathbf{x})$
end for

How to Write a Conditional Distribution? We note $\boldsymbol{\theta}_{-j}$ the set $\{\boldsymbol{\theta}_1, \dots, \boldsymbol{\theta}_{j-1}, \boldsymbol{\theta}_{j+1}, \dots, \boldsymbol{\theta}_M\}$. We need to sample from $p(\boldsymbol{\theta}_j | \boldsymbol{\theta}_{-j}, \mathbf{x})$, $\forall j = 1, \dots, M$. The conditional distribution of $\boldsymbol{\theta}_j$ is defined by

$$p(\boldsymbol{\theta}_j | \boldsymbol{\theta}_{-j}, \mathbf{x}) = \frac{p(\boldsymbol{\theta} | \mathbf{x})}{p(\boldsymbol{\theta}_{-j} | \mathbf{x})} \tag{11.23}$$

Bayes' theorem gives

$$p(\boldsymbol{\theta} | \mathbf{x}) = \frac{p(\mathbf{x} | \boldsymbol{\theta}) \, p(\boldsymbol{\theta})}{p(\mathbf{x})} \tag{11.24}$$

and it follows

$$p(\boldsymbol{\theta}_j | \boldsymbol{\theta}_{-j}, \mathbf{x}) = \frac{p(\mathbf{x} | \boldsymbol{\theta}_j, \boldsymbol{\theta}_{-j}) \, p(\boldsymbol{\theta}_j, \boldsymbol{\theta}_{-j})}{p(\boldsymbol{\theta}_{-j} | \mathbf{x}) \, p(\mathbf{x})} \tag{11.25}$$

$$\propto \underbrace{p(\mathbf{x} | \boldsymbol{\theta}_j, \boldsymbol{\theta}_{-j})}_{likelihood} \; \underbrace{p(\boldsymbol{\theta}_j | \boldsymbol{\theta}_{-j})}_{prior} \tag{11.26}$$

In our case the likelihood is Gaussian and the prior $p(\boldsymbol{\theta})$ factorizes as

$$p(\boldsymbol{\theta}) = p(\mathbf{A}) \, p(\sigma^2) \prod_{i=1}^{n} p(s_i | \boldsymbol{\theta}_{s_i}) \tag{11.27}$$

In the following we only discuss the ways to partition $\boldsymbol{\theta}$ and skip the details about how to calculate the conditional posterior distribution. Since most of the priors used are conjugate, the posteriors are rather straightforward to obtain, and belong to families of distributions easy to sample from, namely unidimensional or multivariate Gaussian and (inverted-)Gamma distributions. Further details can be found in [22].

Update of A and σ^2 Figure 11.1 shows that the conditional posterior distributions of $\mathbf{A}$ and σ^2 merely depend on the source signals $\mathbf{s}$ (and independently of how they are modeled) and $\mathbf{x}$.

When the flat prior $p(\mathbf{a}) \propto 1$ is used, the rows of $\mathbf{A}$ are found to be a posteriori mutually independent with

$$p(\mathbf{r}_i|\sigma^2, \mathbf{s}, \mathbf{x}) = \mathcal{N}(\boldsymbol{\mu}_{\mathbf{r}_i}, \boldsymbol{\Sigma}_{\mathbf{r}}) \tag{11.28}$$

where $\boldsymbol{\Sigma}_{\mathbf{r}} = \sigma^2 \left(\sum_t \mathbf{s}_t \, \mathbf{s}_t^T\right)^{-1}$ and $\boldsymbol{\mu}_{\mathbf{r}_i} = \frac{1}{\sigma^2} \boldsymbol{\Sigma}_{\mathbf{r}} \sum_t x_{i,t} \, \mathbf{s}_t$.[3] The posterior distribution of σ^2 is simply

$$p(\sigma^2|\mathbf{A}, \mathbf{s}, \mathbf{x}) = \mathcal{IG}(\sigma^2|\alpha_{\sigma^2}^{post}, \lambda_{\sigma^2}^{post}) \tag{11.29}$$

with

$$\alpha_{\sigma^2}^{post} = \frac{N\,m}{2} + \alpha_{\sigma^2} \quad \text{and} \quad \beta_{\sigma^2}^{post} = \frac{1}{2} \, \|\mathbf{x} - \mathbf{A}\,\mathbf{s}\|_F^2 + \beta_{\sigma^2} \tag{11.30}$$

Note that $\alpha_{\sigma^2}^{post}$ and $\beta_{\sigma^2}^{post}$ reflects the trade-off between information provided from observation and from the prior. When N is large, α_{σ^2} and β_{σ^2} have little influence in the posterior. As such, Jeffreys noninformative prior $p(\sigma^2) \propto 1/\sigma^2$, corresponding to $\alpha_{\sigma^2} \to 0$ and $\beta_{\sigma^2} \to 0$, can be used in practice.

$\mathbf{A}$ and σ^2 can also be sampled jointly, i.e, from the joint distribution $p(\mathbf{A}, \sigma^2|\mathbf{s}, \mathbf{x})$. This is done by first sampling $\sigma^{2\,(l)}$ from $p(\sigma^2|\mathbf{s}, \mathbf{x})$ and then sampling $\mathbf{A}^{(l)}$ from $p(\mathbf{A}|\sigma^{2\,(l)}, \mathbf{s}, \mathbf{x})$. The first step involves integrating $\mathbf{A}$ out of the conditional distribution $p(\sigma^2|\mathbf{A}, \mathbf{s}, \mathbf{x})$, which is done in [22].

When the dictionary $\boldsymbol{\Phi}$ is orthonormal and the noise $\mathbf{e}$ is iid Gaussian, $\mathbf{A}$ and σ^2 can be updated using $\tilde{\mathbf{s}}$ and $\tilde{\mathbf{x}}$ instead of $\mathbf{s}$ and $\mathbf{x}$.

Update of the Sources Several strategies can be employed to update the sources $s_1, \ldots, s_n$. The most simple one, and more general, is to update the sources one by one, each conditionally upon the others. Indeed, we have, $\forall t = 1, \ldots, N$,

$$\mathbf{x}_t = s_{i,t} \, \mathbf{a}_i + \sum_{j \neq i} s_{j,t} \, \mathbf{a}_j + \mathbf{e}_t \tag{11.31}$$

yielding

$$\underbrace{\frac{\mathbf{a}_i^T \, \mathbf{x}}{\mathbf{a}_i^T \, \mathbf{a}_i} - \sum_{j \neq i} \frac{\mathbf{a}_i^T \, \mathbf{a}_j}{\mathbf{a}_i^T \, \mathbf{a}_i} s_j}_{x_{i|-i}} = s_i + \underbrace{\frac{\mathbf{a}_i^T \, \mathbf{e}}{\mathbf{a}_i^T \, \mathbf{a}_i}}_{e_{i|-i}} \tag{11.32}$$

The update of source s_i conditionally upon s_{-i}, σ^2 and $\mathbf{A}$ can thus be recast into a simple unidimensional sparse regression problem with data $x_{i|-i}$ such that

$$x_{i|-i} = \tilde{s}_i \, \boldsymbol{\Phi} + e_{i|-i} \tag{11.33}$$

[3] In practice the columns of $\mathbf{A}$ are normalized to 1 after each draw to solve the BSS indeterminacy on gain. A rigorous implementation would imply sampling each column of $\mathbf{A}$ from the sphere, which is not straightforward. The proposed scheme, though approximate, proved to be satisfactory in practice.

and where $e_{i|-i}$ is a Gaussian iid residue of variance $\sigma_i^2 = \sigma^2/\|\mathbf{a}_i\|_2^2$. Note that, in this framework, the sources can have different models. For example, speech sources and musical sources can be modeled differently, and be processed iteratively. In some cases, when the sources share the same model, block updates of the sources can also be made, as discussed in the following.

When $\mathbf{\Phi}$ is an orthonormal basis, we simply have

$$\tilde{x}_{i|-i} = \tilde{s}_i + \tilde{e}_{i|-i} \tag{11.34}$$

With the noise and source coefficients iid assumptions, the posterior distribution of $\tilde{s}_i$ factorizes and we simply need to infer $\tilde{s}_{i,k}$ from $\tilde{x}_{i|-i,k}$, $\forall k = 1, \ldots, N$, where $\tilde{s}_{i,k}$ has either a Student t distribution or a hierarchical mixture prior.

Update of Student *t* Source Coefficients We need to sample from $p(\tilde{s}_{i,k}|\alpha_i, \lambda_i, \sigma_i^2, \tilde{x}_{i|-i,k})$. Sampling from this distribution directly is not straightforward, and we instead take advantage of the SMoG formulation of the Student t distribution. As such, a sample $\tilde{s}_{i,k}^{(l)}$ of $p(\tilde{s}_{i,k}|\alpha_i, \lambda_i, \sigma_i^2, \tilde{x}_{i|-i,k})$ can be obtained by sampling $(\tilde{s}_{i,k}^{(l)}, v_{i,k}^{(l)})$ from $p(\tilde{s}_{i,k}, v_{i,k}|\alpha_i, \lambda_i, \sigma_i^2, \tilde{x}_{i|-i,k})$. This is easily done using two Gibbs steps, by alternatively sampling $p(v_{i,k}|\tilde{s}_{i,k}, \alpha_i, \lambda_i)$ and $p(\tilde{s}_{i,k}|v_{i,k}, \sigma_i^2, \tilde{x}_{i|-i,k})$. Conditionally upon $v_{i,k}$, inferring $\tilde{s}_{i,k}$ simply amounts to inferring a Gaussian parameter embedded in Gaussian noise, i.e, Wiener filtering. Thus, we simply have

$$p(\tilde{s}_{i,k}|v_{i,k}, \sigma_i^2, \tilde{x}_{i|-i,k}) = \mathcal{N}(\tilde{s}_{i,k}|\mu_{\tilde{s}_{i,k}}, \sigma_{\tilde{s}_{i,k}}^2) \tag{11.35}$$

with $\sigma_{\tilde{s}_{i,k}}^2 = \left(1/\sigma_i^2 + 1/v_{i,k}\right)^{-1}$ and $\mu_{\tilde{s}_{i,k}} = (\sigma_{\tilde{s}_{i,k}}^2/\sigma_i^2)\, \tilde{x}_{i|-i,k}$.

The sources can alternatively be sampled jointly. Indeed, we have, $\forall k = 1, \ldots, N$

$$\tilde{\mathbf{x}}_k = \mathbf{A}\, \tilde{\mathbf{s}}_k + \tilde{\mathbf{e}}_k \tag{11.36}$$

Conditionally upon $\mathbf{v}_k = [v_{1,k}, \ldots, v_{n,k}]^T$, the vector $\tilde{\mathbf{s}}_k$ is Gaussian with density $\mathcal{N}(\tilde{\mathbf{s}}_k|0, \mathrm{diag}\,(\mathbf{v}_k))$ and can thus be inferred from $\tilde{\mathbf{x}}_k$ again by Wiener filtering, yielding

$$p(\tilde{\mathbf{s}}_k|\mathbf{A}, \sigma^2, \tilde{\mathbf{x}}_k) = \mathcal{N}(\tilde{\mathbf{s}}_k|\boldsymbol{\mu}_{\tilde{\mathbf{s}}_k}, \boldsymbol{\Sigma}_{\tilde{\mathbf{s}}_k}) \tag{11.37}$$

with $\boldsymbol{\Sigma}_{\tilde{\mathbf{s}}_k} = \left(\frac{1}{\sigma^2}\, \mathbf{A}^T\, \mathbf{A} + \mathrm{diag}\,(\mathbf{v}_k)^{-1}\right)^{-1}$ and $\boldsymbol{\mu}_{\tilde{\mathbf{s}}_k} = \frac{1}{\sigma^2}\, \boldsymbol{\Sigma}_{\tilde{\mathbf{s}}_k}\, \mathbf{A}^T\, \tilde{\mathbf{x}}_k$. If block-sampling the sources yields theoretically faster convergence to the stationary distribution, in practice, the two approaches ("one by one" or "full block") involve different computational burdens. With the one by one approach the update of one source $\tilde{s}_i$ requires sampling N univariate independent Gaussian random variables. This does not require any matrix inversion and can be efficiently vectorized in the implementation. The operation has to be repeated n times to update the whole matrix $\tilde{\mathbf{s}}$. On the opposite, with the full block approach, the update of $\tilde{\mathbf{s}}$ requires sampling N times from a n-multivariate Gaussian distribution. This involves inverting a $n \times n$ matrix for each k, and cannot be vectorized, hence requiring to loop over k at each iteration of the Gibbs sampler. The full block approach is thus much more time consuming, and the gain in convergence speed might not be worth the heavier computational burden, as discussed in Sect. 11.4.

$v_{i,k}$ has a conjugate prior and its posterior distribution is also inverted-Gamma, such that

$$p(v_{i,k}|\tilde{s}_{i,k}, \alpha_i, \lambda_i) = \mathcal{IG}(v_{i,k}|\alpha_i^{post}, \lambda_{i,k}^{post}) \tag{11.38}$$

where $\alpha_i^{post} = \frac{1}{2} + \alpha_i$ and $\lambda_{i,k}^{post} = \frac{\tilde{s}_{i,k}^2}{2} + \lambda_i$. Note that here, as opposed to the posterior distribution of σ^2, the influence of the prior in the posterior of $v_{i,k}$ is high because $v_{i,k}$ is observed through only one data point, $\tilde{s}_{i,k}$.

The posterior distribution of the scale parameter is

$$p(\lambda_i|\alpha_i, v_i) = \mathcal{G}\left(\lambda_i|\alpha_{\lambda_i}^{post}, \beta_{\lambda_i}^{post}\right) \tag{11.39}$$

with $\alpha_{\lambda_i}^{post} = N\,\alpha_i + \alpha_{\lambda_i}$ and $\beta_{\lambda_i}^{post} = \sum_k \frac{1}{v_{i,k}} + \beta_{\lambda_i}$. Again, in this case enough data is available to diminish the influence of the prior, and a Jeffrey noninformative prior $p(\lambda_i) \propto 1/\lambda_i$, corresponding to $\alpha_{\lambda_i} = \beta_{\lambda_i} = 0$ can be used for λ_i.

The posterior distribution of α_i is written

$$p(\alpha_i|\lambda_i, v_i) \propto \exp\left(-N \log \Gamma(\alpha_i) + \left(\sum_{k=1}^{N} \log \frac{\lambda_i}{v_{i,k}} - \beta_{\alpha_i}\right)\alpha_i\right) \tag{11.40}$$

This distribution is not straightforward to sample. An exact Metropolis–Hastings (M–H) scheme is described in Appendix (A.2). Alternatively, since the precise value of α_i is unlikely to be important provided it lies within an appropriate small range, this parameter can be sampled from a grid of discrete values with probability mass proportional to (11.40), like in [22]. Though deterministic moves can compromise the theoretical convergence of the sampler, we also found satisfactory in practice to update α_i to the mode of its posterior distribution. The mode cannot be computed analytically, but can be found using a Newton descent. Algorithm 2 recapitulates the Gibbs sampler for the Student t prior.

Update of Source Coefficients with the Mixture Prior As before, the hierarchical mixture prior does not allow to sample the sources

Algorithm 2 Gibbs sampler for the Student t prior

Initialize $\boldsymbol{\theta}$
for $l = 1 : L + L_{bi}$ **do**
  Update A and σ^2
  $\mathbf{A} \sim \prod_i \mathcal{N}(\mathbf{r}_i|\boldsymbol{\mu}_{\mathbf{r}_i}, \boldsymbol{\Sigma}_\mathbf{r})$
  $\sigma^2 \sim \mathcal{IG}(\sigma^2|\alpha_{\sigma^2}^{post}, \beta_{\sigma^2}^{post})$

  Update source coefficients
  for $i = 1 : n$ **do**
    Update $\tilde{s}_i$
    $\tilde{s}_i \sim \prod_k \mathcal{N}(\tilde{s}_{i,k}|\mu_{\tilde{s}_{i,k}}, \sigma^2_{\tilde{s}_{i,k}})$
    Update v_i
    $v_{i,k} \sim \prod_k \mathcal{IG}(v_{i,k}|\alpha_i^{post}, \lambda_{i,k}^{post})$
    Update λ_i
    $\lambda_i \sim \mathcal{G}\left(\lambda_i|\alpha_{\lambda_i}^{post}, \beta_{\lambda_i}^{post}\right)$
    Update α_i
    $\alpha_i = \text{argmax } p(\alpha_i|\lambda_i, v_i)$
  end for
end for

directly from $p(\tilde{s}_{i,k}|\lambda_i, P_i, \sigma_i^2, \tilde{x}_{i|-i,k})$ but rather requires sampling jointly from $p(\tilde{s}_{i,k}, \gamma_{i,k}, v_{i,k}|\lambda_i, P_i, \sigma_i^2, \tilde{x}_{i|-i,k})$. Again, this done by Gibbs sampling, more precisely, alternate sampling from $p(\tilde{s}_{i,k}, \gamma_{i,k}|v_{i,k}, P_i, \sigma_i^2, \tilde{x}_{i|-i,k})$ and $p(v_{i,k}|\tilde{s}_{i,k}, \gamma_{i,k}, \lambda_i)$. The first distribution can be sampled directly, by

1) sampling $\gamma_{i,k}^{(l)}$ from $P(\gamma_{i,k}|v_{i,k}, P_i, \sigma_i^2, \tilde{x}_{i|-i,k})$
2) sampling $\tilde{s}_{i,k}^{(l)}$ from $p(\tilde{s}_{i,k}|\gamma_{i,k}^{(l)}, v_{i,k}, \sigma_i^2, \tilde{x}_{i|-i,k})$

Note that these two latter steps are not Gibbs sampling steps, but constitute an exact draw from the joint distribution.[4] In fact, as pointed out in [23], a Gibbs implementation consisting of sampling alternatively $\tilde{s}_{i,k}|\gamma_{i,k}$ and $\gamma_{i,k}|\tilde{s}_{i,k}$ cannot be used as it leads to a nonconvergent Markov chain (the Gibbs sampler gets stuck when it generates a value $\tilde{s}_{i,k} = 0$). We thus need to evaluate $P(\gamma_{i,k}|v_{i,k}, P_i, \sigma_i^2, \tilde{x}_{i|-i,k})$. This can be seen as an hypothesis testing problem, with

$$(H_1) \iff \gamma_{i,k} = 1 \iff \tilde{x}_{i|-i,k} = \tilde{s}_{i,k} + \tilde{e}_{i|-i,k} \tag{11.41}$$
$$(H_0) \iff \gamma_{i,k} = 0 \iff \tilde{x}_{i|-i,k} = \tilde{e}_{i|-i,k} \tag{11.42}$$

[4] To make things perfectly clear, let us recall that $(x^{(l)}, y^{(l)})$ can be drawn from $p(x, y)$ in two different ways. An exact draw is obtained by sampling $x^{(l)}$ from $p(x)$ and then $y^{(l)}$ from $p(y|x^{(l)})$ (or the other way round). This requires to be able to sample from $p(x)$ directly. If this not possible, the other way is to do Gibbs sampling, i.e sampling alternatively $p(x|y)$ and $p(y|x)$, until the stationary distribution is obtained.

The ratio

$$\tau_{i,k}^{post} = \frac{P(\gamma_{i,k} = 1 | v_{i,k}, P_i, \sigma_i^2, \tilde{x}_{i|-i,k})}{P(\gamma_{i,k} = 0 | v_{i,k}, P_i, \sigma_i^2, \tilde{x}_{i|-i,k})}$$

is thus simply expressed as

$$\tau_{i,k}^{post} = \frac{\mathcal{N}(\tilde{x}_{i|-i,k} | 0, v_{i,k} + \sigma_i^2)}{\mathcal{N}(\tilde{x}_{i|-i,k} | 0, \sigma_i^2)} \frac{P_i}{1 - P_i} \tag{11.43}$$

$$= \sqrt{\frac{\sigma^2}{\sigma^2 + v_{i,k}}} \exp\left(\frac{\tilde{x}_{i|-i,k}^2 \, v_{i,k}}{2\,\sigma^2(\sigma^2 + v_{i,k})}\right) \frac{P_i}{1 - P_i} \tag{11.44}$$

$\gamma_{i,k}$ is thus drawn from the two states discrete distribution with probability masses

$$P(\gamma_{i,k} = 0 | v_{i,k}, P_i, \sigma_i^2, \tilde{x}_{i|-i,k}) = 1/(1 + \tau_{i,k}^{post}) \tag{11.45}$$

$$P(\gamma_{i,k} = 1 | v_{i,k}, P_i, \sigma_i^2, \tilde{x}_{i|-i,k}) = \tau_{i,k}^{post}/(1 + \tau_{i,k}^{post}) \tag{11.46}$$

When a value $\gamma_{i,k} = 0$ is drawn, $\tilde{s}_{i,k}$ is set to zero. Otherwise it is updated through Wiener filtering as before. The posterior distribution of $\tilde{s}_{i,k}$ is thus written as

$$p(\tilde{s}_{i,k} | \gamma_{i,k}, v_{i,k}, \sigma^2, \tilde{x}_{i|-i,k}) = (1 - \gamma_{i,k})\,\delta_0(\tilde{s}_{i,k}) + \gamma_{i,k}\,\mathcal{N}(\tilde{s}_{i,k} | \mu_{\tilde{s}_{i,k}}, \sigma_{\tilde{s}_{i,k}}^2) \tag{11.47}$$

with $\sigma_{\tilde{s}_{i,k}}^2$ and $\mu_{\tilde{s}_{i,k}}$ defined as before.

It is also possible to make block draws of $(\tilde{\mathbf{s}}_k, \boldsymbol{\gamma}_k)$, where $\boldsymbol{\gamma}_k = [\gamma_{1,k}, \ldots, \gamma_{n,k}]^T$. This strategy amounts to solving for every iteration l and every index k a small sized Bayesian variable selection problem, where $\mathbf{A}$ contains the regressors and $\tilde{\mathbf{s}}_k$ contains the regression coefficients [23]. It requires

1) sampling $\boldsymbol{\gamma}_k$ from $p(\boldsymbol{\gamma}_k | \mathbf{v}_k, \boldsymbol{\theta}_\gamma, \sigma^2, \tilde{\mathbf{x}})$, which requires computing 2^n probability masses corresponding to the 2^n values of $\boldsymbol{\gamma}_k$,
2) sampling $\tilde{\mathbf{s}}_k$ from $p(\tilde{\mathbf{s}}_k | \boldsymbol{\gamma}_k, \mathbf{v}_k, \sigma^2, \tilde{\mathbf{x}})$ which is multivariate Gaussian,

where $\boldsymbol{\theta}_\gamma$ contains all the Bernoulli probabilities. The computation of $p(\boldsymbol{\gamma}_k | \mathbf{v}_k, \boldsymbol{\theta}_\gamma, \sigma^2, \tilde{\mathbf{x}})$ involves the integration of $\mathcal{N}(\tilde{\mathbf{x}}_k | \mathbf{A}\,\tilde{\mathbf{s}}_k, \sigma^2\,\mathbf{I}_m)\,\mathcal{N}(\tilde{\mathbf{s}}_k, |0, \mathrm{diag}\,(\mathbf{v}_k))$ over $\tilde{\mathbf{s}}$, which is analytically possible, but involves small size matrix inversions for each value of $\boldsymbol{\gamma}_k$. The full block approach is too computationally demanding, and is still inapplicable in practice.

The conditional posterior distribution of $v_{i,k}$ is

$$p(v_{i,k}|\gamma_{i,k}, \tilde{s}_{i,k}, \lambda_i) = (1-\gamma_{i,k})\,\mathcal{IG}\left(v_{i,k}|\alpha_i, \lambda_i\right) + \gamma_{i,k}\,\mathcal{IG}\left(v_{i,k}|\alpha_i^{post}, \lambda_{i,k}^{post}\right) \quad (11.48)$$

When a value $\gamma_{i,k} = 0$ is generated, $\tilde{v}_{i,k}$ is simply sampled from its prior (no posterior information is available), otherwise, it is inferred from the available value of $\tilde{s}_{i,k}$ as before.

The posterior distribution of the scale parameter λ_i is unchanged, and as given by (11.39). However, because we are looking for sparse representations, most of the indicator variables $\gamma_{i,k}$ take the value 0 and thus most of the variances $v_{i,k}$ are sampled from their prior (see (11.48)). Thus, the influence of the data in the full posterior distribution of λ_i becomes small, and the convergence of λ_i can be very slow. A faster scheme, employed in [24, 25], consists of making one draw from $p(\{v_{i,k} : \gamma_{i,k} = 1\}|\{\tilde{s}_{i,k} : \gamma_{i,k} = 1\}, \lambda_i)$, then one draw from $p(\lambda_i|\{v_{i,k} : \gamma_{i,k} = 1\})$ and finally one draw from $p(\{v_{i,k} : \gamma_{i,k} = 0\}|\lambda_i)$. The posterior $p(\lambda_i|\{v_{i,k} : \gamma_{i,k} = 1\})$ is simply written

$$p(\lambda_i|\alpha_i, v_i) = \mathcal{G}\left(\lambda_i|\alpha_{\lambda_i}^{post'}, \beta_{\lambda_i}^{post'}\right) \quad (11.49)$$

with $\alpha_{\lambda_i}^{post'} = \#\gamma_i\,\alpha_i + \alpha_{\lambda_i}$ and $\beta_{\lambda_i}^{post'} = \sum_{k:\gamma_{i,k}=1} 1/v_{i,k} + \beta_{\lambda_i}$, and where $\#\gamma_i$ is the number of values in γ_i equal to 1.

Finally, the posterior distribution of P_i is

$$p(P_i|\gamma_i) = \mathcal{B}(P_i|\alpha_{P_i}^{post}, \beta_{P_i}^{post}) \quad (11.50)$$

where $\alpha_{P_i}^{post} = \#\gamma_i + \alpha_{P_i}$, $\beta_{P_i}^{post} = N - \#\gamma_i + \beta_{P_i}$.

Algorithm 3 recapitulates the Gibbs sampler for the hierarchical prior.

11.3 Audio-Specific Models

While the previous section described models for sparse sources in general, in this section we focus on audio source separation. As such we bring modifications to the general priors described previously in order to take into account the specificities of audio signals. In particular we lift the iid assumption on $\tilde{s}_i$. In Sect. 11.3.2, the "identically distributed" modeling is replaced by frequency-dependent priors which model the natural nonuniform distribution of energy along frequency. In Sect. 11.3.3, the "independently distributed" modeling is replaced by structured priors which model harmonic properties of sound (persistence of time–frequency coefficients through time). The audio sources are here modeled as a sparse linear combination of Modified Discrete Cosine Transform (MDCT) atoms which is a time–frequency orthonormal basis presented in the next section.

Algorithm 3 Gibbs sampler for the hierarchical prior

Initialize $\boldsymbol{\theta}$
for $l = 1 : L + L_{bi}$ **do**
  Update A and σ^2
  $\mathbf{A} \sim \prod_i \mathcal{N}(\mathbf{r}_i|\boldsymbol{\mu}_{\mathbf{r}_i}, \boldsymbol{\Sigma}_\mathbf{r})$
  $\sigma^2 \sim \mathcal{IG}(\sigma^2|\alpha_{\sigma^2}^{post}, \beta_{\sigma^2}^{post})$

  Update sources
  for $i = 1 : n$ **do**
    Update γ_i **and** $\tilde{s}_i$
    $\gamma_i \sim \prod_k P(\gamma_{i,k}|\tau_{i,k}^{post})$
    $\tilde{s}_i \sim \prod_k (1 - \gamma_{i,k})\,\delta_0(\tilde{s}_{i,k}) + \gamma_{i,k}\,\mathcal{N}(\tilde{s}_{i,k}|\mu_{\tilde{s}_{i,k}}, \sigma_{\tilde{s}_{i,k}}^2)$
    Update $\{v_{i,k} : \gamma_{i,k} = 1\}$
    $v_{i,k} \sim \mathcal{IG}(v_{i,k}|\alpha_i^{post}, \lambda_{i,k}^{post})$
    Update λ_i
    $\lambda_i \sim \mathcal{G}\left(\lambda_i|\alpha_{\lambda_i}^{post'}, \beta_{\lambda_i}^{post'}\right)$
    Update $\{v_{i,k} : \gamma_{i,k} = 0\}$
    $v_{i,k} \sim \mathcal{IG}(v_{i,k}|\alpha_i, \lambda_i)$
    Update P_i
    $P_i \sim \mathcal{B}(P_i|\alpha_{P_i}^{post}, \beta_{P_i}^{post})$
  end for
end for

11.3.1 MDCT Representation

The MDCT basis is a local cosine lapped transform [14], which has proven to give good sparse approximations of audio signals, with many coding applications [26, 27]. It provides an orthonormal decomposition without blocking effects, and has fast implementations based on the FFT. Atoms corresponding to the MDCT transform of a signal of length $N = l_{\text{frame}} \times n_{\text{frame}}$ and a frame length l_{frame}, are defined as

$$\begin{aligned}&\boldsymbol{\Phi}_{(q,p)}(t)\\&= w(t-(p-1)\,l_{\text{frame}})\,\cos\left[\frac{\pi}{l_{\text{frame}}}\left(t-(p-1)\,l_{\text{frame}} + \frac{l_{\text{frame}}+1}{2}\right)\left(q-\frac{1}{2}\right)\right]\end{aligned} \tag{11.51}$$

where $w(t)$ is a window of size $2\,l_{\text{frame}}$, $q = 1, \ldots, l_{\text{frame}}$ is a frequency index and $p = 1, \ldots, n_{\text{frame}}$ is a time frame index. When convenient, we will use in the following the mapping $k = (p-1)\,l_{\text{frame}} + q$ to index the atoms of the dictionary $\boldsymbol{\Phi}$. Figure 11.4 shows several instances of MDCT atoms.

11.3.2 Frequency-Dependent Models

The energy distribution of audio signal is naturally decreasing across frequency. Thus, an identically distributed model for $\tilde{s}_i$ does not best fit audio data. However, the models presented in Sect. 11.2.1 can easily be made

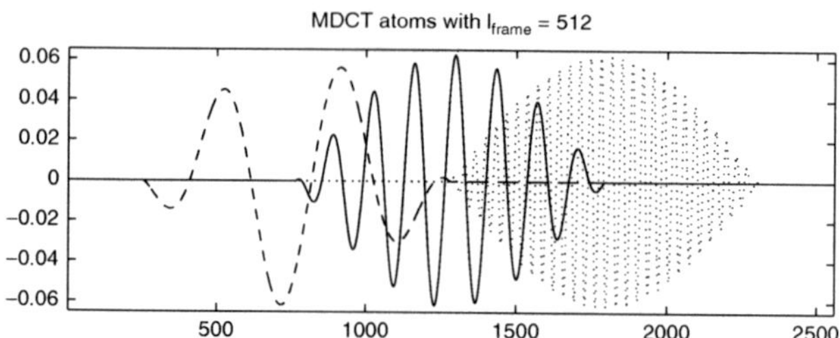

Fig. 11.4. Three MDCT atoms for $l_{\text{frame}} = 512$ and using a sine bell window. The atoms are adjacent in time (50% overlap) and correspond to three different value of frequency q.

frequency dependent by weighting the scale parameter λ_i with a frequency profile. As such, the prior about $v_{i,k}$ becomes

$$p(v_{i,k}|\alpha_i, \lambda_i) = \mathcal{IG}(v_{i,k}|\alpha_i, \lambda_i\, f_k) \tag{11.52}$$

with

$$f_k = f_q = \frac{1}{(1 + ((q-1)/q_c)^R)}, \quad q = 1, \ldots, l_{\text{frame}} \tag{11.53}$$

This frequency shaping is based on the frequency response of a Butterworth lowpass filter, where q_c acts as a cut-off frequency and R acts as the filter order. In practice we set $R = 2$, $q_c = l_{\text{frame}}/3$.

11.3.3 Structural Constraints

The harmonic nature of sound generates spectral lines in the time–frequency plane. The frame length of the MDCT basis being typically 22 ms, a sound event (e.g., a partial of one note) usually lies over several adjacent atoms. Thus, when an atom is selected at frame p and frequency q, it is "likely" that atoms at frame $p-1$ and $p+1$ are selected too, at the same frequency q.[5] This property is not described by the "independently distributed" models of Sect. 11.2.1. Instead, source coefficients should be correlated through time in order to favor horizontal structures in the time–frequency plane. This can be done easily with the mixture prior by replacing the independent Bernoulli prior for γ_i with a *structured prior*, such as Markov chain modeling. As such, for a fixed frequency index q the sequence $\{\gamma_{i,(q,p)}\}_{p=1,\ldots,n_{\text{frame}}}$ is modeled by a two-state first order Markov chain with transition probabilities $P_{i,00}$ and $P_{i,11}$, assumed equal for all frequency indices $q = 1, \ldots, l_{\text{frame}}$. The initial distribution $\pi_i = P(\gamma_{i,(q,1)} = 1)$ of each chain is taken to be its stationary

[5]In fact, it could be at frequency $q-1$, q or $q+1$, but this more general case is not considered here, though it readily fits in the framework we describe.

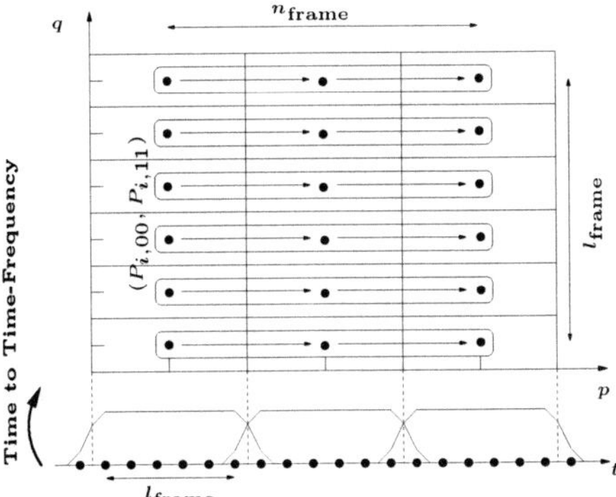

Fig. 11.5. This figure illustrates the structured hierarchical mixture prior. Each square of the time–frequency tiling corresponds to a MDCT atom. To each atom corresponds an indicator variable $\gamma_{i,k}$ which controls whether this atom is selected ($\gamma_{i,k} = 1$) or not ($\gamma_{i,k} = 0$). The set of indicator variables γ_i is modeled as "horizontal and parallel" Markov chains of order 1, with common transition probabilities $P_{i,00}$ and $P_{i,11}$, and with initial probability taken as its equilibrium value.

distribution in order to impose some form of shift-invariance [29], namely

$$\pi_i = \frac{1 - P_{i,00}}{2 - P_{i,11} - P_{i,00}} \quad \text{and} \quad (1 - \pi_i) = \frac{1 - P_{i,11}}{2 - P_{i,11} - P_{i,00}} \tag{11.54}$$

This type of model was introduced for speech denoising in [24]. The model is illustrated in Fig. 11.5, and the corresponding graph for $\tilde{s}_{i,k}$ is given in Fig. 11.6

The Markov transition probabilities are estimated and given Beta priors $\mathcal{B}(P_{i,00}|\alpha_{P_{i,00}}, \beta_{P_{i,00}})$ and $\mathcal{B}(P_{i,11}|\alpha_{P_{i,11}}, \beta_{P_{i,11}})$.

11.3.4 Modified Inference Steps

The use of a weight function with λ_i does not change much in the Gibbs sampler. Basically, the weight function contribution must be mirrored everywhere λ_i appears. As such, the parameters of the posterior distributions of $v_{i,k}$ and λ_i are solely changed as

$$\lambda_{i,k}^{post} = \frac{\tilde{s}_{i,k}^2}{2} + \lambda_i f_k \tag{11.55}$$

and

$$\beta_{\lambda_i}^{post} = \sum_k \frac{f_k}{v_{i,k}} + \beta_{\lambda_i} \tag{11.56}$$

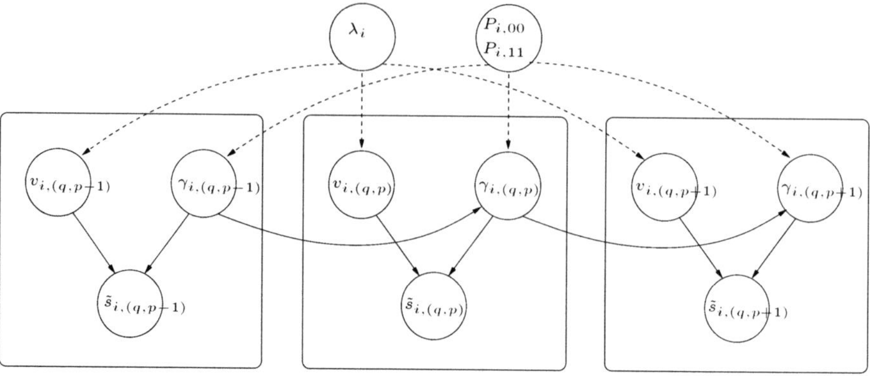

Fig. 11.6. Graph model of the source coefficients with the structured hierarchical mixture prior.

Furthermore, the posterior distribution of α_i writes

$$p(\alpha_i|\lambda_i, v_i) \propto \exp\left(-N \log \Gamma(\alpha_i) + \left(\sum_{k=1}^{N} \log \frac{\lambda_i f_k}{v_{i,k}} - \beta_{\alpha_i}\right) \alpha_i\right) \tag{11.57}$$

and its mode can again be computed using Newton descent rooting.

The update of the source coefficients with the structured hierarchical mixture prior is not much more difficult. As before, starting from (11.33), which writes

$$\tilde{x}_{i|-i} = \tilde{s}_i + \tilde{e}_{i|-i}$$

we wish to sample from $p(\tilde{s}_i, \gamma_i|v_i, \sigma_i^2, \tilde{x}_{i|-i}) = p(\tilde{s}_i|\gamma_i, v_i, \sigma_i^2, \tilde{x}_{i|-i})\, P(\gamma_i|v_i, \sigma_i^2, \tilde{x}_{i|-i})$. Conditionally upon $\gamma_{i,k}$, $\gamma_{i,k'}$, the coefficients $\tilde{s}_{i,k}$, $\tilde{s}_{i,k'}$ are still mutually independent, correlation being introduced through the indicator variables only, see Fig. 11.6. Thus, the distribution $p(\tilde{s}_i|\gamma_i, v_i, \sigma_i^2, \tilde{x}_{i|-i})$ still factorizes as $p(\tilde{s}_i|\gamma_i, v_i, \sigma_i^2, \tilde{x}_{i|-i}) = \prod_k p(\tilde{s}_{i,k}|\gamma_{i,k}, v_{i,k}, \sigma_i^2, \tilde{x}_{i|-i,k})$. The distribution $P(\gamma_i|v_i, \sigma_i^2, \tilde{x}_{i|-i})$ however, does not factorize anymore. Instead, we have to condition the update of $\gamma_{i,k}$ upon the set $\gamma_{i,-k}$, such that

$$\tau_{i,k}^{post} = \frac{P(\gamma_{i,k} = 1|\gamma_{i,-k}, v_{i,k}, \sigma_i^2, \tilde{x}_{i|-i,k})}{P(\gamma_{i,k} = 0|\gamma_{i,-k}, v_{i,k}, \sigma_i^2, \tilde{x}_{i|-i,k})} \tag{11.58}$$

$$= \frac{\mathcal{N}(\tilde{x}_{i|-i,k}|0, v_{i,k} + \sigma_i^2)}{\mathcal{N}(\tilde{x}_{i|-i,k}|0, \sigma_i^2)} \, \frac{P(\gamma_{i,k} = 1|\gamma_{i,-k}, \theta_{\gamma_i})}{P(\gamma_{i,k} = 0|\gamma_{i,-k}, \theta_{\gamma_i})} \tag{11.59}$$

This expression of $\tau_{i,k}^{post}$ only differs from the one given by (11.44) in the ratio $\tau_{i,k} = P(\gamma_{i,k} = 1|\gamma_{i,-k}, \theta_{\gamma_i})/P(\gamma_{i,k} = 0|\gamma_{i,-k}, \theta_{\gamma_i})$, which is given in Appendix A.3. Note that the chosen structure for the coefficients is solely reflected in the inference part through $\tau_{i,k}$. If any other structure was to be considered, such as Markov random fields (like in [24]), one would only need to change $\tau_{i,k}$ accordingly. All the other update steps remain unchanged.

Because we have assumed the initial probability of the chain to be equal to its equilibrium probability, the posterior distributions of $P_{i,00}$ and $P_{i,11}$ do not belong to a family of distributions easy to sample. Their expressions are given in Appendix A.4 where we describe an exact M–H scheme as well as a deterministic scheme to update these variables.

11.4 Separation of a Stereophonic Recording with Three Sources

11.4.1 Experimental Setup

We present results for blind separation of a stereo mixture ($m = 2$) of $n = 3$ musical sources (voice, acoustic guitar, bass guitar).[6] The sources were obtained from the BASS-dB database [31]. They consist of excerpts of original tracks from the song *Anabelle Lee* (Alex Q), published under a Creative Commons Licence. The signals are sampled at $f_s = 22.5$ kHz with length $N = 131072$ ($\approx$6 s). The mixing matrix is given in Table 11.1; it provides a mixture where the voice s_1 is in the middle, the acoustic guitar s_2 originates at 67.5° on the left and the bass guitar s_3 at 67.5° on the right. Gaussian noise was added to the observations with $\sigma = 0.01$, resulting in respectively 25 dB and 27 dB input SNR on each channel. We applied a MDCT to the observations using a sine bell and 50% overlap, with time resolution (half the window length) $l_{\text{frame}} = 512$ (22 ms). We compare the following methods:

(a) The source coefficients are given a Student t iid distribution. The sources are updated with block draws of $\tilde{\mathbf{s}}_k$. Using a MATLAB implementation running on a 1.25 GHz Powerbook G4 with 512 MB RAM, 1000 iterations of the sampler take 6.6 hours.
(b) We apply the approach (a), but the sources are updated one by one, conditionally upon the others. 1000 iterations of the sampler take 1.1 hours.
(c) Same as (b) except that the source coefficients now have a hierarchical mixture frequency-dependent prior with Bernoulli prior on γ_i. 1000 iterations of the sampler take 50 min.
(d) Same as (c) except that the source coefficient now have the structured prior described in Sect. 11.3.3 (horizontal Markov chain modeling of γ_i). The computational burden is nearly unchanged, 1000 iterations take approximately 50 min.

Hyperparameters priors where all chosen so as to yield noninformative priors, i.e, $\alpha_{\lambda_i} = \beta_{\lambda_i} = 0$, $\beta_{\alpha_i} = 0$, $\alpha_{P_i} = \beta_{P_i} = 0$, $\alpha_{P_{i,00}} = \beta_{P_{i,00}} = \alpha_{P_{i,11}} = \beta_{P_{i,11}} = 0$. A Jeffreys noninformative prior was also chosen for σ^2, setting

[6] Part of these results are reproduced from [30].

$\alpha_{\sigma^2} = \beta_{\sigma^2} = 0$. $\mathbf{A}$ was initialized to $[1\ 1\ 1;\ 0\ 0\ 0]$, $\tilde{s}_i$ to $\tilde{x}_1/3$, v_i to ones, λ_i to 0.1, α_i to 0.5, P_i to 0.1, and the Markov transition probabilities to 0.9.

The samplers were run for 2500 iterations in case (a) and for 10000 iterations in the other cases. Approximate convergence was *generally* observed after 1500 iterations in the first case and after 5000 iterations in the second one. In every case σ^2 was annealed to its true posterior distribution during the first hundreds iterations of the sampler. In our framework, simulated annealing basically consists in artificially increasing the values taken by σ^2 through the first iterations of the sampler and gradually decreasing them to their correct expected value. Increasing σ^2 broadens the support of the likelihood and allows the sampler to explore faster the space of parameters. This was empirically proven to accelerate the convergence of the sampler to its stationary distribution in [22]. We anneal the degrees of freedom α_{σ^2} of the input noise by replacing $\alpha_{\sigma^2}^{post}$ in (11.30) with $\alpha_{\sigma^2}^{post'}(l) = (1 - (1 - p_0)\exp(-l/l_0))\ \alpha_{\sigma^2}^{post}$, where l denotes the iteration. In this version of annealing, the degrees of freedom parameter for the input noise is exponentially increased to its correct value from a small starting value. In this way the sampler is more able to explore the probability distribution at earlier iterations, while effectively returning to the true stationary target distribution once $l >> l_0$.

MMSE estimates of the source coefficients were computed in each case by averaging the last 1000 samples. Table 11.1 gives the mixing matrix estimates as well as separation evaluation criteria for the estimated sources in each case. The criteria are described in [32], but basically, the SDR (Source to Distortion Ratio) provides an overall separation performance criterion, the

Table 11.1. Mixing matrix estimates and performance criteria.

Original matrix			
$\mathbf{A} =$	0.7071	0.9808	0.1951
	0.7071	0.1951	0.98079
Method (a)			
$\hat{\mathbf{A}} =$	0.7074 (±0.0003)	0.9811 (±0.0002)	0.1947 (±0.0004)
	0.7067 (±0.0003)	0.1930 (±0.0009)	0.98085 (±0.00008)
Method (b)			
$\hat{\mathbf{A}} =$	0.7074 (±0.0004)	0.9809 (±0.0002)	0.1948 (±0.0004)
	0.7068 (±0.0004)	0.1944 (±0.0008)	0.98084 (±0.00007)
Method (c)			
$\hat{\mathbf{A}} =$	0.7044 (±0.0004)	0.9821 (±0.0002)	0.1943 (±0.0006)
	0.7098 (±0.0004)	0.1881 (±0.0012)	0.9809 (±0.0001)
Method (d)			
$\hat{\mathbf{A}} =$	0.7079 (±0.0003)	0.9811 (±0.0001)	0.1946 (±0.0006)
	0.7064 (±0.0003)	0.1933 (±0.0007)	0.9809 (±0.0001)

	$\hat{s}_1$ (voice)			
Method	SDR	SIR	SAR	SNR
(a)	4.0	14.1	4.6	28.6
(b)	−2.0	14.0	−1.7	22.3
(c)	0.1	5.6	2.6	28.4
(d)	−0.75	12.0	−0.23	28.3

	$\hat{s}_2$ (acoustic guitar)			
Method	SDR	SIR	SAR	SNR
(a)	5.6	17.7	6.0	28.0
(b)	3.2	6.5	6.9	29.3
(c)	0.7	7.7	2.4	27.4
(d)	3.1	10.3	4.5	38.6

	$\hat{s}_3$ (bass guitar)			
Method	SDR	SIR	SAR	SNR
(a)	10.5	20.9	11.0	40.0
(b)	8.0	11.2	11.1	41.6
(c)	5.9	14.4	6.7	51.5
(d)	7.4	15.1	8.3	50.1

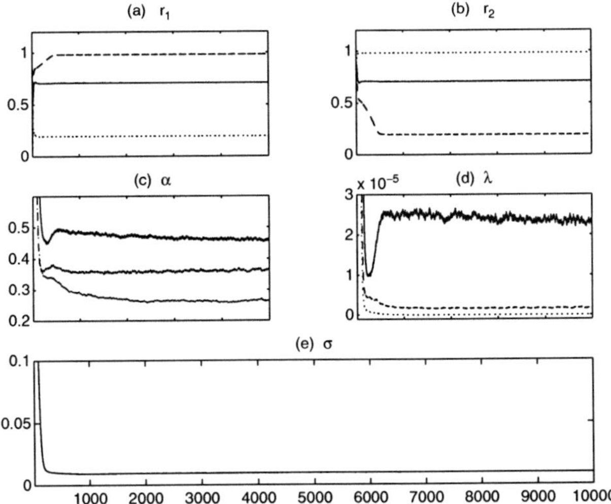

Fig. 11.7. Gibbs samples of the model parameters in the Student t iid case with one by one update of the sources (method (b)); (-): source 1, (- -): source 2, (:) = source 3.

SIR (Source to Interference Ratio) measures the level of interferences from the other sources in each source estimate, SNR (Source to Noise Ratio) measures the error due to the additive noise on the sensors and the SAR (Source to Artifacts Ratio) measures the level of artifacts in the source estimates. Source estimates can be listened to at [33], which is perhaps the best way to assess the quality of the results.

Figures 11.7 and 11.9 respectively represents the sampled values of all the parameters with approach (b) and approach (d). Figure 11.8 represents the histograms of the coefficients of the *original* sources compared with the Student t densities estimated with method (b). Figure 11.10 presents the *significance maps* of the source coefficients, i.e, the MAP estimates of γ_1, γ_2 and γ_3, in the Bernoulli (method (c)) and Markov (method (d)) cases.

11.4.2 Discussion

Separation Quality The sound samples show that best perceptual results are obtained with method (d), that is, when audio-specific structured priors are used. The source estimates with this approach may not be the best in terms of interference rejection, but what prevails is that they sound the *cleanest.* Artifacts remain but may be considered less disturbing than in the other sound samples, originating from isolated time–frequency atoms scattered over the time–frequency plane (as illustrated by Fig. 11.10). These perceptual considerations may not be reflected in the numerical criteria given

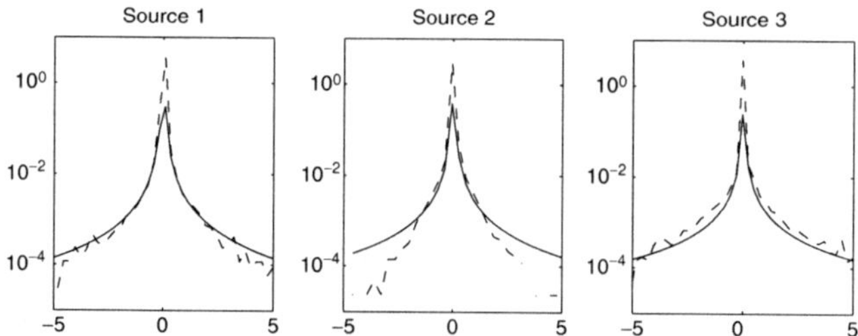

Fig. 11.8. Histograms of the original source coefficients $\tilde{s}_i$ compared to the estimated Student t densities with method (b). The MMSE estimate for α and λ are respectively [0.91, 0.72, 0.52] and [0.0071, 0.0021, 0.0002]. These are close to the ones obtained with method (a), which are respectively [0.84, 0.76, 0.65] and [0.0066, 0.0023, 0.0004].

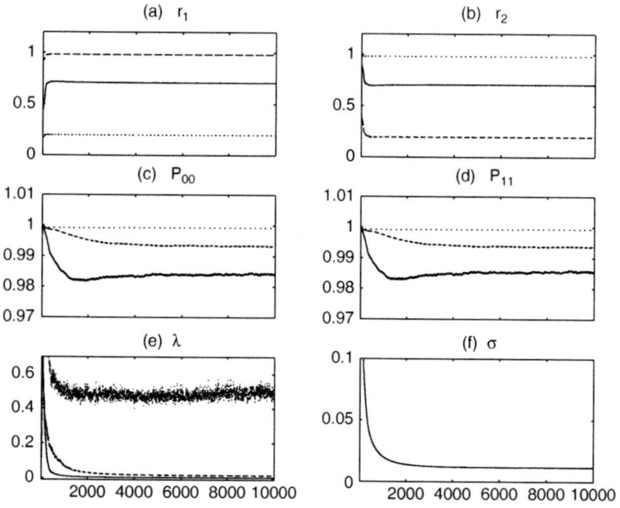

Fig. 11.9. Gibbs samples of the model parameters with the hierarchical frequency dependent mixture prior, with Markov horizontal structure (method (d)); (-): source 1, (- -): source 2, (:) = source 3.

at Table 11.1, which only compares the waveforms themselves. In every case the mixing matrix is very well estimated, with very small error standard deviations.

Strengths and Limits of the MCMC Approach One striking fact about the numerical criteria given in Table 11.1 is the difference in results between methods (a) and (b). Though the two methods should theoretically yield similar source estimates after a "large enough" number of iterations, in practice, over an horizon of 10000 iterations, method (a) still yields better estimates, in

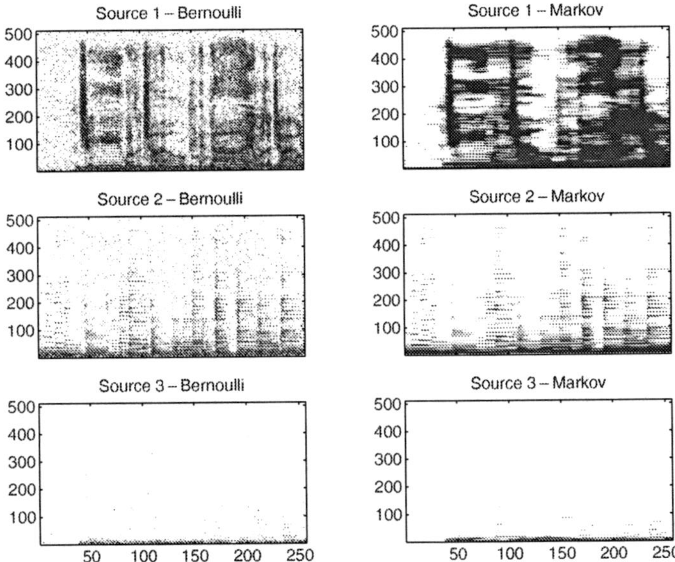

Fig. 11.10. Significance maps of the estimated sources, obtained with Bernoulli priors (left) and horizontal Markov priors (right). Much of the isolated atoms appearing on the left map have been removed in the right map, the latter has been *regularized.*

particular in terms of SIRs. We believe this is because the individual update of each source conditionally upon the others creates some correlation between the sources. If the amount of correlation should theoretically fade away when averaging a large number of samples, well after the burn in period, in practice this seems to be a problem over our limited horizon.[7] We also noticed that, depending on the initializations and the random sequence seeds, method (b), and also (c), could get stuck for long periods in some local modes of the posterior distribution of the mixing matrix, and that full exploration of the posterior could be tedious. This seemed to be a lesser problem with (d), probably because the structure brought in the model regularizes the posterior of the parameters. In contrast, method (a) happened to be more robust to local modes, and convergence is rather fast (in number of iterations) when σ^2 is annealed.

Note however that in every case, the Gibbs sampler fail at exploring *all* the posterior space. Indeed, indeterminacies remain in the model, namely sign and permutation ambiguities between $\mathbf{A}$ and $\mathbf{s}$, corresponding to equivalent modes of the posterior. The Gibbs sampler should explore these modes

[7] Actually, Fig. 11.7 shows that after 10000 iterations, some source hyperparameters seem not to have fully converged to their stationary distribution, when this was not the case with method (a), not shown here.

too, which in practice, over a limited horizon of iterations, was never the case. The sampler converges to one mode, explore this mode, and gets stuck there. This problem is also discussed in [28].

As such, a rigorous use of MCMC methods for practical applications should involve thorough monitoring of the sampler, for example by running several chains with different initializations and comparing the behaviors of each chain [19].

MCMC techniques are nevertheless more robust than EM-like counterparts, though at the expense of higher computation times. They provide a full description of the posterior space, as opposed to point estimates only. Hence, they can be used to compute interval estimates, and also to compute MAP estimates for some parameters and MMSE estimates for others. If MCMC techniques can be considered not yet applicable in practice (but is likely to change owing to the fast evolution of computing facilities), they can at least be used as a diagnostic tool to check the validity of a given source model. Once the model is chosen, lighter optimization techniques can be devised for that particular model.

11.5 Conclusions

In conclusion, let us emphasize that the Bayesian approach provides a framework to gather all information available about the sources into a prior distribution $p(\tilde{s}_i|\boldsymbol{\theta}_{s_i})$. In the case of linear instantaneous mixtures, and more generally convolutive mixtures, each source can have a different model and be updated separately. As mentioned in Sect. 11.2.3, more complex models, involving overcomplete dictionaries can even be considered. As such, musical signals can be modeled as a linear combination of MDCT atoms with short time resolution aiming at representing the transient parts, such the attacks of notes and MDCT atoms with longer time resolution (and finer frequency resolution) aiming at representing tonals (like in this chapter) [29]. As well as models of the sources, other models of mixing systems can be designed. Frequency dependent mixing matrices could be used to model convolution effects, while source models could help solving the frequency permutation problem.

Finally, note that, on a wider scope, these models of sources and mixing systems, generative in essence, propose a semantic, object-based, representation of multichannel sound. Their effectiveness is here evaluated for the source separation problem, but they could be employed for other tasks. As such, the sparsity of the maps shown on Fig. 11.10 suggests efficient coding schemes [34], where each source could be treated separately instead of encoding the mixture directly. Because of the synthesis approach, the sound mixture can also be modified, for example denoised (by simply remixing the estimated sources and mixing system), but also remixed (by changing the parameters of the mixing system), or reshaped (false notes can be corrected by moving or removing the corresponding time–frequency atoms).

A Appendices

A.1 Standard Distributions

Multivariate Gaussian $\mathcal{N}(\mathbf{x}|\boldsymbol{\mu},\boldsymbol{\Sigma}) = |2\pi\,\boldsymbol{\Sigma}|^{-\frac{1}{2}}\exp -\frac{1}{2}(\mathbf{x}-\boldsymbol{\mu})^T\,\boldsymbol{\Sigma}^{-1}\,(\mathbf{x}-\boldsymbol{\mu})$

Gaussian $\mathcal{N}(x|u,\sigma^2) = (2\pi\sigma^2)^{-1/2}\exp -\frac{(x-u)^2}{2\sigma^2}$

Beta $\mathcal{B}(x|\alpha,\beta) = \frac{\Gamma(\alpha+\beta)}{\Gamma(\alpha)\Gamma(\beta)}\,x^{\alpha-1}\,(1-x)^{\beta-1},\quad x\in[0,1]$

Gamma $\mathcal{G}(x|\alpha,\beta) = \frac{\beta^\alpha}{\Gamma(\alpha)}\,x^{\alpha-1}\exp(-\beta\,x),\, x\geq 0$

inv-Gamma $\mathcal{IG}(x|\alpha,\beta) = \frac{\beta^\alpha}{\Gamma(\alpha)}\,x^{-(\alpha+1)}\exp(-\frac{\beta}{x}),\, x\geq 0$

The inverted-Gamma distribution is the distribution of $1/X$ when X is Gamma distributed.

A.2 M–H Update of the Degrees of Freedom Parameter α_i

α_i can be updated using independent M–H sampling,[8] as proposed in [35]. The posterior distribution of α_i, given by (11.40) is written

$$p(\alpha_i|\lambda_i,v_i)\propto g(\alpha_i)^N \tag{11.60}$$

with

$$g(\alpha_i) = \frac{1}{\Gamma(\alpha_i)}\exp(-\beta_{\alpha_i}^{post}\,\alpha_i) \tag{11.61}$$

and

$$\beta_{\alpha_i}^{post} = \frac{1}{N}\left(-\sum_{k=1}^{N}\log\frac{\lambda_i}{v_{i,k}}+\beta_{\alpha_i}\right) \tag{11.62}$$

$g(\alpha_i)$ is approximated by the function $f(\alpha_i)\propto\alpha_i^{\nu_q}\exp(-\beta_q\,\alpha_i)$, by making their respective modes and inflexion points coincide (the computation of the latter involving Newton descent rooting). A proposal distribution is then built as $q(\alpha_i)\propto f(\alpha_i)^N = \mathcal{G}(\alpha_i|N\,\nu_q+1,N\,\beta_q)$.

A.3 Prior Weight of Horizontal Markov Chains

The expression of the prior weight $\tau_{i,k}$ for the model described in Sect. 11.3.3 is given $\forall q=1,\ldots,l_{\text{frame}}$ as follows:

[8]Independent Metropolis–Hasting sampling is another MCMC technique. Given a proposal distribution $q(x)$, at iteration $l+1$, a candidate $x^\star$ is generated from $q(x)$ and accepted with probability $\min\left\{1,\frac{p(x^\star)\,q(x^{(l)})}{q(x^\star)\,p(x^{(l)})}\right\}$. If accepted $x^{(l+1)}=x^\star$, otherwise $x^{(l+1)}=x^{(l)}$.

- $p = 1$

$$\tau_{i,(q,1)} = \frac{P(\gamma_{i,(q,2)}|\gamma_{i,(q,1)} = 1)\ P(\gamma_{i,(q,1)} = 1)}{P(\gamma_{i,(q,2)}|\gamma_{i,(q,1)} = 0)\ P(\gamma_{i,(q,1)} = 0)}$$
$$= \begin{cases} \frac{(1-P_{i,11})\,\pi_i}{P_{i,00}\,(1-\pi_i)} = \frac{(1-P_{i,00})}{P_{i,00}} & \text{if } \gamma_{i,(q,2)} = 0 \\ \frac{P_{i,11}\,\pi_i}{(1-P_{i,00})\,(1-\pi_i)} = \frac{P_{i,11}}{(1-P_{i,11})} & \text{if } \gamma_{i,(q,2)} = 1 \end{cases}$$

- $p = 2, \dots, n_{\text{frame}} - 1$

$$\tau_{i,(q,p)} = \frac{P(\gamma_{i,(q,p+1)}|\gamma_{i,(q,p)} = 1)\ P(\gamma_{i,(q,p)} = 1|\gamma_{i,(q,p-1)})}{P(\gamma_{i,(q,p+1)}|\gamma_{i,(q,p)} = 0)\ P(\gamma_{i,(q,p)} = 0|\gamma_{i,(q,p-1)})}$$
$$= \begin{cases} \frac{(1-P_{i,00})(1-P_{i,11})}{P_{i,00}^2} & \text{if } \gamma_{i,(q,p-1)} = 0 \text{ and } \gamma_{i,(q,p+1)} = 0 \\ \frac{P_{i,11}}{P_{i,00}} & \text{if } \gamma_{i,(q,p-1)} = 0 \text{ and } \gamma_{i,(q,p+1)} = 1 \\ \frac{P_{i,11}}{P_{i,00}} & \text{if } \gamma_{i,(q,p-1)} = 1 \text{ and } \gamma_{i,(q,p+1)} = 0 \\ \frac{P_{i,11}^2}{(1-P_{i,11})(1-P_{i,00})} & \text{if } \gamma_{i,(q,p-1)} = 1 \text{ and } \gamma_{i,(q,p+1)} = 1 \end{cases}$$

- $p = n_{\text{frame}}$

$$\tau_{i,(q,n_{\text{frame}})} = \frac{P(\gamma_{i,(q,n_{\text{frame}})} = 1|\gamma_{i,(q,n_{\text{frame}}-1)})}{P(\gamma_{i,(q,n_{\text{frame}})} = 0|\gamma_{i,(q,n_{\text{frame}}-1)})}$$
$$= \begin{cases} \frac{(1-P_{i,00})}{P_{i,00}} & \text{if } \gamma_{i,(q,n_{\text{frame}}-1)} = 0 \\ \frac{P_{i,11}}{(1-P_{i,11})} & \text{if } \gamma_{i,(q,n_{\text{frame}}-1)} = 1 \end{cases}$$

A.4 Update of Markov Transition Probabilities

We have

$$P(\gamma_i|P_{i,00}, P_{i,11}, \pi_i) = \prod_{q=1}^{l_{\text{frame}}} \prod_{p=2}^{n_{\text{frame}}} P(\gamma_{i,(q,p)}|\gamma_{i,(q,p-1)}, P_{i,00}, P_{i,11}) \times P(\gamma_{i,(q,1)}|\pi_i) \tag{11.63}$$

$$= P_{i,00}^{\#\gamma_i(00)}\ (1 - P_{i,00})^{\#\gamma_i(01)}\ P_{i,11}^{\#\gamma_i(11)} \times (1 - P_{i,11})^{\#\gamma_i(10)} \left(\frac{i - P_{i,00}}{2 - P_{i,00} - P_{i,11}}\right)^{\#\gamma_{i,(q,1)}} \times \left(\frac{i - P_{i,11}}{2 - P_{i,00} - P_{i,11}}\right)^{l_{\text{frame}} - \#\gamma_{i,(q,1)}} \tag{11.64}$$

where $\#\gamma_i(ij)$ is defined as the cardinality of the set $\{\gamma_{i,(q,p)} = j|\gamma_{i,(q,p-1)} = i, q = 1, \dots, l_{\text{frame}}, p = 2, \dots, n_{\text{frame}}\}$ and $\#\gamma_{i,(q,1)}$ is the cardinality of the set $\{\gamma_{i,(q,1)} = 1, q = 1, \dots, l_{\text{frame}}\}$. Hence, we have

$$P(P_{i,00}|\gamma_i, P_{i,11}, \alpha_{P_{i,00}}, \beta_{P_{i,00}}) \propto P(\gamma_i|P_{i,00}, P_{i,11}, \pi_i)\ p(P_{i,00}|\alpha_{P_{i,00}}, \beta_{P_{i,00}})$$
$$\propto \frac{\mathcal{B}(P_{i,00}|\#\gamma_i(00) + \alpha_{P_{i,00}}, \#\gamma_i(01) + \#\gamma_{i,(q,1)} + \beta_{P_{i,00}})}{(2 - P_{i,00} - P_{i,11})^{l_{\text{frame}}}} \tag{11.65}$$

$P_{i,00}$ can be updated using a M–H step, for example using the proposal distribution

$$q(P_{i,00}|\gamma_i, \alpha_{P_{i,00}}, \beta_{P_{i,00}}) = \mathcal{B}(P_{i,00}|\#\gamma_i(00)+\alpha_{P_{i,00}}, \#\gamma_i(01)+\#\gamma_{i,(q,1)}+\beta_{P_{i,00}}) \tag{11.66}$$

The acceptance probability $\alpha(P^\star_{i,00}|P_{i,00})$ of candidate $P^\star_{i,00}$ is simply

$$\alpha(P^\star_{i,00}|P_{i,00}) = \left(\frac{2 - P_{i,00} - P_{i,11}}{2 - P^\star_{i,00} - P_{i,11}}\right)^{l_{\text{frame}}} \tag{11.67}$$

However, because of the exponent l_{frame} in (11.67), the acceptance ratios can stay very low for long periods of time, yielding poorly mixing chains and long burn-in periods. Instead, we found very satisfying in practice to update the transitions probabilities $P_{i,00}$ and $P_{i,11}$ to the modes of their posterior distributions. After calculations of their derivatives, this simply amounts to root polynomials of order two and to choose the root with value lower to one. We favored this latter option in practice. A similar treatment is done for $P_{i,11}$ whose posterior writes

$$p(P_{i,11}|\gamma_i, P_{i,00}, \alpha_{P_{i,11}}, \beta_{P_{i,11}})$$
$$\propto \frac{\mathcal{B}(P_{i,11}|\#\gamma_i(11) + \alpha_{P_{i,11}}, \#\gamma_i(10) + l_{\text{frame}} - \#\gamma_{i,(q,1)} + \beta_{P_{i,11}})}{(2 - P_{i,00} - P_{i,11})^{l_{\text{frame}}}} \tag{11.68}$$

References

1. K. H. Knuth, "Bayesian source separation and localization," in *SPIE'98: Bayesian Inference for Inverse Problems*, San Diego, Jul. 1998, pp. 147–158.
2. ——, "A Bayesian approach to source separation," in *Proc. 1st International Workshop on Independent Component Analysis and Signal Separation*, Aussois, France, Jan. 1999, pp. 283–288.
3. A. Mohammad-Djafari, "A Bayesian approach to source separation," in *Proc. 19th International Workshop on Bayesian Inference and Maximum Entropy Methods (MaxEnt99)*, Boise, USA, Aug. 1999.
4. A. J. Bell and T. J. Sejnowski, "An information-maximization approach to blind separation and blind deconvolution," *Neural Computation*, vol. 7, no. 6, pp. 1129–1159, 1995.
5. J.-F. Cardoso, "Blind signal separation: statistical principles," *Proceedings of the IEEE. Special issue on blind identification and estimation*, vol. 9, no. 10, pp. 2009–2025, Oct. 1998.

6. B. A. Olshausen and K. J. Millman, "Learning sparse codes with a mixture-of-Gaussians prior," in *Advances in Neural Information Processing Systems*, S. A. Solla and T. K. Leen, Eds. MIT press, 2000, pp. 841–847.
7. M. S. Lewicki and T. J. Sejnowski, "Learning overcomplete representations," *Neural Computations*, vol. 12, pp. 337–365, 2000.
8. M. Girolami, "A variational method for learning sparse and overcomplete representations," *Neural Computation*, vol. 13, no. 11, pp. 2517–2532, 2001.
9. T.-W. Lee, M. S. Lewicki, M. Girolami, and T. J. Sejnowski, "Blind source separation of more sources than mixtures using overcomplete representations," *IEEE Signal Processing Letters*, vol. 4, no. 4, Apr. 1999.
10. M. Zibulevsky, B. A. Pearlmutter, P. Bofill, and P. Kisilev, "Blind source separation by sparse decomposition," in *Independent Component Analysis: Principles and Practice*, S. J. Roberts and R. M. Everson, Eds. Cambridge University Press, 2001.
11. M. Davies and N. Mitianoudis, "A simple mixture model for sparse overcomplete ICA," *IEE Proceedings on Vision, Image and Signal Processing*, Feb. 2004.
12. A. Jourjine, S. Rickard, and O. Yilmaz, "Blind separation of disjoint orthogonal signals: Demixing n sources from 2 mixtures," in *Proc. ICASSP*, vol. 5, Istanbul, Turkey, June 2000, pp. 2985–2988.
13. B. D. Rao, K. Engan, S. F. Cotter, J. Palmer, and K. Kreutz-Delgado, "Subset selection in noise based on diversity measure minimization," *IEEE Trans. Signal Processing*, vol. 51, no. 3, pp. 760–770, Mar. 2003.
14. S. Mallat, *A Wavelet Tour of Signal Processing*. Academic Press, 1998.
15. S. Chen, D. Donoho, and M. Saunders, "Atomic decomposition by basis pursuit," *SIAM Journal on Scientific Computing*, vol. 20, no. 1, pp. 33–61, 1998.
16. D. F. Andrews and C. L. Mallows, "Scale mixtures of normal distributions," *J. R. Statist. Soc. Series B*, vol. B, no. 36, pp. 99–102, 1974.
17. H. Snoussi and J. Idier, "Bayesian blind separation of generalized hyperbolic processes in noisy and underdeterminate mixtures," *IEEE Trans. Signal Processing*, vol. 54, no. 9, pp. 3257–3269, Sept. 2006.
18. S. Geman and D. Geman, "Stochastic relaxation, Gibbs distributions, and the Bayesian restoration of images," *IEEE Trans. Pattern Analysis and Machine Intelligence*, vol. PAMI-6, no. 6, pp. 721–741, Nov. 1984.
19. W. R. Gilks, S. Richardson, and D. J. Spiegelhalter, *Markov Chain Monte Carlo in Practice*. Chapman & Hall, 1996.
20. J. S. Liu, "The collapsed Gibbs sampler with applications to a gene regulation problem," *J. Amer. Statist. Assoc.*, vol. 89, no. 427, pp. 958–966, Sept. 1994.
21. J. S. Liu, W. H. Wong, and A. Kong, "Covariance structure of the Gibbs sampler with applications to the comparisons of estimators and augmentation schemes," *Biometrika*, vol. 81, no. 1, pp. 27–40, Mar. 1994.
22. C. Févotte and S. Godsill, "A Bayesian approach to blind separation of sparse sources," *IEEE Transactions on Audio, Speech and Language Processing*, vol. 14, no. 6, pp. 2174–2188, Nov. 2006.
23. J. Geweke, *Variable Selection and Model Comparison in Regression*, 5th ed. Oxford Press, 1996, pp. 609–620, edited by J. M. Bernardo, J. O. Berger, A. P. Dawid, and A. F. M. Swith.
24. P. J. Wolfe, S. J. Godsill, and W.-J. Ng, "Bayesian variable selection and regularisation for time–frequency surface estimation," *J. R. Statist. Soc. Series B*, 2004.

25. C. Févotte and S. Godsill, "Sparse linear regression in unions of bases via Bayesian variable selection," *IEEE Signal Processing Letters*, vol. 13, no. 7, pp. 441–444, July 2006.
26. K. Brandenburg, "MP3 and AAC explained," in *Proc. AES 17th Int. Conf. High Quality Audio Coding*, Florence, Italy, Sept. 1999.
27. L. Daudet and M. Sandler, "MDCT analysis of sinusoids: exact results and applications to coding artifacts reduction," *IEEE Trans. Speech and Audio Processing*, vol. 12, no. 3, pp. 302–312, May 2004.
28. M. Davy, S. Godsill, and J. Idier, "Bayesian Analysis of Polyphonic Western Tonal Music," *Journal of the Acoustical Society of America*, vol. 119, no. 4, pp. 2498–2517, Apr. 2006.
29. C. Févotte, B. Torrésani, L. Daudet, and S. J. Godsill, "Sparse linear regression with structured priors and application to denoising of musical audio," *IEEE Transactions on Audio, Speech and Language*, in press.
30. C. Févotte, "Bayesian blind separation of audio mixtures with structured priors," in *Proc. 14th European Signal Processing Conference (EUSIPCO'06)*, Florence, Italy, Sep. 2006.
31. E. Vincent, R. Gribonval, C. Févotte, et al., "BASS-dB: the blind audio source separation evaluation database," Available on-line, `http://www.irisa.fr/metiss/BASS-dB/`.
32. E. Vincent, R. Gribonval, and C. Févotte, "Performance measurement in blind audio source separation," *IEEE Transactions on Audio, Speech and Language Processing*, vol. 14, no. 4, pp. 1462–1469, July 2006.
33. `http://www.tsi.enst.fr/~fevotte/Samples/book_blind_speech_separation/`.
34. L. Daudet and B. Torrésani, "Hybrid representations for audiophonic signal encoding," *Signal Processing*, vol. 82, no. 11, pp. 1595–1617, 2002, special issue on Image and Video Coding Beyond Standards.
35. S. Moussaoui, D. Brie, A. Mohammad-Djafari, and C. Carteret, "Separation of non-negative mixture of non-negative sources using a Bayesian approach and MCMC sampling," *IEEE Trans. Signal Processing*, vol. 54, no. 11, pp. 4133–4145, Nov. 2006.

Part III

Single Microphone Blind Speech Separation

12 Monaural Source Separation

Gil-Jin Jang and Te-Won Lee

Institute for Neural Computation, University of California, San Diego
9500 Gilman Drive DEPT 0523 La Jolla, California 92093-0523
E-mail: gijang@ucsd.edu, tewon@ucsd.edu

Abstract. This chapterdiscusses source separation methods when only single channel observation is available. The problem is underdeterministic, in that multiple source signals should be extracted from a single stream of observations. To overcome the mathematical intractability, prior information on the source characteristics is generally assumed and applied to derive a source separation algorithm. This chapter describes one of the monaural source separation approach, which is based on exploiting *a priori* sets of time-domain basis functions learned by independent component analysis (ICA). The inherent time structure of sound sources is reflected in the ICA basis functions, which encode the sources in a statistically efficient manner. Detailed derivation of the source separation algorithm is described, given the observed single channel data and sets of basis functions. The prior knowledge given by the basis functions and the associated coefficient densities enables inferring the original source signals. A flexible model for density estimation allows accurate modeling of the observation and the experimental results exhibit a high level of separation performance for simulated mixtures as well as real environment recordings employing mixtures of two different sources.

12.1 Introduction

In natural conversation a speech signal is typically perceived against a background of other sounds carrying different characteristics. The human auditory system processes the acoustic mixture reaching the ears to enable constituent sounds to be heard and recognized as distinct entities, even if these sounds overlap in both spectral and temporal regions with the target speech. The flexibility and robustness of human speech perception is explained by the range of situations in which spoken communication is possible in the presence of competing sound sources [1]. Researchers in signal processing and many other related fields have strived for the realization of this human ability in machines; however, except in limited certain applications, thus far they have failed to produce the desired outcomes.

In order to formulate the problem, we assume that the observed signal y^t is the summation of P independent source signals

$$y^t = \lambda_1 x_1^t + \lambda_2 x_2^t + \ldots + \lambda_P x_P^t \,, \tag{12.1}$$

where x_i^t is the t-th observation of the i-th source, and λ_i is the gain of each source, which is fixed over time. Note that superscripts indicate sample

S. Makino et al. (eds.), Blind Speech Separation, 339–364.

indices of time-varying signals and subscripts identify sources. The gain constants are affected by several factors, such as powers, locations, directions and many other characteristics of the source generators as well as sensitivities of the sensors. It is convenient to assume all the sources to have zero mean and unit variance. The goal is to recover all x_i^t given only a single sensor input y^t. The problem is too ill-conditioned to be mathematically tractable since the number of unknowns is $PT + P$ given only T observations.

Various sophisticated methods have been proposed over the past few years in research areas such as computational auditory scene analysis (CASA) [2, 3] and independent component analysis (ICA) [4–6]. CASA separation techniques are mostly based on splitting mixtures observed as a single stream into different auditory streams by building an active scene analysis system for acoustic events that occur simultaneously in the same spectro-temporal regions. The acoustic events are distinguished according to rules inspired intuitively or empirically from the known characteristics of the sources. Roweis [7] presented a refiltering technique that estimates λ_i in (12.1) as time-varying masking filters that localize sound streams in a spectro-temporal region. In his work, sound sources are supposedly disjoint in the spectrogram and a "mask" divides the mixed streams completely.

Recently nonnegative matrix factorization (NMF) [8] has been focused by many other researchers in solving monaural source separation problem. Although the original NMF algorithm handles only static nonnegative data such as image patches or powerspectral densities of a acoustic signal, introducing convolutive extension to NMF [9, 10], called nonnegative matrix factor deconvolution (NMFD), gives an ability to extract an inherent spectro-temporal structure of a sound source. As a result of NMFD, a dictionary of monotonic trajectories is learned from a course of sound source powerspectral densities, and by classifying the dictionary items into a desired number of elements, the original source signals can be recovered. Although NMFD is successfully applied to several monaural source separation problems such as polyphonic music transcription, it is hard to obtain a reliable dictionary for a complex sources such as speech signals.

ICA has been shown to be highly effective in other aspects such as encoding image patches [11], natural sounds [12, 13], and speech signals [14]. The basis functions and the coefficients learned by ICA constitute an efficient representation of the given time-ordered sequences of a sound source by estimating the maximum likelihood densities, thus reflecting the statistical structures of the sources. The basis filters maximize the amount of information in the transformed domain, constituting an efficient representation of the given sound source and, at the same time, the distributions of the input data are approximated by the pdfs of the output coefficients [5].

The work presented here is motivated by the pdf approximation property involved in the basis filters adapted by ICA learning rules. The intuitive rationale behind the approach is to exploit the ICA basis filters to the separation of mixed source signals observed in a single channel. The basis filters of the source signals are learned a priori from a training data set and these basis filters are used to separate the unknown test sound sources. The algorithm recovers the original auditory streams in a number of gradient-ascent adaptation steps maximizing the log likelihood of the separated signals, computed by the basis functions and the pdfs of their coefficients—the output of the ICA basis filters. We make use of not only the ICA basis filters as strong prior information for the source characteristics, but also their associated coefficient pdfs as an object function of the learning algorithm. The theoretical basis of the approach is **"sparse coding"** [15], once termed **"sparse decomposition"** [16]. Sparsity in this case means that only a small number of coefficients in the representation differ significantly from zero. Empirical observations show that the coefficient histogram is extremely sparse, and the use of generalized Gaussian distributions [17] yields a good approximation.

The chapteris organized as follows. In Sect. 12.2 we present two kinds of generative models for the mixture and the sound sources. We explain in Sect. 12.3 how to derive the adaptation formula for the source signals. We present the experimental results for the synthesized examples in Sect. 12.4, and compare them with Wiener filtering [18]. Finally we summarize our method in comparison to other methods in Sect. 12.5 and draw conclusions in Sect. 12.6.

12.2 Adapting Basis Functions and Model Parameters

The algorithm first involves the learning of the time-domain basis functions of the sound sources that we are interested in separating. This corresponds to the prior information necessary to successfully separate the signals. We assume two different types of generative models in the observed single channel mixture as well as in the original sources. The first one is depicted in Fig. 12.1 (**A**). As described in (12.1), at every $t \in [1, T]$, the observed instance is assumed to be a weighted sum of different sources. In our approach only the case of $P = 2$ is considered. This corresponds to the situation defined in Sect. 12.1: two different signals are mixed and observed in a single sensor.

12.2.1 A Model for Signal Representation

For the individual source signals, we adopt a decomposition-based approach as another generative model. This approach has been formerly employed in analyzing natural sounds [12, 13], speech signals [14], and colored noise [16]. A fixed-length segment drawn from a time-varying signal is expressed as a linear superposition of a number of elementary patterns, called basis functions, with scalar multiples (Fig. 12.1 (**B**)). Continuous samples of length N with $N \ll T$ are chopped out of a source, from t to $t + N - 1$, and the subsequent segment is denoted as an N-dimensional column vector in a boldface

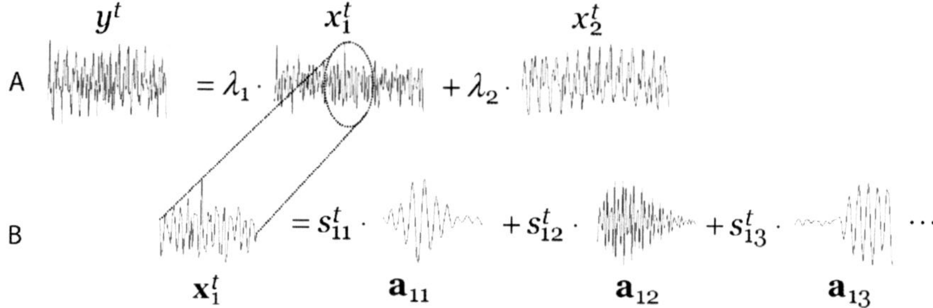

Fig. 12.1. Generative models for the observed mixture and the original source signals. (**A**) The observed single channel input of T samples long is assumed to be generated by a weighted sum of two source signals of the same length: $y^t = \lambda_1 x_1^t + \lambda_2 x_2^t$. (**B**) Decomposition of the individual source signals. The method is to chop x_i^t into blocks of uniform length N starting at t, represented as vectors $\mathbf{x}_i^t = [x_i^t \; x_i^{t+1} \; \ldots \; x_i^{t+N-1}]^T$, which is in turn assumed to be generated by weighted linear superpositions of basis functions: $\mathbf{x}_i^t = \sum_k s_{ik}^t \mathbf{a}_{ik}$.

letter, $\mathbf{x}_i^t = [x_i^t \; x_i^{t+1} \; \ldots \; x_i^{t+N-1}]^T$, attaching the lead-off sample index for the superscript and representing the transpose operator with ^{T}. The constructed column vector is then expressed as a linear combination of the basis functions such that

$$\mathbf{x}_i^t = \sum_{k=1}^{M} \mathbf{a}_{ik} s_{ik}^t = \mathbf{A}_i \mathbf{s}_i^t \,, \tag{12.2}$$

where M is the number of basis functions, $\mathbf{a}_{ik}$ is the k-th basis function of i-th source denoted by an N-dimensional column vector, s_{ik}^t is its coefficient (weight) and $\mathbf{s}_i^t = [s_{i1}^t \; s_{i2}^t \ldots s_{iM}^t]^T$. The r.h.s. is the matrix-vector notation. The second subscript k followed by the source index i in s_{ik}^t represents the component number of the coefficient vector $\mathbf{s}_i^t$. We assume that $M = N$ and $\mathbf{A}$ has full rank so that the transforms between $\mathbf{x}_i^t$ and $\mathbf{s}_i^t$ are reversible in both directions. The inverse of the basis matrix, $\mathbf{W}_i = \mathbf{A}_i^{-1}$, refers to the ICA filters that generate the coefficient vector: $\mathbf{s}_i^t = \mathbf{W}_i \mathbf{x}_i^t$. The purpose of this decomposition is to model the multivariate distribution of $\mathbf{x}_i^t$ in a statistically efficient manner. The ICA learning algorithm is equivalent to searching for the linear transformation that makes the components as statistically independent as possible, as well as maximizing the marginal component densities of the given training data [19],

$$\begin{aligned} \mathbf{W}_i^* &= \arg\max_{\mathbf{W}_i} \prod_t \Pr(\mathbf{x}_i^t; \mathbf{W}_i) \\ &= \arg\max_{\mathbf{W}_i} \prod_t \prod_k \Pr(s_{ik}^t) \,, \end{aligned} \tag{12.3}$$

where $\Pr(a)$ is the probability of a variable a. In contrast, $p(a)$ denotes a realized pdf of a variable a and should be noted distinctively with its conceptual value $\Pr(a)$.

Independence between the components and over time samples factorizes the joint probabilities of the coefficients into the product of marginal component pdf. Thus the important issue is the degree to which the model distribution is matched to the true underlying distribution $\Pr(s_{ik}^t)$. We do not impose a prior distribution on the source coefficients. Instead, we are interested in inferring the distribution that results in maximally independent coefficients for the sources. Therefore we use a generalized Gaussian prior [17] that provides an accurate estimate for symmetric non-Gaussian distributions in modeling the underlying distribution of the source coefficients. The generalized Gaussian prior, also known as exponential power distribution, whose simplest form is $p(s) \propto \exp(-|s|^q)$, can describe Gaussian, platykurtic, and leptokurtic distributions by varying the exponent q. The optimal value of q for given data can be determined from the *maximum a posteriori* value and provides a good fit for the symmetric distributions. In the following sections we present an ICA learning algorithm using a generalized Gaussian function as a flexible prior that estimates the distributions of the sources.

12.2.2 Learning the Basis Functions

In the generative speech model the key parameters are the basis functions. They change the coordinates of the original data so that the resulting coefficients are statistically independent. First we briefly describe the ICA learning rule.

The ICA transformation is performed by the basis filters, the rows of $\mathbf{W}$. Initially, we do not know the structure of the basis filters, and therefore we adapt the filters using a generalized formulation of the ICA cost function. The goal of ICA is to adapt the filters by optimizing $\mathbf{s}$ so that the individual components s_k are statistically independent, and this adaptation process minimizes the mutual information between s_k. A learning algorithm can be derived using the information maximization principle [5] or the maximum likelihood estimation (MLE) method [19], which can be shown to be equivalent to estimating the density functions [20]. In our approach, we use the infomax learning rule with natural gradient extension and update the basis functions by the following learning rule [21]:

$$\Delta\mathbf{W} \propto \left[\mathbf{I} - \varphi(\mathbf{s})\mathbf{s}^T\right]\mathbf{W}\,, \tag{12.4}$$

where $\mathbf{I}$ is the identity matrix, $\varphi(\mathbf{s}) = \partial \log p(\mathbf{s})/\partial \mathbf{s}$ and $\mathbf{s}^T$ denotes the matrix transpose of $\mathbf{s}$. We assume that $\mathbf{W}$ is square; that is, the number of sources is equal to the number of sensors. The coefficient vector $\mathbf{s}$ can be replaced with any of $\mathbf{s}_i^t$ in (12.2). To learn the basis filter for the i-th source, only $\{\mathbf{s}_i^t | t \in [1, T]\}$ are used. We omit the subscripts and the superscripts for compact notations. $\Delta\mathbf{W}$ is the change of the basis functions that is added to $\mathbf{W}$ and will converge to zero once the adaptation process is complete. Calculating $\varphi(\mathbf{s})$ requires a multivariate density model for $p(\mathbf{s})$, which factorizes into component densities: $p(\mathbf{s}) = \prod_k^N p_k(s_k)$. The parametric density estimate $p_k(s_k)$ plays an essential role in the success of the learning rule. Local convergence is assured if $p_k(s_k)$ is an estimate of the true source density [22]. Note that the global shape of $p_k(s_k)$ was fixed in previous works [15, 23, 24].

12.2.3 The Generalized Gaussian Distributions

The success of the ICA learning algorithm for our purpose depends highly on how closely the ICA density model captures the true source coefficient density. The better the density estimation, the better the basis features in turn are responsible for describing the statistical structure. The generalized Gaussian distribution models a family of density functions that is peaked and symmetric at the mean, with a varying degree of normality in the following general form [17, 25]:

$$p_{\mathbf{g}}(s|\theta) = \frac{\omega(q)}{\sigma} \exp\left[-c(q)\left|\frac{s-\mu}{\sigma}\right|^q\right], \quad \theta = \{\mu, \sigma, q\}, \tag{12.5}$$

where $\mu = E[s]$, $\sigma = \sqrt{E[(s-\mu)^2]}$, $c(q) = \left[\frac{\Gamma[3/q]}{\Gamma[1/q]}\right]^{q/2}$, and $\omega(q) = \frac{\Gamma[3/q]^{1/2}}{(2/q)\Gamma[1/q]^{3/2}}$. The exponent q regulates the deviation from normality. The Gaussian, Laplacian, and strong Laplacian—speech signal—distributions are modeled by putting $q = 2$, $q = 1$, and $q < 1$ respectively. Each scalar component of the score function in (12.4) can be computed by substituting the scalar density function $p_{\mathbf{g}}(\cdot)$ for $p_k(\cdot)$ with suitable generalized Gaussian parameters:

$$\varphi(s) = \frac{\partial \log p_{\mathbf{g}}(s)}{\partial s} = -\frac{cq}{\sigma^q}|s-\mu|^{q-1}\operatorname{sign}(s-\mu). \tag{12.6}$$

Gradient ascent adaptation is applied in order to attain the maximal log likelihood. More detailed derivations can be found in [17, 26].

In Fig. 12.2, the coefficient histogram of real data reveals that the distribution has a highly sharpened point at the peak around zero and has heavy and long tails; there is only a small percentage of informative quantities (non-zero coefficients) in the tails and most of the data values are around zero, that is, the data is sparsely distributed. From a coding perspective this implies that we can encode and decode the data with only a small percentage of the coefficients $\mathbf{s}$. For modeling the densities of the source coefficients neither Laplacian nor less kurtotic, logistic functions, are adequate for speech bases. The generalized Gaussian parameter set θ_{ik} approximates the distribution of the coefficient s_{ik}^t, the k-th component of the i-th source coefficient vector:

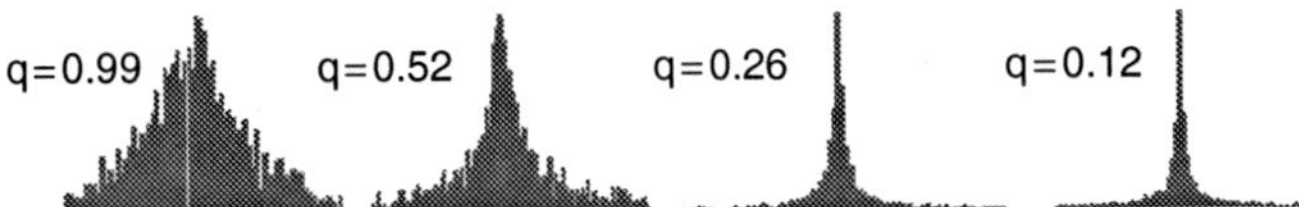

Fig. 12.2. Examples of the actual coefficient distributions and the estimated values of the exponent of the exponential power distributions. The distributions generally have more sharpened summits and longer tails than a Gaussian distribution, and would be classified as super-Gaussian. Generalized Gaussian density functions provide good matches by varying exponents as shown in the equation. From left to right, the exponent decreases, and the distributions become more super-Gaussian.

$\Pr(s_{ik}^t) \approx p_{\mathbf{g}}(s_{ik}^t|\theta_{ik})$. The basis filters $\mathbf{w}_{ik}$, rows of $\mathbf{W}_i$, and the individual parameter set θ_{ik} for the distribution of the filter output are obtained beforehand by the generalized Gaussian ICA learning algorithm [26], and used as prior information for the proposed source separation algorithm.

12.3 Maximum Likelihood Source Inference

We have demonstrated that the learned basis filters maximize the likelihood of the given data. Suppose we know what kind of sound sources have been mixed and we were given the set of basis filters from a training set. Could we infer the learning data? The answer is generally "no" when $N < T$ and no other information is given. In our problem of single channel separation, half of the solution is already given by the constraint $y^t = \lambda_1 x_1^t + \lambda_2 x_2^t$, where x_i^t constitutes the basis learning data $\mathbf{x}_i^t$ (Fig. 12.1 (**B**)). Essentially, the goal of the source inferring algorithm of this chapteris to complement the remaining half with the statistical information given by a set of coefficient density parameters θ_{ik}. If the parameters are given, we can perform *maximum a posteriori* (MAP) estimation simply by optimizing the data likelihood computed by the model parameters.

The separation algorithm has two major features: it is *adaptive* and should perform all relevant adaptation *on a single sample basis*, which means that the solution is achieved by altering a set of unknowns gradually from an arbitrary initial values to a certain goal, and the number of unknowns to be estimated equals the number of samples. In the following subsection we derive a gradient-ascent learning rule for the single filter output given the single sensor input, which is done by the generalized Gaussian expansion of the coefficient pdf. In Sect. 12.3.2, we combine the adaptation rules of the individual filters, with consideration of the interaction between the two source signals. Finally Sect. 12.3.3 explains how to update the scaling factors λ_i. The evaluation of the derived learning rules and practical issues in actual situations are discussed in Sect. 12.4.

12.3.1 For Single Filter and Single Source

Initially, a large amount of probabilistic dependence exists among the nearby components of the source vector $\mathbf{x}_i^t$. When we pass $\mathbf{x}_i^t$ through a set of linear basis filters $\mathbf{W}_i$ obtained by the ICA learning algorithm, a set of statistically independent random variables, $\{s_{ik}^t = \mathbf{w}_{ik}\mathbf{x}_i^t | k = 1, \ldots, N\}$, emerge at the output, where k is a filter index. If $\mathbf{W}_i$ is invertible, it causes no information loss, and $\Pr(\mathbf{x}_i^t)$ factorizes into component densities of the output variables [19]:

$$\begin{aligned}\Pr(x_i^{1\ldots T}; \mathbf{W}_i) &\cong \prod_{t=1}^{T} \Pr(\mathbf{x}_i^t; \mathbf{W}_i) \\ &= \prod_{t=1}^{T} \Pr(\mathbf{s}_i^t) \cdot |\det \mathbf{W}_i| \\ &= \prod_{t=1}^{T} \left\{ \prod_{k=1}^{N} p_{\mathbf{g}}(s_{ik}^t | \theta_{ik}) \right\} |\det \mathbf{W}_i| \,, \end{aligned} \tag{12.7}$$

where $p_{\mathbf{g}}(\cdot)$ is the generalized Gaussian pdf introduced in Sect. 12.2.3, and θ_{ik} is the set of parameters given in advance for the distribution of filter output coefficient s_{ik}^t. The notation '$i \ldots j$' denotes an ordered set of the elements from index i to j. The term $|\det \mathbf{W}_i|$ gives the change in volume produced by the linear transformation [22]. The calculation of the multivariate pdfs is reduced to the product of univariate generalized Gaussian pdfs.

Now we focus on optimizing each component separately. Assuming independence over time, the computation of the log likelihood of the time series $s_{ik}^{1\ldots T}$ amounts to the total sum of the marginal coefficient pdfs over time indices, and we denote it by $\mathcal{L}_{ik}$:

$$\mathcal{L}_{ik} = \log \Pr(s_{ik}^{1\ldots T}) = \log \prod_{t=1}^{T} p_{\mathbf{g}}(s_{ik}^t | \theta_{ik}) = \sum_{t=1}^{T} \log p_{\mathbf{g}}(s_{ik}^t | \theta_{ik}) \,. \tag{12.8}$$

In an ordinary ICA, the learning algorithm optimizes data likelihood by altering $\mathbf{W}_i$. The target of the proposed method is identical, but the values to be altered are the data, not the basis filters. Since (12.8) is the sum of the pdfs of T statistically independent variables along the time samples, any term that is not related to s_{ik}^t—the k-th output coefficient of the i-th source at time t—vanishes while taking the partial derivative of the log likelihood:

$$\begin{aligned}\frac{\partial \mathcal{L}_{ik}}{\partial s_{ik}^t} &= \sum_{m=1}^{T} \frac{\partial \log p_{\mathbf{g}}(s_{ik}^m)}{\partial s_{ik}^t} \\ &= \frac{\partial \log p_{\mathbf{g}}(s_{ik}^t)}{\partial s_{ik}^t} \\ &= \varphi(s_{ik}^t) \,, \end{aligned} \tag{12.9}$$

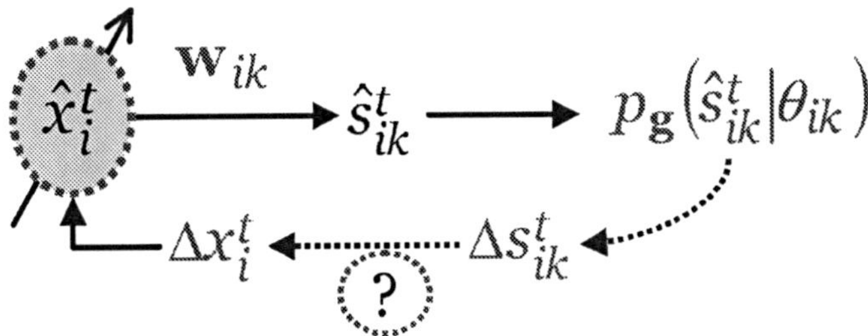

Fig. 12.3. Data flow of the proposed method for a single input and a single filter. The current estimate $\hat{x}_i^t$ is passed through the basis filters $\mathbf{w}_{ik}$, and generates a code $\hat{s}_{ik}^t$ which employs a generalized Gaussian pdf $p_{\mathbf{g}}(\cdot)$ with a set of parameters θ_{ik}. The adjustment in s_{ik}^t, Δs_{ik}^t, is computed to increase the likelihood at the output, then sent back to the input domain by a mapping that specifies Δx_i^t for Δs_{ik}^t, and is finally added to the current estimate $\hat{x}_i^t$. The mapping is discussed in the text. The algorithm iteratively performs all relevant adaptation on a single sample basis.

where $\varphi(\cdot)$ is the component score function defined in (12.6). The parameter set θ_{ik} is omitted but can be identified by the subscript of the filter coefficient. The above equation holds for every $k \in [1, N]$, and $t \in [1, T]$.

Figure 12.3 explains how the proposed method processes the input and the output signals of the filter. It estimates the source signals by iteratively updating them from certain initial values, to maximize the log likelihood computed by the given generalized Gaussian parameters.

An important issue is the mapping of the adjustments: from Δs_{ik}^t to the original input domain Δx_i^t. Figure 12.4 provides a conceptual explanation of the adjustment mapping. Each $\mathbf{w}_{ik}$ takes windows of N continuous samples starting from the t-th sample out of the i-th source signal, $\mathbf{x}_i^t = [x_i^t \;\dots\; x_i^{t+N-1}]^T$, and the filter produces the output coefficient s_{ik}^t. Each sample of the source participates in the generation of N different inputs, and henceforth in the generation of N different output coefficients for each filter. The following matrix-vector expression of the basis filtering highlights positions of a sample x_i^t in all the possible input windows:

$$\begin{aligned}
\left[s_{ik}^{t-N+1} \; s_{i}^{t-N+2} \cdots s_{ik}^{t} \right] &= \mathbf{w}_{ik} \cdot \left[\mathbf{x}_i^{t-N+1} \; \mathbf{x}_i^{t-N+2} \cdots \mathbf{x}_i^{t} \right] \\
&= \begin{bmatrix} w_{ik1} \\ w_{ik2} \\ \vdots \\ w_{ikN} \end{bmatrix}^T \cdot \begin{bmatrix} x_i^{t-N+1} & x_i^{t-N+2} & \cdots & x_i^t \\ x_i^{t-N+2} & \cdots & x_i^t & x_i^{t+1} \\ \vdots & x_i^t & \ddots & \vdots \\ x_i^t & x_i^{t+1} & \cdots & x_i^{t+N-1} \end{bmatrix},
\end{aligned} \tag{12.10}$$

where the scalar w_{ikn} is the n-th component of $\mathbf{w}_{ik}$. The indices of the windows containing x_i^t range from $t-N+1$ to t. We introduce an offset variable $n \in [1, N]$ so that s_{ik}^{t-n+1} may cover the range $[t-N+1, t]$. Then the partial derivative of the output at time $t-n+1$ with respect to the source signal

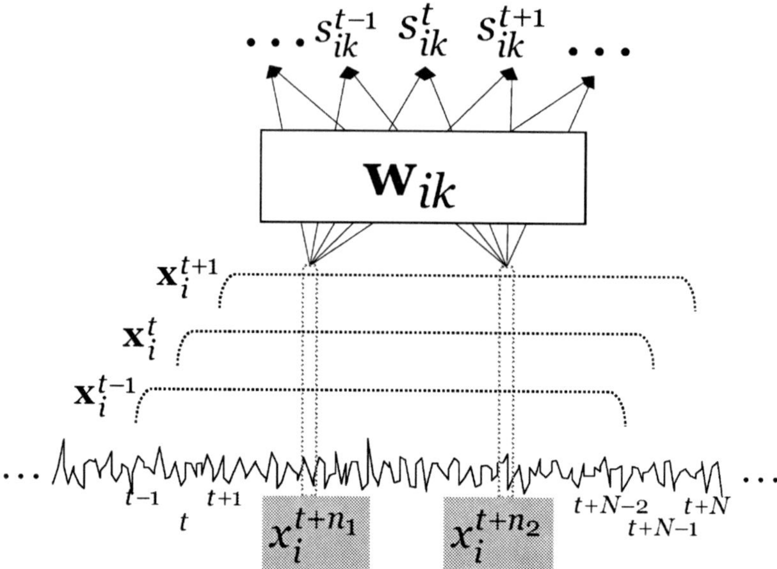

Fig. 12.4. The participation of a sample in the source signal to the generation of each output coefficient. The input $\mathbf{x}_i^t$ is a vector composed of N continuous samples ranging from t to $t+N-1$ in the i-th source. The output coefficient s_{ik}^t is obtained by passing $\mathbf{x}_i^t$ through $\mathbf{w}_{ik}$. The middle of the figure shows that there exist N different possible input covers over a sample, which subsequently participate in the generation of N different output coefficients per filter.

at time t becomes a simple scalar value as

$$\frac{\partial s_{ik}^{t-n+1}}{\partial x_i^t} = \frac{\partial \left(\sum_{m=1}^{N} w_{ikm} \cdot x_i^{t-n+m} \right)}{\partial x_i^t} = w_{ikn} \,. \tag{12.11}$$

The gradient ascent for x_i^t is derived by differentiating the object function $\mathcal{L}_{ik}$ w.r.t. x_i^t, and summing up the gradients over only N relevant output coefficients,

$$\begin{aligned} \frac{\partial \mathcal{L}_{ik}}{\partial x_i^t} &= \sum_{m=1}^{T} \frac{\partial \log p_{\mathbf{g}}(s_{ik}^m)}{\partial x_i^t} = \sum_{n=1}^{N} \frac{\partial \log p_{\mathbf{g}}(s_{ik}^{t_n})}{\partial x_i^t} \\ &= \sum_{n=1}^{N} \frac{\partial \log p_{\mathbf{g}}(s_{ik}^{t_n})}{\partial s_{ik}^{t_n}} \frac{\partial s_{ik}^{t_n}}{\partial x_i^t} = \sum_{n=1}^{N} \varphi(s_{ik}^{t_n}) w_{ikn} \,, \end{aligned} \tag{12.12}$$

where $t_n = t - n + 1$ and the score function $\varphi(\cdot)$ is defined in (12.6). The first multiplier inside the summation, $\varphi(s_{ik}^{t_n})$, is interpreted as a stochastic gradient ascent that gives the direction and the amount of the change at the output of the basis filter, $s_{ik}^t = \mathbf{w}_{ik}\mathbf{x}_i^t$. The second term w_{ikn} accounts for the change produced by the filter between the input x_i^t and the output

$s_{ik}^{t_n}$. The summation implies that the source signal is decomposed to N independent components.

12.3.2 Combining Individual Inference Rules

Now we have the individual learning rules for the source signals in (12.12): the partial derivative of the log likelihood $\mathcal{L}_{ik}$ w.r.t. x_i^t considering each source and each filter separately. However the initial constraint, $y^t = \lambda_1 x_1^t + \lambda_2 x_2^t$, suggests that the sources are definitely no longer independent of each other. The value of the object function $\mathcal{L}_{1k}$ depends on the values of both x_1^t and x_2^t, and $\mathcal{L}_{2k}$ also depends on both. Updating x_1^t and x_2^t separately would violate the constraint. Hence we need to consider the interactions between the two source signals while combining the individual learning rules.

At every time t every source signal can be expressed by the counterpart, $x_2^t = (y^t - \lambda_1 x_1^t)/\lambda_2$ and $x_1^t = (y^t - \lambda_2 x_2^t)/\lambda_1$. We represent the relationships between the sources by the following two equivalent differential equations:

$$\frac{\partial x_2^t}{\partial x_1^t} = -\frac{\lambda_1}{\lambda_2} \Leftrightarrow \frac{\partial x_1^t}{\partial x_2^t} = -\frac{\lambda_2}{\lambda_1} . \tag{12.13}$$

Let us define a unified object function $\mathcal{L}_{\cup}$ as the joint log likelihood of the two source signals, expressed by the sum of the individual object functions $\mathcal{L}_{ik}$

$$\begin{aligned} \mathcal{L}_{\cup} &= \log \left\{ \Pr\left(x_1^{1\dots T}\right) \Pr\left(x_2^{1\dots T}\right) \right\} \\ &\propto \sum_{k=1}^{N} \sum_{t=1}^{T} \log p_{\mathbf{g}}(s_{1k}^t) + \sum_{k=1}^{N} \sum_{t=1}^{T} \log p_{\mathbf{g}}(s_{2k}^t) \\ &\cong \sum_{k=1}^{N} \mathcal{L}_{1k} + \sum_{k=1}^{N} \mathcal{L}_{2k} . \end{aligned} \tag{12.14}$$

In stochastic gradient ascent we differentiate $\mathcal{L}_{\cup}$ w.r.t. the first source x_1^t at every time t, write it in terms of (12.12) and (12.13), and evaluate the gradient to give the subsequent unified learning rule:

$$\begin{aligned} \Delta x_1^t &\propto \frac{\partial \mathcal{L}_{\cup}}{\partial x_1^t} \\ &= \sum_{k=1}^{N} \frac{\partial \mathcal{L}_{1k}}{\partial x_1^t} + \sum_{k=1}^{N} \frac{\partial \mathcal{L}_{2k}}{\partial x_2^t} \frac{\partial x_2^t}{\partial x_1^t} \\ &= \sum_{k=1}^{N} \sum_{n=1}^{N} \varphi(s_{1k}^{t_n}) w_{1kn} + \sum_{k=1}^{N} \sum_{n=1}^{N} \varphi(s_{2k}^{t_n}) w_{2kn} \cdot \left(-\frac{\lambda_1}{\lambda_2} \right) \\ &= \sum_{k=1}^{N} \sum_{n=1}^{N} \left\{ \varphi(s_{1k}^{t_n}) w_{1kn} - \frac{\lambda_1}{\lambda_2} \cdot \varphi(s_{2k}^{t_n}) w_{2kn} \right\} . \end{aligned} \tag{12.15}$$

The second term inside the final summation can be interpreted as a stochastic gradient ascent for x_2^t scaled by $-\lambda_1/\lambda_2$. The denominator λ_2 normalizes the gradient, and the numerator λ_1 scales it to be added to x_1^t. The minus sign implies that adjusting x_2^t affects x_1^t in the opposite direction. Similar reasoning leads to the rule for the second source:

$$\begin{aligned}\Delta x_2^t &\propto \frac{\partial \mathcal{L}_{\cup}}{\partial x_2^t} \\ &= \sum_{k=1}^{N}\sum_{n=1}^{N}\left\{-\frac{\lambda_2}{\lambda_1}\cdot\varphi(s_{1k}^{t_n})w_{1kn}+\varphi(s_{2k}^{t_n})w_{2kn}\right\}. \qquad (12.16)\end{aligned}$$

Updating the sources separately using these learning rules might lead to a violation of the initial constraint. To avoid the violation, the values of the source signals after adaptation must always satisfy

$$\begin{aligned}y^t &= \lambda_1(x_1^t+\Delta x_1^t)+\lambda_2(x_2^t+\Delta x_2^t) \\ &\Leftrightarrow \lambda_1\Delta x_1^t+\lambda_2\Delta x_2^t=0\,. \qquad (12.17)\end{aligned}$$

In the actual application of the adaptation rules, we scale (12.15) and (12.16) appropriately and derive the final learning rules as

$$\begin{aligned}\Delta x_1^t &= \eta\sum_{k=1}^{N}\sum_{n=1}^{N}\left\{\lambda_2^2\cdot\varphi(s_{1k}^{t_n})w_{1kn}-\lambda_1\lambda_2\cdot\varphi(s_{2k}^{t_n})w_{2kn}\right\}, \\ \Delta x_2^t &= \eta\sum_{k=1}^{N}\sum_{n=1}^{N}\left\{-\lambda_1\lambda_2\cdot\varphi(s_{1k}^{t_n})w_{1kn}+\lambda_1^2\cdot\varphi(s_{2k}^{t_n})w_{2kn}\right\}, \qquad (12.18)\end{aligned}$$

where η is a learning gain. The whole dataflow of the proposed method is summarized in four steps in Fig. 12.5. In step **A**, the source signals are decomposed into N statistically independent codes. The decomposition is done by a set of the given ICA filters, $\mathbf{s}_i^t=\mathbf{W}_i\mathbf{x}_i^t$. In step **B**, the stochastic gradient ascent for each filter output code is computed from the derivative of the log likelihood of the code (12.9). In step **C**, the computed gradient is transformed to the source domain according to (12.12). All the filter output codes are regarded as being independent, so all the computations are performed independently. In step **D**, we add up all the gradients and modify them to satisfy the initial constraint according to (12.18). The four steps comprise one iteration of the adaptation of each sample. The solution is achieved after repeating this iteration on the source signal x_i^t at every time t to a convergence from certain initial values.

12.3.3 Updating Scaling Factors

Updating the contribution factors λ_i can be accomplished by finding the maximum *a posteriori* values. To simplify the inferring steps, we force the

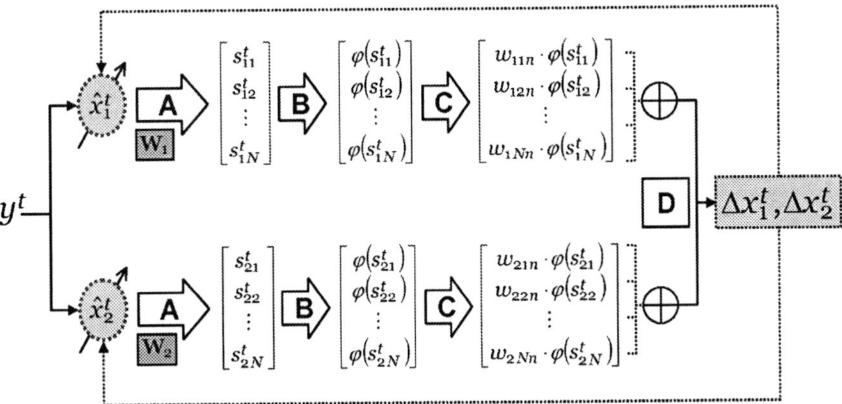

Fig. 12.5. The overall structure and the data flow of the proposed method. In the beginning, we are given single channel data y^t, and we have the estimates of the source signals, $\hat{x}_i^t$, at every adaptation step. **(A)** $x_i^t \Rightarrow s_{ik}^t$: At each time-point, the current estimates of the source signals are passed through a set of basis filters $\mathbf{W}_i$, generating N sparse codes s_{ik}^t that are statistically independent. **(B)** $s_{ik}^t \Rightarrow \Delta s_{ik}^t$: The stochastic gradient for each code is computed from the derivative of the log likelihood of each individual code. **(C)** $\Delta s_{ik}^t \Rightarrow \Delta x_i^t$: The gradient for each code is transformed to the domain of source signal. **(D)** (finalize): The individual gradients are combined and modified to satisfy the given constraints, and added to the current estimates of the source signals.

sum of the factors to be constant: e.g. $\lambda_1 + \lambda_2 = 1$. The value of λ_2 is completely dependent on the value of λ_1, so we need to consider λ_1 only. Given the current estimates of the sources $x_i^{1\ldots T}$, the posterior probability of λ_1 is

$$\Pr(\lambda_1 | x_1^{1\ldots T}, x_2^{1\ldots T}) \propto \Pr(x_1^{1\ldots T}) \Pr(x_2^{1\ldots T}) p_\lambda(\lambda_1), \tag{12.19}$$

where $p_\lambda(\cdot)$ is the prior density function of λ. Performing a log operation on the above equation yields

$$\log \Pr(x_1^{1\ldots T}) \Pr(x_2^{1\ldots T}) p_\lambda(\lambda_1) \cong \mathcal{L}_\cup + \log p_\lambda(\lambda_1), \tag{12.20}$$

where $\mathcal{L}_\cup$ is the unified object function defined in (12.14). If we assume λ_1 to be uniformly distributed in the range $[0, 1]$, $p_\lambda(\cdot)$ is considered as a constant, and vanishes in

$$\begin{aligned} \lambda_1^* &= \arg\max_{\lambda_1} \{\mathcal{L}_\cup + \log p_\lambda(\lambda_1)\} \\ &= \arg\max_{\lambda_1} \mathcal{L}_\cup . \end{aligned} \tag{12.21}$$

We differentiate $\mathcal{L}_{\cup}$ w.r.t. λ_1, and substitute $\lambda_2 = 1 - \lambda_1$:

$$\begin{aligned}
\frac{\partial \mathcal{L}_{\cup}}{\partial \lambda_1} &= \sum_{t=1}^{T}\sum_{k=1}^{N} \frac{\partial \log p_{\mathbf{g}}(s_{1k}^t)}{\partial \lambda_1} + \sum_{t=1}^{T}\sum_{k=1}^{N} \frac{\partial \log p_{\mathbf{g}}(s_{2k}^t)}{\partial \lambda_1} \\
&= \sum_{t=1}^{T}\sum_{k=1}^{N} \varphi(s_{1k}^t)\frac{\partial s_{1k}^t}{\partial \lambda_1} + \sum_{t=1}^{T}\sum_{k=1}^{N} \varphi(s_{2k}^t)\frac{\partial s_{2k}^t}{\partial \lambda_2}\cdot\frac{\partial \lambda_2}{\partial \lambda_1} \\
&= \sum_{t=1}^{T}\sum_{k=1}^{N} \varphi(s_{1k}^t)\frac{\partial s_{1k}^t}{\partial \lambda_1} - \sum_{t=1}^{T}\sum_{k=1}^{N} \varphi(s_{2k}^t)\frac{\partial s_{2k}^t}{\partial \lambda_2} .
\end{aligned} \tag{12.22}$$

From the initial constraint $y^t = \lambda_1 x_1^t + \lambda_2 x_2^t$, we might deduce that the value of $\lambda_i x_i^t$ is unaffected by the change of either λ_i or x_i^t, for all $i \in \{1, 2\}$, $t \in [1, T]$. Because s_{ik}^t is the output of x_i^t, $\lambda_i s_{ik}^t$ is also unaffected by λ_i or s_{ik}^t. Hence we introduce a new variable $c_{ik}^t = \lambda_i s_{ik}^t$, which is constant with λ_i. Differentiating s_{ik}^t w.r.t. λ_i yields

$$\frac{\partial s_{ik}^t}{\partial \lambda_i} = \frac{\partial}{\partial \lambda_i}\left(\frac{c_{ik}^t}{\lambda_i}\right) = -\frac{c_{ik}^t}{\lambda_i^2} . \tag{12.23}$$

(12.22) is rewritten as

$$\begin{aligned}
\frac{\partial \mathcal{L}_{\cup}}{\partial \lambda_1} &= \underbrace{\sum_{t=1}^{T}\sum_{k=1}^{N} \varphi(s_{1k}^t)c_{1k}^t}_{\stackrel{\text{let}}{=}\psi_1}\cdot\left(-\frac{1}{\lambda_1^2}\right) - \underbrace{\sum_{t=1}^{T}\sum_{k=1}^{N} \varphi(s_{2k}^t)c_{2k}^t}_{\stackrel{\text{let}}{=}\psi_2}\cdot\left(-\frac{1}{\lambda_2^2}\right) \\
&= -\frac{\psi_1}{\lambda_1^2} + \frac{\psi_2}{\lambda_2^2} .
\end{aligned} \tag{12.24}$$

Solving equation $\partial \mathcal{L}_{\cup}/\partial \lambda_1 = 0$ subject to $\lambda_1 + \lambda_2 = 1$ and $\lambda_1, \lambda_2 \in [0, 1]$ gives

$$\lambda_1^* = \frac{\sqrt{|\psi_1|}}{\sqrt{|\psi_1|} + \sqrt{|\psi_2|}}, \quad \lambda_2^* = \frac{\sqrt{|\psi_2|}}{\sqrt{|\psi_1|} + \sqrt{|\psi_2|}} . \tag{12.25}$$

These values guarantee the local maxima of $\mathcal{L}_{\cup}$ w.r.t. the current estimates of source signals. The algorithm updates them periodically during the learning steps.

12.4 Evaluations

We now present some examples of single channel separation of artificial mixtures using speech signals and music signals. The performance of the proposed method is compared to the ideal Wiener filtering. We then present the separation results of noise and speech recorded in a real environment.

12.4.1 Simulation Data

We have tested the performance of the proposed method on single channel mixtures of four different sound types. They were monaural signals of rock and jazz music, male and female speech. We used different sets of speech signals for learning basis functions and for generating the mixtures. For the mixture generation, two sentences of the target speakers 'mcpm0' and 'fdaw0', one for each speaker, were selected from the TIMIT speech database. The training sets were designed to have 21 sentences for each gender, 3 each from 7 randomly chosen males and 7 randomly chosen females. The utterances of the 2 target speakers were not included in the training set. Rock music was mainly composed of guitar and drum sounds, and jazz was generated by a wind instrument. Vocal parts of both music sounds were excluded. Half of the music sound was used for training, half for generating mixtures. All signals were downsampled to 8 kHz, from original 44.1 kHz (music) and 16 kHz (speech). The training data were segmented in 64 samples (8 ms) starting at every sample. Audio files for all the experiments are accessible at the web site.[1]

Figure 12.6 displays the actual sources, adapted basis functions, and their coefficient distributions. Music basis functions exhibit consistent amplitudes

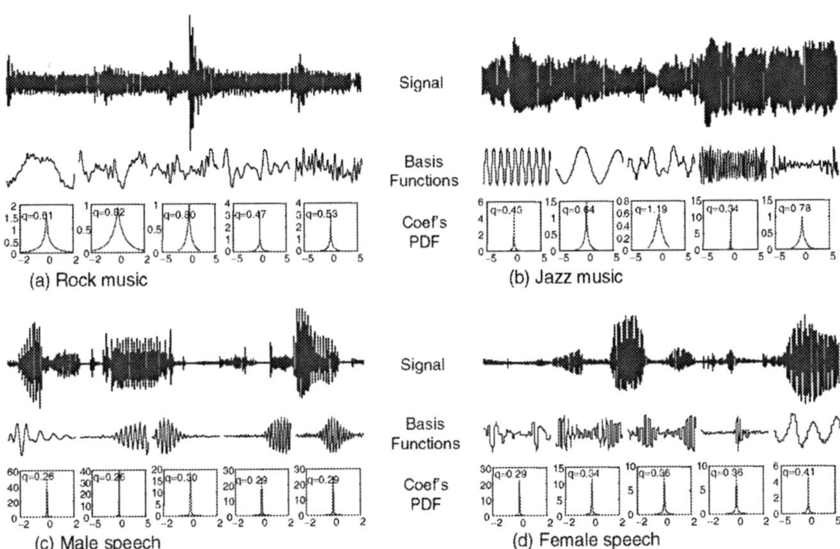

Fig. 12.6. Characteristics of four sound sources. In (a)–(d), the first rows are actual waveforms of the source signals, the second rows are the adapted basis functions $\mathbf{a}_i$, and the third rows show the distributions of the coefficients $p(s_{ik}^t)$ modeled by generalized Gaussians. Only 5 basis functions were chosen out of complete sets of 64. The full set of basis functions is also available at the web site.

[1] `http://home.bawi.org/~jangbal/ch1bss/`

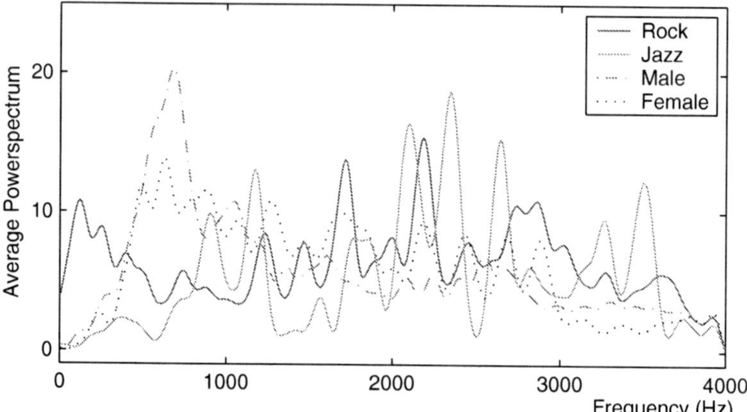

Fig. 12.7. Average powerspectra of the 4 sound sources. Frequency scale ranges in 0∼4 kHz (x-axis), since all the signals are sampled at 8 kHz. The powerspectra are averaged and represented in the y-axis.

with harmonics, and the speech basis functions are similar to Gabor wavelets. Figure 12.7 compares the four sources by the average spectra. Each covers all the frequency bands, although they are different in amplitude. One might expect that simple filtering or masking cannot separate the mixed sources clearly.

12.4.2 Practical Considerations

The learning rules in (12.18) and (12.25) are applied to attain the original source signals from a single channel mixture. Before actual separation, the source signals were initialized to the values of mixture signal: $x_i^t = y^t$. The initial λ_i were both 0.5 to satisfy $\lambda_1 + \lambda_2 = 1$. The single adaptation step was repeated on all the samples, and the scaling factors were updated every 10 steps. The separation converged roughly after 100 steps, depending on the learning rate and other various system parameters. The procedures of the separation algorithm—traversing all the data and computing gradients—are similar to those of the basis learning algorithm, so their time complexities are likewise of the same order. The measured separation time on a 1.0 GHz Pentium PC was roughly 10 min for an 8 sec long mixture.

12.4.3 Separation Results of Simulated Mixtures

We generated a synthesized mixture by selecting two sources out of the four and simply adding them. The proposed separation algorithm was applied to recover the original sources. The similarity between the input and output signals is measured by signal-to-noise ratio (SNR), which is defined by

$$\mathrm{snr}_s(\hat{s})\ [\mathrm{dB}] = 10 \log_{10} \frac{\sum_t s^2}{\sum_t (s - \hat{s})^2} ,$$

where s is the original source signal and $\hat{s}$ its estimate. To qualify a separation result we define a performance measurement function π as the sum of the increases in the SNR values of the two recovered source signals:

$$\pi(\hat{x}_1, \hat{x}_2; y, x_1, x_2) = \sum_{i=1}^{2} \{\mathrm{snr}_{x_i}(\hat{x}_i) - \mathrm{snr}_{x_i}(y)\} ,$$

where y is the mixed signal (initial estimate), $\hat{x}_i$ are the recovered sources, and x_i are the original sources. Table 12.1 reports SNRs and their π's. In terms of π, mixtures containing music were recovered more cleanly than male–female mixtures. Figure 12.8 illustrates the waveforms of the original sources and the recovered results for the mixture of jazz music and male speech, and Fig. 12.9 shows for the mixture of male and female speech. Their π values were 10.5 and 5.6. The separation of speech–speech mixture was much poorer than those of music–speech mixtures.

From the experimental results, we conclude that the demixing performance highly relies on the basis functions. A single channel containing a mixture of different classes of signals is projected on the combined bases sets, and sources are isolated by iteratively approaching the projections on the subsets of bases. This requires that each basis be incomplete in the general space of all possible signals but complete in the subspace of a single class of signal, and that the classes should be disjoint in signal space. This latter requirement is a fundamental limitation of the method. Based on the experimental results, it seems that there is too much overlap in signal space

Table 12.1. SNR results of the proposed method. {R, J, M, F} stand for rock, jazz music, male, and female speech. All the values are measured in dB. '*mix*' column lists the symbols of the sources that are mixed to y, and 'snr_{x_i}' columns are the calculated SNR of the mixed signal (y) and the recovered sources ($\hat{x}_i$) with the original sources (x_i). The last column is the total SNR increase (π). Audio files for all the results are accessible at the web site.

mix	snr_{x1}		snr_{x2}		π
	y	$\hat{x}_1$	y	$\hat{x}_2$	
R + J	−0.1	5.3	0.1	5.4	10.7
R + M	−0.1	4.4	0.1	4.5	8.9
R + F	0.0	3.7	0.0	3.8	7.5
J + M	0.6	5.6	−0.6	4.9	10.5
J + F	0.4	4.0	−0.4	3.6	7.6
M + F	−0.3	2.7	0.3	3.0	5.6

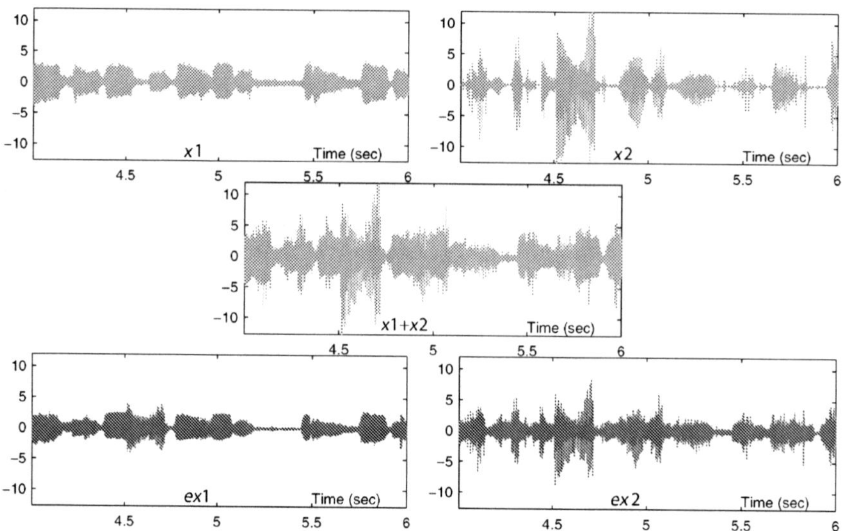

Fig. 12.8. Separation results of jazz music and male speech. In vertical order: original sources (x_1 and x_2), mixed signal ($x_1 + x_2$), and the recovered signals.

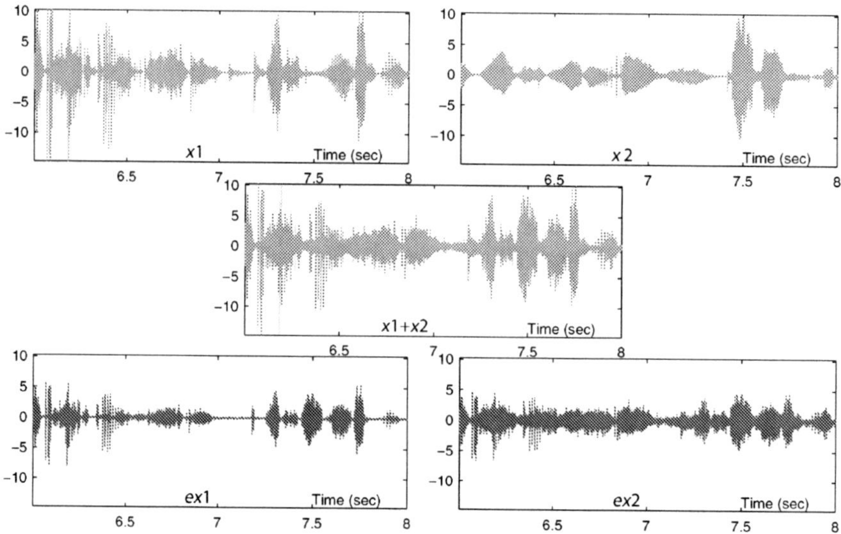

Fig. 12.9. Separation results of male and female speech signals.

between two speakers for this method to ever work for mixtures of speech. ICA found sets of bases that explain the class of the music signals well but performed poorly in explaining the class of the speech signals. Speech basis functions vary in amplitudes frequently in the time domain, and the coefficient distributions are extremely sparse. These characteristics are caused by the unstationary nature of the speech signal. In contrast, as it can be seen in Figs. 12.6 (a) and (b), the amplitudes of the music signals are comparatively stable, and the basis functions cover a longer range in the time axis. The coefficient distributions are less sparse than those of speech basis functions, which is analogous to earlier findings [5].

12.4.4 Comparison to Wiener Filtering

It is very difficult to compare a separation method with other CASA techniques; their approaches are vastly different in many ways such that an optimal tuning of their parameters would be beyond the scope of this chapter. However, we have compared our method with Wiener filtering [18]. The Wiener filter obtains a least squares estimate of the source signals under stationarity assumptions. The construction of the Wiener filter requires an estimate of the powerspectrum of the true source signals:

$$W_i(\omega) = \frac{X_i^*(\omega)}{X_1^*(\omega) + X_2^*(\omega)},$$

where $X_i^*(\omega)$ is the true powerspectum of the i-th source at frequency ω. Although Wiener filtering has the disadvantage that the estimation criterion is fixed and depends on the stationarity assumptions, it provides optimal masking filters if a true spectrogram is given. We regard this as a theoretical limit of the frequency-domain techniques. The filters were computed every block of 0.5, 1.0, 2.0, and 3.0 sec. Their performances are measured by π (the sum of SNR increases after separation) and compared to the proposed method in Table 12.2. In the case of average π, our blind results were comparable

Table 12.2. Comparison of Wiener filtering and the proposed method. '*mix*' column lists the symbols of the sources that are mixed to the input. {R, J, M, F} stand for rock, jazz music, male, and female speech. 'Wiener' columns are the evaluated π values grouped by the block lengths (in seconds). The filters are computed at every block. The last column lists the π values of the proposed method, and the last row is the average π. The performance of the proposed method was closest to Wiener filtering at the block length 2.0s. Audio files for all the results are accessible at the web site.

mix	Wiener				Proposed
	0.5 s	1.0 s	2.0 s	3.0 s	
R + J	11.1	10.3	9.4	9.1	10.7
R + M	8.7	8.1	7.1	7.3	8.9
R + F	10.1	8.9	8.2	7.3	7.5
J + M	13.5	11.9	10.0	10.5	10.5
J + F	14.1	11.0	9.4	8.4	7.6
M + F	9.9	8.5	7.8	6.1	5.6
Average	11.2	9.8	**8.6**	8.1	**8.5**

in SNR with results obtained when the Wiener filters were computed at 2.0 sec.

12.4.5 Experiments with Real Recordings

We have tested the performance of the proposed method on recordings in a real environment. Data were recorded in a diffuse sound room. Four speakers were employed, one in each corner of the sound room. A signal played through these speakers produces a uniform sound field throughout the room. The recorded signals were composed of a male speech utterance on the background of very loud noise. The level of noise was so high that even human listeners could hardly recognize what was spoken. The estimated SNR was about −8 dB. The focus of this experiment is to recover human speech in real recordings, to assess the performance of the proposed separation algorithm in a real environment.

The basis functions of general TIMIT male speakers (Fig. 12.6) are used for the recorded male speaker. Since we do not know exactly the characteristics of the noise source, we assumed two different types of well-known noisy signals: Gaussian white noise and pink noise. White noise has a uniform spectrogram all over frequency axis as well as over time axis. Pink noise is similar but the power decreases exponentially as the frequency increases. The spectrogram of the pink noise resembles the noisy recording in our case. (The noise sources, their basis functions and spectrograms, the recorded sound, and the separated results are available at the provided web site.) The algorithm did not work with the white noise, but it successfully recovered the original sources with the pink noise; the waveforms are displayed in Fig. 12.10. Although a "perfect separation" was not attained, it should be noted that the input was too noisy and we did not have the true basis functions. To achieve better separation in real environments, it is necessary to have a large pool of bases for various kinds of natural sounds and to find the most characterizing basis for a generic problem.

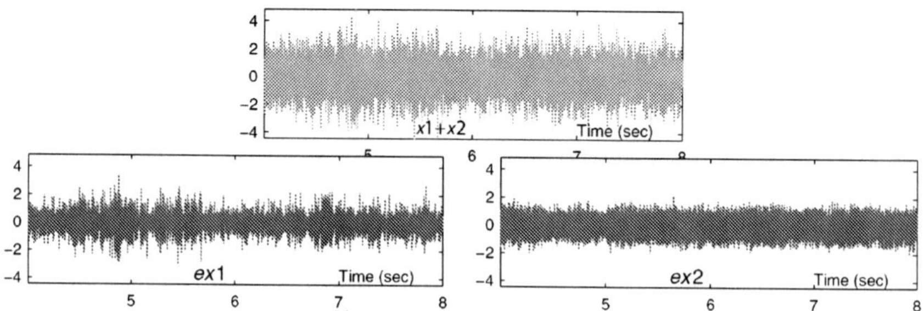

Fig. 12.10. Separation result of real recording. Input signal is at the top of the figure, below are the recovered male speech and the noise.

12.5 Discussions

This section compares the proposed method with other monaural source separation approaches, separability, and more detailed interpretations of experimental results. Future research issues such as dealing with more than two source signals and extensions to temporal continuity modeling are also discussed at the end of the section.

12.5.1 Frequency Domain Approaches to Single Channel Separation Techniques

Traditional approaches to signal separation are classified as either frequency-domain approaches or time-domain nonlinear filtering methods. Frequency-domain approaches assume that source signals are disjoint in the spectrogram, which frequently result in audible distortions of the signal in the regions where the assumption mismatches. Roweis [7] presented a refiltering technique which estimates λ_i in (12.1) as linear time-varying masking filters that localize sound streams in a spectro-temporal region. In his work sound sources are supposedly disjoint in the spectrogram and there exists a "mask" that divides the mixed multiple streams completely. A somewhat similar technique is proposed by Rickard and Balan [27]. They did not try to obtain the "exact" mask but an estimate by a ML-based gradient search. However, being based on the strong assumption in the spectral domain, these methods also suffer from the overlapped spectrogram.

To overcome the assumption of non-overlapping sources, a number of unsupervised adaptation methods have been proposed. They utilize nonnegative matrix factorization (NMF) [8] and its convolutive extensions (nonnegative matrix factor deconvolution; NMFD) [9, 10] to obtain the intrinsic nonnegative sparse basis trajectories in the spectro-temporal region. As a result of NMFD, a dictionary of monotonic trajectories is inferred from a course of sound source powerspectral densities, and by classifying the dictionary items into a desired number of elements, the original source signals can be recovered. Although NMFD is successfully applied to several monaural source separation problems such as polyphonic music transcription, they suffer from the lack of ways to distinguish bases of different and complicated characteristics, and the uncertainty of the outcome due to their nature of unsupervised learning.

Several recent studies show good separation performances with mixtures of natural sound sources that are more complex than monotonic music sources. Although NMF and NMFD exploit the sparseness of the source signals in the spectro-temporal region, Virtanen proposed a new cost function that reflects the reconstruction error and the continuity of the recovered sources, and derived constrained maximization learning formulae [28–30].

With the help of the well-combined cost function, a number of complicated music signals observed in a single channel can be successfully transcribed. The method also has been inserted into a factorial hidden Markov model (HMM) framework and showed good speech recognition performance in noisy environments [31].

12.5.2 Time Domain Approaches to Single Channel Separation Techniques

Most time-domain approaches are based on splitting the whole signal space into several disjoint and orthogonal subspaces that suppress overlaps. The criteria employed by the former time-domain methods mostly involve second-order statistics: least square estimation [32], minimum mean square estimation [33], and Wiener filtering derived from the autocorrelation functions [18]. The use of AR (autoregressive) models on the sources has been successful. In [32] the source signals are assumed to be AR(p) processes, and they are inferred from a monaural input by a least square estimation method. Wan and Nelson [33] used AR Kalman filters to enhance the noisy speech signals, where the filters were obtained from neural networks trained on the specific noise. These methods performed well with input signals well-suited to the AR models, for example speech signals. However that is also a major drawback to applying them to the real applications. Moreover they consider second-order statistics only, which restricts the separable cases to orthogonal subspaces [18].

Our method is also classified as a time-domain method but avoids these strong assumptions by virtue of higher-order statistics. There is no longer orthogonality constraint of the subspaces, as the basis functions obtained by the ICA algorithm are not restricted to being orthogonal. The constraints are dictated by the ICA algorithm that forces the basis functions to result in an efficient representation, i.e. the linearly independent source coefficients; both the basis functions and their corresponding pdfs are key to obtaining a faithful MAP based inference algorithm. The higher-order statistics of the source signal described by a prior set of basis functions capture the inherent statistical structures.

Another notable advantage is that the proposed method automatically generates the prior information. While the other single channel separation methods also require the prior information, their methods to characterize the source signals are dependent on the developer's intuitive knowledge, such as harmonic structures or empirical psycho-acoustics. In contrast, our method exploits ICA for the automation of characterizing source signals. The basis functions can be generated whenever appropriate learning data are available, which may not be identical to the separation data. The training data and the test data are different in our experiments and the mixtures were successfully separated.

12.5.3 Comparison to Blind Signal Separation

The proposed method requires the basis functions of the mixed source signals. In order to expand its applications to real world problems, a 'dictionary' of bases for various kinds of natural sounds and finding the most characterizing basis in the dictionary for a generic case are necessary conditions to achieve good separation performance. These requirements make it more difficult to apply the proposed method to a real world problem than the conventional BSS techniques, which neither make assumptions nor require information about the source signals. However BSS suffers from the necessity of multiple channel observations, at least 2 channel observations are required, while the proposed method deals with single channel observations. The role of the basis functions is in some sense a substitute for extra-channel input.

12.5.4 Separability

The problem stated here is one of finding sets of bases that explain one class of signal well. Then, a single channel containing a mixture of different classes of signals is projected on the combined bases sets, and sources are isolated by considering the projections on the subsets of bases. This requires that each basis be incomplete in the general space of all possible signals but complete in the subspace of a single class of signal, and that the classes are disjoint in signal space. This latter requirement is a fundamental limitation of the method, in that it can do no better than the separation of the classes in signal space. Based on the experimental results, there is too much overlap in signal space between two speakers for this method to ever work for mixtures of speech. One way around this obstacle would be to do the separation in some feature space where there is both better class separation, and the possibility of transformation back to signal space. Future work will be to find more distinguishable subspaces, and develop better criteria for learning the source signals. The current process of training the bases is nondiscriminative. It would seem advantageous to train the sets of bases discriminatively. This would bring the separation results closer to the theoretical limitation.

12.5.5 Extensions

The method can be extended to the case when $P > 2$. We should decompose the whole problem into $P = 2$ subproblems, because the proposed algorithm is defined only in that case. One possible example is a sequential extraction of the sources: if there is a basis that characterizes a generic sound, i.e. which subsumes all kinds of sound sources, then we use this basis and the basis of the target sound that we are interested in extracting. The separation results are expected to be the target source and the mixture of the remaining $P-1$ sources. Repeating this extraction $P-1$ times yields the final results. Another example is merging bases: if there is a method to merge a number of bases and we have all the individual bases, we can construct a basis for Q sources and the other for the remaining $P-Q$ sources. Then we can split the mixture into two submixtures. Likewise repeating the split yields the final separation. In summary, the case $P > 2$ can be handled but additional research such as building a generic basis or merging different bases is required.

12.6 Conclusions

This chapterdescribed one of data-driven single channel source separation techniques. The proposed method is based on the time-domain ICA basis functions. Instead of traditional prior knowledge of the sources, we exploited the statistical structures of the sources that are inherently captured by the basis and its coefficients from a training set. The algorithm recovers original sound streams through gradient-ascent adaptation steps pursuing the maximum likelihood estimate, computed by the parameters of the basis filters and the generalized Gaussian distributions of the filter coefficients. With the separation results, we demonstrated that the proposed method is applicable to real world problems such as blind source separation, denoising, and restoration of corrupted or lost data. Our current research includes the extension of this framework to perform model comparisons to estimate the optimal set of basis functions to use given a dictionary of basis functions, and to reflect temporal continuity of the basis functions. This is achieved by extending the current basis function to cover longer time span, or computing temporal continuity of the basis coefficients. Future work will address the optimization of the learning rules towards real-time processing and the evaluation of this methodology with speech recognition tasks in noisy environments, such as the AURORA database.

References

1. A. S. Bregman, *Auditory Scene Analysis: The Perceptual Organization of Sound.* MIT Press, Cambridge MA, 1990.
2. A. S. Bregman, *Computational Auditory Scene Analysis.* MIT Press, Cambridge MA, 1994.
3. G. J. Brown and M. Cooke, "Computational auditory scene analysis," *Computer Speech and Language*, vol. 8, no. 4, pp. 297–336, 1994.
4. P. Comon, "Independent component analysis, A new concept?" *Signal Processing*, vol. 36, pp. 287–314, 1994.
5. A. J. Bell and T. J. Sejnowski, "An information-maximization approach to blind separation and blind deconvolution," *Neural Computation*, vol. 7, no. 6, pp. 1004–1034, 1995.
6. J.-F. Cardoso and B. Laheld, "Equivariant adaptive source separation," *IEEE Trans. on S.P.*, vol. 45, no. 2, pp. 424–444, 1996.
7. S. T. Roweis, "One microphone source separation," *Advances in Neural Information Processing Systems*, vol. 13, pp. 793–799, 2001.

8. D. D. Lee and S. S. Seung, "Learning the parts of objects by non-negative matrix factorization," *Nature*, vol. 401, pp. 788–791, 1999.
9. P. Smaragdis, "Non-negative matrix factor deconvolution; extraction of multiple sound sources from monophonic inputs," in *Proc. ICA2004*, vol. 3195, pp. 494–501, Sept. 2004.
10. M. N. Schmidt and M. Mørup, "Nonnegative matrix factor 2-D deconvolution for blind single channel source separation," in *Proc. ICA2006*, Apr. 2006.
11. A. J. Bell and T. J. Sejnowski, "The "independent components" of natural scenes are edge filters," *Vision Research*, vol. 37, no. 23, pp. 3327–3338, 1997.
12. A. J. Bell and T. J. Sejnowski, "Learning the higher-order structures of a natural sound," *Network: Computation in Neural Systems*, vol. 7, pp. 261–266, July 1996.
13. S. A. Abdallah and M. D. Plumbley, "If the independent components of natural images are edges, what are the independent components of natural sounds?" in *Proceedings of International Conference on Independent Component Analysis and Signal Separation (ICA2001)*, (San Diego, CA), pp. 534–539, Dec. 2001.
14. T.-W. Lee and G.-J. Jang, "The statistical structures of male and female speech signals," in *Proc. ICASSP*, (Salt Lake City, Utah), May 2001.
15. B. A. Olshausen and D. J. Field, "Emergence of simple-cell receptive-field properties by learning a sparse code for natural images," *Nature*, vol. 381, pp. 607–609, 1996.
16. M. Zibulevsky and B. A. Pearlmutter, "Blind source separation by sparse decomposition," *Neural Computations*, vol. 13, no. 4, 2001.
17. M. S. Lewicki, "Efficient coding of natural sounds," Nature Neuroscience, vol. 5, no. 4, pp. 356–363, 2002.
18. J. Hopgood and P. Rayner, "Single channel signal separation using linear time-varying filters: Separability of non-stationary stochastic signals," in *Proc. ICASSP*, vol. 3, (Phoenix, Arizona), pp. 1449–1452, Mar. 1999.
19. B. Pearlmutter and L. Parra, "A context-sensitive generalization of ICA," in *Proc. ICONIP*, (Hong Kong), pp. 151–157, Sept. 1996.
20. J.-F. Cardoso, "Infomax and maximum likelihood for blind source separation," *IEEE Signal Processing Letters*, vol. 4, pp. 112–114, Apr. 1997.
21. T.-W. Lee, M. Girolami, A. Bell, and T. Sejnowski, "A unifying information-theoretic framework for independent component analysis," *Computers & Mathematics with Applications*, vol. 31, pp. 1–21, Mar. 2000.
22. D. T. Pham and P. Garrat, "Blind source separation of mixture of independent sources through a quasi-maximum likelihood approach," *IEEE Trans. on Signal Proc.*, vol. 45, no. 7, pp. 1712–1725, 1997.
23. A. Hyvärinen, "Sparse code shrinkage: denoising of nongaussian data by maximum likelihood estimation," *Neural Computation*, vol. 11, no. 7, pp. 1739–1768, 1999.
24. J.-H. Lee, H.-Y. Jung, T.-W. Lee, and S.-Y. Lee, "Speech feature extraction using independent component analysis," in *Proc. ICASSP*, vol. 3, (Istanbul, Turkey), pp. 1631–1634, June 2000.
25. G. Box and G. Tiao, *Baysian Inference in Statistical Analysis.* John Wiley and Sons, 1973.
26. T.-W. Lee and M. S. Lewicki, "The generalized Gaussian mixture model using ICA," in *International Workshop on Independent Component Analysis (ICA'00)*, (Helsinki, Finland), pp. 239–244, June 2000.

27. S. Rickard, R. Balan, and J. Rosca, "Real-time time-frequency based blind source separation," in *Proceedings of International Conference on Independent Component Analysis and Signal Separation (ICA2001)*, (San Diego, CA), pp. 651–656, Dec. 2001.
28. T. Virtanen, "Sound source separation using sparse coding with temporal continuity objective," in *Proceedings of International Computer Music Conference*, Oct. 2003.
29. T. Virtanen, "Separation of sound sources by convolutive sparse coding," in *ISCA Tutorial and Research Workshop on Statistical and Perceptual Audio Processing*, 2004.
30. T. Virtanen, *Signal Processing Methods for Music Transcription, Eds. A. Klapuri and M. Davy*, ch. Unsupervised Learning Methods for Source Separation. Springer-Verlag, 2006.
31. T. Virtanen, "Speech recognition using factorial hidden markov models for separation in the feature space," in *Interspeech (ICSLP)*, (Pittsburgh, USA), 2006.
32. R. Balan, A. Jourjine, and J. Rosca, "AR processes and sources can be reconstructed from degenerate mixtures," in *Proceedings of the First International Workshop on Independent Component Analysis and Signal Separation (ICA99)*, (Aussois, France), pp. 467–472, Jan. 1999.
33. E. Wan and A. T. Nelson, "Neural dual extended Kalman filtering: Applications in speech enhancement and monaural blind signal separation," in *Proceedings of IEEE Workshop on Neural Networks and Signal Processing*, 1997.

13 Probabilistic Decompositions of Spectra for Sound Separation

Paris Smaragdis

Mitsubishi Electric Research Laboratories, Cambridge MA 02139, USA

Abstract. In this chapter we present a decomposition algorithm within a probabilistic framework and present some of its extensions which directly manipulate sparsity and introduce invariances. We show that this particular decomposition allows us to use probabilistic analyses that can decompose mixtures of sounds into fundamental building components that facilitate separation. We will present some of these analyses and demonstrate their utility by using them for a variety of sound separation scenarios ranging from the completely blind case, to the case where models of sources are available.

13.1 Spectral Components for Separation

The use of basis functions in the spectral domain to achieve separation has received a fair amount of attention in the last few years. The work described in [1, 4, 5, 8, 11] demonstrates various approaches with which short-time magnitude spectra can be decomposed by basis analyses to extract mixed sounds. The basic principle of this approach can be demonstrated using Fig. 13.1. The spectrogram shown in Fig. 13.1 (**b**) is that of a scene containing various percussive instruments which can be easily identified by eye. The two plots at top left and bottom right display a set of spectral bases (Fig. 13.1 (**a**)) and their corresponding weights (Fig. 13.1 (**d**)). The basis decomposition is of the form $\mathbf{F} = \mathbf{W} \cdot \mathbf{H}$. Where $\mathbf{F}$ is the input magnitude spectrogram in the form of a matrix, $\mathbf{W}$ is a matrix containing the basis functions in its columns, and $\mathbf{H}$ is a matrix containing the basis weights across time. The bottom left plot shows the approximation of the input from the displayed bases and weights.

It was discovered early on that seeking such low rank approximations to magnitude spectrograms results into decompositions which are semantically informative. Considering the example in Fig. 13.1, one can see that the basis functions approximate the spectral shape of the instruments in the mixture, whereas their corresponding weights position these bases in time and present us with the temporal location (and energy) of these sources. If we wish to separate the component corresponding to base i then we simply have to perform $\hat{\mathbf{F}}_{(i)} = \mathbf{w}^{(c_i)} \cdot \mathbf{h}^{(r_i)}$, where $\mathbf{w}^{(c_i)}$ is the i-th column of $\mathbf{W}$, and $\mathbf{h}^{(r_i)}$ is the i-th row of $\mathbf{H}$. The resulting reconstruction $\hat{\mathbf{F}}_{(i)}$ will be a magnitude spectrogram containing the energy attributed to basis i. We can then modulate the phase of the original mixture spectrogram with $\hat{\mathbf{F}}_{(i)}$ and invert the frequency

S. Makino et al. (eds.), Blind Speech Separation, 365–386.

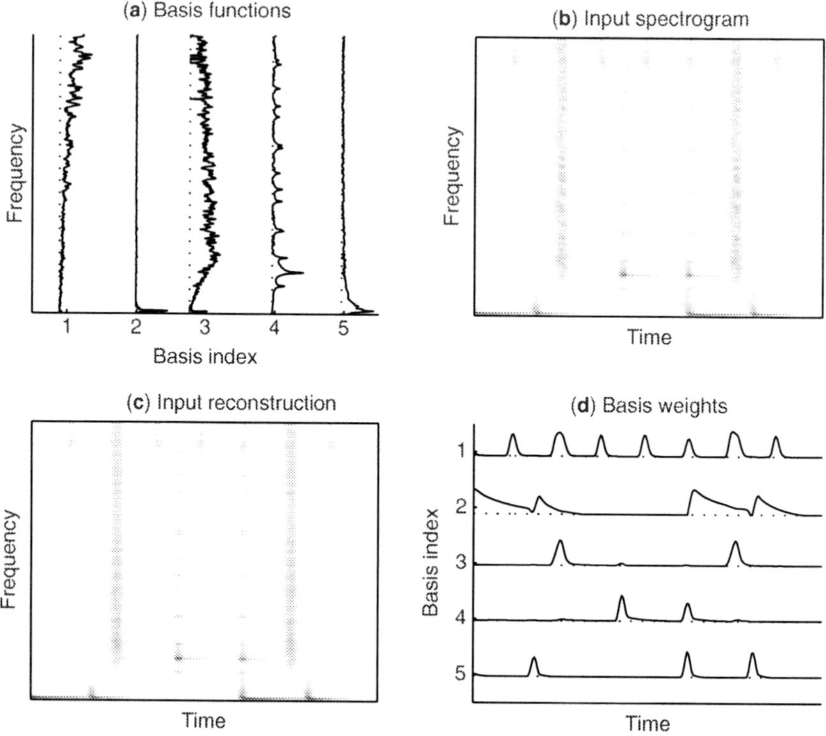

Fig. 13.1. Basis decomposition of a spectrogram. Plot (**b**) shows the input spectrogram which represents a recording of a few percussive instruments. Plot (**a**) diplays the basis functions extracted from the input and subfigure (**d**) displays their corresponding weights. It should be observed that the basis functions match the spectral character of the instruments in the mixture and the basis weights, which appropriately position these bases in time, describe the temporal positioning of the present sources. Plot (**c**) shows the approximation of the input from the basis decomposition. One can see that all relevant information has been maintained.

transform to obtain the separated source. If basis i appropriately describes a source in the mixture then the separation is often satisfying.

The way that basis functions are computed in this process can be varied. One of the most early examples [2] employed Principal Component Analysis (PCA) for this task. Although the resulting decomposition was useful for certain tasks it hardly provided the level of separation that was desired. Futher work employing Independent Component Analysis provided superior results [1, 3] since it was actively striving for a set of bases which were employed as independently as possible. Nonnegative Matrix Factorization (NMF) was also used [13] to further improve results by ensuring that the basis approximation would not produce negative magnitude values which can result into noise artifacts and poor analysis. Subsequent work in [5, 6, 9, 11] worked on more elaborate convolutive basis decomposition models which accounted for time structure and extracted more accurate basis functions.

In this chapter we will introduce yet another way that spectra can be decomposed which exhibits very good performance and some very desirable mathematical characteristics that allow easy formulations of important concepts such as sparsity, invariances and integration in a probabilistic learning framework. In the next few sections we will describe this decomposition method and some common extensions. We will then demonstrate how this decomposition it can be used for source separation in various ways.

13.2 Probabilistic Latent Component Analysis (PLCA)

In this section we will introduce the PLCA decomposition and demonstrate how it can be applied on magnitude spectra. Before we move into the formulation of the decomposition we will have to use a different interpretation of magnitude short-time spectra than the one we are accustomed to. For the purposes of this chapter we will be treating spectra as distributions or histograms. The intuitive explanation behind this concept is as follows. We can assume that a spectrum is a distribution of energy across frequencies. In fact one can make a more explicit analogy when we consider sounds as being composed by a collection of time/frequency localized 'wavelets'[1]. The distribution of these basic sound elements in the time/frequency plane will effectively be the spectrogram of the analyzed sound. Using this representation we can use a wealth of probabilistic analysis techniques directly on the spectral distributions without having to worry about enforcing nonnegativity (lack of which results into non-interpretable outputs) and also providing a clear way to incorporate these techniques in learnings framework.

To start with let us consider the spectrogram in Fig. 13.2. The top right Fig. 13.2 (**b**) displays a magnitude spectrogram of a short sound. We can reinterpret this as a probability distribution $P(f,t)$ spanning across frequency and time. This being a 2-dimensional distribution it will have two marginals, one over frequency $P(f)$ and one over time $P(t)$. These distributions can be straightforwardly computed by:

$$P(f) = \int P(f,t)dt \qquad \text{and} \qquad P(t) = \int P(f,t)df$$

The acute reader will notice that the computation of the two marginals is the same as the computation of the power spectrum and the signal energy over time (in the case of discrete data summations should substitute the integrations). We note therefore that the marginals of a time/frequency distribution

[1]The use of the term wavelets should not constrict our representation to the traditional wavelet format. Our implied meaning of this term is that of a sound atom which can assume an arbitrary form as long as it is localized in time and frequency.

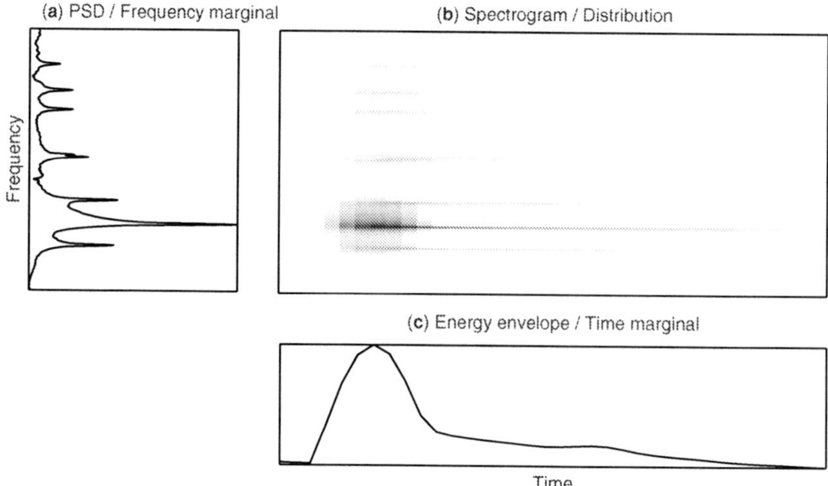

Fig. 13.2. A spectrogram as a distribution and its marginals across both dimensions. Plot (**b**) shows the input spectrogram which we also interpret as a time/frequency distribution. Plot (**a**) displays the marginal of that distribution across the frequency axis. This distribution also coincides with the power spectrum of the original signal. Likewise in plot (**c**) we plot the marginal distribution across time. Once again we note that this coincides with the energy of the signal over time.

relate to the power spectrum and the energy envelope of the signal. This is a perfectly logical outcome since the marginals themselves are distributions describing energy across either frequency and time.

The analysis procedure we just described is not particularly insightful for analyzing sound mixtures, nor is it new information. We will therefore extend it so that it becomes a more useful decomposition. We will still use the concept of a marginal distributions although this time we will introduce a latent variable which will allow us to extract multiple marginals. The model we will use is called Probabilistic Latent Component Analysis (PLCA) and is defined for the general case as:

$$P(\mathbf{x}) = \sum_z P(z) \prod_{j=1}^N P(x_j|z) \tag{13.1}$$

where $P(\mathbf{x})$ is an N-dimensional distribution of the random variable $\mathbf{x} = \{x_1, x_2, ..., x_N\}$. The z is a latent variable (which for our purposes assumes a finite set of discrete values), the $P(x_j|z)$ are one dimensional distributions on each of x's dimensions, and $P(z)$ is a weighting factor. Effectively this model represents a mixture of marginal distribution products that construct an N-dimensional distribution. Our objective is to discover the most appropriate distributions $P(x_j|z)$ and $P(z)$ so that $P(\mathbf{x})$ is best approximated.

The estimation of the marginals $P(x_j|z)$ is performed using a variant of the EM algorithm. In short this algorithm contains an expectation and a maximization step which we alternate between in an iterative manner. In the expectation step we estimate the contribution of each value of the latent variable z:

$$R(\mathbf{x}, z) = \frac{P(z)\prod_{j=1}^{N} P(x_j|z)}{\sum_{z'} P(z')\prod_{j=1}^{N} P(x_j|z')} \tag{13.2}$$

and in the maximization step we reestimate the marginals and the priors using the contribution weighting to obtain new and more accurate estimates $P^*(z)$ and $P^*(x_j|z)$:

$$P^*(z) = \int P(\mathbf{x})R(\mathbf{x}, z)d\mathbf{x} \tag{13.3}$$

$$P^*(x_j|z) = \frac{\int \cdots \int P(\mathbf{x})R(\mathbf{x}, z)dx_k, \forall k \neq j}{P^*(z)} \tag{13.4}$$

$P(x_j|z)$ will contain a latent marginal distribution across the dimension of variable x_j, relating to the latent variable z, and $P(z)$ will contain the prior of that latent variable. Repeating the above steps successively for multiple iterations produces a converging solution for the marginals and the latent variable priors. For practical realizations of this algorithms the input distribution $P(\mathbf{x})$ and all the estimated quantities will have to be discrete. In this case we will have to substitute the integrations with summations and impose an extra step after each iteration which ensures that all estimated distributions sum to one. $P(\mathbf{x})$ itself does not need to be a true distribution and can instead be a histogram (i.e. there is no need for it to be normalized).

Applying this model on spectra means that the input will be a 2-dimensional distribution over frequency and time and that the extracted marginals will be a collection of multiple frequency distributions and multiple time distributions. Interestingly the 2-dimensional version of this algorithm is numerically identical to the Nonnegative Matrix Factorization (NMF) algorithm [14] which has been used extensively in the past for source separation work. For a 2-dimensional input PLCA will extract two sets of distributions $P(x_1, z)$ and $P(x_2, z)$ which map respectively to the $\mathbf{W}$ and $\mathbf{H}$ matrices of NMF. The additional distribution $P(z)$ is there for normalization purposes and in the case of NMF is absorbed in the two matrix factors. The connection between these two approaches stems from the fact that one of the formulations of NMF optimizes a pseudo-Kullback–Leibler divergence between the input and its factor product approximation. The EM training we employ for PLCA results into minimizing the Kullback–Leibler divergence between the input distribution and its approximation by our model. Extra operations in the NMF algorithm normalize the inputs so that they behave as distributions in the training process. Since it is out of scope we will not elaborate on this subject in detail, we note however that it is fairly straightforward to observe the equivalency between the two algorithms after some simple algebraic manipulation.

The result of performing PLCA on a spectrogram and a description of the results was already shown in Fig. 13.1. The marginal distributions over frequency are shown in Fig. 13.1 (**a**), and their counterparts over time are shown in Fig. 13.1 (**a**). The fact that PLCA and NMF are identical means that we can exactly reproduce any NMF work for source separation using PLCA.

There might be no computational advantage of PLCA as compared to NMF, however its nature allows us to perform NMF-type analysis for higher dimensional inputs, which in our case would most likely be multichannel or higher-order spectra and representations similar to correlograms. Another significant advantage of PLCA is that we can easily incorporate the PLCA model in statistical frameworks since its outputs can be interpreted and used in a probabilistic setting. An example of this advantage is shown in the following section.

13.2.1 Sparsity

Sparsity manipulation has always been an optimization target in work relating to source separation. Sadly component models such as PCA, ICA and NMF do not provide a convenient and principled way to directly optimize sparseness. In this section we show how we can apply arbitrary sparsity constraints on the PLCA model which we will use later on for better separation performance.

Let us denote with $\boldsymbol{\theta}$ an arbitrary distribution in the PLCA model whose sparsity we wish to manipulate. Just like before we will be using the EM algorithm to discover all the distributions in the PLCA model. In order to manipulate the sparsity of $\boldsymbol{\theta}$ we will be using its entropy $\mathcal{H}(\boldsymbol{\theta}) = -\sum_i \theta_i log\theta_i$ as a sparsity metric. We therefore modify the estimation procedure to incorporate it. In order to do that we will have to switch from a maximum likelihood (ML) estimation that we have used in the previous section, to a maximum a posteriori (MAP) estimation which includes a prior on $\boldsymbol{\theta}$ in the form of an entropic distribution $P(\boldsymbol{\theta}) = e^{-\beta\mathcal{H}(\boldsymbol{\theta})}$. The β parameter is used to weight the importance of the entropy prior as well as tune it towards maximization or minimization by changing its sign.

The E-step under this new formulation will remain the same as in (13.2). During the M-step the entropic prior will come into play. Even though this time we have switched from ML to MAP estimation, the update equations for the PLCA distributions remain the same as shown in before. For $\boldsymbol{\theta}$, whose entropy we wish to manipulate, we will have an extra step as a result of its prior. Estimation of $\boldsymbol{\theta}$ during the M-step will result in the following set of simultaneous transcendental equations:

$$\frac{\omega}{\theta_i} + \beta + \beta log\theta_i + \lambda = 0 \tag{13.5}$$

where λ is the Lagrange multiplier enforcing the constraint $\sum_i \theta_i = 1$. As shown in [16] these equations can be solved by using Lambert's $\mathcal{W}$ function. The resulting estimator is

$$\boldsymbol{\theta} = \frac{-\omega/\beta}{\mathcal{W}(-\omega e^{1+\lambda/\beta}/\beta)} \tag{13.6}$$

where $\omega = \int \cdots \int P(\mathbf{x})R(\mathbf{x}, z)dx_k, \forall k \neq j$ for $\boldsymbol{\theta} = P(x_j|z)$ (for the definition of R refer to (13.2)). Equations (13.5) and (13.6) are used in a 2 iteration fixed-point loop after the M-step which refines the estimate of $\boldsymbol{\theta}$ to accommodate the entropy constraint.

Application of this constraint on the PLCA model can come in many forms. One can optimize the entropy of $P(x_j|z)$ for any combination of j and z, or even optimize $P(z)$ (in which case $\omega = \int P(\mathbf{x})R(\mathbf{x}, z)d\mathbf{x}$). To illustrate the effect of these optimizations we can revisit the example shown previously in Fig. 13.1. Figure 13.3 displays the effect of manipulating the entropy constraints of the estimated time distributions. Each column in the figure displays the estimated set of time marginals for high sparsity, no sparsity optimization and low sparsity. It is easy to see that the high sparsity constraint is creating much more peaky distributions, which are poorly representing the instruments in the input. The low sparsity marginals on the other hand are much smoother and describe the input more accurately than the other versions (without enforcing sparsity the bass drum instrument is usually described by two components and the snare drum and claps are consolidated into one. The low sparsity version overcomes this and resolves each instrument as one component).

Although we won't elaborate with additional examples, it is easy to encourage or discourage sparsity on any of the estimated distributions. It is

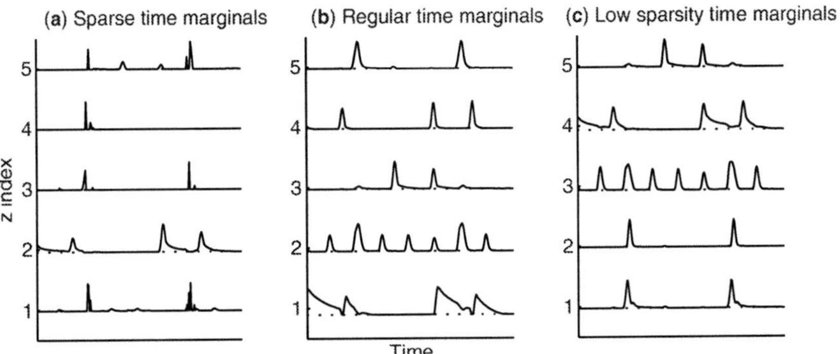

Fig. 13.3. Effect of entropy constraints on time distributions. Plot (**a**) shows time marginals from the input in Fig. 13.1 with the constraint of high sparsity. Plot (**b**) displays the same time distributions without any sparseness optimization. Plot (**c**) displays the same distributions once more but with a low sparseness constraint. The order of the distributions is arbitrary in every run.

also possible to use this approach to impose sparseness across values that do not belong in the same distribution, or amongst groups of distributions. More details and examples of this approach can be found in [17]. As we move on to more elaborate models and specialized applications this flexibility to manipulate sparseness becomes increasingly useful and in some cases necessary.

13.2.2 Shift-Invariant Form

The NMF model was extended in [9] and [6] to extract components that exhibit time structure. Likewise it is straightforward to extend PLCA to perform a similar decomposition. The PLCA model in this case is reformulated as:

$$\begin{aligned} P(\mathbf{x}) &= \sum_z P(z)P(\mathbf{w},\boldsymbol{\tau}|z) * P(\mathbf{y}|z) \\ &= \sum_z P(z)\int P(\mathbf{w},\boldsymbol{\tau}|z)P(\mathbf{y}-\boldsymbol{\tau}|z)d\boldsymbol{\tau} \end{aligned} \tag{13.7}$$

where this time the input distribution $P(\mathbf{x})$ is modeled as a sum of convolutions as opposed to a sum of outer products. The distributions to be estimated for this model are $P(z)$, $P(\mathbf{w},\boldsymbol{\tau}|z)$ and $P(\mathbf{y}-\boldsymbol{\tau}|z)$. $P(\mathbf{w},\boldsymbol{\tau}|z)$ will be a set of distributions which will be convolved with the distributions $P(\mathbf{y}|z)$ and summed according to the weighting given from $P(z)$. For semantic convenience we will refer to $P(\mathbf{w},\boldsymbol{\tau}|z)$ as the kernel distributions and $P(\mathbf{y}|z)$ as the impulse distributions. Just as before we can derive an EM-like algorithm to estimate all these quantities. The resulting steps are once again an expectation step which estimates the contribution of each convolution pair at all delays:

$$R(\mathbf{w},\mathbf{y},\boldsymbol{\tau},z) = \frac{P(z)P(\mathbf{w},\boldsymbol{\tau}|z)P(\mathbf{y}-\boldsymbol{\tau}|z)}{\sum_{z'} P(z')\int P(\mathbf{w},\boldsymbol{\tau}'|z')P(\mathbf{y}-\boldsymbol{\tau}'|z')d\boldsymbol{\tau}'} \tag{13.8}$$

And the M-step which updates the desired distributions by:

$$P^*(z) = \iiint P(\mathbf{x})R(\mathbf{w},\mathbf{y},\boldsymbol{\tau},z)d\mathbf{x}d\mathbf{y}d\boldsymbol{\tau} \tag{13.9}$$

$$P^*(\mathbf{y}|z) = \frac{\iint P(\mathbf{w},\mathbf{y}+\boldsymbol{\tau})R(\mathbf{w},\mathbf{y}+\boldsymbol{\tau},\boldsymbol{\tau},z)d\mathbf{w}d\boldsymbol{\tau}}{\iiint P(\mathbf{w},\mathbf{y}'+\boldsymbol{\tau})R(\mathbf{w},\mathbf{y}'+\boldsymbol{\tau},\boldsymbol{\tau},z)d\mathbf{y}'d\mathbf{w}d\boldsymbol{\tau}} \tag{13.10}$$

$$P^*(\mathbf{w},\boldsymbol{\tau}|z) = \frac{\int P(\mathbf{x})R(\mathbf{x},\mathbf{y},\boldsymbol{\tau},z)d\mathbf{y}}{P^*(z)} \tag{13.11}$$

The above formulation gives us full control over the nature of the estimated components and is a probabilistic generalization of the convolutive NMF [6, 9] and tensor deconvolution [10] work.

As one might observe the decomposition presented here is redundant having too many degrees of freedom for the given input. This is addressed by limiting the space where $P(\mathbf{y}|z)$ and $P(\mathbf{w},\boldsymbol{\tau}|z)$ are defined. In general we will often require that $P(\mathbf{w},\boldsymbol{\tau}|z)$ is smaller than $P(\mathbf{x})$ so that we extract small repeating components in the input. Additionally we also employ the sparsity framework described in the preceding section to impose the constraint that the kernel distributions should be high entropy. Using this constraint we ensure that the kernel distributions are more informative and the impulse distributions are sparser and are used mainly for positioning the kernels. Not imposing this constraint does not have an adverse effect in performance, it does however obscure the difference between kernel and impulse distributions. Ideally we wish to have information-rich kernels that are positioned by sparse impulses, so that we can identify salient components that appear repeatedly throughout the input. Not imposing this requirement can result in equally optimal results where information is split between the two distributions making it hard to pinpoint the repeating patterns.

To better illustrate how this transform could be used in our spectrogram analysis framework consider the plots in Fig. 13.4. We use the same input as in the past experiments only this time instead of extracting frequency and time distributions, we are extracting kernels which will be 'mini time/frequency distributions' and impulses which will be their distribution across the

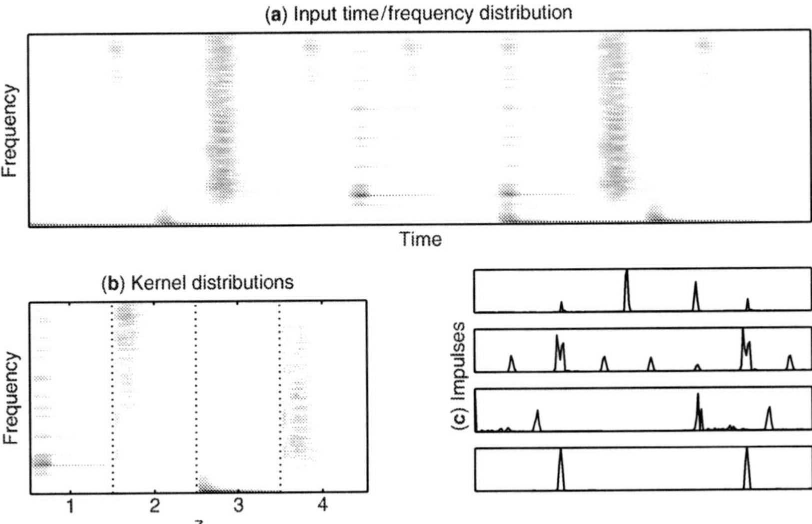

Fig. 13.4. Kernel and impulse distributions of a spectrogram as derived from PLCA. Plot (**a**) shows the input spectrogram. Plot (**b**) displays the kernel distributions which represent small time/frequency distributions that repeat within the input. Plot (**c**) displays the impulse distributions which place the kernel distributions at the appropriate position to properly approximate the input. Note how the kernel distributions describe the spectral evolution of the percussion instruments in the input, and the impulse distributions describe their positions in time.

time axis. As shown in Fig. 13.4 the kernel distributions that are discovered correspond to the instruments in the mixture, only this time in addition to obtaining spectral information for each instrument we also we obtain how that information changes with time. The impulse distributions position the kernels appropriately to approximate the input, and therefore denote the point in time where each instrument takes place.

In this example we demonstrated shift-invariance only across the time axis. The shift-invariant formulation of PLCA is quite general can be used for an arbitrary number of dimensions. It is also well suited for analysis of constant-Q transforms where we observe shifting of harmonic series on the frequency axis as well as on the time axis. It is easy to use this formulation of PLCA to replicate results along the lines of [15].

Additional invariances, such as invariance to shearing, rotation, etc., are also straightforward to implement in this framework and depending on the problem at hand they could be very valuable. Since we do not have any use of these invariance properties in the context of this chapter, in the interest of space we will not present them. Their derivation however is straightforward and similar to the one we used for shift-invariance.

13.3 Source Separation using PLCA

In the following sections we will show how the PLCA model and its extensions can be used to separate sources from a monophonic mixture. We will consider three cases. First the case where we have no prior information on the sources which requires unsupervised learning of our model. The second case will be a supervised scenario in which we know the type of sounds that are to be separated. Finally we also demonstrate the semi-supervised method in which we know some information about one source but not the interference.

13.3.1 Unsupervised

In the case of unsupervised source separation we expect to separate sources as components of PLCA. This approach has been extensively reported in the context of NMF. In this section we will repeat some of these experiments and show how PLCA can be used in this process.

We start with a spectrogram representing the mixture we wish to analyze. As we noted previously this transform can be interpreted as a probability of acoustic energy on a time/frequency plane. We therefore denote it as a distribution $P(f, t)$ and proceed to analyze it using PLCA. As we have already demonstrated analysis of a mixture using PLCA will result into a component-wise description. By attempting to analyze the input with a small number of components we effectively produce an information bottleneck. This forces the components to be maximally informative and to converge to the elements that

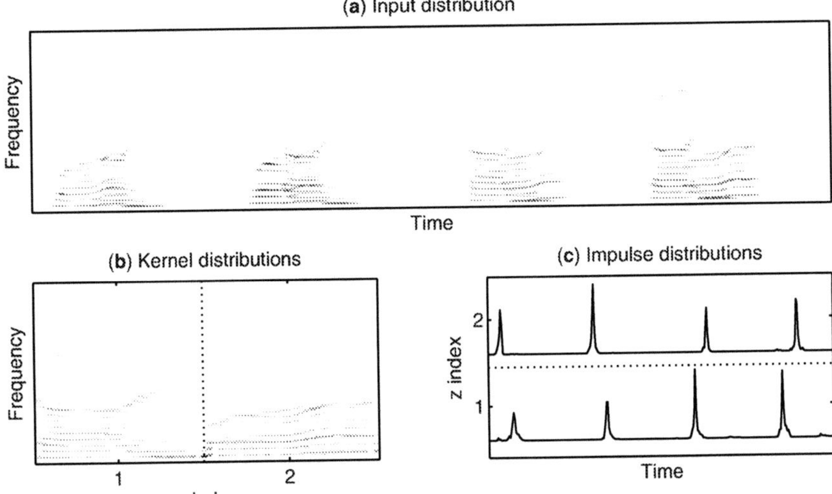

Fig. 13.5. A mixture of words and its analysis using shift-invariant PLCA. Plot (**a**) displays the input time/frequency distribution containing two repeating and variably overlapping words. Plot (**b**) displays the kernel distributions that we recovered after analysis of that input. Plot (**c**) displays their corresponding impulse distributions. Note how the kernel distributions resemble the two words that exist in the mix.

occur the most in the input. As we have demonstrated in the preceding sections the components we extract from spectrograms often coincide with what we perceive as the elements in the mixture. To demonstrate this in a more challenging situation consider the time/frequency distribution in Fig. 13.5 (**a**). This is composed out of the same speaker repeatedly uttering the words "what" with an upward pitch inflection and "time" with a downwards pitch inflection. The words are repeated four times and as is evident they mostly overlap making their separation a nontrivial problem. In Figs. 13.5 (**b**) and (**c**) we show the results of a shift-invariant PLCA with two components. Note that the kernel distributions have converged to the spectrogram patterns of the two words uttered (one can easily verify that from the different pitch inflections). As expected the impulse distributions position these words at the proper time locations, denoting their energy at each instance as well.

Once this kind of analysis is performed it is then easy to extract the sources by performing a selective reconstruction. Each component will be defined as the convolution of each kernel distribution with its corresponding impulse distribution. So to reconstruct the two words we would do:

$$P(f,t|z_1) = P(z_1)P(f,\tau|z_1) * P(t|z_1) \tag{13.12}$$

$$P(f,t|z_2) = P(z_2)P(f,\tau|z_2) * P(t|z_2) \tag{13.13}$$

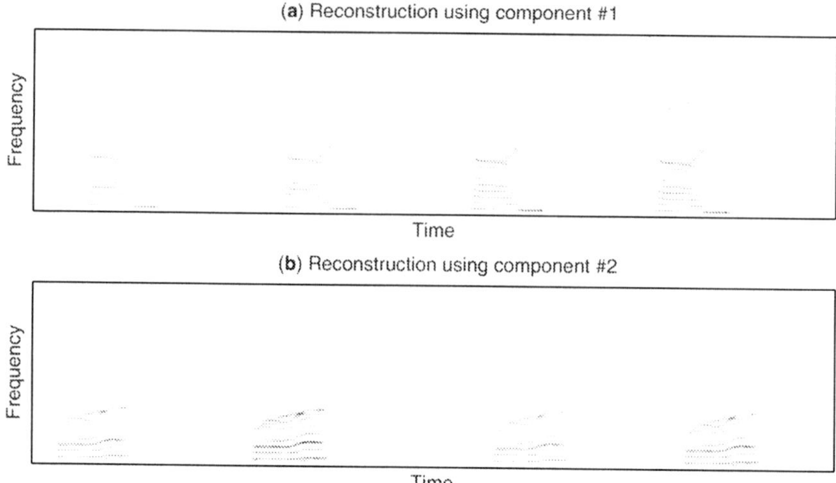

Fig. 13.6. Component-wise reconstruction of the mixture in Fig. 13.5 (**a**). The top subplot (**a**) displays the reconstruction using the first component shown in Fig. 13.5 (**b**), and the bottom subplot (**b**) displays the reconstruction using the second component shown in Fig. 13.5 (**b**). The two reconstruction effectively separate the two word series that were part of the original mixture.

where $P(f, \tau|z_i)$ are the kernel distributions, $P(t|z_2)$ are the impulse distributions and $P(z_i)$ are the component priors as estimated by the PLCA learning algorithm. The resulting reconstructions $P(f, t|z_i)$ correspond to the portions of the input that contain only one of the two spoken words in isolation. The result of this operation is shown in Fig. 13.6 in which we see the two words separated as the two individual time/frequency distributions $P(f, t|z_i)$. The sum of these two is our model's approximation of the input. In order to obtain an audible representation of these results we can use the time/frequency energy patterns provided by $P(f, t|z_i)$ to modulate the phase of the mixture at the corresponding time/frequency points and invert the resulting spectrograms to the time domain using overlapped Fourier transforms.

Note that in order to make this example work we had to provide an appropriate selection of PLCA parameters. We had to constrain this analysis to two components since we knew there were two words present. We also had to ensure that the size of the kernel distributions would be approximately the size of the two words in the input. Larger sizes would also produce similar results when aided by a sparsity constraint on the kernels. Smaller sizes however would result into an inadequate representation which cannot capture the complex structure of the input. At the extreme case of small kernels we would have kernel distributions of unit width where the model becomes the plain non-shift-invariant PLCA. In that case we will extract the average spectral shape of each word as a marginal distribution over frequency, but it will obviously not be enough to provide a sufficient audible reconstruction.

This method closely resembles the work in [9] and [6] who introduced a similar analysis method based on NMF. The results are equivalent in quality and many of the insights and extensions of these techniques are directly applicable in this case as well.

13.3.2 Supervised

As one might imagine the utility of the unsupervised method presented in the preceding section has a limited application range and is most useful with mixtures containing repeating sources. However the method by which the sources are separated is very powerful and can be reformulated as a supervised learning problem in order to address more challenging situations.

Let us see what happens when we analyze a long speech segment from a single speaker using shift-invariant PLCA. The extracted kernels of this analysis are shown in Fig. 13.7 (**a**). Anyone well acquainted with speech signals would observe that the resulting kernels possess the structure of basic speech elements (such as phonemes or phones). This is to be expected since this analysis produces patterns that are commonly repeating along the time axis. It is natural for a single speaker to reuse the same phonetic patterns throughout time. These patterns are what characterizes the timbre and elements of that speaker's unique phrasing and accent. Likewise note the extracted kernels from a recording of another speaker in Fig. 13.7 (**b**). Just as in the case of the first speaker we see that salient features of the second speaker

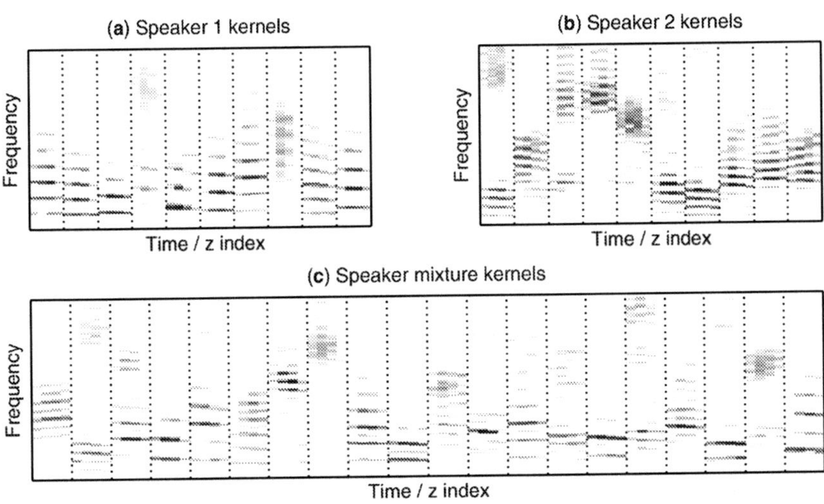

Fig. 13.7. Kernel distributions extracted from two speakers and their mixture. Plot (**a**) shows the kernels of a female speaker, plot (**b**) the kernels of a male speaker and plot (**c**) the kernels of a mixture of the two speakers. Note how the kernels from the mixture are very similar to the kernels of the individual speakers and are not mixtures of the speaker kernels.

are described in the kernels derived from analyzing his signal. Upon closer inspection it is easy to see that one of the speakers exhibits a lower pitch which results into closely clustered harmonics, whereas the other speaker exhibits a higher pitch which results into more spaced out harmonics. This is to be expected since one speaker is male, and the other is female.

We perform this operation once more only this time on a mixture of these two speakers. It is safe to assume that the time/frequency distribution of the mixture will be approximately that of the sum of the time/frequency distributions of the two included sounds. In that case since the model we use is linear we can expect to recover both speakers' components by performing PLCA analysis on the mixture. Indeed as Fig. 13.7 (**c**) shows, the kernels extracted from the mixture distribution belong primarily to an individual speaker and seldom contain elements of both.

If we were to know which kernels belonged to the which or the original speakers, we could reconstruct the input using only kernels from one speaker at a time to reconstruct the original speech segments. Suppose we somehow knew the values of z of the kernels that belong to the female speaker (let's call the set z_f)and likewise for the male speaker (set z_m). Denoting the extracted kernels as $P(f,\tau|z_f)$ we could then do:

$$P_f(f,t) = \sum_{z \in z_f} P(z)P(f,\tau|z) * P(t|z) \tag{13.14}$$

$$P_m(f,t) = \sum_{z \in z_m} P(z)P(f,\tau|z) * P(t|z) \tag{13.15}$$

The resulting $P_f(f,t)$ and $P_m(f,t)$ will be reconstructions of the mixture which are dependent on elements of only a single speaker, and will effectively represent the separated sources.

However sorting the kernels in groups that split the input into the constituent sources is a very hard problem to solve accurately. Instead we will use a simpler procedure which employs prior knowledge about the sources to separate them. The underlying assumption we will make is that already learned kernels from a sound source can adequately describe new instances of sounds from that source. Effectively we assume that the union of the kernels in Figs. 13.7 (**a**) and (**b**) is very similar to the kernels in Fig. 13.7 (**c**). If that is the case we can obtain properly segmented sets of kernels by training in advance from clean data and then using them to extract each speaker from the mixture.

To illustrate this process in action let us consider the example in Fig. 13.7. This time we will not train on the exact sounds that are part of the mixture but rather on different and isolated speech segments from the two speakers and apply the learned statistics to resolve the mixture. By training on clean speech segments of the two speakers we obtain two sets of kernels $P_f(f,\tau)$ for the female speaker and $P_m(f,\tau)$ for the male speaker. We then proceed to analyze the mixture distribution $P(f,t)$. We do so by consolidating the two kernel sets in $P(f,\tau|z) = \{P(f,\tau|z_f), P(f,\tau|z_m)\}$ and then performing shift-invariant PLCA with $P(f, \tau|z)$ kept fixed with the already known values so that we only estimate $P(t|z)$ and $P(z)$. This will result into decomposing the mixture by the already known speaker dependent kernels. Assuming that the training data adequately describes the two speakers in the mixture we can hope that the already known kernels can produce a good approximation of the input. Given reasonable training data this always turns out to be the case. During training on the mixture we obtain the appropriate $P(t|z)$ and $P(z)$ to approximate the input with the already known kernels. Note now that some of the kernels are best suited for data from one speaker, and some for the other. This means that the values of $P(t|z)$ and $P(z)$ that correspond to $P(f|z_f)$ should approximate the parts of the mixture that can best be explained from the kernels of the female speaker, and similarly for the male speaker. By selective reconstruction using kernels from only one speaker at a time (equation (13.15)) we can extract the spectral energy attributed to each speaker in the mixture. We can then transform the speaker dependent time/frequency distributions to time series as described in the previously. Figure 13.8 shows an example of this analysis. The mixture is shown in Fig. 13.8 (**a**), and the two separated speakers in Figs. 13.8 (**b**) and (**c**).

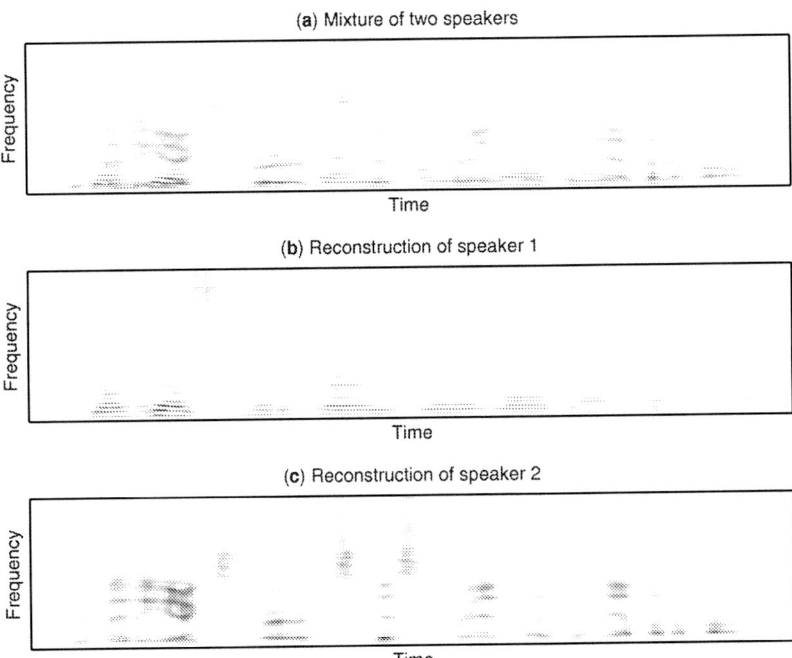

Fig. 13.8. Decomposition of a mixture to its sources. Figure (**a**) shows the input time/frequency distribution. Figures (**b**) and (**c**) show the two extracted distributions derived from using kernels already trained on the two speakers in the mixture. Although it is hard to judge the quality of separation from the figures, one can clearly see that in each of the separated spectrograms we have a different type of dominant speaker.

This techniques is equivalent to the NMF based approach presented in [4], where a lengthy discussion on limitations and parameter selection for this task is also presented. Due to the numerical equivalence between shift-invariant PLCA and convolutive NMF we refer the interested reader to that publication. Here we will note that performance on average is around 5–6 dB improvement for 0 dB speech mixtures. Additional separation quality gains can achieved when making good use of sparsity and more sophisticated estimation as shown in [7, 8]. As one might suspect this approach would fail if the two mixed signals are very similar (as is the case with mixtures of the same speaker), but excels in the case where the two mixed sounds are spectrally very different. In that case an improvement of 20 dB or more is often common. This makes this approach more suited for denoising as opposed to separation of concurrent speakers.

Although not presented here, it is very easy to use this approach to separate more than just two sources, as well as adapting this to multiple channel inputs (where we would learn multichannel spectra instead). An exhaustive demonstration of these options is beyond the scope of this paper, it is however very simple to implement.

13.3.3 Semi-Supervised

The supervised approach showed in the preceding section works fine for some constrained scenarios. It is not however not very convenient for some common denoising situations. Knowing in advance the nature of the sounds to be separated is often difficult. A more likely case is that of knowing only the target source or only the interfering noise. In this section we will explore that scenario.

Let us assume a mixture containing multiple sources. One of these sources is known to us and is our target of separation. Referring back to the approach in the previous section we note that we can obtain the kernels that belong to the target, but we would have no information about the additional interfering sources. Had we had the set of kernels that describe all the interfering sources, we could then revert back to the separation process shown in the previous section. It therefore beneficial if we somehow estimate the kernels for all the sources present in the mixture other than our target. Due to the flexibility of the EM training procedure this is easy to do, and requires only a minor change from the procedure just shown.

Since we already have training data of the target sound we can easily extract its kernels $P(f,\tau|z_t)$ using PLCA. The task now is to estimate the kernels $P(f,\tau|z_i)$ of the interfering sources. Since the only instance we have of the interfering sources is the mixture itself, we perform PLCA on it using the set of kernels $\{P(f\tau|z_t), P(f,\tau|z_i)\}$ and during training we only update the kernels $P(f,\tau|z_i)$ and keep $P(f,\tau|z_t)$ fixed to the already known values. This results into an analysis which explains as much of the mixture as possible from the known target kernels and then uses the interference kernels to explain the nontarget elements in the mixture. Upon conclusion of training we obtain the desired $P(f,\tau|z_i)$ as well as $P(t|z)$ and $P(z)$. Just as before we can now selectively reconstruct the input mixture using the kernels of either the target or the interfering sources and obtain both target and background separated.

Once again, the more dissimilar these two sources are the better the separation will be. Note that in this formulation we only need a binary segmentation of target and background. Although the background can be composed out of multiple sources in terms of our analysis it is treated as one source. Obviously the definition of the target and the background are arbitrary and can be switched. We can assume that we only know a model of the background and we wish to separate the extra signal present instead. Either way we decide to define the sources the training procedure is the same and the results are often of comparable quality.

An example of this approach is shown in Fig. 13.9. In this example we have a piece of music which contains a piano and a soprano Fig. 13.9 (**a**). The first half of the segment contains solo piano and the second half contains

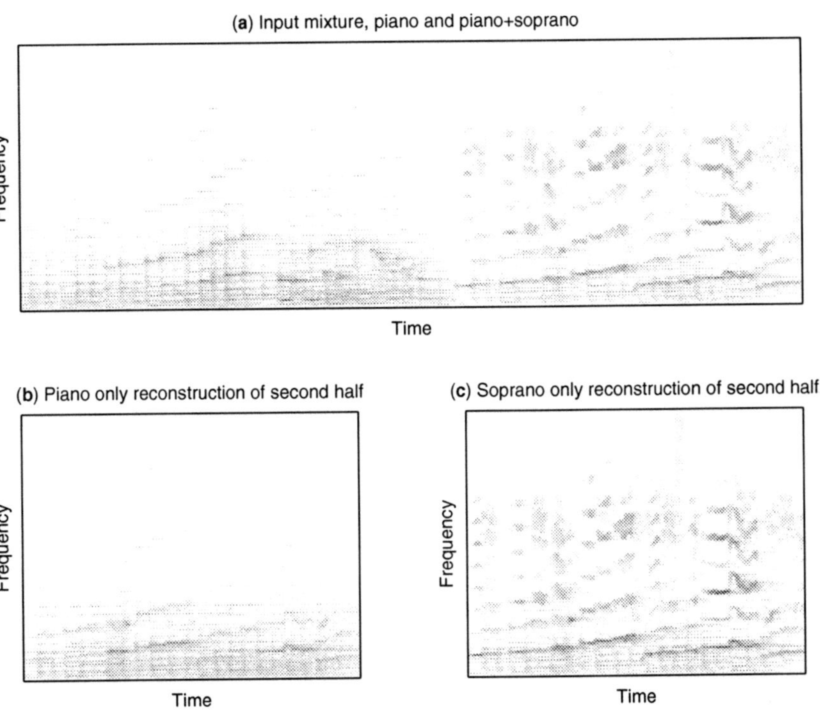

Fig. 13.9. Decomposition of a piano/soprano mixture. Plot (**a**) shows the input time/frequency distribution, the first half contains just the piano and the second half contains a mixture of the piano and a soprano. Plots (**b**) and (**c**) show the two extracted distributions derived from using kernels trained on the piano and estimated kernels from the soprano.

the piano mixed with the soprano. In this case we can learn a model of the piano by training on the first half but we cannot learn a model of the soprano since she is never isolated as a source. To extract the soprano we learn the piano kernels from the first half and then use them in conjunction with some untrained kernels and train on the second part. Since the piano kernels will do a good job in describing the piano parts in the mixture the untrained kernels will be updated to explain the remaining parts of the signal. These parts are the soprano who is the only other source present aside from the piano. Carefully examining the untrained kernels after they converge we can see that they resemble spectra similar to these generated by the soprano. We can then reconstruct the second half of the input once using just the piano kernels and once more using just the soprano kernels. These reconstructions are shown in Figs. 13.9 (**b**) and (**c**). One can see that the reconstruction according to the piano kernels produces a signal with steady tones which look very much like piano notes, whereas the reconstruction using the soprano kernels looks very much like vocal spectrogram with severe pitch fluctuations. Reverting these reconstructions back to the time domain results into an very good separation despite the fact that the two sources have significant harmonic and temporal overlap since they are engaged in the same piece of music.

Although it is hard to judge performance using the figures, separation using this techniques has produced excellent results often yielding more that 20 dB of separation between the target and the background. This approach is described in more detail using its equivalent NMF formulation in [12]. It has been used successfully for a variety of applications ranging from removal of vocals from music, removal of mechanical noises from microphone equipped robots and denoising of speech signals where either noise or speech models are known in advance.

13.4 Discussion

In the preceding section we have described some techniques to extract sources from mixtures when we wish to operate on the spectrogram of a mixture. Although the formulation of our approach is new to these applications, it is also for the basic part identical to prior work using NMF. We have tried to present bibliographical references which can provide the interested reader with all the relevant details in the above approaches and we kept the tone of this chapter in the conceptual level.

That said there are a lot of parameters that the above techniques share which need to be carefully selected in order to obtain reasonable results. The main things one needs to be concerned about are the frequency analysis parameters and the PLCA parameters. The optimal use of the frequency analysis parameters depends of course on the type of analysis. The most common is that of the short-time Fourier transform (STFT), where successive windowed

Fourier transforms are taken on the input signal to compose its spectrogram. When using the STFT for the approaches in this chapter one needs to address two issues, that of adequate temporal analysis and that of adequate frequency analysis. One must always observe the mixture spectrogram to ensure that the features of the targeted sources are visually distinct. Quite often poor frequency resolution or a bad choice of a hop size results into a transform which cannot be segmented to its constituent sources.

When it comes to the PLCA transform the parameters one needs to adjust properly are the number of desired components and their shape. The optimal number of desired components depends on the approach we use for separation. For the unsupervised case it is obvious that the number of components should be close to the number of actual sources in the mixture. For the other cases a good number of components needed to describe a source is around 50 or 100 components per source. Fewer components than that will result into a coarse description of the source which will not overlap as much with competing sources, whereas too many components will result into a very high quality representation which will unfortunately be fundamental enough to explain other sources as well. Of course when dealing with simple almost stationary sounds, fewer components should be used. Likewise the size of the extracted kernels is also an important parameter. The kernels can be of unit width in which case they are equivalent to marginals, or they can be much wider and describe entire sections. The choice of the kernel size depends on the type of source one has to separate. If there is a strong repetition of almost identical elements (such as the case of percussive instruments), wider kernels are preferred since they will capture a lot of the structure of the sources. On the other hand when we have sources that are dynamic and don't have very strict repetitions (such as speech), it is best to avoid very wide kernels since the probability that they will generalize well and fit at multiple parts of the analyzed input decreases with larger sizes. An extensive description of the effects that these parameters (as well as others) have on the separation of concurrent speech can be found in [4]. A lot of the results in that paper are directly relevant to the presented approaches.

Although in all sections we did refer the reader to similar work using NMF, as we have mentioned before there are significant advantages when using PLCA. The most significant one shown here is the ability to manipulate the sparsity of the estimated distributions. In the PLCA model this allows us to fine-tune the expected results and bias them towards our preferences. We have already demonstrated the effect of sparsity on the regular PLCA model in Fig. 13.3, similar gains can be achieved with the shift-invariant version as well. One would often impose the constraint that the kernels are nonsparse and the impulses are. This results into a more informative decomposition by pushing most information in the kernels and making the impulses mere placeholders. Another possibility that we have not presented in this chapter is to impose sparsity across distributions. The regular sparsity formulation allows us to make each individual distribution in a set sparse but it does not directly manipulate their relationships. It is often desirable to have the constraint that distributions should be sparse in relation to each other. That means that when one of them has a high value the rest should not. Imposing this constraint is not as straightforward since we would have to sparsify a nondistribution. In these cases we usually have to revert to the Polya Urn model [8] which is a simpler formulation of PLCA that allows us to have that flexibility. We should also note that it is possible to use the sparsity framework to enforce manipulation of the Kullback–Leibler divergence between arbitrary distributions within the PLCA model. This brings this model closer to independent component analysis methods, although it operates on a different domain.

Finally it should be noted that the probabilistic nature of this technique allows us to integrate PLCA in a variety of statistical models in ways that NMF could not be used. A notable example is that of a hidden Markov model which can use PLCA as a state model (as opposed to the more traditional Gaussian mixture). Under this formulation we can model mixture sounds and use the techniques mentioned in this chapter to make use of temporal information as well. Other extensions which are directly applicable are Latent Dirichlet Allocation [7], and other probabilistic constraints on the model distributions, as well as meta-learning models within which PLCA can be incorporated.

13.5 Conclusions

In this chapter we introduced the Probabilistic Latent Component Analysis method in its plain form, and with extensions to deal with shift-invariance and to manipulate sparsity. We then demonstrated some of its applications for source separation from monophonic mixtures. We have shown how we can use PLCA to describe spectrograms in terms of constituent parts and how to learn models of sources and use that information to extract them from monophonic mixtures. In our experience this approach can yield very good separation performance. However to achieve that performance there needs to be information to guide the separation process. We have shown that it is possible to extract simple components using PLCA on spectrograms, and then introduced procedures to incorporate prior information about the sources to guide extraction. Obviously these are only a few straightforward approaches and there is plenty of investigation that needs to be carried out in search for more efficient schemes. It is our hope that this chapter can serve as a compact introduction to this type of separation and foster some additional research in this area.

The author wishes to thank Bhiksha Raj and and Madhusudana Shashanka for their collaboration on this research and for providing invaluable insights during the preparation of this chapter.

References

1. M.A. Casey and W. Westner, "Separation of Mixed Audio Sources by Independent Subspace Analysis," *International Computer Music Conference (ICMC)*, Aug. 2000.
2. M.A. Casey, "Auditory Group Theory: with Applications to Statistical Basis Methods for Structured Audio," *Ph.D. Dissertation*, Massachusetts Institute of Technology, MA, USA. Feb. 1998.
3. P. Smaragdis, "Redundancy Reduction for Computational Audition, a Unifying Approach," *Ph.D. Dissertation*, Massachusetts Institute of Technology, MA, USA. June 2001.
4. P. Smaragdis, "Convolutive speech bases and their application to supervised speech separation," *IEEE Transaction on Audio, Speech and Language Processing*, Jan. 2007.
5. T. Virtanen, "Monaural Sound Source Separation by Non-Negative Matrix Factorization with Temporal Continuity and Sparseness Criteria," *IEEE Transactions on Audio, Speech, and Language Processing*, vol. 15, no. 3, Mar. 2007.
6. T. Virtanen, "Separation of sound sources by convolutive sparse coding," *Workshop on Statistical and Perceptual Audio Processing (SAPA)*, Oct. 2004.
7. B. Raj, M.V. Shashanka, and P. Smaragdis, "Latent Dirichlet decomposition for single channel speaker separation," *IEEE International Conference on Acoustics, Speech and Signal Processing (ICASSP)*, May 2006.
8. B. Raj and P. Smaragdis, "Latent variable decomposition of spectrograms for single channel speaker separation," *IEEE Workshop on Applications of Signal Processing to Audio and Acoustics (WASPAA)*, Oct. 2005.
9. P. Smaragdis, "Discovering auditory objects through non-negativity constraints," *Workshop on Statistical and Perceptual Audio Processing (SAPA)*, and Oct. 2004.
10. M. Mørup and M.N. Schmidt, "Sparse non-negative tensor 2D deconvolution (SNTF2D) for multi channel time-frequency analysis," *DTU Informatics and Mathematical Modeling Technical Report* 2006. Available at: `http://www2.imm.dtu.dk/pubdb/views/edoc_download.php/4659/pdf/imm4659.pdf`
11. M.N. Schmidt and M. Mørup, "Nonnegative Matrix Factor 2-D Deconvolution for Blind Single Channel Source Separation," *6th International Conference on Independent Component Analysis and Blind Signal Separation*, Mar. 2006.
12. P. Smaragdis, "From learning music to learning to separate," *Forum Acusticum*, Aug. 2005.
13. P. Smaragdis and J.C. Brown, "Non-negative matrix factorization for polyphonic music transcription," *IEEE Workshop on Applications of Signal Processing to Audio and Acoustics (WASPAA)*, Oct. 2003.
14. D. D. Lee and H. S. Seung, "Algorithms for non-negative matrix factorization," *Advances in Neural Information Processing 13*, 2001.
15. D. FitzGerald, M. Cranitch, and E. Coyle, "Sound source separation using shifted non-negative tensor factorisation," *IEEE International Conference on Acoustics, Speech, and Signal Processing*, May 2006.

16. M.E. Brand, "Structure learning in conditional probability models via an eutropic prior and parameter extinction," *Neural Computation*, vol. 11, no. 5, pp. 1155–1182, July 1999.
17. M.V. Shashanka "A Unified Probabilistic Approach to Modeling and Separating Single-Channel Acoustic Sources," *Ph.D. Dissertation*, Department of Cognitive and Neural Systems. Boston University, 2007.

14 Sparsification for Monaural Source Separation

Hiroki Asari,[1] Rasmus K. Olsson,[2] Barak A. Pearlmutter,[3] and Anthony M. Zador[4]

[1] Watson School of Biological Sciences, Cold Spring Harbor Laboratory
One Bungtown Road, Cold Spring Harbor, NY 11724, USA
E-mail: asari@cshl.edu
[2] Informatics and Mathematical Modelling, Technical University of Denmark
2800 Lyngby, Denmark
E-mail: rko@imm.dtu.dk
[3] Hamilton Institute, NUI Maynooth, Co. Kildare, Ireland
E-mail: barak@cs.nuim.ie
[4] Cold Spring Harbor Laboratory
One Bungtown Road, Cold Spring Harbor, NY 11724, USA
E-mail: zador@cshl.edu

Abstract. We explore the use of sparse representations for separation of a monaural mixture signal, where by a sparse representation we mean one where the number of nonzero elements is smaller than might be expected. This is a surprisingly powerful idea, as the ability to express a signal sparsely in some known, and potentially overcomplete, basis constitutes a strong model, while also lending itself to efficient algorithms. In the framework we explore, the representation of the signal is linear in a vector of coefficients. However, because many coefficient values could represent the same signal, the mapping from signal to coefficients is nonlinear, with the coefficients being chosen to simultaneously represent the signal and maximize a measure of sparsity. This conversion of the signal into the coefficients using L_1-optimization is viewed not as a preprocessing step performed before the data reaches the heart of the algorithm, but rather as *itself* the heart of the algorithm: after the coefficients have been found, only trivial processing remains to be done. We show how, by suitable choice of overcomplete basis, this framework can use a variety of cues (*e.g.*, speaker identity, differential filtering, differential attenuation) to accomplish monaural separation. We also discuss two radically different algorithms for finding the required overcomplete dictionaries: one based on nonnegative matrix factorization of isolated sources, and the other based on end-to-end optimization using automatic differentiation.

14.1 Introduction

This chapter reviews the concept of sparsity in the context of single-channel signal separation. The key idea is to impose restrictions on the decompositions of data: while the codebooks/dictionaries are allowed to have a very large number of components, the encodings are constrained to be sparse, *i.e.*, to contain only a small number of nonzero values. Enforcing sparsity helps

S. Makino et al. (eds.), Blind Speech Separation, 387–410.

ensure a unique decomposition, and, more importantly, can be used to extract the individual source signals from the mixture. In a learning phase, dictionaries are adapted for each source class. By concatenating such dictionaries for all sources in a given mixture, separation can then be achieved in the joint sparse encoding if the sources are exclusively distributed and thus discriminable enough in the adapted dictionaries of each source class. Detailed prior knowledge is sometimes available on the sources, in which case sparse coding is sometimes able to isolate the sources even without going through the learning phase. However, here we will give a special focus on full adaptive methods, which includes learning from data a set of dictionaries that allows for sparse representations.

The inclusion of sparsity objectives in a machine learning task is very much biologically inspired. A striking feature of many sensory processing problems in humans is that by far more neurons appear to be engaged in the internal representations of the signal than in its transduction. The auditory (or visual) cortex has in fact orders of magnitude more neurons than the cochlear (or optic nerve), and thus the neural representation of an acoustic (or visual) stimulus is *overcomplete* in the sense that many more neurons are available than are needed to represent the stimulus with high fidelity. How does the brain then choose a unique representation if many different patterns of auditory (or visual) cortical activity could all faithfully represent any given pattern of cochlear (or optic nerve) activity? It is biologically appealing to sparsely encode the patterns because such representations are metabolically efficient [1, 2], and the principle of sparse (or "efficient") coding has been used to predict receptive field properties of both auditory and visual neurons [3–7].

In single-channel separation of musical and speech signals, mimicking the sparsity of neural representations has yielded good results [8–11]. The methods fail, however, when the signals are too similar, *e.g.*, in the cases of similarly sounding voices or two trumpets in a duet. In such adverse conditions, improved performance can be expected from exploiting grouping cues such as time-continuity and common-onset known to be employed by the auditory system [12]. For instance, Asari et al. [13] use the information provided by the differential filtering imposed on a source by its path from its origin to the cochlea (the head-related transfer function, or HRTF).

In Sect. 14.2, we formulate a general framework for monaural source separation using dictionary methods. We also discuss learning algorithms for finding such dictionary elements suitable for sparse representations of given sources [14]. In Sect. 14.3, we describe methods to achieve a sparse representation for given signals in an overcomplete basis, *i.e.*, L_1-norm minimization by linear programming, and demonstrate an application of the framework in Sect. 14.4. We then introduce one particular monaural segregation cue, the HRTF, and reformulate the model accordingly in Sect. 14.5. In Sect. 14.6 it is demonstrated that the HRTF cues lead to improved separation when the source signals originate from different directions. Note that in contrast to much previous work, the HRTF is used here to separate auditory streams rather than to localize them in space; the model assumes that the locations of the sources have already been determined by other mechanisms. Finally, we close with a brief comment on the perspectives of sparse overcomplete representations in Sect. 14.7.

14.2 Problem Formulation

We consider a situation, where the observable is the sum of the source signals. For instance, this is a reasonable assumption in an acoustic setup, where sound waves from various emitters superpose at the microphone. While it is a common theme in techniques for blind source separation (BSS) to exploit the strong information provided by multiple sensors, here, only a single sensor is available for the estimation of the sources. Hence,

$$y(t) = \sum_{i=1}^{P} x_i(t) \tag{14.1}$$

where $y(t)$ and $x_i(t)$ are the time-domain mixture and source signals, respectively. While the problem cannot be solved in general for all classes of sources, solutions may be obtained for certain types of source distributions. For instance, humans at large possess the ability to isolate what is being said by a single speaker in a cocktail party situation, whereas a special training is required to listen out a single instrument from a musical piece, say, transcribe the bass from a rock 'n roll track [15]. Hence, the key to achieve the separation of the sources lies in learning features of the source distributions that are sufficiently discriminative to achieve separation and invertible such that the source signals can be reconstructed.

Inspired by the human auditory system, we will proceed to work in a time–frequency representation, $\mathbf{Y} = \mathrm{TF}\{y(t)\}$, since a number of advantages are associated with performing the computations in the transformed domain. We restrict TF such that $\mathbf{Y}$ is a real-valued matrix with spectral vectors, $\mathbf{y}$, as columns. The result of such a mapping is that certain classes of sources will become less overlapped in the transformed domain, which in turn facilitates the separation of the signals. More generally, if the sources can be assumed *sparsely* distributed in the frequency domain, additivity is approximately preserved in the transformed mixture,

$$\mathbf{y} = \sum_{i=1}^{P} \mathbf{x}_i \tag{14.2}$$

where $\mathbf{x}_i$ is the transformed source signal.

A class of algorithms, here denoted "dictionary methods," generally relies on learning factorizations of $\mathbf{x}_i$ from a training set of isolated source ensembles in terms of dictionaries $\mathbf{d}_{ij}$ and its encodings c_{ij},

$$\mathbf{x}_i = \sum_{j=1}^{N_i} \mathbf{d}_{ij} c_{ij} = \mathbf{D}_i \mathbf{c}_i \tag{14.3}$$

where the j-th column of $\mathbf{D}_i$ consists of $\mathbf{d}_{ij}$, and the j-th element of $\mathbf{c}_i$ holds the corresponding coefficient c_{ij}. Combining models (14.2) and (14.3) results in

$$\mathbf{y} = \sum_{i=1}^{P} \mathbf{D}_i \mathbf{c}_i = \mathbf{D}\mathbf{c} \tag{14.4}$$

We allow the number of dictionary elements, $\sum_i N_i$, to be larger than the dimensionality of $\mathbf{y}$, meaning that $\mathbf{D}$ is potentially overcomplete, *i.e.*, many possible decompositions exist. This has been shown to result in more natural and compact representations [4–7, 16–18].

The application of a factorization in terms of dictionaries to the problem of signal separation fundamentally consists of two steps: first, a set of dictionaries, $\mathbf{D}_i$, is learned from a training set of unmixed $\mathbf{x}_i$. Second, the combined encoding, $\mathbf{c}$, is mapped onto the concatenation of the prelearned source dictionaries, $\mathbf{D}$. Finally, the sources are estimated, resynthesizing according to (14.3). In Sect. 14.4, we provide examples of applications.

The method relies on the premise that the dictionaries of the sources in the mixture are sufficiently different such that $\mathbf{D}_1$ almost exclusively encode $\mathbf{x}_1$ but not $\mathbf{x}_2$, etc. Alternatively, it has been shown that source signals from identical distributions can be separated provided that information about the signal path is available [13]. This is described in more detail in Sect. 14.5.

Different matrix factorization methods can be conceived based on various a priori assumptions of the dictionaries and encodings. Since computing $\mathbf{c}$ (given $\mathbf{D}$) from (14.4) is generally ill-posed, the model should at least impose sufficient constraints for the inversion to produce a well-defined solution. In Sect. 14.2.1, we will proceed to describe criteria for learning dictionaries from training data. An important tool in this regard is linear programming, which can be employed to (i) learn the dictionaries, and (ii) compute the sparse decomposition required in (14.4) for the separation of the sources. The relevant aspects of linear programming are covered in Sect. 14.3.

14.2.1 Dictionary Learning

We are concerned with devising a machine learning solution to acquire a set of dictionaries that can be used for source separation as formalized in (14.4). In order to be relevant in this regard, a dictionary should easily encode its class of signal, but at the same time be discriminative, meaning that encodings of other signals can be attributed with a low likelihood. In the following, it is described how to exploit inherent properties of the source signals to derive a learning algorithm producing dictionaries of the mentioned sort.

Nonnegativity The particular choice of time–frequency representation determines the set of algorithms that can be employed. In audio applications, TF is often selected in a way so to mimic features of the early processing performed by the human auditory system. A common choice is to use a compressed version (*e.g.*, cube-root) of the power spectrogram, as computed by the short-time Fourier transform. This is often motivated by the fact that loudness perception in hearing can be approximated by a power law model [19]. An important feature of this representation is that it is nonnegative, reflecting a property of neuronal signaling in terms of spike rates, which by definition are nonnegative.

Lee and Seung [20] derived an efficient algorithm which minimizes the Euclidean distance between the data and the factorization, subject to nonnegativity constraints. In terms of learning the dictionaries, $\mathbf{D}_i$, the objective is to optimize the function,

$$E_{\mathrm{NMF}} = \|\mathbf{X}_i - \mathbf{D}_i\mathbf{C}_i\|_{\mathrm{F}}^2 \quad \text{for} \quad \mathbf{D}_i \geq \mathbf{0}, \quad \mathbf{C}_i \geq \mathbf{0} \tag{14.5}$$

where $\|\cdot\|_{\mathrm{F}}$ is the Frobenius norm, and $\mathbf{X}_i$ and $\mathbf{C}_i$ are matrices with data points and corresponding encodings as columns, respectively. From a probabilistic point of view, we can interpret the optimizer of (14.5) as a maximum posterior (MAP) estimator, assuming additive i.i.d. Gaussian noise and heaviside/uniform nonnegative (improper) a priori distributions.

A closed-form solution is not available, but an effective gradient descent method emerges when the step-size is associated with a certain function of $\mathbf{X}_i$, $\mathbf{D}_i$, and $\mathbf{C}_i$. Starting from random nonnegative matrices,

$$\mathbf{C}_i \leftarrow \mathbf{C}_i \bullet \frac{\mathbf{D}_i^\top \mathbf{X}_i}{\mathbf{D}_i^\top \widetilde{\mathbf{X}}_i} \tag{14.6}$$

$$\mathbf{D}_i \leftarrow \mathbf{D}_i \bullet \frac{\mathbf{X}_i \mathbf{C}_i^\top}{\widetilde{\mathbf{X}}_i \mathbf{C}_i^\top} \tag{14.7}$$

converges to a local minimum of (14.5), where $\widetilde{\mathbf{X}}_i = \mathbf{D}_i\mathbf{C}_i$ and the operators $\bullet$ and $\div$ indicate elementwise multiplication and division, respectively.[1] Non-negativity constraints have been used to learn signal dictionaries for single-channel separation of audio signals [9], and a convolutive extension

[1]The NMF updates are derived from a steepest descend starting point, *i.e.*, $\mathbf{C}_i \leftarrow \mathbf{C}_i - \Delta_C \bullet \nabla_E$, where $\nabla_E = -2\mathbf{D}_i^\top(\mathbf{X}_i - \widetilde{\mathbf{X}}_i)$ is the gradient with respect to $\mathbf{C}_i$ and Δ_C is a step-size matrix. Setting $\Delta_C = \frac{\mathbf{C}_i}{2\mathbf{D}_i^\top \widetilde{\mathbf{X}}_i}$, we arrive at the stated learning rule for $\mathbf{C}_i$. The derivation for the rule regarding $\mathbf{D}_i$ is similar.

of NMF has been particularly effective in this regard [21, 22], a technique reviewed by Smaragdis in Chapter 13 in this volume. Virtanen [23] provides a comprehensive review of NMF and related methods in audio analysis.

Sparsity In the following we describe how to apply the principle of sparsity to the learning of a decomposition in terms of a dictionary and its encoding as formalized in (14.3). Insisting on the sparsity of the encodings can be viewed as applying the principle of Occam's razor to the model, which states that the simplest explanation in some sense is to be preferred. The implication in a factorization setup is that $\mathbf{c}_i$ should be optimized as to contain as few nonzero entries as possible. In mathematical terms, minimize the L_0-norm, $\|\mathbf{u}\|_0 = \sum_i u_i^0$ if we define $0^0 \overset{\text{def}}{=} 0$.

Furthermore, sparseness is motivated from a perceptional, neural computational point of view. In neural terms, we could interpret c_{ij} as the neural activities (*e.g.*, spike rates) of the corresponding neurons characterized by their features $\mathbf{d}_{ij}$. The sparseness assumption then corresponds to representing the acoustic stimulus $\mathbf{y}$ in terms of the minimum number of spikes (Fig. 14.1),

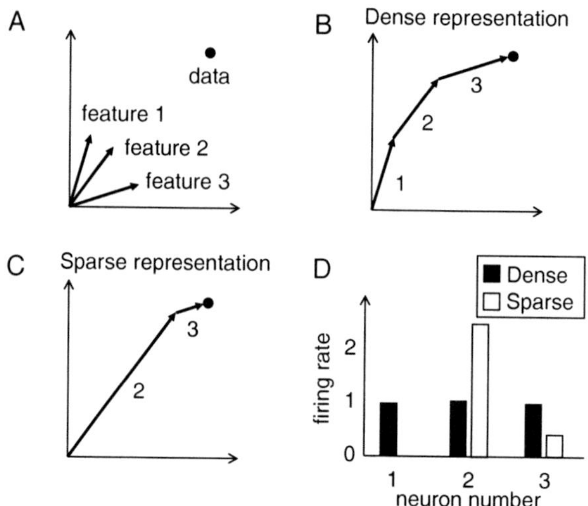

Fig. 14.1. Overcomplete representation in two dimensions. (A) Three non-orthogonal basis vectors (neural features) $\mathbf{d}_{ij}$ in two dimensions constitute an overcomplete representation, offering many possible ways to represent a data point $\mathbf{y}$ with no error. **(B)** The conventional solution is given by the pseudoinverse, yielding a *dense* neural representation where the squared sum of the coefficients (neural activities), $\|\mathbf{c}\|_2^2 = \sum_{ij} c_{ij}^2$, is minimized. This representation invokes all neural features about evenly. **(C)** The sparse solution invokes at most two neural features because it minimizes $\|\mathbf{c}\|_1 = \sum_{ij} |c_{ij}|$. **(D)** Comparison of neural activity for the two cases. For the dense representation, all three neurons participate about equally, whereas for the sparse representation activity is concentrated in neuron 2. From [13], with permission.

a biologically appealing constraint which leads to an energy-efficient representation [1, 2]. Also note that the sparse coding is compatible with the "efficient coding hypothesis" [24], according to which the goal of sensory processing is to construct an efficient representation of the sensory environment.

The problem of finding an overcomplete signal dictionary tuned to a given stimulus ensemble, so that signals drawn from that ensemble will have sparse representations in the constructed dictionary, has received increasing attention, due to applications in both neuroscience and in the construction of efficient practical codes [25, 26]. Unfortunately, it is not computationally tractable to optimize directly the L_0-norm of the encoding. In fact, the problem is NP-complete [27]. As an alternative to the L_0-norm, the L_1-norm, defined $\|\mathbf{u}\|_1 = \sum_i |u_i|$, can be applied. In many situations, the L_1-norm solution approximates the L_0-norm solution, leading to equally sparse solutions [28], particularly in the presence of a noise model. The objective function balances the norm of the encoding with the accuracy of the fit,

$$E_{\mathrm{L1}} = \|\mathbf{c}_i\|_1 \quad \text{for} \quad \|\mathbf{x}_i - \mathbf{D}_i \mathbf{c}_i\|_p \leq \beta \tag{14.8}$$

where β is proportional to the noise level and with $p = 1, 2$, or ∞. The optimization of (14.8) with respect to $\mathbf{c}_i$ can be viewed as a MAP estimator where an i.i.d. exponential a priori distribution is assumed for $\mathbf{c}_i$, and additive i.i.d. noise whose distribution is specified by β and p. Letting $\beta \to 0$ is equivalent to assuming that the noise is very small, and the solution converges to the zero-noise solution. The Gaussian noise case, $p = 2$, can be solved by semidefinite programming methods [29]. Both $p = 1$ and $p = \infty$ can be solved using linear programming, the details of which are covered in Sect. 14.3.

When the objective is to learn dictionaries, *i.e.*, learning $\mathbf{D}$ from training data, one option is to optimize (14.8) with respect to $\mathbf{c}$ *and* $\mathbf{D}$ [14]. This is likewise described in Sect. 14.3. Benaroya et al. [9] combined sparsity and nonnegativity constraints in order to learn audio dictionaries, whereas Jang and Lee [11] applied independent component analysis (ICA) which can also be seen as sparsity-inducing, depending on the source prior distribution.

14.3 Sparse Representation by Linear Programming

Linear programming solvers (LP) are often used as subroutines within larger systems, in both operations research and machine learning [30, 31]. One very simple example of this is in sparse signal processing, where it is common to represent a vector as sparsely as possible in an overcomplete basis; this representation can be found using LP, and the sparse representation is then used in further processing [25, 32–37]. In this section we explain how to use linear programming for two related tasks, namely (i) performing a sparse decomposition, as defined by the L_1-norm, of (14.8), and (ii) learning dictionaries by optimizing on a training set the L_1 sparsity of the decomposition.

In order to do so, we develop in Sect. 14.3.1 a useful notation and formulate the sparse decomposition in terms of linear programming. Then, in Sect. 14.3.2, we describe how to efficiently compute derivatives of functions where linear program solution is used as inputs. Finally, these theoretical foundations allow us to formulate a learning rule for obtaining dictionaries optimized for sparsity in Sect. 14.3.3.

14.3.1 Basics

In order to develop a notation for LP, consider the general LP problem

$$\arg\min_{\mathbf{z}} \mathbf{w}^\top \mathbf{z} \text{ s.t. } \mathbf{A}\mathbf{z} \leq \mathbf{a} \text{ and } \mathbf{B}\mathbf{z} = \mathbf{b} \tag{14.9}$$

We will denote the linear program solver lp, and write the solution as $\mathbf{z} = \mathsf{lp}(\mathbf{w}, \mathbf{A}, \mathbf{a}, \mathbf{B}, \mathbf{b})$. It is important to see that $\mathsf{lp}(\cdot)$ can be regarded as either a mathematical function which maps LP problems to their solutions, or as a computer program which actually solves LP problems. Our notation deliberately does not distinguish between these two closely related notions.

Assuming feasibility, boundedness, and uniqueness, the solution to this LP problem will satisfy a set of linear equalities consisting of a subset of the constraints: the *active* constraints [38–40]. An LP solver calculates two pieces of information: the solution itself, and the identity of the active constraints. We will find it convenient to refer to the active constraints by defining some very sparse matrices that extract the active constraints from the constraint matrices. Let $\alpha_1 < \cdots < \alpha_n$ be the indices of the rows of $\mathbf{A}$ corresponding to active constraints, and $\beta_1 < \cdots < \beta_m$ index the active rows of $\mathbf{B}$. Without loss of generality, we assume that the total number of active constraints is equal to the dimensionality of the solution, $n + m = \dim \mathbf{z}$. We let $\mathbf{P}_\alpha$ be a matrix with n rows, where the i-th row is all zeros except for a one in the α_i-th column, and $\mathbf{P}_\beta$ similarly have m rows, with its i-th row all zeros except for a one in the β_i-th column. So $\mathbf{P}_\alpha\mathbf{A}$ and $\mathbf{P}_\beta\mathbf{B}$ hold the active rows of $\mathbf{A}$ and $\mathbf{B}$, respectively. These can be combined into a single matrix

$$\mathbf{P} \equiv \begin{bmatrix} \mathbf{P}_\alpha & \mathbf{0} \\ \mathbf{0} & \mathbf{P}_\beta \end{bmatrix}$$

Using these definitions, the solution $\mathbf{z}$ to (14.9), which presumably is already available having been computed by the algorithm that identified the active constraints, must be the unique solution of the system of linear constraints

$$\mathbf{P} \begin{bmatrix} \mathbf{A} \\ \mathbf{B} \end{bmatrix} \mathbf{z} = \mathbf{P} \begin{bmatrix} \mathbf{a} \\ \mathbf{b} \end{bmatrix}$$

or

$$\mathsf{lp}(\mathbf{w}, \mathbf{A}, \mathbf{a}, \mathbf{B}, \mathbf{b}) = \mathsf{lq}\left(\mathbf{P} \begin{bmatrix} \mathbf{A} \\ \mathbf{B} \end{bmatrix}, \mathbf{P} \begin{bmatrix} \mathbf{a} \\ \mathbf{b} \end{bmatrix}\right) \tag{14.10}$$

where lq is a routine that efficiently solves a system of linear equations, $\mathsf{lq}(\mathbf{M}, \mathbf{m}) = \mathbf{M}^{-1}\mathbf{m}$. For notational convenience we suppress the identity of the active constraints as an output of the lp routine. Instead we assume that it is available where necessary, so any function with access to the solution $\mathbf{z}$ found by the LP solver is also assumed to have access to the corresponding $\mathbf{P}$.

14.3.2 Automatic Differentiation

Automatic differentiation (AD) is a process by which a numeric calculation specified in a computer programming language can be mechanically transformed so as to calculate derivatives (in the differential calculus sense) of the function originally calculated [41]. There are two sorts of AD transformations: forward accumulation [42] and reverse accumulation [43]. (A special case of reverse accumulation AD is referred to as backpropagation in the machine learning literature [44].) If the entire calculation is denoted $\mathbf{y} = h(\mathbf{x})$, then forward accumulation AD arises because a perturbation $d\mathbf{x}/dr$ induces a perturbation $d\mathbf{y}/dr$, and reverse accumulation AD arises because a gradient $dE/d\mathbf{y}$ induces a gradient $dE/d\mathbf{x}$. The Jacobian matrix $\mathbf{J}$ whose i, j-th entry is dh_i/dx_j plays a dominant role in reasoning about this process: forward AD calculates $\acute{\mathbf{y}} = \mathbf{J}\acute{\mathbf{x}} = \overrightarrow{h}(\mathbf{x}, \acute{\mathbf{x}})$, and reverse AD calculates $\grave{\mathbf{x}} = \mathbf{J}^\top \grave{\mathbf{y}} = \overleftarrow{h}(\mathbf{x}, \grave{\mathbf{y}})$. The difficulty is that, in high dimensional systems, the matrix $\mathbf{J}$ is too large to actually calculate. In AD the above matrix-vector products are found directly and efficiently, without actually calculating the Jacobian.

The central insight is that calculations can be broken down into a chained series of assignments $v := g(u)$, and transformed versions of these chained together. The transformed version of the above internal assignment statement would be $\acute{v} := \overrightarrow{g}(u, \acute{u}, v)$ in forward mode [42], or $\grave{u} := \overleftarrow{g}(u, v, \grave{v})$ in reverse mode [43]. The most interesting property of AD, which results from this insight, is that the time consumed by the adjoint calculations can be the same as that consumed by the original calculation, up to a small constant factor. (This naturally assumes that the transformations of the primitives invoked also obey this property, which is in general true.)

We will refer to the adjoints of original variables introduced in forward accumulation (perturbations) using a forward-leaning accent $v \mapsto \acute{v}$; to the adjoint variables introduced in the reverse mode transformation (sensitivities) using a reverse-leaning accent $v \mapsto \grave{v}$; and to the forward- and reverse-mode transformations of functions using forward and reverse arrows, $h \mapsto \overrightarrow{h}$ and $h \mapsto \overleftarrow{h}$, respectively. A detailed introduction to AD is beyond the scope of this chapter, but one form appears repeatedly in our derivations, *i.e.*, $\mathbf{V} := \mathbf{A}\mathbf{U}\mathbf{B}$ where $\mathbf{A}$ and $\mathbf{B}$ are constant matrices and $\mathbf{U}$ and $\mathbf{V}$ are matrices as well. This transforms to

$$\acute{\mathbf{V}} := \mathbf{A}\acute{\mathbf{U}}\mathbf{B} \tag{14.11}$$

$$\grave{\mathbf{U}} := \mathbf{A}^\top \grave{\mathbf{V}} \mathbf{B}^\top \tag{14.12}$$

AD of a Linear Equation Solver We first derive AD equations for a simple implicit function, namely a linear equation solver. We consider a subroutine lq which finds the solution $\mathbf{z}$ of $\mathbf{Mz} = \mathbf{m}$, written as $\mathbf{z} = \mathsf{lq}(\mathbf{M}, \mathbf{m})$. This assumes that $\mathbf{M}$ is square and full-rank, just as a division operation $z = x/y$ assumes that $y \neq 0$. We will derive formulae for both forward mode AD (the $\acute{\mathbf{z}}$ induced by $\acute{\mathbf{M}}$ and $\acute{\mathbf{m}}$) and reverse mode AD (the $\grave{\mathbf{M}}$ and $\grave{\mathbf{m}}$ induced by $\grave{\mathbf{z}}$).

For forward propagation of perturbations, we will write $\acute{\mathbf{z}} = \overrightarrow{\mathsf{lq}}(\mathbf{M}, \acute{\mathbf{M}}, \mathbf{m}, \acute{\mathbf{m}}, \mathbf{z})$. Using (14.11), we have that $(\mathbf{M} + \acute{\mathbf{M}})(\mathbf{z} + \acute{\mathbf{z}}) = \mathbf{m} + \acute{\mathbf{m}}$ which reduces to $\mathbf{M}\acute{\mathbf{z}} = \acute{\mathbf{m}} - \acute{\mathbf{M}}\mathbf{z}$. Hence, we conclude that

$$\overrightarrow{\mathsf{lq}}(\mathbf{M}, \acute{\mathbf{M}}, \mathbf{m}, \acute{\mathbf{m}}, \mathbf{z}) = \mathsf{lq}(\mathbf{M}, \acute{\mathbf{m}} - \acute{\mathbf{M}}\mathbf{z})$$

Note that lq is linear in its second argument, where the perturbations enter linearly. For reverse propagation of sensitivities, we will write

$$\left[\grave{\mathbf{M}}\ \grave{\mathbf{m}}\right] = \overleftarrow{\mathsf{lq}}(\mathbf{M}, \mathbf{m}, \mathbf{z}, \grave{\mathbf{z}}) \qquad (14.13)$$

First observe that $\mathbf{z} = \mathbf{M}^{-1}\mathbf{m}$ and hence $\grave{\mathbf{m}} = \mathbf{M}^{-\top}\grave{\mathbf{z}}$ so

$$\grave{\mathbf{m}} = \mathsf{lq}(\mathbf{M}^{\top}, \grave{\mathbf{z}})$$

For the remaining term we start with our previous forward perturbation $\acute{\mathbf{M}} \mapsto \acute{\mathbf{z}}$, namely $\acute{\mathbf{z}} = -\mathbf{M}^{-1}\acute{\mathbf{M}}\mathbf{z}$, and note that the reverse must be the transpose of this linear relationship (*i.e.*, using (14.11) and (14.12)), $\grave{\mathbf{M}} = -\mathbf{M}^{-\top}\grave{\mathbf{z}}\mathbf{z}^{\top}$, which is the outer product

$$\grave{\mathbf{M}} = -\grave{\mathbf{m}}\mathbf{z}^{\top}$$

AD of Linear Programming We apply (14.13) followed by some book-keeping, yields

$$\begin{aligned} \begin{bmatrix} \grave{\mathbf{A}} & \grave{\mathbf{a}} \\ \grave{\mathbf{B}} & \grave{\mathbf{b}} \end{bmatrix} &= \overleftarrow{\mathsf{lp}}(\mathbf{w}, \mathbf{A}, \mathbf{a}, \mathbf{B}, \mathbf{b}, \mathbf{z}, \grave{\mathbf{z}}) \\ &= \mathbf{P}^{\top}\,\overleftarrow{\mathsf{lq}}(\mathbf{P}\begin{bmatrix} \mathbf{A} \\ \mathbf{B} \end{bmatrix}, \mathbf{P}\begin{bmatrix} \mathbf{a} \\ \mathbf{b} \end{bmatrix}, \mathbf{z}, \grave{\mathbf{z}}) \\ \grave{\mathbf{w}} &= \mathbf{0} \end{aligned}$$

Forward accumulation is similar, but is left out for brevity.

Constrained L_1 Optimization We can find AD equations for linearly constrained L_1-norm optimization via reduction to LP. Consider

$$\arg\min_{\mathbf{c}} \|\mathbf{c}\|_1 \text{ s.t. } \mathbf{Dc} = \mathbf{y}$$

Although $\|\mathbf{c}\|_1 = \sum_i |c_i|$ is a nonlinear objective function, a change in parametrization allows optimization via LP. We name the solution $\mathbf{c} = \mathsf{L1opt}(\mathbf{y}, \mathbf{D})$ where

$$\mathsf{L1opt}(\mathbf{y}, \mathbf{D}) = \begin{bmatrix}\mathbf{I} & -\mathbf{I}\end{bmatrix} \mathsf{lp}(\mathbf{1}, -\mathbf{I}, \mathbf{0}, \mathbf{D}\begin{bmatrix}\mathbf{I} & -\mathbf{I}\end{bmatrix}, \mathbf{y})$$

in which $\mathbf{0}$ and $\mathbf{1}$ denote column vectors whose elements all contain the indicated number, and each $\mathbf{I}$ is an appropriately sized identity matrix. The reverse-mode AD transformation follows immediately

$$\begin{aligned}
&\overleftarrow{\mathsf{L1opt}}(\mathbf{y}, \mathbf{D}, \mathbf{c}, \grave{\mathbf{c}}) \\
&\quad = \begin{bmatrix}\grave{\mathbf{D}} & \grave{\mathbf{y}}\end{bmatrix} = \begin{bmatrix}\mathbf{0}' & \mathbf{I}\end{bmatrix} \overleftarrow{\mathsf{lp}}\left(\mathbf{1}, -\mathbf{I}, \mathbf{0}, \mathbf{D}\begin{bmatrix}\mathbf{I} & -\mathbf{I}\end{bmatrix}, \mathbf{y}, \mathbf{z}, \begin{bmatrix}\mathbf{I} \\ -\mathbf{I}\end{bmatrix} \grave{\mathbf{c}}\right) \begin{bmatrix}\mathbf{I} & \mathbf{0} \\ -\mathbf{I} & \mathbf{0} \\ \mathbf{0}^\top & 1\end{bmatrix}
\end{aligned}$$

where $\mathbf{z}$ is the solution of the internal LP problem and $\mathbf{0}'$ is an appropriately sized matrix of zeros.

14.3.3 Dictionaries Optimized for Sparsity

A major advantage of the LP differentiation framework, and more specifically the reverse accumulation of the constrained L_1-norm optimization, is that it provides directly a learning rule for finding sparse representations in overcomplete dictionaries.

We assume an overcomplete dictionary in the columns of $\mathbf{D}$, which is used to encode a signal represented in the column vector $\mathbf{y}$ using the column vector of coefficients $\mathbf{c} = \mathsf{L1opt}(\mathbf{y}, \mathbf{D})$ where each dictionary element has unit L_2 length. We will update $\mathbf{D}$ so as to minimize $E = \langle \|\mathsf{L1opt}(\mathbf{y}, \mathbf{D})\|_1 \rangle$ while keeping the columns of $\mathbf{D}$ at unit length. This can be regarded a special case of ICA [45], where measures of independence across coefficients are optimized. We wish to use a gradient method so we calculate $\nabla_{\mathbf{D}} E_{\mathbf{y}}$ where $E_{\mathbf{y}} = \|\mathsf{L1opt}(\mathbf{y}, \mathbf{D})\|_1$ making $E = \langle E_{\mathbf{y}} \rangle$. Invoking AD

$$\begin{aligned}
\nabla_{\mathbf{D}} E_{\mathbf{y}} &= \grave{\mathbf{D}} = \begin{bmatrix}\grave{\mathbf{D}} & \grave{\mathbf{y}}\end{bmatrix} \begin{bmatrix}\mathbf{I} \\ \mathbf{0}^\top\end{bmatrix} \\
&= \overleftarrow{\mathsf{L1opt}}(\mathbf{y}, \mathbf{D}, \mathbf{c}, \mathrm{sign}(\mathbf{c})) \begin{bmatrix}\mathbf{I} \\ \mathbf{0}^\top\end{bmatrix}
\end{aligned} \tag{14.14}$$

where $\mathrm{sign}(x) = +1/0/-1$ for x positive/zero/negative, and applies element-wise to vectors.

We are now in a position to perform stochastic gradient optimization [46], modified by the inclusion of a normalization step to maintain the columns of $\mathbf{D}$ at unit length and nonnegative.

1: Draw $\mathbf{y}$ from signal distribution.
2: Calculate $E_{\mathbf{y}}$.
3: Calculate $\nabla_{\mathbf{D}} E_{\mathbf{y}}$ by (14.14).
4: Step $\mathbf{D} := \mathbf{D} - \eta \nabla_{\mathbf{D}} E_{\mathbf{y}}$.
5: Set any negative element of $\mathbf{D}$ to zero.
6: Normalize the columns $\mathbf{d}_i$ of $\mathbf{D}$ to unit L_2-norm.
7: Repeat to convergence of $\mathbf{D}$.

This procedure can be regarded as an efficient exact maximum likelihood treatment of the posterior calculated by Lewicki and Sejnowski [25] using a Gaussian approximation. It is interesting to note that the formulation here can be easily and mechanically generalized to other objectives.

14.4 Source Separation Example

We will now demonstrate an application of the framework laid out in Sects. 14.2 and 14.3. More specifically, we will attempt to separate two speech signals of equal power from a single mixture as was proposed in Pearlmutter and Olsson [14]. This is a relevant task in hearing aids, as a speech recognition preprocessor, and in other applications which might benefit from better noise reduction. For this reason, there has been a flurry of interest in the problem.

A common trait of many approaches is that speaker-dependent models have been learned from a training set of isolated recordings and subsequently a combination of these have been applied to the mixture. Roweis [47] learned hidden Markov models (HMM) of individual speakers and combined them in a factorial HMM, separating a mixture. The high dimensionality of the combined state space prohibited direct inference, but an approximate solution was obtained. A Bayesian solution to inference in the factorial HMM, applying a set of milder assumptions, was provided by Kristjansson [48], achieving a very good (super-human) performance on a word recognition task. Bach and Jordan [49] devised a clustering algorithm based on specific features of speech, which does not learn models for each speaker. Dictionary methods, which do not require combinatorial searches, have been based on a priori assumptions of sparsity and/or nonnegativity (see Sect. 14.2).

In the following, we will twice evoke the assumption of L_1 sparsity: first in order to learn the dictionaries, *i.e.*, inverting (14.3); second, to compute the separating decomposition of (14.4).

14.4.1 Dictionary Learning

A set of personalized speech dictionaries was learned by sparsity optimization using the method described in Sect. 14.3.[2] Defining the time–frequency

[2]The GRID corpus was used [50]. It contains 1000 short sentences recorded for each of 34 speakers.

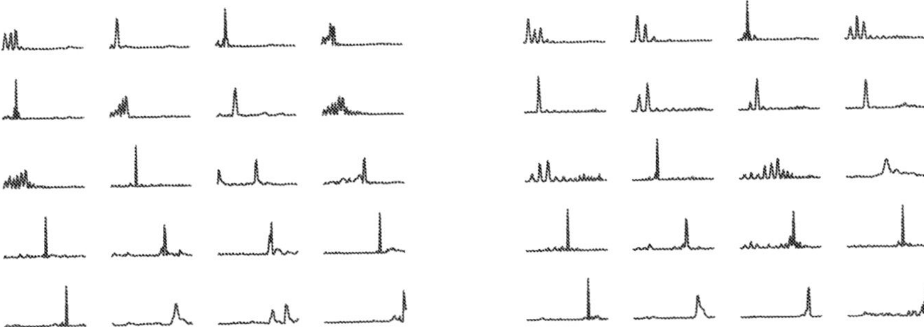

Fig. 14.2. A sample of learnt dictionary entries for male (left) and female (right) speech in the Mel spectrum domain. Harmonic features have clearly been found from the data, but broad and narrow noise spectra are also visible.

transformation (TF), the speech was preprocessed and represented to (essentially) transform the audio signals into an amplitude Mel time–frequency representation [51]. The stochastic gradient optimization of the linearly constrained L_1-norm was run for 40,000 iterations. The step-size η was decreased throughout the training. The $N = 256$ columns of the dictionaries were initialized with narrow pulses distributed evenly across the spectrum and nonnegativity was enforced following each iteration. In Fig. 14.2 is displayed a randomly selected sample of learnt dictionary elements of one male and one female speaker. The dictionaries clearly capture a number of characteristics of speech, such as quasi-periodicity and dependencies across frequency bands.

14.4.2 Source Separation

In order to separate the sources, we assume the additive mixture model of (14.4) and perform a sparse decomposition by minimizing the L_1-norm. Thus, a linear program is used to compute $\mathbf{c} = \texttt{L1opt}(\mathbf{y}, \mathbf{D})$, where $\mathbf{y}$ is the mixture, $\mathbf{D} = [\mathbf{D}_1 \quad \mathbf{D}_2]$ is the concatenated dictionary, and $\mathbf{c} = \begin{bmatrix} \mathbf{c}_1 \\ \mathbf{c}_2 \end{bmatrix}$ is the joint source encoding. Assuming that the $\mathbf{D}_1$ and $\mathbf{D}_2$ are different in some sense, it can be expected that a sparse representation in the basis $\mathbf{D}$ coincides with the separation of the sources. The degree of success depends on the level to which the signals (and dictionaries) are different. The source estimates in the Mel spectrum domain are then resynthesized according as (14.3): $\hat{\mathbf{x}}_1 = \mathbf{D}_1\mathbf{c}_1$ and $\hat{\mathbf{x}}_2 = \mathbf{D}_2\mathbf{c}_2$. The conversion back to the time-domain consists of mapping to the amplitude spectrogram and subsequently reconstructing the time-domain signal using the noisy phase of the mixture. Due to the sparsity of speech in the transformed domain, the degree of overlap of the sources is small, which causes the approximation to be fairly accurate.

Table 14.1. Monaural two-speaker signal-to-noise separation performance (mean±stderr of SNR), by speaker gender. The simulated test data consisted of all possible combinations, $T = 6$ sec, of the 34 speakers. Clearly, it is an easier task to separate the speech signals in the case of opposite-gender speakers. This indicates the required level of contrast between the source signals for the method to work.

Genders	SNR (dB)
M/M	4.9±1.2
M/F	7.8±1.3
F/F	5.1±1.4

The quality of the source estimates was evaluated in the time-domain simply as the ratio of powers of the target to reconstruction error, here termed the signal-to-noise ratio (SNR). Table 14.1 lists the performance of the method on the GRID database.

14.5 Convolutional Mixing and Head-Related Transfer Function

One limitation of the BSS model as described in (14.1)–(14.4) is that source signals from identical distributions (or from different distributions with the same statistics) can hardly be separated because the performance depends on the "personalized" dictionaries that exclusively encode one source signal but not the others. In this section, we will then describe how we could extend the model to exploit additional separation cues that "tag" the dictionaries so they can be assigned to the appropriate sources in the framework of sparse overcomplete representations. Specifically, we will consider the monaural source separation of convolutive sources, *i.e.*, separating multiple *prefiltered* signals combined at a single sensor, using biologically inspired spectral cues that segregate the auditory scene based on the source locations [13].

14.5.1 Head-Related Transfer Function

The auditory system uses a wide variety of psychophysical cues to segregate auditory streams [12], including both binaural and monaural cues. Many monaural cues have been identified, such as common onset time or comodulation of stimulus power in different parts of the spectrum.

For simplicity, here we focus on just one set of cues: those provided by the differential filtering imposed on a source by its path from its origin in space to the cochlea. This filtering or "spectral coloring" is caused both by the head and the detailed shape of the ear (the head-related transfer function, or HRTF), and by the environment on sources at different positions in space. The HRTF depends on the spatial position—both the relative azimuth and elevation—of the source. At some frequencies, the HRTF can attenuate sound from one location by as much as 40 dB more than from another, and such HRTF cues, when present, help in source separation [52].

The HRTF is also important for generating a three-dimensional experience of sound, so that acoustic sources that bypass the HRTF (*e.g.*, those presented with headphones) are typically perceived unnaturally, as though arising inside the head [53, 54]. Note however that the HRTF is used here to *separate* auditory streams rather than to *localize* them in space, in contrast to much previous work on the role of the HRTF in sound localization [53–57].

It is often reasonable to assume that sound arriving from different locations should be treated as arising from distinct sources. We thus assume that all sounds from a given position are *defined* to belong to the same source, and any sounds from a different position are defined to belong to different sources. We emphasize that although sound localization (the process by which an animal determines where in space a source is located) is related to source separation (the process by which an animal extracts different auditory streams from a single waveform), the two computations are distinct; neither is necessary nor sufficient for the other. Here we focus on the separation problem, and assume that source localization occurs by other mechanisms.

14.5.2 Reformulation

Here we will reformulate the BSS model in (14.1)–(14.4) for the monaural source separation problem of convolutive sources. Suppose there are P acoustic sources located at known distinct positions in space, with $x_i(t)$ being the time course of the stimulus sound pressure of the i-th source at its point of origin. Associated with each position is a known filter given by $h_i(t)$. In what follows we will refer to $h_i(t)$ as the HRTF, but in general $h_i(t)$ will include not just the filtering of the head and external ear, but also the filter function of the acoustic environment (reverberation, etc.)

The signal $y(t)$ at the ear is then the sum of the filtered signals,

$$y(t) = \sum_{i=1}^{P} h_i(t) * x_i(t) = \sum_{i=1}^{P} \widetilde{x}_i(t) \qquad (14.1')$$

where $*$ indicates convolution and $\widetilde{x}_i(t) = h_i(t) * x_i(t)$ is the i-th source in isolation following filtering. (We can say that $x_i(t)$ is the i-th source measured in source space, while $\widetilde{x}_i(t)$ is the same source measured in sensor space.) The goal is then to recover the underlying sources $x_i(t)$ from the signal $y(t)$, using knowledge of the directional filters $h_i(t)$. Note that the actual spatial locations of the sources are not computed during the separation in this model but we assume the locations (and thus associated directional filters $h_i(t)$) have already been identified by other mechanisms.

In the TF domain, we have

$$\mathbf{y} = \sum_{i=1}^{P} \mathbf{h}_i \bullet \mathbf{x}_i = \sum_{i=1}^{P} \widetilde{\mathbf{x}}_i \tag{14.2$'$}$$

where $\bullet$ indicates elementwise multiplication. As in (14.3), we then assume that each source $\mathbf{x}_i$ can be expressed as a linear combination of dictionary elements $\mathbf{d}_j$:

$$\mathbf{x}_i = \sum_{j=1}^{N_i} \mathbf{d}_j c_{ij} = \mathbf{D}\mathbf{c}_i \tag{14.3$'$}$$

Note that we no longer have to use "personalized" dictionaries for each source but we could use any dictionary set $\mathbf{D}$ that captures the spectral correlations in the sources and permits sparse representations, *i.e.*, where only a small number of coefficients c_{ij} are significantly nonzero. By further assuming that the dictionaries in sensor space, $\widetilde{\mathbf{d}}_{ij}$, are related to the dictionaries in source space, $\mathbf{d}_j$, by convolution with each filter $\mathbf{h}_i$:

$$\widetilde{\mathbf{d}}_{ij} = \mathbf{h}_i \bullet \mathbf{d}_j \tag{14.15}$$

the signal $\mathbf{y}$ received at the ear can be expressed as a linear combination of the dictionary elements in sensor space:

$$\begin{aligned}
\mathbf{y} &= \sum_{i=1}^{P} \mathbf{h}_i \bullet \mathbf{x}_i && \text{by (14.2$'$)} \\
&= \sum_{i=1}^{P} \mathbf{h}_i \bullet \left(\sum_{j=1}^{N_i} \mathbf{d}_j c_{ij} \right) && \text{by (14.3$'$)} \\
&= \sum_{i,j} \widetilde{\mathbf{d}}_{ij} c_{ij} && \text{by (14.15)} \\
&= \widetilde{\mathbf{D}}\mathbf{c} && \text{(14.4$'$)}
\end{aligned}$$

As before, the BSS model in (14.1′)–(14.4′) consists of two steps: first, a set of dictionary in source space $\mathbf{D}$ is learned from a training set of unmixed signals $\mathbf{x}_i$. Second, given a convolutional mixture $\mathbf{y}$ and position-dependent filters $\mathbf{h}_i$, appropriate coefficients c_{ij} are obtained for (14.4′) under a sparseness prior (*i.e.*, by computing $\mathbf{c} = \texttt{L1opt}(\mathbf{y}, \widetilde{\mathbf{D}})$), and a given source i can be reconstructed by summing over all dictionary elements associated with position i using (14.3′). Note that separation and deconvolution are simultaneously achieved here by estimating the coefficients by using a post-HRTF (sensor space) dictionary $\widetilde{\mathbf{D}}$ but reconstructing the signals by using a pre-HRTF (source space) dictionary $\mathbf{D}$ (Fig. 14.3).

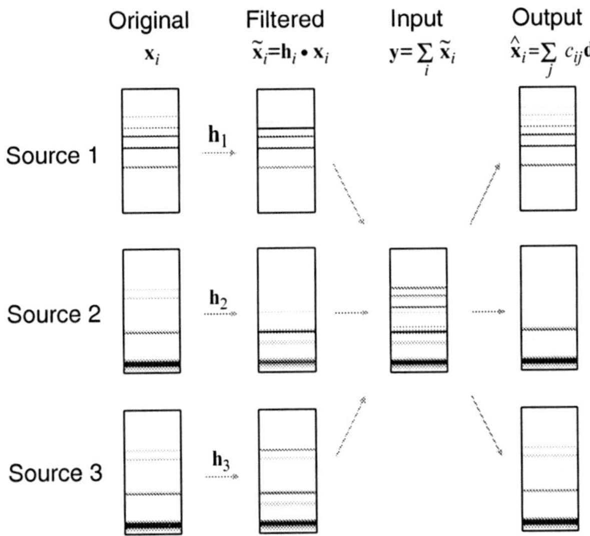

Fig. 14.3. Separation of three musical sources. Three musical instruments at three distinct spatial locations were filtered (by $\mathbf{h}_1, \ldots, \mathbf{h}_3$, corresponding to the HRTFs for azimuth $-90°$, $0°$, and $90°$ with zero elevation, respectively) and summed to produce the *input* $\mathbf{y}$, and then separated using a sparse overcomplete representation to produce the *output*. Note that two of the sources (a harp playing the note "D", *center* and *bottom*) were chosen to be identical; this example is thus particularly challenging, since the only cue for separating the sources is the filtering imposed by the HRTF. Nevertheless, separation was good as seen by comparing the left (*Original*) and right (*Output*) columns. From [13], with permission.

14.6 Separation of Convolutive Sources

Successful source separation for the BSS models described in Sect. 14.5 requires that two conditions are satisfied. First, the sources must be sparsely representable, as is the case with natural auditory stimuli [4, 5, 18, 58]. Second, the sources must have spectral correlations matched to the HRTF. In the following, we will demonstrate that the model is able to separate acoustic sources consisting of mixtures of music, natural sounds, and speech.[3]

14.6.1 Dictionary Learning

Here we used nonnegative matrix factorization (NMF; see also Sect. 14.2.1) to generate a set of complete dictionaries from spectrograms obtained from unmixed samples of solo instrumental music, natural sounds, and speech

[3]Sound data were taken from commercially available audio CDs, and the HRTF data for a representative left human pinna were downloaded from http://www.itakura.nuee.nagoya-u.ac.jp/HRTF/ [59].

($\mathbf{D}_{\rm ms}$, $\mathbf{D}_{\rm ns}$, and $\mathbf{D}_{\rm sp}$, respectively), and concatenated them to form an overcomplete source space dictionary: $\mathbf{D} = [\mathbf{D}_{\rm ms} \quad \mathbf{D}_{\rm ns} \quad \mathbf{D}_{\rm sp}]$.

The ability of the NMF dictionaries to represent sounds in a sparse model can be quantified in terms of the "sparseness index," defined as $\|\mathbf{c}_i\|_0 / \dim \mathbf{x}_i$ in the presence of a single (unmixed) source $\mathbf{x}_i$ (see also (14.3′)). The distribution of the index was 0.61±0.27, 0.64±0.17, 0.49±0.13 (mean±SD) for $\mathbf{D}_{\rm ms}$, $\mathbf{D}_{\rm ns}$, and $\mathbf{D}_{\rm sp}$, respectively, over 10,000 test samples. This suggests that the NMF dictionaries generally led to sparse representations of the ensembles, satisfying the first condition for the model to work.

When applied to music, NMF typically yielded elements suggestive of musical notes, each with a strong fundamental frequency and weaker harmonics at higher frequencies. In many cases, listeners could easily use timbre to identify the instrument from which a particular element was derived. When applied to sounds from other ensembles (natural sounds and speech), NMF yielded elements that had rich harmonic structure, but it was not in general easy to "interpret" the elements (*e.g.*, as vowels). Nonetheless these elements captured aspects of the statistical structure of the underlying ensemble of sounds, and thus satisfied the second condition as well.

It should be mentioned that the choice of NMF was merely a matter of convenience; we could have used any basis that satisfies the two conditions. Finding good overcomplete dictionaries from samples of a stimulus ensemble is a subject of ongoing research [26] (see also Sect. 14.3.3). NMF is then not necessarily the best algorithm in this context, but is simply good enough for our monaural BSS model.

14.6.2 Separation with HRTF

To demonstrate the model's ability to separate sources, we generated digital mixtures of three sources positioned at three distinct positions in space (Fig. 14.3). On the *left column* are the spectrograms of the sources at their origin. Two of the sources (a harp playing the note "D", *center* and *bottom*) were chosen to be identical; this example is thus particularly challenging, since the only cue for separating the sources is the filtering imposed by the HRTF.

Separation was nevertheless quite successful (compare *left* and *right* columns). These results were typical: whenever the underlying assumptions about the sparseness of the stimulus were satisfied, sources consisting of mixtures of music, natural sounds or speech were separated well (Fig. 14.4 (A)). Separation worked particularly well for mixtures of sparsely representable sources (*i.e.*, smaller sparseness index values), whereas it did not work for sources that were not sparsely represented (*i.e.*, larger sparseness index values). Figure 14.4 (B) shows that separation without differential prefiltering by the HRTF was unsuccessful, as was separation using the Gaussian prior instead of the sparseness prior (dense representation: L_2-norm minimization).

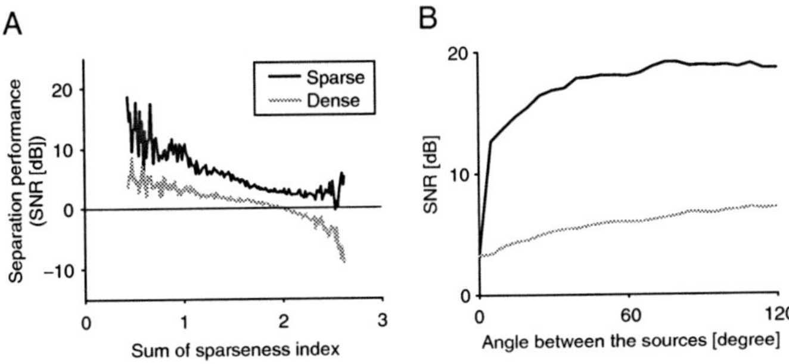

Fig. 14.4. Separation performance with three sources. **(A)** The separation performance (SNR in the TF domain averaged across the sources) is shown as a function of the sum of the "sparseness index" of the three sources (average over 20,000 sets of test sample mixtures). The sparse prior (*black*) always outperforms the dense prior (*gray*), and excellent separation was achieved especially when the sources are sparsely representable. The source locations were randomly chosen but 90° apart from each other with zero elevation. **(B)** Using a typical example of three novel stimuli (trumpet and two same harp), separation performance (y-axis) was examined with all the possible combinations of the three sources (from 0° to 120° apart; x-axis). The average performance is shown here under either the sparse (*black*) or the dense (*gray*) prior. Note that separation was unsuccessful at angle zero since we cannot exploit *differential* filtering, whereas the performance gets better as the sources get further apart. From [13], with permission.

The procedure for source separation in the BSS model conceptually consists of two distinct steps (although in practice the two steps occur simultaneously). In the first step, the stimuli are decomposed into the appropriate dictionary elements. In the second step, the dictionary elements are tagged and bundled together with other elements from the same source. It is for this bundling or "tagging" step that the HRTF along with the prior knowledge of source locations is essential.

The failure of the dense representation to separate sources (Fig. 14.4) results from a failure of the first step. Instead of decomposing the sources into a small number of dictionary elements, the dense representation assumes that each element contributed about equally to the received signal, and so finds a representation in which a large fraction of dictionary elements are involved. That is, instead of "explaining" the sources in terms of two harps and a trumpet, the dense representations also finds some clarinet, some cello, etc. at all positions. This is intrinsic to the dense solution, since it finds the "minimum power" solution in which neural activity is spread among the population (Fig. 14.1 (B)).

The failure of even the sparse approach when the spectral cues induced by the HRTF are absent (Fig. 14.4 (B), leftmost point showing 0° separation) results from a failure at the second step. That is, the sparse approach finds a useful decomposition at the first step even without the HRTF, but in the absence of HRTF cues the active elements are not tagged, and so the dictionary elements cannot be assigned appropriately to distinct sources. Other psychophysical cues relevant for source separation, such as common onset time, might provide alternative or additional tags in this same framework. A more general formulation of source separation might allow tagging on longer time scales, so that a set of dictionaries active at one moment might be more (or less) likely to be active the next, reflecting the fact that sources tend to persist, but we do not pursue that approach further here.

14.7 Conclusion

Sparse overcomplete representations can monaurally separate a mixture of sound sources into its constituent auditory streams. In our framework it is critical to use an appropriate overcomplete basis in order to achieve acceptable separation performance, and we described one way to exploit inherent properties of source signals for finding discriminative or "personalized" dictionaries that allow sparse representations of particular sound ensembles. We also modified the separation model to instead exploit one type of monaural separation cues that animals use, the HRTF, to "tag" dictionary elements so they can be assigned to the appropriate sources. We expect that other psychophysical cues important for acoustic stream segregation, such as common onset time, could be used in a similar way.

Recent advances in ICA have emphasized the utility of sparse overcomplete representations for source separation problems in acoustic, visual and other domains [25, 28, 35, 36, 60–62]. Our formulation of the source separation problem has been built on these ideas, generalizing the framework to allow cues other than differential attenuation. (It does not, however, sacrifice the ability to use binaural cues like differential attenuation, since such cues can also be incorporated by simply replacing the single-input-single-output HRTF filters by single-input-two-output filters, doubling the size of the dictionary elements by leaving the algorithm otherwise unchanged.) We have demonstrated the power and flexibility of the framework by applying it to two difficult monaural separation problems, one using as its sole cue differential low-level source models for the speakers, the other using as its sole cue the differential filtering of different sources by the HRTF.

Sparseness provides a powerful and useful constraint for choosing a unique representation in an overcomplete basis. We think that sparse representations can be a generic model for signal processing even in control theory or statistics as well as in neuroscience, and further advances in optimization and learning algorithms will find out its practical usages in many aspects, including the cocktail party problem in more general settings.

References

1. S. B. Laughlin and T. J. Sejnowski, "Communication in neuronal networks," *Science*, vol. 301, no. 5641, pp. 1870–1874, 2003.
2. W. B. Levy and R. A. Baxter, "Energy efficient neural codes," *Neural Computation*, vol. 8, no. 3, pp. 531–543, 1996.
3. E. P. Simoncelli and B. A. Olshausen, "Natural image statistics and neural representation," *Annual Review of Neurosci.*, vol. 24, no. 1, pp. 1193–1216, 2001.
4. M. S. Lewicki, "Efficient coding of natural sounds," *Nature Neuroscience*, vol. 5, no. 4, pp. 356–363, 2002.
5. E. C. Smith and M. S. Lewicki, "Efficient auditory coding," *Nature*, vol. 439, pp. 978–982, 2006.
6. A. J. Bell and T. J. Sejnowski, "The 'independent components' of natural scenes are edge filters," *Vision Research*, vol. 37, no. 23, pp. 3327–3338, 1997.
7. B. A. Olshausen and D. J. Field, "Emergence of simple-cell receptive field properties by learning a sparse code for natural images," *Nature*, vol. 381, pp. 607–609, 1996.
8. M. D. Plumbley, S. A. Abdallah, J. P. Bello, M. E. Davies, G. Monti, and M. B. Sandler, "Automatic music transcription and audio source separation," *Cybernetics and Systems*, vol. 33, no. 6, pp. 603–627, 2002.
9. L. Benaroya, L. M. Donagh, F. Bimbot, and R. Gribonval, "Non negative sparse representation for wiener based source separation with a single sensor," in *Acoustics, Speech, and Signal Processing*, 2003, pp. 613–616.
10. T. Virtanen, "Sound source separation using sparse coding with temporal continuity objective," in *ICMC*, 2003.
11. G. J. Jang and T. W. Lee, "A maximum likelihood approach to single channel source separation," *Journal of Machine Learning Research*, vol. 4, pp. 1365–1392, 2003.
12. A. S. Bregman, *Auditory Scene Analysis: The Perceptual Organization of Sound*, Cambridge, Massachusetts: MIT Press, 1990.
13. H. Asari, B. A. Pearlmutter, and A. M. Zador, "Sparse representations for the cocktail party problem," *Journal Neuroscience*, vol. 26, no. 28, pp. 7477–7490, 2006.
14. B. A. Pearlmutter and R. K. Olsson, "Linear program differentiation for single-channel speech separation," in *International Workshop on Machine Learning for Signal Processing*, Maynooth, Ireland: IEEE Press, 2006.
15. S. W. Hainsworth, "Techniques for the automated analysis of musical audio," Ph.D. dissertation, Department of Engineering, University of Cambridge, 2004.
16. B. A. Olshausen and D. J. Field, "Sparse coding with an overcomplete basis set: A strategy employed by V1?" *Vision Research*, vol. 37, no. 23, pp. 3311–3325, 1997.
17. O. Schwartz and E. P. Simoncelli, "Natural signal statistics and sensory gain control," *Nature Neuroscience*, vol. 4, no. 8, pp. 819–825, 2001.
18. D. Klein, P. Konig, and K. P. Kording, "Sparse spectrotemporal coding of sounds," *Journal on Applied Signal Processing*, vol. 7, pp. 659–667, 2003.
19. S. S. Stevens, "On the phychophysical law," *Psychological Review*, vol. 64, pp. 153–181, 1957.

20. D. D. Lee and H. S. Seung, "Learning the parts of objects with nonnegative matric factorization," *Nature*, vol. 401, pp. 788–791, 1999.
21. P. Smaragdis, "Non-negative matrix factor deconvolution; extraction of multiple sound sources from monophonic inputs," in *Fifth International Conference on Independent Component Analysis*, ser. LNCS 3195. Granada, Spain: Springer-Verlag, 2006, pp. 494–499.
22. P. Smaragdis, "Convolutive speech bases and their application to supervised speech separation," *IEEE Transactions on Speech Audio Processing*, vol. 15, no. 1, pp. 1–12, 2007.
23. T. Virtanen, "Techniques for the automated analysis of musical audio," Ph.D. dissertation, Institute of Signal Processing, Tampere University of Technology, 2006.
24. H. B. Barlow, "Possible principles underlying the transformations of sensory messages," in *Sensory Communication*, W. A. Rosenblith, Ed. MIT Press, 1961, pp. 217–234.
25. M. S. Lewicki and T. J. Sejnowski, "Learning overcomplete representations," *Neural Computation*, vol. 12, no. 2, pp. 337–365, 2000.
26. K. Kreutz-Delgado, J. F. Murray, B. D. Rao, K. Engan, T.-W. Lee, and T. J. Sejnowski, "Dictionary learning algorithms for sparse representation," *Neural Computation*, vol. 15, no. 2, pp. 349–396, 2003.
27. D. L. Donoho and M. Elad, "Maximal sparsity representation via $l1$ minimization," *Proceeding of the National Academy of Sciences USA*, vol. 100, pp. 2197–2202, 2003.
28. Y. Li, A. Cichocki, and S. Amari, "Analysis of sparse representation and blind source separation," *Neural Computation*, vol. 16, no. 6, pp. 1193–1234, 2004.
29. R. Fletcher, "Semidefinite matrix constraints in optimization," *SIAM Journal on Control and Optimization*, vol. 23, pp. 493–513, 1985.
30. O. L. Mangasarian, W. N. Street, and W. H. Wolberg, "Breast cancer diagnosis and prognosis via linear programming," *Operations Research*, vol. 43, no. 4, pp. 570–577, 1995.
31. P. S. Bradley, O. L. Mangasarian, and W. N. Street, "Clustering via concave minimization," in *Advances in Neural Information Processing Systems 9*. MIT Press, 1997, pp. 368–374.
32. I. F. Gorodnitsky and B. D. Rao, "Sparse signal reconstruction from limited data using FOCUSS: A re-weighted minimum norm algorithm," *IEEE Transactions on Signal Processing*, vol. 45, no. 3, pp. 600–616, 1997.
33. M. Lewicki and B. A. Olshausen, "Inferring sparse, overcomplete image codes using an efficient coding framework," in *Advances in Neural Information Processing Systems 10*. MIT Press, 1998, pp. 815–821.
34. S. S. Chen, D. L. Donoho, and M. A. Saunders, "Atomic decomposition by basis pursuit," *SIAM Journal on Scientific Computing*, vol. 20, no. 1, pp. 33–61, 1998.
35. T.-W. Lee, M. S. Lewicki, M. Girolami, and T. J. Sejnowski, "Blind source separation of more sources than mixtures using overcomplete representations," *IEEE Signal Processing Letters*, vol. 4, no. 5, pp. 87–90, 1999.
36. M. Zibulevsky and B. A. Pearlmutter, "Blind source separation by sparse decomposition in a signal dictionary," *Neural Computation*, vol. 13, no. 4, pp. 863–882, 2001.
37. B. A. Pearlmutter and A. M. Zador, "Monaural source separation using spectral cues," in *Fifth International Conference on Independent Component Analysis*, ser. LNCS 3195. Granada, Spain: Springer-Verlag, 2004, pp. 478–485.

38. G. B. Dantzig, "Programming in a linear structure," 1948, uSAF, Washington D.C.
39. S. I. Gass, *An Illustrated Guide to Linear Programming.* McGraw-Hill, 1970.
40. R. Dorfman, "The discovery of linear programming," *Annals of the History of Computing*, vol. 6, no. 3, pp. 283–295, 1984.
41. A. Griewank, *Evaluating Derivatives: Principles and Techniques of Algorithmic Differentiation*, ser. Frontiers in Appl. Math. Philadelphia, PA: SIAM, 2000, no. 19.
42. R. E. Wengert, "A simple automatic derivative evaluation program," *Communications of the ACM*, vol. 7, no. 8, pp. 463–464, 1964.
43. B. Speelpenning, "Compiling fast partial derivatives of functions given by algorithms," Ph.D. dissertation, Department of Computer Science, University of Illinois, Urbana-Champaign, 1980.
44. D. E. Rumelhart, G. E. Hinton, and R. J. Williams, "Learning representations by back–propagating errors," *Nature*, vol. 323, pp. 533–536, 1986.
45. A. J. Bell and T. J. Sejnowski, "An information-maximization approach to blind separation and blind deconvolution," *Neural Computation*, vol. 7, no. 6, pp. 1129–1159, 1995.
46. H. Robbins and S. Monro, "A stochastic approximation method," *Annals of Mathematical Statistics*, vol. 22, pp. 400–407, 1951.
47. S. T. Roweis, "One microphone source separation," in *Advances in Neural Information Processing Systems*, 2001, pp. 793–799.
48. T. Kristjansson, J. Hershey, P. Olsen, S. Rennie, and R. Gopinath, "Super-human multi-talker speech recognition," in *ICSLP*, 2006.
49. F. Bach and M. I. Jordan, "Blind one-microphone speech separation: A spectral learning approach," in *Advances in Neural Information Processing Systems 17*, 2005, pp. 65–72.
50. M. P. Cooke, J. Barker, S. P. Cunningham, and X. Shao, "An audio-visual corpus for speech perception and automatic speech recognition," *Journal of the Acoustical Society of America*, vol. 120, pp. 2421–2424, 2006.
51. D. P. W. Ellis and R. J. Weiss, "Model-based monaural source separation using a vector-quantized phase-vocoder representation," in *ICASSP*, 2006.
52. W. A. Yost, R. H. Dye, Jr., and S. Sheft, "A simulated "cocktail party" with up to three sound sources," *Perception and Psychophysics*, vol. 58, no. 7, pp. 1026–1036, 1996.
53. F. L. Wightman and D. J. Kistler, "Headphone simulation of free-field listening. II: Psychophysical validation," *Journal of the Acoustical Society of America*, vol. 85, no. 2, pp. 868–878, 1989.
54. A. Kulkarni and H. S. Colburn, "Role of spectral detail in sound-source localization," *Nature*, vol. 396, no. 6713, pp. 747–749, 1998.
55. E. I. Knudsen and M. Konishi, "Mechanisms of sound localization in the barn owl," *Journal of Comparative Physiology*, vol. 133, pp. 13–21, 1979.
56. E. M. Wenzel, M. Arruda, D. J. Kistler, and F. L. Wightman, "Localization using nonindividualized head-related transfer functions," *Journal of the Acoustical Society of America*, vol. 94, no. 1, pp. 111–123, 1993.
57. P. M. Hofman and A. J. V. Opstal, "Bayesian reconstruction of sound localization cues from responses to random spectra," *Biological Cybernetics*, vol. 86, no. 4, pp. 305–316, 2002.
58. H. Attias and C. Schreiner, "Temporal low-order statistics of natural sounds," in *Advances in Neural Information Processing Systems*, 1997.

59. T. Nishino, Y. Nakai, K. Takeda, and F. Itakura, "Estimating head related transfer function using multiple regression analysis," *IEICE Transactions A*, vol. J84-A, no. 3, pp. 260–268, 2001, in Japanese.
60. H. Farid and E. H. Adelson, "Separating reflections from images by use of independent components analysis," *Journal of Optical Society of America*, vol. 16, no. 9, pp. 2136–2145, 1999.
61. S. T. Rickard and F. Dietrich, "DOA estimation of many W-disjoint orthogonal sources from two mixtures using DUET," in *Proceedings of the 10th IEEE Workshop on Statistical Signal and Array Processing (SSAP2000)*, Pocono Manor, PA, 2000, pp. 311–314.
62. A. Levin and Y. Weiss, "User assisted separation of reflections from a single image using a sparsity prior," in *Proceedings of the European Conference on Computer Vision (ECCV)*, Prague, 2004.

15 Monaural Speech Separation by Support Vector Machines: Bridging the Divide Between Supervised and Unsupervised Learning Methods

Sepp Hochreiter[1] and Michael C. Mozer[2]

[1] Institute of Bioinformatics
Johannes Kepler University
4040 Linz, Austria

[2] Department of Computer Science
University of Colorado
Boulder, CO 80309

Abstract. We address the problem of identifying multiple independent speech sources from a single signal that is a mixture of the sources. Because the problem is ill-posed, standard independent component analysis (ICA) approaches which try to invert the mixing matrix fail. We show how the unsupervised problem can be transformed into a supervised regression task which is then solved by support-vector regression (SVR). It turns out that the linear SVR approach is equivalent to the sparse-decomposition method proposed by [1, 2]. However, we can extend the method to *nonlinear ICA* by applying the "kernel trick." Beyond the kernel trick, the SVM perspective provides a new interpretation of the sparse-decomposition method's hyperparameter which is related to the input noise. The limitation of the SVM perspective is that, for the nonlinear case, it can recover only whether or not a mixture component is present; it cannot recover the strength of the component. In experiments, we show that our model can handle difficult problems and is especially well suited for speech signal separation.

15.1 Introduction

Independent component analysis (ICA) [3–5] attempts to recover multiple source signals from one or more observed signals that are mixtures of the sources. ICA algorithms rely on the assumption that the sources are statistically independent, and most make the further assumption that mixtures are linear combinations of the sources. Well known ICA methods like "infomax" [6, 7], maximum likelihood approaches [8], entropy and cumulant based methods [9–11] have the restriction that the number of source and mixture signals must be equal. The restriction stems from the fact that these methods recover the sources by inverting the mixing matrix. To avoid ill-posed problems, the inversion requires at least as many observations as sources.

In many real world applications, only one observation is available. Speech processing is an important example in which this situation is often true. However, many nonspeech acoustic signals, such as bird songs, music, and

S. Makino et al. (eds.), Blind Speech Separation, 411–428.

traffic, are recorded with only one or two microphones. Other technological applications are based on a single mixture time series, such as mobile communication signals with direct sequence code division multiple access (DS-CDMA) [12]. Standard ICA approaches cannot be used in these cases.

The human auditory system is unmatched in its ability to robustly distinguish among multiple sound sources from two mixtures—the ears. It can even extract sources from monaural recordings. In some cases, separation of speech signals is easy because the signals occur in different frequency bands. However, when simple physical characteristics do not distinguish the sources, the task is extremely difficult.

As one realizes when listening to an orchestra, the human auditory system is able to separate complex mixtures from just two sources. The conductor is able to isolate individual melody lines, instruments, or even musicians from the ensemble, whereas a naive audience member may not. The difference between the conductor and the audience member is the conductor's knowledge and familiarity with the sound patterns that constitute the performance. One could even imagine that the conductor has a dictionary of sound *atoms*—canonical or prototypical musical phrases and timbres—and identification of components comes by isolating the atoms from the mixture. This argument carries over to speech processing. Our conjecture is that people can separate speech so well because of the accumulation of speech experience since birth. These experiences allow for the construction of a dictionary of sounds corresponding to vowels, words, etc. The dictionary contains not only one entry per vowel or word but multiple entries which allow to distinguish female from male and help to identify words spoken in a dialect or foreign accent.

Several ICA approaches have adopted the idea of using a dictionary to extract multiple sources from a smaller number of mixtures, or even one mixture [1, 2, 13, 14]. The dictionary can be composed of primitive functions (e.g., Fourier bases, wavelet packages, or Gabor functions) [1, 2], it can be predefined based on prior knowledge, or it can be trained to fit the problem [14–17]. Zibulevsky and Pearlmutter [1, 2] specify not only a dictionary, but also a prior that enforces sparseness—i.e., an expectation as to how many sources will be present simultaneously. All these approaches are restricted to mixtures consisting of linearly superimposed dictionary atoms; this restriction is necessary to avoid ambiguity in the problem.

In this chapter we show that the *sparse-decomposition method* of Zibulevsky and Pearlmutter can be reinterpreted as ϵ-support vector regression (ϵ-SVR) [18–23] when there is a single mixture and a Laplacian prior.

By drawing the connection between ICA—an unsupervised learning task—and support-vector regression—traditionally used for supervised learning—we show how an unsupervised task can be transformed into a supervised framework. This transformation may be useful to other problems as well. The key idea of the transformation is to use a predefined set of examples—the dictionary atoms—for generating target values for the supervised learning framework. In our approach, each unsupervised example is paired with each atom, and a fixed function is used to generate a "target" value for the example, i.e., the target for the SVR. Typically, the target value might be a measure of similarity between the dictionary atom and the example, such as a dot product. Using this approach for speech signal processing, which may contain a mixture of multiple speakers, we break the input signal into short segments which serve as the unsupervised examples. Using a prespecified dictionary of atomic speech signals [16], we use our method to determine, for each segment, which combination of the atoms is present. Our approach may also be useful for DS-CDMA mobile communication, where the dictionary consists of spreading sequences of the users.

The ϵ-SVR framework allows us to obtain sparse solutions. The sparsity constraint reflects the fact that at any point in the input signal, only a few mixture components (i.e., a few speakers) should be present. Within the SVR framework, support vectors will correspond to exactly those dictionary atoms that are correlated with the input signal, but which are mutually decorrelated. The support-vector machine determines which target values are spurious, which are produced by superimposition, and which indicate that a dictionary entry is present in the input signal.

The idea of applying SVR to unsupervised problems was also suggested for filtering [24], and recently for estimating missing values in cDNA microarrays [25].

By drawing an analogy between ϵ-SVR and the sparse-decomposition method for ICA, we obtain a new interpretation to the sparse-decomposition method's hyperparameter that determines the degree of sparseness. Further, the analogy yields a generalization of the sparse-decomposition method to allow for nonlinear transformations of the sources before they are mixed, and to allow for a further nonlinear transformation in the process of identifying dictionary atoms in the mixture.

We demonstrate our approach with experiments using noisy mixtures of speech with a single microphone. As we show, our approach incorporates nonlinear transformations of the dictionary atoms that achieve a degree of robustness and invariance to irrelevant characteristics of the speech signal. Examples of transformations and the corresponding invariants that we consider include: taking the local variance of the absolute value or square of the waveform to reduce sensitivity to the sign of the waveform combined with adding a constant value to the waveform (a nonlinear transformation, and a nonlinear invariant); or the power spectrum which is invariant to temporal shifts of the waveform (linear transformation and nonlinear invariant). These transformations are highly relevant for speech separation where, for example, shift invariance avoids the need to segment the speech signal.

15.2 Viewing Sparse Decomposition as ϵ-SVR

In this section, we review the sparse-decomposition method introduced by Zibulevsky and Pearlmutter [1], focusing on the case of a single mixture signal. We will also describe the relation between sparse decomposition and ϵ-support vector regression (ϵ-SVR).

15.2.1 The Sparse-Decomposition Method

Denote the mixture signal by $\boldsymbol{x} \in \mathbb{R}^L$. In the case of speech, the signal might correspond to a window of L consecutive samples from a longer input stream. We assume a dictionary matrix, $\boldsymbol{S} \in \mathbb{R}^{L \times P}$, whose columns consist of the P atomic signals of length L. We assume a generative process in which the mixture is created by first choosing a set of dictionary atoms and then combining them linearly with noise:

$$\boldsymbol{x} = \boldsymbol{S}\,\boldsymbol{c} + \boldsymbol{\nu} = \sum_{i=1}^{P} c_i\, \boldsymbol{s}^i + \boldsymbol{\nu}, \tag{15.1}$$

where $\boldsymbol{c} \in \mathbb{R}^P$ is a vector of weighting coefficients, one per atom, $\boldsymbol{\nu} \sim N\left(0, \sigma^2\right)$ is an L-dimensional i.i.d. additive Gaussian noise vector, and $\boldsymbol{s}^i$ the ith atom in the dictionary: $\boldsymbol{s}^i = [\boldsymbol{S}]_i$. Figure 15.1 illustrates the generative process that produces the mixture.

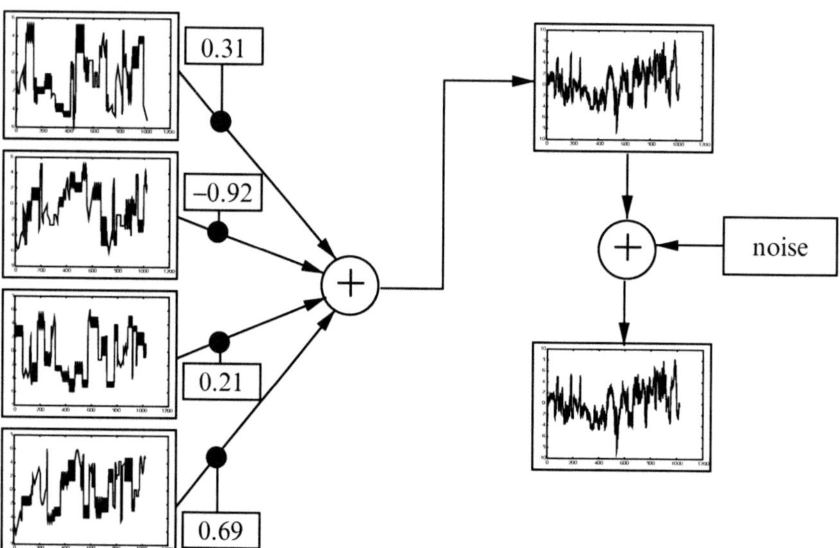

Fig. 15.1. The data generation process. Four atoms from the dictionary $\boldsymbol{S}$ are weighted by real nonzero coefficients and are added together with noise resulting in the mixture. The goal is to recover the nonzero weighting factors, or at least to detect whether an atom is present in the mixture.

The goal of the sparse-decomposition method is to determine the coefficient vector, $\hat{\boldsymbol{c}}$, that satisfies two properties: (I) $\boldsymbol{x}$ must be well approximated by $\hat{\boldsymbol{c}}$, and (II) $\hat{\boldsymbol{c}}$ is sparse, i.e., it has elements of small magnitude. These two properties are achieved by a Bayesian approach in which property 1 is characterized by the likelihood, $p(\boldsymbol{x} \mid \boldsymbol{c}, \boldsymbol{S})$, and property 2 is characterized by the prior, $p(\boldsymbol{c})$. Thus, the approach tries to maximize the posterior

$$p(\boldsymbol{c} \mid \boldsymbol{x}, \boldsymbol{S}) \propto p(\boldsymbol{x} \mid \boldsymbol{c}, \boldsymbol{S})\ p(\boldsymbol{c}),$$

where we use "$\propto$" because we omit the constant normalization factor in the denominator of Bayes rule. Given the Gaussian noise model, the likelihood is

$$p(\boldsymbol{x} \mid \boldsymbol{c}, \boldsymbol{S}) \propto \exp\left(-\frac{1}{2\,\sigma^2}\left(\boldsymbol{x} - \boldsymbol{S}\,\boldsymbol{c}\right)^2\right).$$

To enforce sparseness of the coefficients, a Laplacian prior for $\boldsymbol{c}$ is used with $\|\boldsymbol{c}\|_1 = \sum_{i=1}^{P} |c_i|$ we have:

$$p(\boldsymbol{c}) \propto \exp\left(-\frac{\epsilon}{\sigma^2}\,\|\boldsymbol{c}\|_1\right).$$

Consequently, the posterior is

$$p(\boldsymbol{c} \mid \boldsymbol{x}, \boldsymbol{S}) \propto \exp\left(-\frac{1}{\sigma^2}\left[\frac{1}{2}\left(\boldsymbol{x} - \boldsymbol{S}\,\boldsymbol{c}\right)^2 + \epsilon\,\|\boldsymbol{c}\|_1\right]\right).$$

The solution, $\hat{\boldsymbol{c}}$, is obtained by maximum a posteriori (MAP) search. Taking the log of the posterior, flipping its sign, and ignoring irrelevant constant terms and factors, we obtain the minimization problem

$$\hat{\boldsymbol{c}} = \arg\min_{\boldsymbol{c}}\ \frac{1}{2}\left(\boldsymbol{x} - \boldsymbol{S}\,\boldsymbol{c}\right)^2 + \epsilon \sum_{i=1}^{P} |c_i|.$$

This unconstrained optimization problem can be turned into a constrained optimization problem in which $\boldsymbol{c}$ is split into two vectors such that $\boldsymbol{c} = \boldsymbol{c}^+ - \boldsymbol{c}^-$, where $\boldsymbol{c}^+$ and $\boldsymbol{c}^-$ contain the magnitudes of the positive and negative coefficients of $\boldsymbol{c}$, respectively. The MAP solution $\{\hat{\boldsymbol{c}}^+, \hat{\boldsymbol{c}}^-\}$ is

$$\begin{aligned}\arg\min_{\boldsymbol{c}^+,\boldsymbol{c}^-}\ & \frac{1}{2}\left(\boldsymbol{c}^+ - \boldsymbol{c}^-\right)^T \boldsymbol{S}^T\boldsymbol{S}\left(\boldsymbol{c}^+ - \boldsymbol{c}^-\right) - \\ & \boldsymbol{x}^T\boldsymbol{S}\left(\boldsymbol{c}^+ - \boldsymbol{c}^-\right) + \epsilon\,\boldsymbol{1}^T\left(\boldsymbol{c}^+ + \boldsymbol{c}^-\right) \\ \text{s.t.}\ & 0 \le c_i^+, c_i^- \le C, \end{aligned} \tag{15.2}$$

where T is the transposition operator, $\boldsymbol{1}$ is the vector of ones, and C is an upper bound that can serve as an additional constraint on the solution (which was not part of the original formulation by Zibulevsky and Pearlmutter).

We will show that *this formulation is ϵ-support vector regression (ϵ-SVR)* [18], but in order to do so, we must first give a brief overview of ϵ-SVR.

15.2.2 ϵ-Support Vector Regression

ϵ-SVR is a supervised approach to regression in which we are given training data $\{(\boldsymbol{s}^i, y_i), \ldots, (\boldsymbol{s}^P, y_P)\}$, where $\boldsymbol{s}^i \in \mathbb{R}^L$ and y_i is a scalar. The goal is to produce a function, h, such that $h(\boldsymbol{s}^i)$ closely approximates y_i. In the linear formulation of ϵ-SVR, $h(\boldsymbol{s}^i) = \langle \boldsymbol{w}, \boldsymbol{s}^i \rangle$, where $\boldsymbol{w} \in \mathbb{R}^L$, and $\langle .,. \rangle$ denotes the dot product. In the nonlinear formulation, $h(\boldsymbol{s}^i) = \langle \boldsymbol{w}, \boldsymbol{\theta}(\boldsymbol{s}^i) \rangle$, where $\boldsymbol{\theta}$ is a mapping from $\mathbb{R}^L$ to $\mathbb{R}^L$.

Note we consider regression functions through the origin. Consequently, the formulation excludes a constant offset term. The ϵ-SVR attempts to obtain a "flat" function by minimizing $\frac{1}{2}\|\boldsymbol{w}\|^2$, but subject to the constraint that the fit is good enough, as quantified by the constraint

$$\left| y_i \ - \ h(\boldsymbol{s}^i) \right| \ < \ \epsilon \ + \ \xi_i \tag{15.3}$$

for all i. The parameter ϵ is a measure of how accurate the fit needs to be, or intuitively, a measure of the noise in the data. The *slack variables* $\xi_i \geq 0$ allow for the fact that it may not be possible to find an h that satisfies the ϵ accuracy criterion.

$$\xi_i \ = \ \max\{0, \ \left| y_i \ - \ h(\boldsymbol{s}^i) \right| \ - \ \epsilon\}.$$

However, to ensure that the deviations are minimal, the optimization attempts to minimize the magnitude of the slack variables as well. Specifically, the constrained optimization is over the objective function

$$\frac{1}{2}\|\boldsymbol{w}\|^2 + C \ \|\boldsymbol{\xi}\|_1,$$

where C determines the trade off between the flatness of the function and the tolerance of prediction errors and

$$\|\boldsymbol{\xi}\|_1 = \sum_i \xi_i = \sum_i \max\{0 \ , \ \left| y_i - h(\boldsymbol{s}^i) \right| - \epsilon\} = \left| y_i - h(\boldsymbol{s}^i) \right|_\epsilon . \tag{15.4}$$

We divide (15.3) into an inequality set for which $\left| y_i - h(\boldsymbol{s}^i) \right| = y_i - h(\boldsymbol{s}^i)$, where ξ_i in (15.3) is denoted by ξ_i^+, and into an inequality set for which $\left| y_i \ - \ h(\boldsymbol{s}^i) \right| \ = \ h(\boldsymbol{s}^i) \ - \ y_i$, where ξ_i in (15.3) is denoted by ξ_i^-. We obtain for the linear case $h(\boldsymbol{s}^i) \ = \ \boldsymbol{w}^T \boldsymbol{s}^i$ as ϵ-SVR optimization problem

$$\begin{aligned} \min_{\boldsymbol{w}, \boldsymbol{\xi}^+, \boldsymbol{\xi}^-} \quad & \frac{1}{2} \ \|\boldsymbol{w}\|^2 \ + \ C \sum_{i=1}^{l} (\xi_i^+ \ + \ \xi_i^-) \\ \text{s.t.} \quad & y_i \ - \ \boldsymbol{w}^T \boldsymbol{s}^i \ \leq \ \epsilon \ + \ \xi_i^+, \\ & \boldsymbol{w}^T \boldsymbol{s}^i \ - \ y_i \ \leq \ \epsilon \ + \ \xi_i^-, \\ & \xi_i^+ \ \geq \ 0 \ \text{ and } \ \xi_i^- \ \geq \ 0, \end{aligned} \tag{15.5}$$

It is known that the primal ϵ-SVR optimization problem (15.5) possesses an alternative but entirely equivalent formulation, the *dual* formulation:

$$\begin{aligned} \min_{\boldsymbol{c}^+,\boldsymbol{c}^-} \; & \frac{1}{2} \left(\boldsymbol{c}^+ \; - \; \boldsymbol{c}^-\right)^T \boldsymbol{S}^T \boldsymbol{S} \left(\boldsymbol{c}^+ \; - \; \boldsymbol{c}^-\right) \; - \\ & \boldsymbol{y}^T \left(\boldsymbol{c}^+ \; - \; \boldsymbol{c}^-\right) \; + \; \epsilon \; \boldsymbol{1}^T \left(\boldsymbol{c}^+ \; + \; \boldsymbol{c}^-\right) \\ \text{s.t.} \;\; & 0 \leq c_i^+, c_i^- \leq C, \end{aligned} \tag{15.6}$$

where $\boldsymbol{S}$ is the matrix formed by $\boldsymbol{s}^i$: $\boldsymbol{s}^i \;=\; [\boldsymbol{S}]_i$. The coefficient c_i are the Lagrange multipliers for the primal constraints in (15.5) which are split into positive and negative components, c_i^+ and c_i^-: $c_i = c_i^+ - c_i^-$. The $\boldsymbol{s}^i$ for which $c_i \neq 0$ are called *support vectors* or *support sources* or *support signals.* Note that

$$\left(\boldsymbol{c}^+ \; - \; \boldsymbol{c}^-\right)^T \boldsymbol{S}^T \boldsymbol{S} \left(\boldsymbol{c}^+ \; - \; \boldsymbol{c}^-\right) \;=\; \frac{1}{2} \sum_{i,j=1}^{P} c_i \, c_j \left\langle \boldsymbol{s}^i, \boldsymbol{s}^j \right\rangle .$$

For the nonlinear formulation we define the Gram matrix $\boldsymbol{K}$ as

$$K_{ij} = k(\boldsymbol{s}^i, \boldsymbol{s}^j) \;=\; \left\langle \boldsymbol{\theta}\left(\boldsymbol{s}^i\right), \boldsymbol{\theta}\left(\boldsymbol{s}^j\right)\right\rangle .$$

The Gram matrix can be written as $\boldsymbol{K} \;=\; \boldsymbol{\theta}(\boldsymbol{S})^T \boldsymbol{\theta}(\boldsymbol{S})$, where $\boldsymbol{\theta}(\boldsymbol{S})$ is the matrix formed by $\boldsymbol{\theta}(\boldsymbol{S})_i \;=\; \boldsymbol{\theta}(\boldsymbol{s}^i)$. The nonlinear dual formulation is therefore

$$\begin{aligned} \min_{\boldsymbol{c}^+,\boldsymbol{c}^-} \; & \frac{1}{2} \left(\boldsymbol{c}^+ \; - \; \boldsymbol{c}^-\right)^T \boldsymbol{K} \left(\boldsymbol{c}^+ \; - \; \boldsymbol{c}^-\right) \; - \\ & \boldsymbol{y}^T \left(\boldsymbol{c}^+ \; - \; \boldsymbol{c}^-\right) \; + \; \epsilon \; \boldsymbol{1}^T \left(\boldsymbol{c}^+ \; + \; \boldsymbol{c}^-\right) \\ \text{s.t.} \;\; & 0 \leq c_i^+, c_i^- \leq C. \end{aligned} \tag{15.7}$$

It can be seen that in (15.6) only $\boldsymbol{S}^T \boldsymbol{S}$ is replaced by $\boldsymbol{K}$ to obtain a nonlinear formulation.

As a consequence of the transformation from the primal to the dual formulation, the primal vector $\boldsymbol{w}$ can be expressed through the coefficients c_i and the training data $\boldsymbol{s}^i$:

$$\boldsymbol{w} = \sum_{i=1}^{P} c_i \; \boldsymbol{s}^i$$

for the linear version and

$$\boldsymbol{w} = \sum_{i=1}^{P} c_i \; \boldsymbol{\theta}\left(\boldsymbol{s}^i\right)$$

for the nonlinear version. Consequently, for the linear formulation

$$h(\boldsymbol{s}) = \sum_{i=1}^{P} c_i \left\langle \boldsymbol{s}^i, \boldsymbol{s}\right\rangle \;=\; \boldsymbol{s}^T \boldsymbol{S} \; \mathbf{c}.$$

For the nonlinear formulation we obtain

$$h(\boldsymbol{s}) = \sum_{i=1}^{P} c_i \left\langle \boldsymbol{\theta}\left(\boldsymbol{s}^i\right), \boldsymbol{\theta}\left(\boldsymbol{s}\right)\right\rangle = \boldsymbol{\theta}(\boldsymbol{s})^T \boldsymbol{\theta}(\boldsymbol{S})\ \boldsymbol{c} \tag{15.8}$$

which can reformulated with a kernel as

$$h(\boldsymbol{s}) = \sum_{i=1}^{P} c_i\ k\left(\boldsymbol{s}^i, \boldsymbol{s}\right) = \boldsymbol{k}(s, \boldsymbol{S})^T \boldsymbol{c},$$

where $\boldsymbol{k}(s, \boldsymbol{S})$ is the vector

$$\boldsymbol{k}(s, \boldsymbol{S}) = \boldsymbol{\theta}(\boldsymbol{s})^T \boldsymbol{\theta}(\boldsymbol{S}) = \left(k(s, s^1), k(s, s^2), \ldots, k(s, s^P)\right).$$

15.2.3 The Relationship Between the Sparse-Decomposition Method and ϵ-SVR

The sparse-decomposition method and ϵ-support vector regression are equivalent: Optimization problem (15.2) is *identical* to optimization problem (15.6). To spell out the mapping, consider framing the sparse-decomposition method as ϵ-SVR. The data for the ϵ-SVR consists of P training examples, where the input for example i is the dictionary atom $\boldsymbol{s}^i$, $\boldsymbol{s}^i \in \mathbb{R}^L$, and the target for the example, $\boldsymbol{y}_i$, is the dot product of the mixture $\boldsymbol{x}$ and dictionary atom $\boldsymbol{s}^i$: $y_i = \boldsymbol{x}^T \boldsymbol{s}^i = \left\langle \boldsymbol{x}, \boldsymbol{s}^i \right\rangle$.

The ϵ-SVR formulation gives an interpretation to the hyperparameter ϵ in the sparse-decomposition method. It is a measure of the noise level in the data, and it indirectly affects the number of $\hat{c}_i$ that are significantly non-zero. As depicted in Fig. 15.2, each example will have a target, y_i, that either

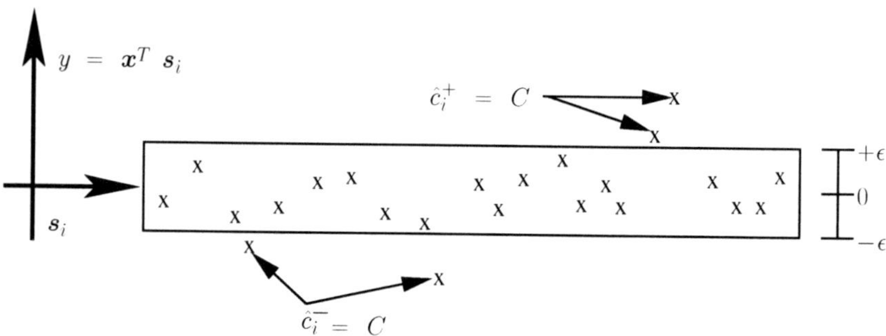

Fig. 15.2. The linear ϵ–support vector regression corresponding to the sparse-decomposition method. Each "x" in the figure denotes a single training example in the ϵ-SVR model. The horizontal axis is a one-dimensional depiction of the input space, and the vertical axis is the target output. The grey area, the ϵ-tube, specifies the range of target outputs that are not significantly different from zero. The examples i that lie outside the ϵ-tube will have $|\hat{c}_i| = C$.

lies inside or outside the ϵ-*tube*. The closer a target y_i is to zero, the more nearly orthogonal is the mixture $\boldsymbol{x}$ to atom $\boldsymbol{s}^i$, and the less likely atom i is to be present in the mixture. Thus, the ϵ-tube distinguishes atoms that are likely to be relevant from those likely to be irrelevant. It turns out that any example i lying outside the ϵ-*tube* will have either $\hat{c}_i = C$ or $\hat{c}_i = -C$. In the sparse-decomposition formulation, c_i indicates the degree to which a dictionary atom i is present.

15.3 Nonlinear Formulation

The ϵ-SVR framework also provides for a nonlinear approximation of $\boldsymbol{y}$ by introducing a nonlinear kernel $k(\boldsymbol{a}, \boldsymbol{b})$, where $\boldsymbol{a}, \boldsymbol{b} \in \mathbb{R}^L$. The dot products $\left\langle \boldsymbol{s}^i, \boldsymbol{s}^j \right\rangle$ in the ϵ-SVR are replaced by

$$k\left(\boldsymbol{s}^i, \boldsymbol{s}^j\right) = \left\langle \boldsymbol{\theta}\left(\boldsymbol{s}^i\right), \boldsymbol{\theta}\left(\boldsymbol{s}^j\right)\right\rangle,$$

or in matrix notation $\boldsymbol{S}^T\boldsymbol{S}$ is replaced by the kernel matrix $\boldsymbol{K}$ with $K_{ij} = k\left(\boldsymbol{s}^i, \boldsymbol{s}^j\right)$. The interpretation of this kernel is that the P vectors $\boldsymbol{s}^i$ are mapped by a function $\boldsymbol{\theta}$ into a feature space. The kernel is the dot product in the feature space: $k(\boldsymbol{a}, \boldsymbol{b}) = \langle \boldsymbol{\theta}(\boldsymbol{a}), \boldsymbol{\theta}(\boldsymbol{b})\rangle$.

Exploiting the correspondence we have established between the sparse-decomposition method and the ϵ-SVR, we can generalize the linear sparse-decomposition method to a nonlinear method by analogy to the nonlinear ϵ-SVR. We present two different nonlinear extensions of the sparse-decomposition method. In both extensions, the dictionary atoms $\boldsymbol{s}^i$ are mapped into a feature space using a nonlinear kernel. In the first extension we assume that the mixture $\boldsymbol{x}$ arises from a nonlinear combination of atoms $\boldsymbol{s}^j$. In the second extension we assume that mixture is linear but prior to being mixed, the atoms undergo a nonlinear transformation. We consider two subclasses of the second extension: (a) we directly approximate the nonlinear transformation by applying this transformation to the atom waveform sources as well, and (b) we extract invariances of the nonlinear transformation through the feature map which maps into a space of components which are invariant under the nonlinear transformation. In this space of invariances, the dot products extract similarities between the mixture and the dictionary atoms.

15.3.1 Nonlinear Mixtures

In our first extension, we assume that the mixture $\boldsymbol{x}$ results from a nonlinear combination of the atoms $\boldsymbol{s}^1, \boldsymbol{s}^2, \ldots, \boldsymbol{s}^P$:

$$\boldsymbol{x} = g\left(\boldsymbol{s}^1, \boldsymbol{s}^2, \ldots, \boldsymbol{s}^P\right).$$

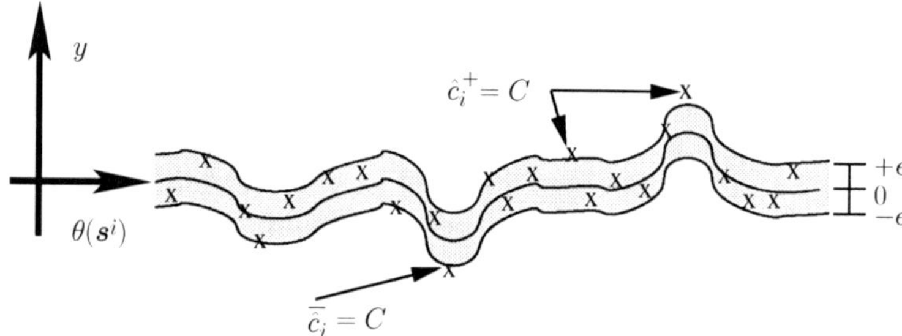

Fig. 15.3. The nonlinear ϵ-SVR. The dictionary atoms s^i are mapped by $\boldsymbol{\theta}$ into a feature space.

We further assume that the nonlinear mixture g can be expressed as a linear mixture in some feature space:

$$k\left(\boldsymbol{s}^i, \boldsymbol{x}\right) = k\left(\boldsymbol{s}^i, g\left(\boldsymbol{s}^1, \boldsymbol{s}^2, \ldots, \boldsymbol{s}^P\right)\right) = \sum_{j=1}^{P} k\left(\boldsymbol{s}^i, \boldsymbol{s}^j\right) c_j .$$

The feature space that allows for the linear mixing is defined by the mapping $\boldsymbol{\theta}(.)$, yielding the kernel k already mentioned above:

$$k\left(\boldsymbol{s}^i, \boldsymbol{s}^j\right) = \left\langle \boldsymbol{\theta}\left(\boldsymbol{s}^i\right), \boldsymbol{\theta}\left(\boldsymbol{s}^j\right)\right\rangle .$$

The target values for the ϵ-SVR are $y_i = k\left(\boldsymbol{s}^i, \boldsymbol{x}\right)$.

Finally we apply nonlinear support-vector regression to obtain $\hat{\boldsymbol{c}}$. The presence of dictionary atom i in the mixture is indicated by $\hat{c}_i = C$ or $\hat{c}_i = -C$ (see Fig. 15.3). Optimization of the kernel parameters for an SVR-kernel may be used to obtain $\boldsymbol{\theta}$. However, $\boldsymbol{\theta}$ cannot be estimated explicitly, because it is represented only implicitly via its correspondence with a kernel function.

15.3.2 Nonlinear Transformation of Atomic Sources

Another generalization of the sparse-decomposition method is obtained by allowing each atomic source $\boldsymbol{s}^i$ to be explicitly transformed by $\boldsymbol{\theta}$—a mapping from $\mathbb{R}^L$ into $\mathbb{R}^L$—before the sources are linearly superimposed to produce the mixture. In contrast to the previous section, where $\boldsymbol{\theta}$ was implicitly specified via a kernel function, here we must specify $\boldsymbol{\theta}$ explicitly. Given a choice of $\boldsymbol{\theta}$, the atomic sources are first transformed by $\boldsymbol{\theta}$, and the linear theory is applied to the transformed sources.

With the transformation $\boldsymbol{\theta}$, (15.1) becomes

$$\boldsymbol{x} = \sum_{j=1}^{P} c_i \, \boldsymbol{\theta}\left(\boldsymbol{s}^i\right) = \boldsymbol{\theta}\left(\boldsymbol{S}\right) \, \boldsymbol{c}.$$

Multiplying the equation from left with $\boldsymbol{\theta}(\boldsymbol{s})^T$ gives

$$\boldsymbol{\theta}(\boldsymbol{s})^T \boldsymbol{x} = \boldsymbol{\theta}(\boldsymbol{s})^T \boldsymbol{\theta}(\boldsymbol{S})\ \boldsymbol{c}.$$

This equation is exactly the definition of the nonlinear ϵ-SVR regression function, $h(\boldsymbol{S})$—see (15.8)—with $\boldsymbol{\theta}$ as the feature space mapping and the target

$$y_i = h(\boldsymbol{s}^i) = \boldsymbol{\theta}\left(\boldsymbol{s}^i\right)^T \boldsymbol{x}.$$

15.3.3 Nonlinear Transformed Atomic Sources: Inhomogeneous Case

In the previous section, we assumed that each atomic source $\boldsymbol{s}^i$ is transformed by the same function $\boldsymbol{\theta}$. Extending this *homogeneous* case, we now consider a more general *inhomogeneous* case in which each atomic source, $\boldsymbol{s}^i$, can be transformed by a distinct function, f_i (it will become clear shortly why we use the notation f_i rather than θ_i), where the $\{f_i\}$ represent a family of functions parameterized by a single variable τ_i, i.e., $f_i(\boldsymbol{s}) = g(\boldsymbol{s}; \tau_i)$. For example, if $\boldsymbol{x}$ represents a time series, g might shift the input in time by τ_i elements, and f_i would then correspond to a particular shift.

In the situation we consider, we wish to extract the *invariant feature*, e.g., a representation of the time series independent of its particular shift. Suppose that we can determine a function of the input space, $\boldsymbol{\theta}(.)$, that is invariant under and linear in f_i, i.e.,

$$\boldsymbol{\theta}(a\ f_i(\boldsymbol{s})) \propto \boldsymbol{\theta}(\boldsymbol{s}).$$

For example, if $\boldsymbol{\theta}$ computes the power spectrum, it would be invariant under shifts in the signal and $q(a) = a$. This function $\boldsymbol{\theta}$ then serves to extract the invariant feature from the signal.

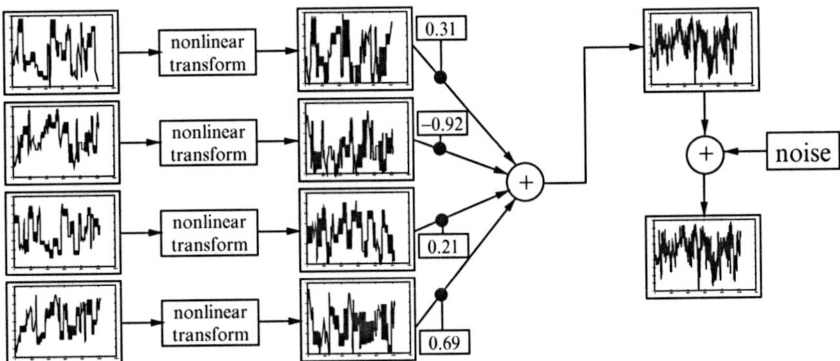

Fig. 15.4. The data generation process allowing for transformations of the atomic sources prior to mixing, in contrast with the original sparse-decomposition method (see Fig. 15.1).

Our goal is to approximate the feature of the mixture, $\boldsymbol{\theta}(\boldsymbol{x})$, in terms of a linear combination of the features of the atoms, i.e.,

$$\boldsymbol{\theta}(\boldsymbol{x}) = \sum_{i=1}^{P} t(c_i)\, \boldsymbol{\theta}\left(f_i\left(\boldsymbol{s}^i\right)\right) = \sum_{i=1}^{P} c_i\, \boldsymbol{\theta}\left(\boldsymbol{s}^i\right).$$

Under this formulation, the unknown transformations f_i are eliminated from the approximation problem. $\hat{c}_i$ indicates whether mixture $\boldsymbol{x}$ and atom $\boldsymbol{s}^i$ share the same features or not: large $\hat{c}_i$ implies that $\boldsymbol{x}$ and $\boldsymbol{s}^i$ are mapped to similar (correlated) feature vectors.

The approach in this subsection is equivalent to the approach in Sect. 15.3.1 if we set

$$k\left(\boldsymbol{s}^i, \boldsymbol{s}^j\right) = \left\langle \boldsymbol{\theta}\left(\boldsymbol{s}^i\right), \boldsymbol{\theta}\left(\boldsymbol{s}^j\right)\right\rangle.$$

However, here $\boldsymbol{\theta}$ is designed for the f_i and via $t(c_i)$, the magnitudes of the components c_i can be computed. In contrast, in Sect. 15.3.1, the c_i indicate only the presence of source atoms.

15.4 Experiments

15.4.1 Nonlinear Approximation of the Linear Correlations

We use a dictionary consisting of 64 atoms of length 128. The atoms vary in their frequencies and their shape (e.g., sinuidal, triangular, rectangular, and asymmetric triangular). On average, we chose 4.5 dictionary atoms to be present in one mixture. The nonzero coefficients c_i are randomly chosen from $[0.1, 1.0] \cup [-1.0, -0.1]$. To generate training data, we then (1) added Gaussian noise with variance 1.0 to each mixture component, (2) added noise frequencies to the whole mixture (sinuidal with amplitude form $[0; 0.2]$), and (3) shifted the phase of each atom by a random amount chosen from a uniform distribution corresponding to a phase shift of 0–20% of the period. The values for hyperparameters ϵ and C are chosen through a validation set. They are adjusted so that the average number of sources which are not recognized is below 0.5. In doing so, we impose an upper bound on the error that results from failing to detect sources that are present. To evaluate the performance, we count the number of sources which were wrongly detected.

The linear sparse-decomposition method leads to an error of 35.32. Table 15.1 shows the result for the nonlinear kernel $k(\boldsymbol{a}, \boldsymbol{b}) = (\gamma + \boldsymbol{a} \cdot \boldsymbol{b})^d$ with different values for γ and d. The nonlinear kernel obtains results comparable to the linear sparse-decomposition method. This simulation demonstrates that nonlinear kernels work even for linear problems. We assume the reason for the good performance of the nonlinear kernels is that they are more robust against noise or against the specific noise which is present in the data if appropriate parameters are chosen.

Table 15.1. Average over 100 trials of wrongly detected atoms (false positives) for kernels $k(\boldsymbol{a}, \boldsymbol{b}) = (\gamma + \boldsymbol{a} \cdot \boldsymbol{b})^d$ (the linear result was 35.32).

d	$\gamma = 10^4$	$\gamma = 10^5$	$\gamma = 10^6$	$\gamma = 10^7$
2	35.14	35.21	35.23	35.23
5	35.22	35.20	35.23	35.23
10	35.30	35.20	35.22	35.23
20	35.25	35.14	35.21	35.23
30	35.40	35.19	35.21	35.23

15.4.2 Transformed Atomic Sources

Artificial Data We consider a case where the mixture components have been corrupted by a class of componentwise transformations that produce nonnegative values: $[f(\boldsymbol{s})]_j = |s_j|^w$, where w is a scalar that parameterizes f. We wish to extract an invariance that mitigates the effect of w for any w. The *local variance* is a measure that achieves this goal. The local variance of $\boldsymbol{s}$ is defined as:

$$(2\ l\ +\ 1)^{-1} \sum_{t=j-l}^{j+l} (s_t - \bar{s}_j)^2,$$

where

$$\bar{s}_j := (2\ l\ +\ 1)^{-1} \sum_{t=j-l}^{j+l} s_t$$

and l is a parameter that characterizes the neighborhood locality.

The local variance serves as an invariance extractor ($\boldsymbol{\theta}$). For $w = 1$ the invariance is exact, because the curve is only mirrored at the $y = 0$ axis. The mirroring is just a shift of the mean signal, a base-line-shift, which occurs for many measurement devices (it is a huge problem for EEG measurements). The variance as a central moment corrects the means to zero and eliminates base-line-shifts. However in one signal the base-line-shifts may occur multiple, therefore the variance should be taken locally. Also local frequency measures are invariant to base-line-shifts but the local variance combines the frequency of a whole frequency band.

For $w > 1$, the local variance is very robust against values of w, and serves to define a similarity measure (signals which have high local variance in the same position may be similar). For our tests, we used three different neighborhood sizes: $l = 8$ (referred to as AV1), $l = 10$ (AV2), $l = 20$ (AV3).

We generated 100 dictionary atoms of length 1024. To produce an atom we segmented the 1024 length vector into random segments of length between 1 and 64. Each segment consists of a scaled (from $[-0.8, -0.2] \cup [0.2, 0.8]$) periodic function from previous experiment. To each segment component a

Table 15.2. Best linear results out of 100 trials.

Task	False negative	False positive
1: s_i^2	0.94	9.20
2: $\|s_i\|$	1.03	20.97
3: $\|s_i\|^w$	1.00	20.02

Table 15.3. Average over 100 mixtures of the number of wrongly detected atoms (false positives) for the linear sparse-decomposition method ("linear") and three different methods measuring the local variance ("AV1"–"AV3"). "Failed" means that we were not able to push the average number of undetected sources (false negatives) below 0.4.

Task	Linear	AV1	AV2	AV3
1: s_i^2	Failed	0.63	0.72	0.99
2: $\|s_i\|$	Failed	5.41	7.38	Failed
3: $\|s_i\|^w$	Failed	0.55	0.84	2.37

constant between $[-ac, +ac]$ is added. Figure 15.4 depicts the data generation process.

Task 1 uses $w = 1$, task 2 uses $w = 2$, and task 3 uses $|s_j|^w$ with w randomly chosen from $[0.5; 2.0]$ as transformation. We set $ac = 5.0$ for task 1 and $ac = 0.5$ for task 2 and 3. The transformations are mixed as in previous experiment and Gaussian noise has $\sigma = 0.01$. As in the previous experiment, we ensure that the average number of undetected sources falls below a bound which is here set to 0.4. The best linear results are shown in Table 15.2 and the average results of our new method are shown in Table 15.3. The nonlinear mapping by the local variance formulas was able to extract the invariant and, therefore, could classify an atom as being present or not. The linear model failed at the task.

The results show that extracting invariant features (here the local variance) and measuring similarities between these invariant features instead of the original waveforms can help to robustly detect sources in a one-dimensional mixture. However prior knowledge is necessary to extract appropriate features.

Speech Data We considered transformations that shift the dictionary atoms, where each atom can have a different shift. As an invariant we use the power spectrum. The dictionary entries are 5 spoken words ("hid", "head", "had", "hud", and "hod") spoken by 20 different speakers, yielding a dictionary size of 100 atoms. The data was obtained from `areas/speech/database/hvd/` in the AI-Repository at `cs.cmu.edu`. The speech is sampled at 10 kHz.

Table 15.4. Average number of wrongly detected atoms for the linear sparse-decomposition method ("linear"), and three nonlinear transformation into the power spectrum ("PS1" to "PS3"). The error values are an average of 100 mixtures. "Failed" means that we were not able to push the missed atom signals below a threshold (see text for details).

	Linear	PS1	PS2	PS3
Task 1	Failed	1.82	1.72	1.50
Task 2	Failed	5.06	4.82	5.10

We did not restrict the shifts of the atoms. The coefficients c_i are chosen from $[0.2, 0.8]$. The power spectrum is obtained by using fast Fourier transformation with shifting Hanning window of size 256. The power spectrum is often used in speech processing and it is suited to define various "speech kernels". The lowest twenty frequencies were set to zero. The additive Gaussian noise had standard deviation of $\sigma = 0.05$ for task 1 (T1) and $\sigma = 0.2$ for task 2 (T2).

We compared three methods: PS1 is the power spectrum of the original mixture, PS2 is the power spectrum of the mixture where absolute mixture values smaller than 0.05 are set to zero, and PS3 is the power spectrum of the mixture where absolute mixture values smaller that 0.1 are set to zero. As in previous experiments we keep the average number of not detected atoms below a certain bound: 0.64 for T1 and 0.74 for T2.

The linear approach completely failed to solve the task. The results for the nonlinear transformation (power spectrum) are given in Table 15.4. Figure 15.5 shows an example of atomic source detection for PS3.

15.5 Conclusions

In this chapter, we reinterpreted the sparse-decomposition method for a single mixture as ϵ-support vector regression (ϵ-SVR). The ϵ-SVR analogy supplied a new view on the sparse-decomposition method's hyperparameter and allowed us to introduce family of similar algorithms of which the sparse-decomposition method is one member. This family includes methods that allow for nonlinear transformations of the sources before they are mixed, and other nonlinear transformations in the process of identifying dictionary atoms in the mixture. One benefit of the nonlinearities incorporated into the approach is that they can be used to achieve some degree of invariance to irrelevant characteristics of signals. We demonstrated our approach with experiments using noisy single mixtures and speech datasets.

At a broader level, this chapter provided a means of mapping an unsupervised ICA problem into a supervised framework. To achieve this mapping, we introduced the notion of a dictionary of predefined atoms. In our approach, each unsupervised example is paired with each atom, and a fixed function

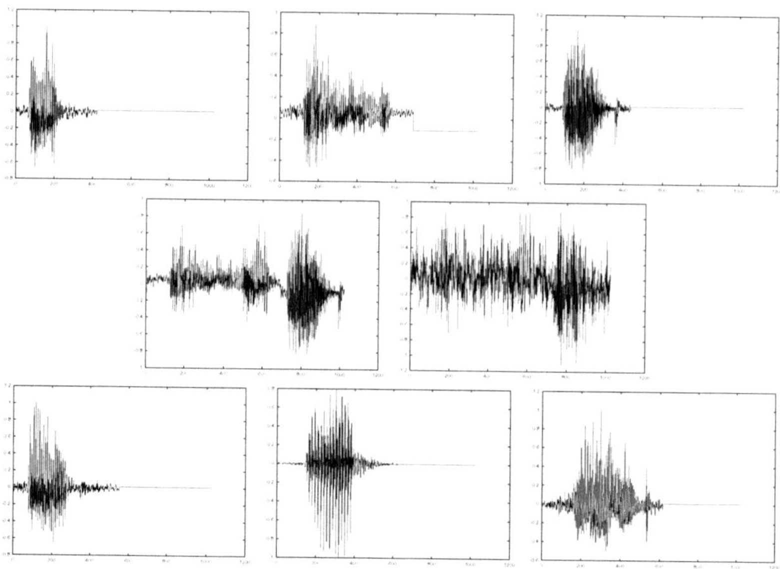

Fig. 15.5. Example for the method PS3. It detected 6 dictionary atoms in mixture: three were correctly and three wrongly detected. First line: three dictionary entries which are present (but shifted) in the mixture. Second line: (left) the mixture without noise and (right) the mixture. Third line: wrongly detected dictionary entries out of 100.

is used to generate a "target" value for the example. This target value can then be used within a supervised learning framework. Typically, the target value might be a measure of similarity between the dictionary atom and the example, such as a dot product. This mapping idea may well be useful in other unsupervised learning problems, where the target values from different unsupervised examples may be combined.

References

1. M. Zibulevsky and B. A. Pearlmutter, "Blind source separation by sparse decomposition," *Neural Computation*, vol. 13, no. 4, pp. 863–882, 2001.
2. B. A. Pearlmutter and A. M. Zador, "Monaural source separation using spectral cues," in *Proc. of the Fifth International Conference on Independent Component Analysis and Blind Signal Separation*, C. G. Puntonet and A. Prieto, Eds. Springer Berlin/Heidelberg, 2004, pp. 478–485.
3. A. Cichocki, R. Unbehauen, L. Moszczynski, and E. Rummert, "A new on-line adaptive algorithm for blind separation of source signals," in *Proc. Int. Symposium on Artificial Neural Networks, ISANN-94*, 1994, pp. 406–411.
4. A. Hyvärinen, "Survey on independent component analysis," *Neural Computing Surveys*, vol. 2, pp. 94–128, 1999.

5. C. Jutten and J. Herault, "Blind separation of sources, part I: An adaptive algorithm based on neuromimetic architecture," *Signal Processing*, vol. 24, no. 1, pp. 1–10, 1991.
6. A. J. Bell and T. J. Sejnowski, "An information-maximization approach to blind separation and blind deconvolution," *Neural Computation*, vol. 7, no. 6, pp. 1129–1159, 1995.
7. B. A. Pearlmutter and L. C. Parra, "Maximum likelihood blind source separation: A context-sensitive generalization of ICA," in *Advances in Neural Information Processing Systems 9*, M. C. Mozer, M. I. Jordan, and T. Petsche, Eds. MIT Press, Cambridge, MA, 1997, pp. 613–619.
8. H. Attias and C. E. Schreiner, "Blind source separation and deconvolution: The dynamic component analysis algorithm," *Neural Computation*, vol. 10, no. 6, pp. 1373–1424, 1998.
9. S. Amari, A. Cichocki, and H. Yang, "A new learning algorithm for blind signal separation," in *Advances in Neural Information Processing Systems 8*, D. S. Touretzky, M. C. Mozer, and M. E. Hasselmo, Eds. MIT Press, Cambridge, MA, 1996, pp. 757–763.
10. P. Comon, "Independent component analysis – a new concept?" *Signal Processing*, vol. 36, no. 3, pp. 287–314, 1994.
11. J.-F. Cardoso and A. Souloumiac, "Blind beamforming for non Gaussian signals," *IEE Proceedings-F*, vol. 140, no. 6, pp. 362–370, 1993.
12. T. Tanaka, "Analysis of bit error probability of direct-sequence CDMA multiuser demodulators," in *Advances in Neural Information Processing Systems 13*, T. K. Leen, T. G. Dietterich, and V. Tresp, Eds. MIT Press, Cambridge, MA, 2001, pp. 315–321.
13. G. Cauwenberghs, "Monaural separation of independent acoustical components," in *Proceedings of the 1999 IEEE International Symposium on Circuits and Systems (ISCAS'99)*, vol. 5. IEEE, 1999, pp. 62–65.
14. T.-W. Lee, M. S. Lewicki, M. Girolami, and T. J. Sejnowski, "Blind source separation of more sources than mixtures using overcomplete representations," *IEEE Signal Processing Letters*, 1998.
15. S. T. Roweis, "One microphone source separation," in *Advances in Neural Information Processing Systems 13*, T. K. Leen, T. G. Dietterich, and V. Tresp, Eds. MIT Press, Cambridge, MA, 2001, pp. 793–799.
16. M. S. Lewicki and T. J. Sejnowski, "Learning overcomplete representations," *Neural Computation*, vol. 12, no. 2, pp. 337–365, 2000.
17. ——, "Learning nonlinear overcomplete representations for efficient coding," in *Advances in Neural Information Processing Systems 10*, M. I. Jordan, M. J. Kearns, and S. A. Solla, Eds. MIT Press, Cambridge, MA, 1998, pp. 556–562.
18. V. Vapnik, *The Nature of Statistical Learning Theory.* Springer-Verlag, New York, 1995.
19. C. Cortes and V. N. Vapnik, "Support vector networks," *Machine Learning*, vol. 20, pp. 273–297, 1995.
20. B. Schölkopf and A. J. Smola, *Learning with kernels – Support Vector Machines, Reglarization, Optimization, and Beyond.* MIT Press, Cambridge, 2002.
21. B. Schölkopf, P. L. Bartlett, A. J. Smola, and R. Williamson, "Support vector regression with automatic accuracy control," in *Proceedings of ICANN'98*, ser. Perspectives in Neural Computing, L. Niklasson, M. Bodén, and T. Ziemke, Eds. Berlin: Springer Verlag, 1998, pp. 111–116.

22. ——, "Shrinking the tube: a new support vector regression algorithm," in *Advances in Neural Information Processing Systems 11*, M. S. Kearns, S. A. Solla, and D. A. Cohn, Eds. Cambridge, MA: MIT Press, 1999, pp. 330–336.
23. A. J. Smola and B. Schölkopf, "A tutorial on support vector regression," *Statistics and Computing*, vol. 14, pp. 199–222, 2004. Also: NeuroCOLT Technical Report NC-TR-98-030.
24. R. Vollgraf, M. Scholz, I. Meinertzhagen, and K. Obermayer, "Nonlinear filtering of electron micrographs by means of support vector regression," in *Advances in Neural Information Processing Systems 16*. MIT Press, Cambridge, Massachusetts, 2004, pp. 717–724.
25. X. Wang, A. Li, Z. Jiang, and H. Feng, "Missing value estimation for DNA microarray gene expression data by support vector regression imputation and orthogonal coding scheme," *BMC Bioinformatics*, vol. 7, p. 32, 2006.

Index

Printed in the United States
89833LV00001B/58-96/A

9 781402 064784